THE CLIMATE CONNECTION

The Climate Connection highlights the influence of saltatory evolution and rapid climate change on human evolution, migration and behavioural change. Growing concern over the potential impacts of climate change on our future is clearly evident. In order to better understand our present circumstances and deal effectively with future climate change, society needs to become more informed about the historical connection between climate and humans. The authors' combined research in the fields of climate change, evolutionary biology, Earth sciences and human migration and behaviour complement each other, and have facilitated an innovative and integrated approach to the human evolution–climate connection.

   The Climate Connection provides an in-depth text linking 135 000 years of climate change with human evolution and implications for our future, for those working and interested in the field and those embarking on upper-level courses on this topic.

RENÉE HETHERINGTON obtained a BA (Business and Economics) from Simon Fraser University, Canada in 1981 and a Master's of Business from the University of Western Ontario, Canada in 1985 and an Interdisciplinary PhD (Anthropology, Biology, Geography and Geology) from the University of Victoria, British Columbia, Canada in 2002. She was awarded a National Science and Engineering Research doctoral fellowship for her work reconstructing the paleogeography and paleoenvironment of the Queen Charlotte Islands/Haida Gwaii. The Social Sciences and Humanities Council subsequently awarded her a postdoctoral fellowship for her research relating climate change to human evolution and adaptability over the last 135 000 years. She is co-leader of the International Geological Correlation Programme Project 526 'Risks, Resources and Record of the Past on the Continental Shelf.' She and her husband Bob are partners in RITM Corp., a consulting company whose expertise is committed to helping organizations, especially in the resource sector, reach their potential while recognizing we are in a changing world. She has recently been nominated as candidate for Member of Parliament with the Federal Liberal Party of Canada. She lives with her family in Canada on Vancouver Island.

ROBERT G. B. REID is Emeritus Professor of Biology at the University of Victoria, British Columbia, Canada. He holds BSc and PhD degrees from Glasgow University. His major fields of professional interest are digestive physiology, malacology and evolutionary theory. He is the author of *Evolutionary Theory: The Unfinished Synthesis* (1985) and *Biological Emergences: Evolution by Natural Experiment* (2007). Robert Reid has taught in the fields of marine biology and comparative physiology, his major source of experience, as well as the history of biology. He has presented seminars at the Konrad Lorenz Institute workshop on Environment, Development and Evolution in Altenberg, Austria. Throughout his career he has worked closely with Environment Canada, and the Department of Fisheries and Oceans, Canada. He was Chair of the Pacific Aquaculture Centre of Excellence Shellfish Committee and Chair of the Shellfish Research Group of British Columbia. He was a member of the University of Victoria Arts and Science Committee on Liberal Arts Programme Implementation from 1974 to 1976 and on the University of Victoria, Arts and Science Dean's Advisory Committee from 1977 to 1980. He was a member of the University of Victoria Senate Committee on University Extension from 1981 to 1983 and was University of Victoria Biology Honours Director from 2000 to 2002.

THE CLIMATE CONNECTION

Climate Change and Modern Human Evolution

RENÉE HETHERINGTON
RITM Corp., Canada

ROBERT G. B. REID
University of Victoria, British Columbia, Canada

CAMBRIDGE
UNIVERSITY PRESS

CAMBRIDGE
UNIVERSITY PRESS

Shaftesbury Road, Cambridge CB2 8EA, United Kingdom

One Liberty Plaza, 20th Floor, New York, NY 10006, USA

477 Williamstown Road, Port Melbourne, VIC 3207, Australia

314–321, 3rd Floor, Plot 3, Splendor Forum, Jasola District Centre, New Delhi – 110025, India

103 Penang Road, #05–06/07, Visioncrest Commercial, Singapore 238467

Cambridge University Press is part of Cambridge University Press & Assessment,
a department of the University of Cambridge.

We share the University's mission to contribute to society through the pursuit of
education, learning and research at the highest international levels of excellence.

www.cambridge.org
Information on this title: www.cambridge.org/9780521147231

First published 2010

A catalogue record for this publication is available from the British Library

| ISBN | 978-0-521-19770-0 | Hardback |
| ISBN | 978-0-521-14723-1 | Paperback |

Contents

The colour plates can be found between pages 174 and 175.

Foreword

Whenever people stop to think about it, there can scarcely be a more obvious and common-sense idea or awareness than that humans, and all their activities, have a relation to climate. The clothes we wear, the houses we live in, the food we eat and where it is produced, our perceptions of the rest of the world and all its living creatures, of the changes in weather and of the seasons – all are influenced by, or are expressions of, the climate of our immediate surroundings and of the whole planet. Each of us, wherever she or he may live, 'knows', instinctively, and through experience, that the climate is a vital and sometimes dominant component of our environment. Also, through our family and collective memories, as well as by simple observation of the natural world around us, through the many stories that are parts of our cultures in most of our societies, and acceptance that there was in the past something called the 'Ice Age', most of us are aware that the climate in the past was somehow different from what we are experiencing today. But just what is the relationship between humans, as intelligent living beings on the dynamically changing planet on which we depend, and the climate which itself appears to be changing? How is the relationship expressed, and how do humans respond? A greater understanding of this relationship may be important, indeed may be vital, in helping people in all parts of the world and their institutions, to understand the problems and to take more effective actions in the years ahead.

This book delves deeply and comprehensively into the background and underlying factors of evolving climatic conditions on the surface of planet Earth and the evolution of what has now become the modern human. It tells the fascinating story of how, again and again, changes in climatic conditions presented groups or populations of the evolving pre-human and human species, wherever they were at the time, with difficulties, limitations and challenges, so that the survivors – our ancestors – used their ingenuity, or ability to migrate, to cope with the challenges; and in so doing they moved a step closer to what today we rather gratuitously call 'behaviourally modern humans'. Clearly, the fact that today we – *Homo sapiens* – are the dominant large animal on the planet, that we are the only living species of our genus, and that we have planet-wide distribution and have recently been or can be found not only on all parts of all continents but have ventured over and under the seas, into the air, and into outer space, has a lot to do with the fact that changes in climate have obliged our ancestors, or given them the opportunity, to develop the capacity for behavioural change, the flexibility, and the tools to cope with changes in our external environment. In a very real sense, modern humankind in all its present successes is a product of severe climate changes in the past.

However, those same evolved abilities that have brought us to our present pre-eminent position in the global biosphere now have resulted in the situation that our activities are seriously affecting the processes and chemical characteristics that produce the dynamic climate equilibrium and the biological productivity of the Earth itself. To understand what is happening, and to understand whether, and in what way, humans can take action to prevent this situation from being self-destructive, we need to examine carefully our inherited and evolved ability as a species to act within our environment. This book not only tells us where we have come from and what we are today, but calls for sobering reflection on what we can do in light of where we appear to be going. Just at this time in our current history, when our economy, our politics, and the popular concern in many parts of the world is focused on the prospect of impending severe climate change, this book, with its thorough, logical analyses and broad perspective on the connections between climate and humans, is very valuable.

The authors, an eminent evolutionary biologist and an experienced palaeogeographer-ecologist, have teamed up to examine the climate–human evolution connections in a comprehensive and thorough way. To do so, they have taken us back to the story of our planet, the evolutionary and chemical changes in our atmosphere and oceans as the setting in which organic life developed and which, in due course, resulted in living cells and organisms that had the capacity to respond and adapt to changes – changes in the climate as it then was – and to pass characteristics for survival to the next generation. We are reminded that the main units into which the geological timescale is divided, identified a century ago by the disappearance or emergence of distinctive fauna, are testimony to periodic severe climatic disturbances, or to other catastrophes such as bolide impacts or volcanism, which not only have constrained, but have also stimulated the evolution of life.

The story of the growing knowledge of the geologically earliest hominins, and of the discoveries, the interpretations, the premature and disputed conclusions that led to scientific recognition of the genus *Homo* and its several species; of the influence of the dominant personalities who led the thinking in this subject, is well told, in the light of current discoveries and new techniques of analysis. After a century and a half of controversy, the studies of human fossils and of the evidences of human behaviour and expression – tools, habitations, art, indications of language – together with the evidence of dispersion of early humans to widely scattered locations in Africa, Asia, Australia, Europe, the Americas and Polynesia, especially in the last 135 000 years, make a coherent picture. The issues of whether 'out of Africa' or multiple origins explains the sources of humankind and of the role and distribution of other species *(Homo erectus, Homo ergaster, Neanderthal Man*, and the others including our late little relative *Homo floresiensis)* fall convincingly into place in this larger perspective.

Having reviewed the evidence and our progressive understanding of early humans, the authors address the question of the changes of climate during the last 135 000 years, since the time when our direct ancestors, *Homo sapiens*, left Africa and began forays which would take them throughout the world. To do this, they first bring to our attention the variations of solar energy received by the planet – the Milankovic cycles and modifications – the movements of the continents, variations in three-dimensional ocean circulation, and the progressive changes in atmospheric chemical composition which have affected climate, climate stability and the support of biological life during the past two billion years. They then follow the climate story through planetary evolution, including haphazard but important events such as volcanic outpourings and cycles of glaciation that have affected the conditions and stability of the climate, and thus the development and distribution of vegetation and all animal life, up until the spread of *Homo sapiens* 'out of Africa' to the rest of the world about at the time of the onset of the last global glacial cycle. That cycle resulted not only in glaciers on land and ice in the oceans of northern regions, but also severely affected climate and thus vegetation and food supplies for primitive humans in subtropical regions.

The story of the changes in climate during the last 135 000 years, the evidence for changes in landscape and vegetation, and the progressive evolution, sophistication and distribution of *Homo sapiens* in response to different climates and climatic disturbances, up to the beginning of relative stability about 11 500 years ago – the Holocene epoch – is important. With the compressed timescale made possible by computer modelling, the tale reads somewhat like an adventure story – 'just one damn climate change after another' (though the incidents are separated by thousands of years) – presenting our ancestors with stresses and challenges from which the survivors, rather like Hercules after each of his labours, were better equipped to meet the next challenge. The markers for this remarkable chronicle are the successive 'marine isotope stages' that are well identified in geological and oceanographic data and allow computer simulations of climate that can be correlated with non-biological observations and measurements. The story thus brings human history and various branches of the natural sciences together to set the stage, during the Holocene, for humans to reside in most parts of the world and then, themselves, to have the ability and capacity to make significant changes in the world environment.

The development of agriculture, with the consequent growth of hierarchical societies, domestication of animals, larger-scale fisheries, fixed habitations with people in
large numbers, trade between communities, etc., marked a quite new dimension in the
human–climate connection. The authors summarize significant climate or climate-
related events in different parts of the world and relate them to societal and technological developments in favoured locations or progressive groups, through processes
that can be hypothesized as 'catastrophe–communication–collaboration'. That process is still going on. Through it, humankind has developed various civilizations,
spread throughout the world in drastically increasing numbers, learned how to exploit
living and non-living resources for short-term human ends, and succeeded in mana-
ging, for better or for worse, many environmental processes. A new epoch, the
Anthropocene, is upon us. But the effects of, and vulnerability to changes of climate
are by no means lessened.

While the story of climate and its changes, and its connections to the evolution of
humankind over the last two million years or so is fascinating and of great portent as
a basis for assessing our ability to cope with the severe climatic changes that are
impending, the masterful review and analyses presented in this book also have a
significant value as a record of scientific thought and research during the past two
hundred years. The book provides a commentary on how understanding of a topic of
intense interest to a large number of scholars and investigators, as well as to the
general public, grows through careful thought and meticulous work, strong opinions
and philosophies in one direction or another, and through new discoveries and new
technologies which may overturn or replace established ideas. It presents a unique
story of the scientific method itself.

Anyone reading the book cannot help but be impressed with how tempting it is
for rational, knowledgeable scholars to draw sweeping interpretations and conclusions from very sparse data or observations. The find of a single human skull, very
carefully examined for cranial characteristics which give indications of anatomy,
diet, intelligence, etc., is for better or worse, interpreted to represent many generations of a whole human race, until some other find is made. Ideas which seemed
sensible deductions or conclusions by the originator sometimes become fixed by the
followers, and defended with almost religious inflexibility. Some tools made of
stone, which happened not to perish through millennia, are surmised to represent
and define a whole way of life of a people, when it is likely that the people who made
them mostly used tools of materials that have since disappeared, and the objects
found today, while genuine, may not have been representative at all. The authors of
this book, through their very comprehensive review (more than 1100 quoted
refereed papers) and their careful, generous and yet critical analyses, bring to light
the many controversies and different schools of thought, and put them into perspective in the light of recent discoveries and new technologies of dating chronology and

genetics. They lay many misconceptions to rest. Where major uncertainty still exists, the authors do not hesitate to give their personal interpretation, for the reader or future researchers to resolve. Thus this comprehensive book puts into perspective and simplifies what are surely whole libraries of strongly held but often conflicting ideas about the history of our own species and our biological relatives. And the unifying perspective, rarely brought out until now, is the effects of, and responses to, climate changes. This in itself is a very great service to the scientific community and to the advance of knowledge.

The Climate Connection also admonishes us to be more careful about the use of common words, and behind them be careful of the distinctions between concepts often loosely undifferentiated. The 'adaptation' of our ancestors to the climate where our species developed half a million or more years ago is a genetic characteristic that each of us carries within us and which cannot be changed within a few millennia. It is why most of us wear clothes wherever we go, to keep a warm tropical environment next to our skin. Donning clothes is however not adaptation, but a form of 'adaptability'. Adaptability is a characteristic that enables humans to change our behaviour relatively quickly in any number of ways (by developing new tools, etc.) to cope with, or escape from, changes in the environment. It is our adaptability that has enabled us to survive through, and in the long run benefit from, rapid climate changes in the past. And it is our adaptability that we must call upon to meet the challenges ahead.

While presenting a balanced and comprehensive review of who we are as modern humans, the only surviving species of our genus, and how we got to where we are today as an animal shaped by successive encounters with climatic challenges, the book does not pull any punches with respect to the challenges ahead. That the impending rapid change in climate, with reduction of planetary biological productivity on land and oceans, and geologically rapid change in sea level, may be in large measure due to human actions and our own short-sightedness does not make it any less real. The authors do a commendable job in outlining the evidence for the challenges ahead. They note a selection of indicators and plausible speculations about what may lie in store for the living resources and environmental conditions upon which we all depend, and describe how modern complex societies that have apparently become locked into increasing numbers of people, increasing use of energy, and use of biological resources far beyond the ability of the planet to produce them, are increasingly vulnerable to failure because of changes in climate, hydrological systems or sea level. Whether the *adaptability* built into the make-up of *Homo sapiens* will enable us to survive, and ultimately to benefit from the coming climate change, and whether terrestrial and ocean ecosystems can ever recover from the severe damage that human actions in the past century have done, is an open question that this book leaves the reader to ponder. We have no knowledge of the losses that were suffered by human groups in each or any of the climate change challenges in the past, so aptly described herein; but we do know that some individuals, enough to carry on our story, survived and progressed. Their story at least is encouraging.

Fred Roots
Science Advisor Emeritus, Environment Canada, and Chair of the
Canadian National Committee for the UNESCO Man
and Biosphere (MAB) Programme

Preface

Another beautiful sunny summer day has dawned on Canada's west coast. Gulls glide across the cloud-speckled blue sky. Eagles dive, plucking their dinner from the cool ocean. Children play on the beach, clams squirting their bare legs. It is an idyllic paradise. Striations on the bedrock are evidence of the two-kilometre-thick ice sheet that blanketed this land fifteen thousand years ago. Soon after, aboriginal people lived and played on these beaches leaving behind deep clam-laden middens and stone tools.

Yet, storm clouds brew on the horizon. When we look back on Earth's climate over the last 135 000 years it is clear that climate change is not new but is part of the natural change ubiquitous in Earth's history. Long, long before land- and sea-ice covered the northern hemisphere palm trees grew in the Arctic. Yet even within these cycles, that oscillate between warm and cold, there have been occasions when climate changed rapidly. These events often coincided with the extinction of once-dominant species and rapid, saltatory evolution of others. Thus, it is important to understand the role climate change has and continues to play in the evolution of our species, particularly as we face future climate change.

Adaptation and adaptability are hopelessly confused in the public mind, as well as in the writings of anthropologists and archaeologists. The adaptations with which neo-Darwinists work are random, genetically fixed mutations that require the approval of natural selection to become general species characteristics. Adaptability is what the individual organism can do to respond physiologically and behaviourally to change. In the case of humans, the proper application of intelligence is part of our adaptability. The distinction is particularly important in the context of this book, since the process of adaptation and natural selection in the strict sense is much too slow to respond to sudden environmental alteration. In contrast, the adaptable organism can do something about it instantly. Unfortunately for humans, tradition, ritual, and 'sticking to tried and true ways' can obstruct effective action, despite our potential ability to respond effectively to change.

While the twentieth century stands out in history for two world wars, many local wars, genocides and political revolutions, there were little-considered developments that present us in the twenty-first century with even more menace. Many of us tend not to notice them, or to discount their effects. We do so at our own peril. We refer to the environmental consequences of global warming, deforestation, soil erosion, the degradation of ocean fisheries, expanding populations and inflated economies. We humans have periodically mapped roads to our own downfall, without taking significant action to forestall such fates. In the case of the present danger, it will affect the entire population of Earth.

Our aim in this book is to link climate change with modern human evolution. Our global climate simulations provide 135 000 years of climate change data that are combined with geological proxy data and archaeological evidence to illuminate connections between climate change and key events in human history. We identify the relevance of saltatory evolution to rapid climate change. We introduce the 3 C's syndrome – catastrophe, communication and cooperation – as an impetus for human social evolution. We seek a new way of understanding our past evolutionary relationship with climate, one that makes the connection between human history and climate history; one that will generate a present that does not borrow on our children's future.

Acknowledgements

This book grew from our desire to relate climate change with human evolution, to explain how rapid environmental changes have affected us in the past, and to find ways to cope with a changing future. For years, the authors met weekly to discuss evolution, the environment and humans. Our discussions have frequently benefited from the contributions of Richard A. Ring, Professor Emeritus, Department of Biology, University of Victoria, Rodney Roche, Professor Emeritus, Department of Biosciences, University of Calgary, the late Bill Livant, Professor Emeritus, Department of Psychology, University of Regina, Dawna Brand, PhD candidate, Department of Biology, University of Victoria, and numerous other faculty and students from the University of Victoria Biology Department who joined in our discussions. Gareth Nelson, School of Botany, University of Melbourne has provided much appreciated encouragement at all levels for Robert's evolutionary ideas. Also valued are Elizabeth Vrba's contributions at the Konrad Lorenz Institute Workshop on Environment, Development and Evolution and their influence on our evolutionary ideas. Renée Hetherington's research into climate change and human evolution at the University of Victoria Climate Modelling Lab benefited from Andrew J. Weaver's enthusiasm, vision and support. The authors strayed into unchartered territory and were particularly supported by the UVic Climate Modelling Lab staff – Michael Eby, Wanda Lewis, and Ed Wiebe, postdoctoral fellows – Jeff Lewis and Kirsten Zickfeld, and other staff and students. The last glacial land-ice was interpreted by Shawn Marshall of the University of Calgary. Roger MacLeod has provided graphics and GIS input and advice throughout this project. Kathleen W. Matthews has provided research assistance.

The duration of the writing of this book makes it extremely difficult to thank all those who have contributed. Too numerous to mention are the many colleagues and students with whom we have worked, spoken and to whom we have lectured. They have provided insights and evidence. They have asked questions that provoked us to search for clear, insightful responses. Their assistance has been much appreciated.

E. Fred Roots, Science Advisor Emeritus with Environment Canada, dug up and presented us with climate policy research papers and notes extending back over 25 years, providing a perspective on the current climate debate. We thank him for the foreword to this book and valuable suggestions for improvement of the text. We also thank our editors Audrey McClellan, West Coast Editorial Associates, Joseph Bottrill and Polly Wolf, Out of House Publishing Solutions, and particularly Matt Lloyd, Laura Clark and the rest of the editing team at Cambridge University Press for their thoughtful and expert advice. We very much appreciate the assistance of Beverly Duthie with the index and Robert Thompson, Richard Ring and Rodney Roche for correcting the final proofs. We thank reviewers of various chapters including Nathan Gillett, Research Scientist, Canadian Centre for Climate Modelling, University of Victoria, Richard A. Ring, E. Fred Roots, Andrew J. Weaver, and three anonymous reviewers. For personal communications, we thank Gizelle Rhyon-Berry, Research Associate, The Foundation for Shamanic Studies, Úrsula Oswald Spring, Professor, Regional Centre of Multidisciplinary Research at the National University of Mexico (CRIM-UNAM), Anwar A. Abdullah, Senior Advisor to the Prime Minister on Sustainable Development, Kurdistan Regional Government, Ursula Franklin, University Professor Emeritus, University of Toronto, and E. Fred Roots.

Our friends and loved ones have been particularly supportive during the extended researching and writing of this book. Renée's dear friend Marion Farrant offered encouragement, advice and support from the outset and was instrumental in its inception. Margaret Fulton has provided friendship, advice and encouragement. Robert's daughter, Clio Reid, has provided valuable personal support and editorial assistance. Renée's sons, John and Ryley, gave willingly their love and support and inspired us to see the world through the next generation's eyes and to recognize the responsibility we all share for our future. Robert I. Thompson, Renée's husband, has provided ongoing love, support and encouragement; without him, this book would never have been written.

The Canadian Foundation for Climate and Atmospheric Sciences and the Natural Science and Engineering Research Council of Canada indirectly funded part of the research via their operating grant programmes. Renée was also supported by a Social Sciences and Humanities Research Council of Canada postdoctoral fellowship.

This book is a contribution to UNESCO and the International Geological Correlation Programme Project No. 526 'Risks, Resources, and Record of the Past on the Continental Shelf.' Any errors, omissions or inconsistencies are solely the authors' responsibility and in no way reflect on the contributions made by all the people we have mentioned and the many we have not named here, but appreciate nonetheless.

1

Introduction

1.1 The climate connection

Until the early nineteenth century, there was a widespread belief that the Biblical Flood was the greatest climatic catastrophe in human history. The deluge was supposedly God's punishment for the sins of man, and the story of Noah is the account of a family that in the face of mockery took drastic steps to ensure its survival and the preservation of life on Earth. This allegory was found to be an oversimplification when Baron Cuvier's studies of the Paris basin suggested that, for whatever reason, a *series* of deluges had altered the environment, so that it was, in turn, dry land, covered with fresh water, or inundated by the sea. However, beginning in 1836, studies of glaciers by Cuvier's student, Louis Agassiz, provided an alternative explanation for phenomena thought to be caused by flooding. Agassiz attributed glaciation to the divine hand, referring to it as 'God's Plough'. About 20 years later, Charles Darwin gave Agassiz's work an evolutionary interpretation, but Agassiz could not see the connection between climate and evolution.

Darwin's own work subsumed adaptation to climatic conditions as part of evolution, but he did not expand on the effects that the environment has on evolution. *The Origin of Species* (1872) is particularly weak on the evolution of physiological and behavioural adaptability; factors that are crucial for the survival of individuals during severe environmental changes. Furthermore, because he was a gradualist, Darwin paid little or no attention to catastrophic events in his theory. Although he assumed that human evolution was affected by climate, he did not explore the ways in which humans can change the climate.

In this book we examine both these connections: the impact of climate on evolution, with an emphasis on human evolution, and the impact of humans on climate. We are in the midst of a new era of climatic change, and modern society needs to assess its vulnerability and adaptability to such change. While technology has obviously advanced since the time of Noah, the task of adjusting to future

EON	ERA	PERIOD	EPOCH	AGE IN MILLIONS OF YEARS
PHANEROZOIC	CENOZOIC (TERTIARY)	QUATERNARY	Holocene / Pleistocene	0.01 / 1.8
		NEOGENE		23
		PALEOGENE	Oligocene Epoch / Eocene Epoch / Paleocene Epoch	34 / 56 / 65
	MESOZOIC	CRETACEOUS		146
		JURASSIC		200
		TRIASSIC		250
	PALEOZOIC	PERMIAN		299
		CARBONIFEROUS		360
		DEVONIAN		416
		SILURIAN		444
		ORDOVICIAN		488
		CAMBRIAN		542
PRECAMBRIAN	PROTEROZOIC	NEOPROTEROZOIC		1000
		MESOPROTOEROZOIC		1600
		PALEOPROTEROZOIC		2500
	ARCHAEAN			4500

Figure 1.1 Geologic timescale

climate change will be no less challenging as we move down the path towards the eventual decarbonization of our global energy system.

1.2 Earth's changing climate

Climate change has been ubiquitous in Earth's 4.5-billion-year history (Figure 1.1). During the Archaean era there was little or no oxygen in the Earth's atmosphere, hence no ozone and no protection from solar radiation. However, after single-celled microbes invented photosynthesis, one of the photosynthetic by-products, free oxygen, gradually accumulated in the oceans and atmosphere.

In the late pre-Cambrian, most organisms were aquatic unicells, and the land-masses were cold, with a high albedo – bare rock, snow and ice reflected much of the Sun's heat back into space. A series of intense glaciations froze the surface water on the planet, creating what is known as 'snowball Earth'. The effect on life and its evolution at the surface of the oceans, except perhaps in a narrow equatorial zone, was devastating. Over the next several million years, the process of volcanic eruption and outgassing is thought to have slowly increased the atmospheric concentration of carbon dioxide until the greenhouse effect was sufficiently strong to melt the ice and bring the Earth out of its 'snowball' state. Atmospheric oxygen remained low as plants invaded the land, dispersed, evolved and increased in biomass. Gradually the amount of oxygen in the atmosphere increased until it eventually reached its present concentration of 21%, the rest being mainly nitrogen gas. Vegetation darkened the previously snow-covered landscape, decreasing albedo, and as a result the Earth became warmer.

The big bang of animal evolution occurred about 542 million years ago, at the beginning of the Cambrian period. Prior to that time, marine animals had been diversifying for maybe 100 million years. Land plants evolved during the Silurian about 444 million years ago. Plants, in the form of mats of fungi, bacteria and algae, inhabited the intertidal zone ever since the climate was propitious – within the range of supporting life. By 50 million years ago, Earth's climate was sufficiently warm that the Arctic experienced a Mediterranean-like climate with temperatures as high as 24 °C (Brinkhuis *et al.*, 2006).

Five major extinction events are believed to be related to catastrophic climate events that were associated with bolide impacts or other geological events, including large volcanic eruptions that spewed forth huge lava flows (see, for example, Rampino *et al.*, 1988; Rampino and Stothers, 1988). These occasioned some of the most sudden historical changes in Earth's climate, including mass extinctions of many species. The five major extinction events occurred at the ends of the Ordovician period (444 million years ago), Devonian period (360 million years ago), Permian period (250 million years ago), Triassic period (200 million years ago) and Cretaceous period (65 million years ago). The most extreme occurred at the end of the Permian and resulted in the extinction of 95% of the marine species known from the earlier Permian fossil record. While the impact of a bolide is only one of several mechanisms hypothesized as the cause of most extinctions, the impact of the Chicxulub comet in the Yucatan Peninsula is considered to be a potential cause of the Cretaceous event associated with extinction of the dinosaurs. In the North American splash zone of Chicxulub, higher-level plants are thought to have been exterminated, and the continent was populated by hardy ferns for millions of years. However, the work of Gerta Keller and associates found that the worldwide extinction of dinosaurs occurred some 150 000 to 350 000 years after the impact of the Chicxulub comet and suggest instead it may be due to volcanic outpourings in the Deccan trap terrain in India (see, for example, Keller *et al.*, 2009).

Although these climate catastrophes wreaked havoc on the previously dominant species, their subsequent demise opened new environments for the lucky survivors. Those survivors that were adaptable enough to take advantage of the changed environment moved into the newly vacated territories. As a result, novel or previously

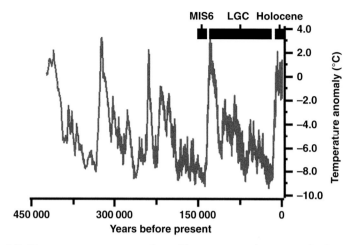

Figure 1.2 Temperature reconstruction of lower atmosphere over the last 450 000 years, expressed as an anomaly relative to present-day values. The data, described in Petit *et al.* (1999), is derived from analysis of historical stable isotopes from the Vostok ice core. Also indicated are marine isotope stage 6 (MIS6), last glacial cycle (LGC) and Holocene intervals.

rare species became widely distributed. After catastrophic changes, climatic stability allowed the surviving plants and animals to enter a prolonged period of evolutionary stability. This presented barriers to evolution, through competition by specialists, although a prolonged period of warm, stable climates was probably at the foundation of the emergence of homeothermy (warm-bloodedness) in birds and mammals (see 'Human adaptability', Appendix C, for a full account).

Continental drift, caused by the movement of crustal plates (i.e., plate tectonics), broke up the largest land masses, slowly affecting climate because the sea acts as a heat buffer for smaller bodies of land, and because of the redistribution of ocean currents. Furthermore the mountain building caused by movements of the crustal plates influenced global atmospheric winds, with a marked effect on precipitation, causing, for example, the monsoons of South East Asia and the drying of East Africa. By the time of the Pleistocene (1.8 million years ago), continental drift had produced a topography similar to what we see at present and had ceased to be a major factor in climate-induced evolution, except in areas where the plates crash together, causing earthquakes and volcanic eruptions.

Cycles of glacial and interglacial events have provided the largest changes in the climate of the Pleistocene. Interpretations of data from the Antarctic Vostok ice core by Petit and others (1999) and more recently from the Antarctic Dome Concordia ice core (Spahni *et al.*, 2005) reveal large variations in the atmospheric temperature and carbon dioxide of the Antarctic during the past 650 000 years (Figure 1.2 and Figure 1.3). Six cycles of cold glacial and warm interglacial phases are evident during this interval.

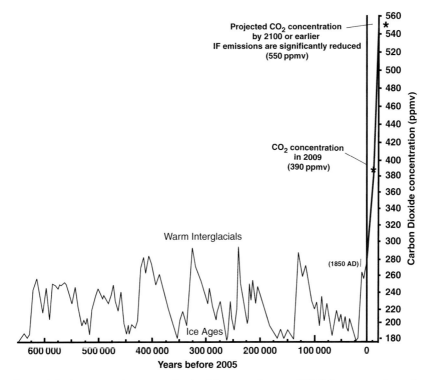

Figure 1.3 A composite atmospheric CO_2 record over 650 000 years – six and a half ice-age cycles – based on a combination of CO_2 data from three Antarctic ice cores: Dome C, Vostok, and Taylor Dome. Adapted from Siegenthaler *et al.* (2005, Figure 1.04). Also indicated are atmospheric CO_2 levels in 2009 and levels projected to be reached by 2100 or earlier.

1.3 Climate and humans

The global expansion and migration of *Homo sapiens* occurred during the last glacial cycle (LGC), which began 135 000 years ago (Figure 1.4). It was a time of significant global climatic change. The strength of the thermohaline circulation in the oceans, which directly influences the climate of the northern hemisphere, was highly variable. Global sea level rose and fell. Continental shelves were exposed for hundreds of kilometres in some areas and were submerged in others. Vegetation and animal distribution changed rapidly, sometimes decreasing biomass and sometimes increasing food resources for humans. Yet despite the variable climate and their prehistoric culture and technology, humans not only persisted but also spread throughout the world.

At the height of the last ice age, *Homo sapiens* had migrated out of Africa, colonized Asia and Australia, and begun moving towards the Americas. After the demise of the Neanderthals, *Homo sapiens* was the sole *Homo* species in Europe. More recently, a dwarf species of *Homo* lived on Flores Island. *Homo floresiensis*, a

PERIOD	EVENT	AGE IN THOUSANDS OF YEARS
QUATERNARY	*H. sapiens* reach New Zealand	0.8
	H. sapiens reach Hawaii & Easter Island	1.1
	H. sapiens reach Pacific Islands New Caledonia & Samoa	3
	Aboriginal peoples inhabit Canada's Arctic	6
	Short cold event	8.2
	Holocene warm interval & first agriculture begin, megafauna extinction in N.America	10
	H. floresiensis disappears	18–12
	Younger Dryas cold spell	12.7–11.7
	H. sapiens appear in the Americas	>20–10
	Coldest time of last ice age	21
	H. sapiens in NE Russia	32–18
	Neanderthals disappear	30
	H. sapiens in Tasmania	35
	H. sapiens appear in Europe, New Guinea	~40
	H. sapiens appear in Australia	~45
	H. erectus disappears in Java, cultural renaissance in Europe	after 50
	Behaviourally modern *H. sapiens* appear in China	~67–30
	Neanderthals in W. Russia until	~73–36
	H. floresiensis appears on Flores Island	74–38
	H. sapiens migrate out of Africa	~119
	Beginning of last glacial cycle	~135
	Modern humans appear in Africa	~195
	H. erectus disappears in China	230
	Neanderthals appear; *H. erectus* in Japan	by 500
	H. erectus on Flores Island	840
	Wooden javelins in Germany	900
	Homo erectus dispersal out of Africa	Prior to 1000
		1800
	Early stone tools	2300

Fig 1.4 Quaternary timeline

small-brained hominin, was only one-metre tall, walked on two legs, and was found associated with fossils of the Komodo dragon and a stone tool kit that was used to hunt the dwarf *Stegodon*, an elephant thought to have become extinct 840 000 years ago. *Homo floresiensis* survived on Flores Island from before 38 000 years ago until at least 18 000 years ago (P. Brown *et al.*, 2004), making them our last known living *Homo* relative.

1.4 Climate and species dominance

During the last 4 billion years there has been a series of dominant species on Earth. Although they were well suited to the stable environment in which they became ascendant, circumstances changed, and in the wake of major climate disruptions, the formerly dominant species' near or total extinction left an opening for adaptable survivors who could take advantage of the changed environment. For example, the climate catastophe at the end of the Cretaceous–Tertiary boundary 65 million years ago resulted in the extinction of large numbers of previously common species and an expanded distribution of new or previously rare species. Mammals and flowering plants were liberated from dynamic stasis that had restricted their distribution.

Homo sapiens has been the dominant mammalian species on Earth during only the last 10 000 years. In that time, humans have developed agriculture and domesticated animals; new civilizations and concentrated population centres have risen and fallen.

Generally, the biological response to a major climate disruption is a decrease in interspecific and intraspecific competition, usually because of the reduction in numbers, or the extinction, of previously dominant species. The result may be a per capita surplus of resources/biomass for the surviving species, although overall biomass may be severely reduced. New resource opportunities are created for use by previously non-dominant but adaptable species. Then, as climate begins to restabilize, populations of the most adaptable types expand.[1] The new dominant type diversifies as different populations specialize or adjust to the new conditions of life. These new changes begin to proliferate through speciation and behaviour modifications. As the climate restabilizes, circumstances become conducive to dynamic equilibrium or stability, which is strongly influenced by competition. This leads to the dominance of new species, expanded use of innovative behaviours, and an increase in populations of dominant species, which may, in turn, cause increased complexity of communities and ecosystems, and, therefore, larger biomasses and an increase in efficient use of resources.

For humans, the disruption of a stable climate puts a premium on physiological and behavioural adaptability, leading to the application of novel behaviours and new ideas that might have been obstructed as 'too revolutionary' or too much of a change from traditional practices in previously stable social groups. Some anatomical forms may become less common or disappear – e.g., gracile bodies in cold conditions; squat bodies in hot conditions. Migration from formerly supportive environments to a small number of refugia may result in the intermingling of previously isolated groups that have already developed novel tools and ideas. The combination of *crisis*, *communication* and *collaboration* is a powerful generator of emergent social novelty. New social wholes are greater than the sum of their parts.

As climate restabilizes, new human behaviours and techniques become traditional behaviours and techniques that are resistant to change. Populations expand, which may create economic stress (i.e., competition for resources) that brings internecine stress. This results in the emergence of new political organizations, such as social hierarchies involving division of labour and, consequently, a new distribution of resources. These developments put a premium on intergenerational transmission of knowledge (i.e., education).

During the last glacial maximum about 21 500 years ago, glacial ice covered much of the northern hemisphere. Subsequent climatic warming was interjected by a series of cold intervals caused by the injection of fresh water into the Atlantic Ocean, which interrupted its overturning circulation. The most renowned cold spike, the Younger Dryas, began about 12 700 years ago when the ice dam blocking massive Lake Agassiz in North America collapsed. It lasted just over a thousand years. Another cold event 8200 years ago resulted in a 4 °C drop in temperatures in Greenland for about 200 years. This short cold period interrupted the Holocene interval, an otherwise remarkably stable climate interval that has lasted nearly 12 000 years.

Coincident with the termination of the Younger Dryas and the onset of the relatively stable Holocene was the expansion of agriculture, which began between 12 000 and 8000 years ago at different locations around the world. This was the first time in Earth's history that a species consciously and systematically controlled its environment, generating additional stability in order to guarantee access to resources and increase the biomass of food plants and animals. The development of agriculture created a unique circumstance. Normally, climate stability would be conducive to dynamic stability and competition; biological communities would remain in a steady state for a prolonged period until the climate cycle disequilibrated things again. And in some sense this was true; the agriculturalists became dominant and killed off or otherwise outcompeted the hunter–gatherers. However, the onset of agriculture altered what was otherwise a stable state of resource availability, previously influenced only by climate.

In this case, climate stability facilitated the expansion of new ideas and technology (agriculture) that led to an increasing per capita surplus for *Homo sapiens*, a situation that previously occurred only when climate instability resulted in population declines. This surplus increased the region's carrying capacity for the native population, which in turn facilitated an increase in *Homo sapiens'* agriculturalist population and resulted in further control/manipulation of the environment and additional production increases. Population continued to grow, as did the number of population centres. This stimulated the development of innovative ideas in isolated population centres.

The cycle repeated. Self-induced states of disequilibrium, a consequence of the soil's increasing salt content and the depletion of natural fertilizers, led to conflict, disease and migrations away from environments that were no longer supportive. Migrating populations encroached upon fringe populations of non-agriculturalist aboriginal hunter–gatherers. The development of new civilizations in agriculturally rich regions again led to the acceptance of new behaviours and ideas, such as medicine and the use of fossil fuels. Populations increased again, further expanding the number of population centres. Groups placed increasing emphasis on accepted and successful industrialist practices. Migration was curtailed by political boundaries and the occupation of all available territories. Exploitation of fuels, minerals, timber and monoculture resources created a 'third world'. The population of this region had restricted access to technology and natural and medical resources, and its ability to migrate was limited to regions possessing less favourable conditions. Political instability was the common symptom of such limitations.

A noticeable effect of these later developments in human civilization has been climate change resulting from the use of fossil fuels and consequent greenhouse warming. Fossil fuels have been predominantly used by the industrialized nations as they developed their economies. However, the impact of climate change is global; it is not limited to 'first world' nations that may have the technological or economic ability to adjust to a changing climate. The seeds of discontent are therefore sown. These seeds could grow to resentment and hostility, or, if we take appropriate international steps to both mitigate climate change and assist non-industrialized nations in adjusting to its effects, they could blossom into a mechanism for creating global stability. In short, dealing with climate change is about dealing with domestic and global security.

1.5 What can be learned from evolutionary history?

Although this book is largely about climatic aspects of human history and the interaction between human activities and climate change, we must remember that we as a species are a recent step in the evolutionary journey and there were many pre-human interactions between environment and evolution. These early interactions provide us with principles that may help us understand the climate connection.

In the appendices on evolution, we discuss three arenas of evolutionary causation: developmental evolution; functional evolution, which incorporates physiological and behavioural causes; and association, which takes the form of symbiosis and socialization. All these evolutionary causes interact. They also operate within, and interact with, the physicochemical and biotic environment. In addition, all of them have molecular biological components.

We contrast two theories of evolution. The first, Darwinian theory, progenitor of the modern synthesis of evolution, emphasizes gradual change through natural selection over geological time. The second, emergence theory, treats evolutionary change as a saltatory (or rapid) process, combining internal organismal causes with environmental causes. Emergence theory therefore accepts the idea that the history of life takes the form of punctuated equilibria, with rapid emergent evolution punctuating the bulk of geological time, which is made up of states of dynamic stability in communities and ecosystems. The equilibrium phase is characterized by the operation of the syndrome of causes and effects generally referred to as 'natural selection'. However, natural selection is not evolutionarily creative, and emergent novelty is resisted by specialist organisms that are already highly adapted to prevailing conditions. Therefore for innovative evolutionary experiments to succeed, they must be able to either significantly outcompete the local specialists, hang on at the fringes where competition is reduced, penetrate new environments where resources are rich but competition low or benefit from natural catastrophes that either reduce the survivorship of dominant types or make them extinct. It follows that the greatest potential advantage of novel organisms is to be more adaptable than the older specialists. Emergent evolution can be shown to increase adaptability, often accompanied by multifunctional properties. But it does not follow that emergent types will enjoy immediate competitive success. This is illustrated by the coexistence of the mammals and dinosaurs through the Jurassic and the Cretaceous. Although more adaptable, the mammals could not become the dominant type until the dinosaurs had been wiped out at the end of the Cretaceous.

Emergence theory implies that with each major emergence there are not only novel qualities and interactions, but also a potential for further evolutionary diversification. However, emergences are largely unpredictable, and since they change the rules of the game, we can only gain a complete understanding of their evolutionary significance at the new emergent level. In our survey of evolutionary causes – developmental, functional and associative – we discuss the historical foundations of biological adaptability, which provides some general principles that apply to human evolution, despite the unpredictable nature of emergences. Adaptability is at a premium under conditions of stress and climatic instability, or on occasions where an unexploited new environment exists that cannot be entered by the specialists – only by organisms that are appropriately adaptable. The movement of physiologically adaptable animals from the sea to fresh water, or from fresh water to the land illustrates this point. Each migration removes competition and predation and makes new resources available. This mix of adaptability and migration applies to human evolution too. Even earlier in the history of life, the congregation of different types of microorganism potentiated symbioses that had greater adaptability and could exploit new environments. Indeed, all of the major ecosystems were founded on symbiosis. This too may be applied to human evolution: under stress conditions, formerly isolated groups might communicate and pool their resources, if they are wise enough not to kill each other off! Thus, crisis or catastrophe potentiates congregation, collaboration and the complexification of ideas. For a time the old ways are disregarded and novelty has a chance to succeed.

With reservations regarding the unpredictability of emergences and the limitations of using reductionist epistemology to take us from simple systems to complex systems, we have largely applied our evolutionary section to human evolution. Here, evolution in the broad sense is not only biological but also cultural and social. There are emergent novelties of cognition, language, tool manufacture, socialization, aesthetic and religious awareness, education and agriculture. In the main body of this book, we apply ideas from biological evolution to human cultural evolution, treating such emergences as rapid changes followed by periods of dynamic equilibrium where evolutionary changes are rare or are resisted by prevailing traditions. In particular, we have emphasized the significance of climatic instability and social crisis as conduits for experiment, novelty and adaptability.

Referring to what we have just surmised about the nature of evolution and the importance of crisis for releasing innovation, we might have predicted that the relatively stable climate of the last 11 650 years would have resulted in little real change on Earth, evolutionarily speaking. However, the development of agriculture after the Younger Dryas and the consequent success of the agriculturalists at the expense of the hunter–gatherers were important evolutionary stimuli, especially since crisis in the form of agricultural collapse and 'the depredations of warring savages' were common.[2] By controlling the environment, humans have created their own mechanisms for increasing per capita surplus. The problem is that this is unsustainable. We are living far beyond Earth's present capacity to provide, living on borrowed surplus, and in doing so we are altering Earth's delicate balance and throwing it into rapid climate change.

Currently, those least affected by climate instability are the wealthiest nations, which are best able to protect themselves from climate change, political unrest and destabilized economies by maintaining the status quo. However, evolutionary principles indicate that dynamic equilibrium obstructs novelty and the ability to adjust to change. It is those societies in which differences are established, encouraged and celebrated that will foster an attitude of respect for disparate cultures and worldviews. While a dominant culture narrows the number of acceptable behaviours and limits eccentric variations, thereby decreasing potential available alternatives during a period of change, a tolerant culture stimulates variety and increases the number of potential options available to implement change.

1.6 Back to the future

Despite or, more correctly, because of agriculture, technology and population expansion, Earth is once again in a state of climatic instability. However, in this case, *Homo sapiens* has instigated a climate warming unprecedented in our history, with scientific evidence suggesting carbon dioxide and methane levels are greater than they have been for at least 650 000 years.

The global human population has reached approximately 6.75 billion people. Recent research indicates it will reach over 9 billion by 2050. The consequential burning of fossil fuels now generates an additional input of 7 gigatons (billion tons) of carbon dioxide per year into the atmosphere. Given the current state of economic and technological expansion, this amount is expected to reach 12 gigatons by AD 2100.

Between the time of the coldest period of the last global ice age and the Industrial Revolution in 1850, the level of carbon dioxide in Earth's atmosphere increased by less than the increase experienced between the Industrial Revolution and today; the increase in the concentration of atmospheric methane is even greater. Yet, because of the delay in climatic response associated with the thermal inertia of the oceans, we have only just begun to feel the consequences of our burgeoning energy consumption.

Through technological innovation and behavioural adaptability, humans have managed a tremendous increase in population. We have mitigated climate change and maintained global dominance by stabilizing and controlling our environment. However, that dominance and environmental manipulation are now stimulating climate change that will increasingly result in economic, social and cultural instability.

The practices of presently dominant human societies are unsustainable, especially with a rapidly expanding global population. Climate models indicate that even if we rapidly reduce the level of greenhouse gas emissions, sea levels will continue to rise, global average surface-air temperatures will continue increasing and the incidence of severe droughts and floods will escalate. Although Earth's climate has continually changed in the past, we are now influencing and exacerbating that change, leading us to a place that humans may never before have experienced and to levels of carbon dioxide that have not occurred on Earth since the Mesozoic era, known as the Age of the Reptiles. All our cousin *Homo* species are now extinct; global biodiversity is plummeting; and non-dominant languages and cultures are disappearing at an alarming rate.

If we can better understand our past evolutionary relationship with climate, perhaps we can better prepare for our future – because a drastic change is needed in our behaviour to generate an immediate and global reduction in greenhouse gas emissions.

Notes

1. We use the word 'type' rather than 'species' to indicate, for example, the original placental mammals as a group rather than one particular species.
2. This is a quote from Thomas Robert Malthus's *Essay on Population* (1798). He believed that population expansion in the New World would be repressed by the native hunter–gatherers!

Part I

Early human history

2

From ape to human: the emergence of hominins

2.1 Introduction

Humans are not descended from apes. However, we do share a common ancestor and must accept that our place in the world has been, and continues to be, a part of the natural world. It was Aristotle's *scala naturae* that first placed humans at the top of the Ladder of Life, followed by apes, cetaceans, and other animals, in descending order of complexity, down to lower plants on the bottom rung. Aristotle perceived the ladder as a series; what he did not grasp was an evolutionary process that would see one type change into another. This concept of a primordial form that progresses in complexity and diverges in form, although frequently attributed to Charles Darwin, must be ascribed to Georges Louis Leclerc Buffon (1707–88), who stated: 'If, for example, it could be once shown that the ass was but a degeneration from the horse – then there is no further limit to be set to the power of nature, and we should not be wrong in supposing that with sufficient time she could have evolved all other organized forms from one primordial type' (quoted in Butler, 1879, pp. 90–1).

Charles Darwin's grandfather Erasmus postulated in *Zoonomia* (1794) that all life could have evolved from a primordial fibre. Lamarck (1744–1829) later wrote in his *L'Histoire naturelle des animaux sans vertèbres* (1815–22) that evolution was progressionistic – moving in the direction of complexity at a fixed rate. But there are many who would advocate Charles Darwin's 'notion of a steady progression of more and more complex organisms as a result of natural selection … [as] the greatest intellectual and philosophical revolution in human history' (Leakey and Lewin, 1977, pp. 16–17). Darwin's book *On the Origin of Species by Means of Natural Selection, or the Preservation of Favoured Races in the Struggle for Life* (1859) was published at a time when the overarching paradigm – the Judeo-Christian story – set Creation at 4004 BC and credited the immediate manifestation of humans in the fully 'modern' form to the appearance of Adam and Eve. This view is still strongly held by some. However, fossils of animals no longer alive, combined with geological evidence,

15

including that provided by Charles Lyell's *Principles of Geology* (1830–33), made it unavoidably clear that humans lived on a planet of great antiquity.

Darwin's supporter T. H. Huxley, impatient with Darwin's reluctance to get into arguments about human evolution, published *Evidence as to Man's Place in Nature* in 1863. It was not until 1871 that Darwin, in *The Descent of Man, and Selection in Relation to Sex*, made explicit his views on 'man's' place in the history of evolution. Darwin inferred that Africa was the cradle of humankind and depicted humans as descendants of apes, similar in form, physiology, susceptibility to disease, instinct, emotion and social habits. 'Toolmaking, terrestriality, and social life composed the vital distinctions, the foundation on which Darwin's scenario of human emergence was built' (Potts, 1996a, p. 8). The more threatening existence on the treeless savannah required that the ancestors of humans, lacking large eyeteeth, create tools to protect themselves. Toolmaking on the savanna, then, was the primary adjustment, generating the conditions for walking upright, facilitating brain enlargement and encouraging the development of culture and social contracts.

Today, we rely on genetic theory to explain diversification in organisms. But despite many recent scientific achievements we possess little detailed knowledge of what genetic or developmental variations (e.g., language, cognitive abilities, longevity, modes of learning communicating and aging) set archaic humans apart from apes (Pääbo, 2003). There are no genes for these characteristics; there are only genes that participate in events at the cellular and organismal level. The degree to which apes possess culture – 'rich behavioural complexity' (Whiten *et al.* 1999, p. 685) – and language has only recently been described (see Tomasello, 1999; Tomasello and Call, 1997; Whiten *et al.*, 1999). As a consequence, Pääbo (2003) states, 'We have come to realize that almost all features that set humans apart from apes may turn out to be differences in grade rather than absolute differences.' He adds that 'many genes that enable humans to perform tasks of interest may exert their effects during early development where our ability to study their expression both in apes and humans is extremely limited' (p. 411). However, this statement makes the genes the driving force of development, which is not true. Begging the question of developmental causation, and hiding its proof under technical difficulty, does not advance our understanding of the matter.

For over a century, the fossil record of human evolution was assumed to suggest that Darwin's progression of speciation from ape to 'modern' humans was plausible. At each stage of progression in the fossil record, there were to have appeared so-called transitional forms displaying the morphological transitions that spanned the long evolutionary distance between apes and humans. Sometime along this progressive ladder leading to human modernity, a defining moment occurred that marked the beginning of the history of 'modern' humans – the appearance of stone tools.

Although it is common practice to recognize species' diversity in other animals, palaeoanthropology maintained a narrow linear perspective of human evolution; it lacked any species' diversity. This unilinear progressionistic perspective dominated the interpretation of human evolution until the 1960s, when evidence of more than one, and potentially three, contemporaneous hominid species began to appear. The discovery of multiple hominid species necessitated the reformulation of the definition of what it was and is to be human. The criteria became absolute brain size (600 cm^3), perceived capacity for language ability (as inferred from endocranial casts), hand function (precision grip and opposable thumb) and stone toolmaking. However, these criteria remain unsatisfactory for a variety of reasons: the biological significance of cranial capacity is questionable; the function of language cannot be reliably inferred from brain appearance, nor are language-related parts of the brain clearly localized; the 'modern' human-like hand grip cannot be restricted to *Homo*; early stone tools dating to more than 2.3 million years ago are contemporaneous with both *Homo* and *Paranthropus*, from the Australopith genera (Wood and Collard, 1999).

Stedman and colleagues published a paper in *Nature* magazine (2004) that describes what appears to be the first genetic mutation in the human genome that roughly coincides with 'human-like' characteristics: smaller jaw muscles, which potentially 'removed an evolutionary constraint on encephalization' (p. 418) (see also Currie, 2004). A mutation in the human gene MYH16 prevents the high level of protein build-up in the jaw muscles of humans. However, it is not clear whether the build-up of that particular protein is entirely limited to the jaw. If it is, then there have to be other mechanisms that confine it there. In any event, according to Stedman *et al.*, all non-human primates have an intact copy of the gene and, as a result, their jaw muscles grow larger, potentially constraining their brain size. Stedman *et al.* place the appearance of the mutation at about 2.4 million years ago, just before the evolution of the 'relatively gracile masticatory apparatus [that] appears in *Homo erectus/ergaster* by 1.8–2.0 Myr [million years]' (pp. 417–8). Based on these findings, they link humanness to the appearance of the mutation in the MYH16 gene.

But has this reductionist approach truly led to an understanding of what it is to be human? Genomists fundamentally believe that natural selection affects both the results of gene shuffling and the results of non-synonymous point mutation. Correlated with that, they believe that the establishment of 'successful' point mutations is slow but constant (averaged over millions of years). One of the problems that genomists are beginning to recognize is that mutation rates per se cannot explain the major phenotypic differences between related species such as the bonobo chimpanzee (*Pan paniscus*) and the human. These differences can only be explained in terms of epigenetics – the regular process of development, including evolutionary events that change that path of development – of which gene expression is only a part, along with gene and chromosome duplication and differentiation and physiological factors. In any case, there is only a 1.2% difference in the DNA at the nucleotide level between the two species (Britten, 2002), little more than the average 0.2% range of variation within *Homo sapiens* (Chakravarti, 2001; Sachidanandam *et al.*, 2001) and less than the variation among chimpanzees. Thus, looking for explanations of divergence in terms of either point mutations or recombination may in fact be futile. That intraspecific and interspecific differences at those levels suggest drift or neutrality (Hellmann *et al.*, 2003) is not a surprise. The most distinct differences (divergences) between say bonobo and human – hair loss, increased bipedalism, new hand anatomy, reduction in facial bone growth and relative expansion of the neocortex – are all epigenetic, mainly involving differences in the distribution in space and time of hormones and their receptors. Those molecules need not change at all in their structure to produce the effects. That means that you will not find explanations by looking for structural gene mutations and gene shuffling.

As Pääbo (2003) suggests, new approaches are required that study the interaction of genes and their influence on developmental and physiological systems. Reid (1985) and Oyama (1985) have long argued that we cannot have a theory of evolution without theories of development and physiology. Evolution results in increased levels of organismal complexity. But, if natural selection creates stable equilibria both in development and physiology, what makes complexification possible? The answer, Oyama and Reid argue, lies in a system of interacting factors, including environment, behaviour, physiology, development and gene expression (see Appendix A, Sections 2 and 3, for a discussion).

Palaeoanthropologists' insistence on maintaining an evolutionary continuum from ape to human, and the fact that early hominid fossils were discovered after comparative anatomists had decided on human diagnostic features, has made it necessary to adjust the 'human-indicative' criteria as new evidence is uncovered. Further, this monophyletic, progressive preoccupation has been intrinsically linked to, and perhaps limited by, Darwinistic selectionism and the notion of gene-determined, sequential adaptations from ape to human. According to palaeoanthropologist Jeffrey Schwartz (1999), 'While other disciplines in evolutionary science have expanded their appreciation of different possible mechanisms and processes of evolutionary change, the study of human evolution had been firmly planted in the traditional dogmas of Darwinism, including the belief that evolutionary change manifests itself only through an insensible series of infinitesimally small modifications' (p. 33).

It is important to keep in mind that 'modern' human origin theories are tested based on the presence or absence of a suite of 'modern' behavioural traits. The implications of these theories are dependent on a series of assumptions about behavioural modernity, and their validity is only as good as the assumptions made (for an interesting discussion see Henshilwood and Marean, 2003). Let us look more closely at the emergence of 'modern' humans and the traits used to define 'modern' human behaviour.

2.2 The emergence of anatomically 'modern' humans

2.2.1 Where and when did 'modern' humans (Homo sapiens) first appear?

Significant scientific debate centres on whether living humans (*Homo sapiens sapiens*) are the direct descendants of *Homo erectus*. Until the mid-1980s it was believed that a single hominin species, *H. erectus*, left Africa for Europe and Asia around 1 million years ago, taking with it early Acheulean stone tools, a prerequisite of dispersal. (Note: the term 'hominid' here refers to all the great apes, whereas hominin refers to the bipedal apes, including all the fossil species and living humans.) However, a second, more recent theory suggests that *H. erectus* split into two species: *H. erectus* and *H. ergaster*. According to the second theory, *H. erectus* went east to Asia and later became extinct. The earliest evidence of hominins in Asia dates to Pakistan by 2.0 million years ago (Dennell *et al.*, 1988a; Rendell *et al.*, 1987); to Dmanisi, Georgia, in westernmost Asia, between about 1.7 and 1.9 million years ago (Gabunia, Vekua, Lordkipanidze, Swisher *et al.*, 2000; Gabunia, Antón *et al.*, 2001); and to Java, South East Asia, at least 1.5 million years ago (Larick *et al.*, 2001). The two partial hominin crania found at Dmanisi have been assigned to *H. erectus*; they are relatively small compared with the early African *H. erectus* and the larger Javanese *H. erectus* fossils. Unfortunately, these early fossil dates have been fraught with controversy. However, in relatively cold northern China, recently discovered stone tools have been carefully dated to 1.36 million years ago and demonstrate not only the early presence of *H. erectus* in Asia, but also the capacity of *H. erectus* to deal with challenging and seasonal environments (Gibbons, 2001; Zhu *et al.*, 2001).

Many archeologists believe the second species, *H. ergaster*, is our direct ancestor. *H. ergaster* lived along the shores of Lake Turkana, Kenya, up to 1.9 million years ago (Wood and Collard, 1999). But, surprisingly, the first evidence of an ancestral species that combined *H. ergaster/H. erectus* traits with 'modern' traits has been found outside Africa. The potential common ancestor of Neanderthals and 'modern' humans has been excavated from the Gran Dolina site in the Sierra de Atapuerca of north-central Spain and possibly from the Ceprano site in southern Latium, Italy; this proposed new species, *Homo antecessor*, dates to more than 780 000 years ago at Atapuerca and between 800 000 and 900 000 years ago at Ceprano (Bermudéz de Castro *et al.*, 2004; Falguères *et al.*, 1999; Manzi *et al.*, 2001). The Gran Dolina fossil hominins possess a 'modern' face with primitive dentition; the Ceprano fossil hominin likewise possesses characteristics intermediate between *H. erectus/ H. ergaster* and 'modern' humans. If these interpretations are correct, it is possible that *H. antecessor* represents a speciation event that occurred in Africa or Eurasia coincident with a major dispersal event out of Africa at around 900 000 years ago.

Further complicating the separation of early *Homo* spp. are hominin finds dating to approximately 1.0 million years ago recently found in association with Acheulean stone tools from Middle Awash, Ethiopia. These fossil hominins are intermediate between earlier and later African fossils. Measurements on the fossil crania indicate an overlap between Asian and African sample ranges, making it difficult to distinguish these hominins as belonging to either an Asian or African sample. In light of these findings, Asfaw *et al.* (2002) suggest that separating *H. erectus* into separate species is 'misleading' and that *H. erectus* is the most likely ancestor of *H. sapiens*. They further suggest that by about 1 million years ago, *H. erectus* had colonized much of the Old World without speciating, but by around 500 000 years ago, hominin speciation may have begun. They recommend additional research to determine the rate of morphological change and to ascertain whether linkages exist between their interpreted rapid hominin speciation and large-magnitude global climatic oscillations.

Irrespective of whether *H. erectus* or *H. ergaster* was the ancestor of 'modern' humans, researchers continue their search for the origin of the first anatomically 'modern' humans. That is, based on the shape of their bones, when and where did hominins first develop into the *H. sapiens* species? Geneticists, using mutation rates to set the human 'molecular clock', suggest that 'modern' humans are the descendants of an ancestral African population living between 100 000 and 200 000 years ago (Balter, 2002; Bräuer, 1989; Stringer, 1989, 1990). Fossil experts are generally in agreement, basing their interpretation on the crania from Herto, Middle Awash, Ethiopia – dated to between 160 000 and 154 000 years old, which is intermediate between archaic and anatomically 'modern' humans – as well as on the anatomically 'modern', ~195 000-year-old *Homo sapiens*' fossil remains found at Omo Kibish in Ethiopia, and 100 000-year-old fossils from Klasies River Mouth, South Africa (Clark *et al.*, 2003; McDougall *et al.*, 2005). Alternatively, some anthropologists, including Wolpoff (1989), argue that humans evolved simultaneously across the globe beginning some 2 million years ago.

These contrary perspectives form the basis of the two main competing theories explaining the origins of 'modern' humans. One is the 'multi-regional evolution' or regional continuity hypothesis (see Wolpoff, 1989); the other is the 'out-of-Africa' or 'Garden of Eden' hypothesis (see Stringer and Andrews, 1988).

M. H. Wolpoff championed the multi-regional hypothesis and suggested that human populations evolved more or less in parallel in all the major regions of the

Old World (i.e., Africa, Asia and Europe) over at least the last million years. One of the most significant points of conflict between the two theories is the timing and character of the evolutionary divergence of modern regional diversity among present-day populations. The multi-regional hypothesis implies long time-scale evolutionary processes and assumes that adaptation to environment is the same, whether in China, Australia or Europe. The differences observed between Old World and New World monkeys make it difficult to substantiate such a hypothesis biologically. Different environments elicit different behaviours and different evolutionary tracks. New World monkeys are anatomically distinct from Old World monkeys, having flat faces, prehensile tails, and, hence, different behaviours. Geographically divergent types might have distinct differences to begin with, but their encounters with different environments surely affect their evolution.

C. Stringer first defined the "out-of-Africa" model, proposing that anatomically and genetically 'modern' humans originated in Africa sometime after 200 000 years ago, with most dates converging on 130 000 years ago, and subsequently dispersed throughout the world over the last 100 000 years. The out-of-Africa model assumes short-term evolutionary processes, and the strictest version theorizes that there was no genetic contribution from any other early humans (e.g., the Neanderthals) to the modern human gene pool. The limited genetic diversity apparent in modern humans is considered the result of a bottleneck in our recent evolutionary past, when the total human population dropped to just over 10 000 individuals. Prior to the bottleneck, the population is estimated to have numbered around 40 000 individuals, suggesting to some that such limited diversity of mitochondrial DNA could not have facilitated the genetic diversity apparent between Neanderthal and modern humans (Lahr and Foley, 1998; Sherry *et al.*, 1997).

Proponents of the multi-regional model emphasize continuity, with populations transformed gradually through processes of natural selection. Yet the congruent neo-Darwinistic idea of allopatric speciation is based on different adaptations to different environments, particularly continuous variation, progressive adaptive change and gene flow. This is not consistent with the multi-regionalists' assumption of coincident evolution of *Homo sapiens* in various regions of the world. In contrast, the out-of-Africa model assumes extensive migrations of small groups, emphasizing divergence and subsequent replacement of species, paying little attention to gene flow or natural selection mechanisms. We concur with Lahr and Foley (1998) that these fundamental problems are the basis of the difficulties with the human origin hypotheses and that a third position is required that addresses a single origin in Africa and considers 'broader evolutionary issues relating to human diversity and evolutionary mechanisms' (p. 143). But it is not just the lack of empirical support for the multi-regional model, despite its consistency with population genetics, nor the 'outmoded evolutionary ideas' (p. 143) of the out-of-Africa model that we see as the problems. Whether blindly accepting the mantra of natural selection and the modern synthesis of biology in the case of the multi-regional model, or simply ignoring it in the case of the out-of-Africa model, neither theory addresses the questions left unanswered by the modern synthesis and natural selection, which include: what causes emergence in the first instance, and how does emergent evolution occur in the face of change-resistant stasis generated by natural selection? It is our thesis that the environment, including rapid changes in climate, generates the conditions of disequilibrium that overcome periods of stasis. Environmental change affects species' epigenesis – embryonic and juvenile development, physiology and behaviour. The *Homo* species is no exception. But before delving into these issues further, it is necessary to clarify what we mean by the 'modern' human.

2.2.2 What is the definition of a 'modern' human?

To assist in resolving the ancestral history of 'modern' humans and to ascertain whether Neanderthals were truly a different species, researchers have analysed hominin fossils in order to identify morphological characteristics, calculate brain size and analyse brain structure and development processes. Fossils contain ancient genes that are used by geneticists to perform genetic comparisons between species and to piece together the history of genetic changes or adaptations. Archeologists seek evidence from stone tools, structures and art discovered in archeological deposits and use these findings to make interpretations about the behaviour of 'modern' humans, including symbolic expression, general intelligence or cognitive abilities, language and foresight and social behaviour. However, these interpretations remain fraught with controversy. Often archeologists assume that behaviour is gene determined, so the most adaptive patterns are naturally selected. However, such behaviours could not be changed according to circumstances except through generations of random gene mutation and selection. Human emergent properties, such as intelligence, foresight and planning, require that behaviour patterns be flexible, rather than genetically fixed.

2.2.2.1 Adaptation and adaptability

In the arena of human action, we commonly use the word 'adaptation' when we are talking about behavioural change. We 'adapt to the cold' by putting on more clothes and lighting fires. However, the word 'adaptation' can apply to two distinct phenomena. The first is the Darwinian natural selection of genetic change appropriate to particular conditions. For example, 'adaptation to the cold' in such a case might refer to hereditary changes in subcutaneous adipose tissue, hair structure and general anatomy. The second is adaptability: the physiological and behavioural self-modification made by individuals in the face of changing conditions – donning clothes and lighting fires when the temperature drops. Common usage does not often distinguish between adaptation and adaptability. This lack of discrimination carries over into the literature of biology and anthropology, so the two distinct categories tend to remain confused. The result is that adaptability is falsely assumed to be a subset of Darwinian adaptation. Consequently, in biological writing, the evolution and importance of adaptability are underplayed. In some anthropological writing, the importance of physiological, behavioural and cognitive adaptability is recognized. However, it is often assumed to have evolved through an accumulation of Darwinian genetic adaptations. There are no sound grounds for that assumption, yet its application by evolutionary psychologists to anthropology further assumes that contemporary humans are adapted, in the Darwinian sense, to the particular environmental conditions of the Pleistocene.

Darwin himself excluded adaptability from the category of adaptation, referring to it as an 'innate flexibility of constitution' (Darwin 1872, p. 107) that had *not* been the product of natural selection. Yet the progressive improvement of that innate flexibility is what has made organisms in the vertebrate lineage more *evolvable*, in the sense of potentiating the more visible diversification of functional anatomy. For example, the emergence of a placenta rounded off a series of advances in physiological adaptability that allowed the placental mammals to diversify into tigers, horses, bears, bats, whales and humans. The lifelong ability to maintain a dynamic internal homeostasis under changing conditions, coupled with exploratory behaviour, permitted placental mammals to penetrate cold, hot, wet and dry environments.

When selecting criteria that distinguish 'modern' humans from other species and thus ascertaining species phylogenies, the assumption is made that 'modern' humans are genetically distinct and that offspring are reproductively viable. Obviously, tens of thousands of years later, it is difficult to determine whether this is true, for instance, between Neanderthals and 'modern' humans. Scientists use similarities in fossil morphology, brain size and structure, genetics and artifacts to identify what makes us unique. The interpretations are not always consistent. Further, some of that uniqueness is the result of genetic adaptation, while some is the result of developmental, behavioural and physiological adaptability. However, this distinction is not always recognized, further complicating the issue.

Genetic adaptation is the result of slow gradual genetic change as defined by Darwin. It is slow to respond to environmental change, whereas developmental, behavioural and physiological adaptability can respond rapidly to sudden climate change. During a period of climate stability, the importance of this distinction may not be particularly noticeable or relevant to our species, but during a period of rapid climate change, adaptability is the characteristic that gave humans an opportunity to respond quickly.

2.2.3 *Morphology*

Hominin fossil bones provide information about body shape and design that anthropologists use to understand locomotor behaviour, temperature regulation, water balance and habitat. Body size is used as an indication of primate population density, social organization, range size and initial breeding age. But making inter-pretations based on fossil evidence is complex. The human fossil record is limited in its capacity to provide information about early human genetic adaptations and developmental, physiological and behavioural adaptability.

Fossils show a similarity between Neanderthal and anatomically 'modern' human physiques, indicating that they were both bipedal and lived in an open habitat. Fossils also show that the only *Homo* species that is similar in brain size to *H. sapiens* is *H. neanderthalensis*. Interestingly, when scientists group hominin fossils by brain size the resultant groups are different from the groups that result when they gather hominin fossils by body size and shape, skeletal concomitants of locomotor behaviour, rate and pattern of development and relative size of masticatory apparatus. These findings suggest that grouping fossil hominins into species is a multi-faceted process and, further, that the correlation between brain size and adaptive zone is complex (Wood and Collard, 1999; also Deacon, 1990).

Neanderthal crania differ from anatomically 'modern' *H. sapiens* in a suite of features. For example, Neanderthal fossils possess a low, elongated and broadened braincase. The large faces of the Neanderthal contained rounded eye orbits, a wide nasal aperture, large brow ridges and a receding chin. Alternatively, *H. sapiens'* fossil crania from the same period are high and rounded, possessing a chin, a small orthognathic face, reduced masticatory apparatus and small brow ridges (Department of Informatics, 2003; McBrearty and Brooks, 2000; Ponce de León and Zollikofer, 2001). Neanderthal skeletal fossils also differ from anatomically 'modern' *H. sapiens* in a number of ways, including longer collar bones and a stocky, barrel chest; broad elbow and knee joints compared to the main bone shafts; a long pubic bone, with the bridge at the front tall from top to bottom, but thin from front to back; and robust hands with broad fingertips (Dolan DNA Learning Center, 2003).

These combinations of characteristics has led some researchers to suggest that Neanderthals were a separate species. Alternatively, it has led others to argue that these features – for example, the stocky body build – would have allowed Neanderthals to better retain heat during the cold of the last glacial age. Contrary to early suggestions that Neanderthals were apelike in their stance, scientists now suggest that their posture, muscle and limb function were similar to 'modern' humans.

Some researchers suggest that most of the skeletal differences are a result of increased muscularity of Neanderthals and do not represent significant beha-vioural differences, particularly locomotory.[1] Others note that the partial *H. sapiens'* skeletons uncovered at the sites of Skhul and Qafzeh, Israel, and dated to between 119 000 and 85 000 years ago (Valladas *et al.*, 1998), possess elongated limbs and linear physique combined with a low estimated body mass, which resembles the characteristics of recent humans from hot, dry climates and contrasts with the heavier, shorter-limbed Neanderthals, who more resemble recent humans from the Arctic (Holliday, 1997a, 1997b; Pearson, 2001; Ruff, 1994; Trinkaus, 1981). Anthropologist Osbjorn Pearson (2001) suggests that any definition of anatomically 'modern' form should recognize the fact that some present human populations have evolved different body forms in response to climate and that those forms are substantially different from those of early 'modern' humans. Further, analysis of fossil postcranial skeletons indicates that the 'modern' postcranial skeletal form appeared surprisingly recently, about the time of the last glacial maximum. According to Pearson, changes in skeletal morphology can arise from genetic variation and result from adaptations to climate, as well as variations in degree or kind of habitual activity.

According to Lieberman *et al.* (2002), brain shape, specifically the change from a prognathously thrust-forward face and eyes to one which is more tucked under the braincase, and the shift from an oblong to a globular-shaped head, are the key features that distinguish 'modern' (*H. sapiens)* skulls from early hominin skulls. Further, he suggests that these two features would have been produced by small evolutionary changes that occurred relatively abruptly rather than through gradual evolution (Lieberman *et al.*, 2002; see also Balter, 2002). If so, an explanation other than slow, gradual genetic evolution based on natural selection is required. But a recently discovered *H. erectus* skull in Java may obligate a modification of this theory (Lieberman *et al.*, 2003). The new Java skull has morphology intermediate between skulls from central Java, dating between 1.0 and 1.5 million years ago, and the youngest fossils from eastern Java, dating to between 400 000 and 50 000 years ago and even earlier (Swisher *et al.*, 1996), suggesting not only that *H. erectus* lived continuously in Java for around one million years, but also that because the base of the cranium looks 'modern', with a strongly flexed cranial base similar to that of *H. sapiens*, other developmental changes may have led to the 'modern' human brain.

Computerized fossil reconstructions that distinguish between unique craniofacial features of Neanderthals and *H. sapiens* suggest Neanderthals were a separate evolutionary lineage, distinct from *H. sapiens* for at least 500 000 years. This implies a divergence prior to the emergence of 'modern' humans at about 130 000 years before present. Ponce de León and Zollikofer (2001) suggest that characteristic cranial and mandibular differences arose early during development, possibly prenatally, were maintained during postnatal ontogeny, and show hypermorphosis – through faster rates of growth and development. They further suggest many aspects of Neanderthal mid-facial prognathism may be the result of more rapid rates of fetal growth and/or subsequent reversal of growth fields, and that the differences observed between Neanderthal and 'modern' human skulls is the result of ontogenetically early differences in the rates and timing of action at particular growth fields. Thus, these differences may be better understood as developmental adaptabilities as opposed to genetic adaptations (see discussion of allometry in Appendix A, Section 2). Unfortunately, Ponce de León and Zollikofer do not explain what they mean by 'growth fields' or how growth fields are differentiated in their activities, making interpretations about developmental evolution problematical.

Skeletal remains found at archeological sites throughout Europe have further stimulated debate on the morphological and genetic similarity between Neanderthal and anatomically 'modern' humans. Evidence in favour of the out-of-Africa model, in which 'modern' humans originated in Africa and subsequently displaced all archaic humans in other regions, includes anatomically 'modern' skeletal remains uncovered from two sites in northern Israel, Mugharet es Skhul and Jebel Qafzeh, that date to between 119 000 and 85 000 years ago. These appeared at a time when more 'archaic' forms, including the Neanderthals, were living in adjacent or over-lapping areas. Additional 'anatomically modern' finds in Border Cave and Klasies River Mouth in South Africa, dating between 80 000 and 118 000 years ago (McBrearty and Brooks, 2000), and Omo in Ethiopia, dating to 195 000 years ago (McDougall *et al.*, 2005), contrast with a typically Neanderthal skeleton found at Saint-Césaire in western France, dated to about 41 400 years ago (Lévêque and Vandermeersch, 1980; Mercier *et al.*, 1991), and appear, at least initially, to support the out-of-Africa model for the dispersal of 'modern' humans.

However, archeologists disagree over the identity of the Saint-Césaire skeletal remains. Some argue that researchers who imply the Saint-Césaire skeletal remains are Neanderthal do not take sufficient account of anatomically 'modern' human variability within populations. They suggest that recent hominin finds in South East Asia, Australasia and central Europe indicate morphological and genetic continuity between 'archaic' and 'anatomically modern' populations (Thorne and Wolpoff, 1992; Wolpoff, 1989). In the same vein, the recent discovery of a four-year-old child from Lagar Velho in Portugal's Lapedo Valley that dates to approximately 24 500 years ago, several thousands of years after Neanderthals are thought to have disappeared, indicates the child possessed morphological characteristics potentially inherited from Neanderthal ancestors (Duarte *et al.*, 1999). The child skeleton has the short arms and broad trunk typical of the Neanderthal, combined with a 'modern'-looking chin and pubic bone. The skeleton is interpreted as evidence that Neanderthals and anatomically 'modern' humans interbred and produced fertile off-spring for many generations. Similar claims are made about Neanderthal remains found in Vindija Cave, Croatia, which show assimilation of early 'modern' features (F. H. Smith *et al.*, 1999; Wong, 2003). Such hybridization would imply that the *H. sapiens* and Neanderthals were the same species and that the genetic traits of the more populous 'modern' humans overwhelmed those of Neanderthals.

These interpretations support a rejection of the extreme out-of-Africa model. The Lagar Velho child evidence interpreted as just indicated suggests that although anatomically 'modern' humans may have come out of Africa, they mixed with archaic populations encountered in Europe. Further, Neanderthals and anatomically 'modern' humans would have had similar social behaviour and means of communication, including language, making Neanderthals simply another group of Pleistocene hunter–gatherers (Trinkaus and Duarte, 2003). However, not everyone accepts this interpretation. Others argue that the Lagar Velho skeleton lacks any indication of Neanderthal morphology and is simply a more chunky 'modern' human child, whose body structure might reflect an adaptation to the cold climate in Portugal at that time (Tattersall and Schwartz, 1999; Wong, 2003).

One of the reasons why archeologists' opinions vary so drastically in their attribution of fossil hominins to species is because those attributions are frequently based on cladistic analysis of cranial and dental evidence. Although it is true that a reliable phylogeny gives us confidence in hypothesizing our own and other species' ancestors, and provides an opportunity to link human evolution with changes in the environment and ecological influences, unless those hypothesized phylogenies are reliably validated, we risk generating unfounded species and relationships and species-based interpretations. Craniodental characteristics of fossil hominins have typically been used to generate hominin phylogenetics in the past, but based on research on hominoids (gibbons, apes – chimpanzees, gorillas, and orangutans, and humans) and papionins (baboons), it is apparent that phylogenetic hypotheses based on craniodental data can be incompatible with phylogenies based on molecular data for the same groups. Broadening the set of characteristics used so they also include postcranial and soft-tissue characters has yielded more positive results, indicating a need to pay greater attention to nonmorphological characteristics (Collard and Wood, 2000).

2.2.4 The brain

2.2.4.1 Brain size

It has long been argued that human brain size is directly correlated with intelligence, and intelligence is likewise associated with the progressive evolution from ape to 'modern' human. In 1861, Paul Broca concluded: 'In general, the brain is larger in mature adults than in the elderly, in men than in women, in eminent men than in men of mediocre talent, in superior races than in inferior races' (p. 304). 'Other things equal, there is a remarkable relationship between the development of intelligence and the volume of the brain' (p. 118).

Later Broca nearly abandoned his thesis that brain size records intelligence when he discovered that the brain size of peoples of the Mongolian type could surpass that of the 'civilized' peoples of Europe. Although he concluded that 'lowly' races could also have large brains, he nevertheless insisted that small brain size was a mark of inferiority (Broca, 1873 cited in Gould, 1981, p. 87). Modern female brains weighed on average 14% less than modern male brains, substantiating a similar argument about the intellectual capacity of women (Broca, 1873 cited in Gould, 1981, p. 103). As Gould (1981) states, 'An unbeatable argument. Deny it at one end where conclusions are uncongenial; affirm it by the same criterion at the other. Broca did not fudge numbers, he merely selected among them or interpreted his way around them to favored conclusions' (p. 87). In fact, as Gould states, brain size 'increases with body size, decreases with age, and decreases during long periods of poor health' (p. 89).

Broca also described a sample of Cro-Magnon skulls, only to find that their cranial capacity exceeded that of modern Frenchmen, a finding that did not concur with his thesis. However, the sutures between the skull bones of the Cro-Magnon skulls closed earlier in the anterior region of the brain – the region where higher mental functions were localized – restricting the development of the frontal cortex and thus limiting the developmental period of the more complicated intellectual parts of the brain. Broca reasoned that the posterior regions of the brain were associated with more mundane, primitive and emotive roles. This explained the larger Cro-Magnon cranial capacities (quoted in Gould, 1981, p. 98).

It is widely accepted that the transition from *H. erectus* to *H. sapiens* involved an increase in brain size, although the underlying morphogenetic basis for that change is only beginning to be explored (Baba *et al.*, 2003; Lieberman *et al.*, 2003). However, the correlation between brain size and the intelligence of anatomically 'modern' humans versus archaic humans has drawn significantly more contentious debate.

In his discussion about the difficulties in drawing conclusions about the Neanderthal mind, archeologist Paul Mellars (1996) makes two basic assumptions. First, that 'we should assume that there were not significant contrasts between the mental capacities of Neanderthal and "modern" human populations, and that such contrasts should be inferred only after a rigorous and essentially skeptical analysis of the available archaeological evidence makes this conclusion at least highly plausible if not inescapable' (p. 366). Second, that 'we cannot rule out the possibility that there were significant differences in the mental or cognitive capacities of Neanderthal populations' (p. 366). This is because

… it is inherent in evolutionary theory that the brain has been no less subject to adaptive [improvements] over the time-span of human evolution than have other physical and biological features of early populations. There is no doubt that modern human brains are significantly more complex, highly structured, efficient and intelligent than those of even the most intelligent great ape … and this increase in intelligence has clearly taken place at various stages during the last five million years or so.

(pp. 366–7)

These opinions do not, however, define the nature of the progressive evolution of the human brain. It certainly cannot be simplistically assumed that the process was the natural selection of point mutations of genes for brain growth, intelligence, language etc.

Holloway (1985 cited in Mellars, 1996, p. 368) suggests that although Neanderthal brains appear to be slightly larger (~1400 cm^3) than those of most modern human populations (1300–1400 cm^3), much of this variability is related to body size and other environmental facts including metabolic costs of maintaining brain functions under various climatic conditions (pp. 319–24). But Mellars (1996) says that, by itself, brain size is a poor indicator of overall intelligence levels. Martin concurs, indicating that the biological significance of absolute cranial capacity is open to question (Martin, 1983; see also Wood and Collard, 1999).

But all this toing and froing about brain size may be moot after the recent discovery at Flores, Indonesia, of what is believed to be a new species of *Homo* from the late Pleistocene. *Homo floresiensis* dates from before 38 000 years ago to at least 18 000 years ago (P. Brown *et al.*, 2004; Morwood *et al.*, 2004). This endemically dwarfed species of *Homo* walked on two legs, was only about one metre in height, and possessed an endocranial volume of 380 cm^3, which is outside the previously accepted range of *Homo* species and equal to the smallest estimates for *Australopithecus*, which lived over 3 million years ago. Yet this small-brained hominin was found with stone artifacts, including a formal tool component associated solely with the dwarf *Stegodon* (an extinct elephant-like mammal). *H. floresiensis* disappeared about the same time the *Stegodon* disappeared. Associated deposits possess animal remains, including the Komodo dragon. The earliest archeological evidence of hominins on Flores dates to approximately 840 000 years ago (Morwood *et al.*, 1998). Morwood *et al.* (2004) suggest that *H. floresiensis* is a descendant of *H. erectus* that survived in refuge on the island of Flores and experienced subsequent dwarfing or, alternatively, it may be an unknown small-bodied, small-brained hominin that arrived on the island from elsewhere on the Sunda Shelf (P. Brown *et al.* 2004).

Irrespective of where it came from or how it got there, *H. floresiensis* overlapped with the first arrival (between 55 000 and 35 000 years ago) and continued presence of *H. sapiens* in the region. Further, this small-brained hominin, or its predecessor, was able to migrate across the sea and colonize Flores during the early Pleistocene, a time when other land mammals were unable to make the crossing, even when sea levels were lower during glacial periods.

The *H. floresiensis* find raises serious questions about our previous commitment to correlating large brain size with intelligence and toolmaking, as well as questions about our definition of human modernity. As P. Brown *et al.* (2004) state, these small-bodied *Homo* sp. fossils force us to recognize that 'the genus *Homo* is morphologically more varied and flexible in its adaptive responses than is generally recognized' (p. 1061) and that 'body size is not a direct expression of phylogeny' (p. 1060). In this case, 'adaptive responses' are equivalent to adaptability, not genetic adaptation. Further, the tiny brain size of these obviously capable tool-makers suggests that scientists need to consider neurological organization, as opposed to focusing strictly on brain size, when ascertaining behavioural and intellectual complexity. In other words, this unusual and relatively rapid change in morphology and brain size may be a result not of slow and gradual genetic adapta-tion, but of anatomical and physiological adaptability in response to a rapidly changing environment. And if brain size is a problematic indicator for intelligence and therefore, by suggestion, for 'humanness', do other brain-related features exist, including brain structure and development, which may make better indicators?

2.2.4.2 *Brain structure and development*

Research by Gannon *et al.* (1998) indicates that the anatomical left-hemisphere predominance of the planum temporale (PT) – a region of the brain critical in auditory association, as well as in the production and comprehension of language, both spoken and signed – is clearly not unique to humans. Of the 18 chimpanzee brains they studied, 17, or 94%, possessed significantly larger left PT than right PT. As suggested by Deacon (1992), communication and cognitive abilities may have evolved in both chimpanzees and humans, and such independent PT evolution served to develop 'species-specific repertoires', which characterize human and chimpanzee cognition and communication. Stated this way, it implies that 'language both spoken and signed' is characteristic of both species. If this is the case, the difference in degree is large enough as to require further explanation.

Research disclosed by Zilles and Falk at a meeting of the American Association of Physical Anthropologists (AAPA) in 2002 showed that by using a new imaging technique they could identify which parts of the brain expanded during human evolution. A comparison between the modern chimp and human brain showed a larger right frontal lobe and a decreased right temporal lobe in the human brain. According to Falk, this is surprising because the left hemisphere – the language-bearing side of the brain – was generally thought to have expanded the most. Instead, their research indicated the most substantial changes occurred in the right frontal lobe, particularly the bottom of the right frontal lobe, which expanded even more in archaic *H. sapiens* and Neanderthals. Falk thinks that this area of the brain may be responsible for understanding the patterns of stress and intonation in ordinary speech, including rhythm, tone and emotional content (for a discussion see Gibbons, 2002b).

Also during the 2002 AAPA conference, researchers identified nutritional constraints on brain evolution, and generally concurred that a fish or shellfish diet must have been required to ensure sufficient omega fatty acids were available to support the dramatic expansion of the early human brain. According to Stephen C. Cunnane (cited by Gibbons, 2002b) a lack of omega-3 and mega-6 fatty acids during foetal brain development is 'catastrophic'. According to C. Leigh Broadhurst even after birth, depletion of these essential acids may be linked to cognitive disorders – dyslexia, senile dementia, schizophrenia – and vision problems (Broadhurst *et al.*, 2002). The research suggests that our ancestors could not support an expanding brain even by eating the brains of large fauna. To sustain the kind of brain development seen in early humans, generations of women must have lived near lakes, rivers and the sea, where they could have obtained the necessary marine resources. South African Cape sites, where large shell middens dating from 100 000 to 18 000 years ago, and Rift Valley lake sites, as well as sites along the Nile corridor into the Middle East, would have facilitated the exploitation of littoral resources including fish, shellfish, sea bird nestlings and eggs that would have contained the high levels of long-chain poly-unsaturated fatty acids that facilitate multi-generational brain development.

Terrestrial meat sources, particularly bone marrow, may also have played a key role in human encephalization, facilitating the advancement of behavioural traits that potentially allowed *H. erectus* to migrate into cool northern climates (Chamberlain, 1996). Understanding the impact of diet on human development and behaviour has important implications to our understanding of the connection between humans and their environment, particularly during periods of rapid climate change.

2.2.5 Genetics

Although fossils record the evolution of the morphological features of an organism and the timing of their first appearance, their prevalence is extremely limited and highly fragmented. If we study solely fossils and neglect genetic, physiological, behavioural and evolutionary ecological aspects of human evolution and behaviour, we risk negatively restricting our perspective. In the early 1940s, when scientists gathered to develop an evolutionary theory of biology and Huxley's *Evolution: The Modern Synthesis* (1942) amalgamated Darwinism, Mendelian genetics, classical genetics, theoretical population genetics, ecology and physiology, a new era of evolutionary science began. 'Genetics became the language of evolution'

(Schwartz, 1999, p. 8). Ernest Mayr, a member of the committee that had formulated the synthesis and who became a bastion of neo-Darwinism, defined species as a group of individuals capable of breeding and producing offspring that could themselves reproduce (see Mayr, 1963). He also promoted the idea that a new species could be generated by reproductively isolating members of the parental species. Separating individuals from one another geographically could interrupt gene flow. Mayr's concept of species' formation perpetuated the concept of linear transformation and was generally accepted by evolutionary biologists. Yet Mayr still remained intrigued by the possibility of 'explosive' evolution resulting from epigenetic events – anything that reflects the development of the organism (Schwartz, 1999, p. 42).

Increasingly, over the last 60 years, genetics has shaped the focus of evolutionary biology, in particular human evolutionary biology. Genetic models have been used to analyze blood groups, serum proteins, enzymes, and, more recently, saliva and hair samples of both modern and ancient DNA to interpret population structure and human dispersal events. The idea that each human has distinctive characteristics that are passed on from generation to generation became popular when the 5300 year-old 'Iceman' was found in the Austrian Alps in 1991 and the identification of his modern relatives undertaken through mitochondrial DNA (mtDNA) analysis. It was further spread by the book *The Seven Daughters of Eve: The Science that Reveals our Genetic Ancestry* written by Bryan Sykes, Director of Molecular Medicine at Oxford. Additional interest was generated with the discovery of Kwaday Dän Ts'inchi, the name given to the 500 year-old body found in the ice in Tatshenchini-Alsek Park, Canada in 1999, along with the identification of the human groups to which he belonged.

Geneticists working on patterns of mtDNA variation in modern human populations in different regions of the world have argued that the similarity in mtDNA within the modern human population must be the product of an independent evolution, spanning a period of about one million years, which diverged from a single common female ancestor who lived somewhere in Africa about 200 000 years ago. Descendants from this initial population then dispersed around the world between about 30 000 and 100 000 years ago (see, for example, Cann *et al.*, 1987; Stoneking and Cann, 1989; Wilson and Cann, 1992). Recent research suggests that although mtDNA evidence has consistently shown the greatest diversity of within-group variation is found in the African population, this variation 'can be explained by divergence from an initial source (perhaps Africa) into a number of small isolated populations, followed by later population expansion throughout our species' (Relethford and Harpending, 1994, p. 249). Cavalli-Sforza *et al.* (1988) and others have performed genetic research based on patterns of genetic marker variation in blood groups and proteins. Their findings, plus those of researchers studying variations in nuclear DNA of modern populations, have made similar claims about the African origin of 'modern' humans (see also Cavalli-Sforza, 1991; Lucotte, 1989; Mountain *et al.*, 1993; Wainscoat *et al.*, 1989).

mtDNA studies indicate a lack of diversity within the Neanderthal mtDNA sequences and large differences between Neanderthal and 'modern' human mtDNA (Höss, 2000; Ingman *et al.*, 2000; Knight, 2003; Krings, Capelli *et al.*, 2000; Krings, Stone *et al.*, 1997; Ovchinnikov *et al.*, 2000; Schmitz *et al.*, 2002; Stoneking and Cann, 1989; Wilson and Cann, 1992). Recent research by Serre *et al.* (2004) compared mtDNA from fossil remains of four Neanderthals and five early 'modern' humans found at archeological sites throughout Western Europe. The Neanderthal mtDNA contained sequences similar to previously determined Neanderthal mtDNA sequences, whereas the early 'modern' human sequences failed to yield any such combinations of mtDNA. Serre *et al.* suggest that this excludes any possibility of a large genetic contribution from Neanderthals to early humans. Modelling of 'modern' human range expansion into Europe indicates that despite a likely cohabitation of more than 12 000 years, the absence of Neanderthal mtDNA in Europe is consistent with no more than 120 'admixture events' between Neanderthals and 'modern' humans. These findings imply that Neanderthals and anatomically 'modern' humans developed separately and did not interbreed. Further, it supports the 'replacement' view, that Neanderthals became extinct without passing their genes on to 'modern' humans. Offspring of a Neanderthal and an anatomically 'modern' human mating, if any, were infertile (Currat and Excoffier, 2004).

However, because mtDNA is only passed down through the mother, these findings do not take into consideration the contribution made by Neanderthal males. Recent Y-chromosome studies suggest that the forefathers of modern men originated in Africa between 35 000 and 89 000 years ago (Ke *et al*, 2001; Underhill *et al.*, 2000). In response to a critique of this work by Hawks (2001), Jin and Su (2001) further suggest that the significant gap that appears in the East Asian fossil record between 100 000 and 40 000 thousand years ago is consistent with the hypothesis that archaic humans became extinct during the last Ice Age in East Asia. However, using non-recombining Y chromosome (NRY) studies in 50 human populations, Hammer *et al.* (2001) have found patterns of NRY diversity that indicate an early out-of-Africa model followed by migrations out of Asia. This latter study provides the possibility that Neanderthal males, or, for that matter, *H. erectus* outside Africa, may have passed genes to the *H. sapiens'* nuclear genome. Thus, there remains the possibility that some of us may yet possess nuclear DNA inherited from Neanderthal or *H. erectus* ancestors. Stringer (quoted in Hopkin, 2004) speculates that, given a stable climate, Neanderthals and 'modern' humans could have overlapped in time, and Neanderthals were just unlucky.

Yet despite advances in genetical investigative methods, the application of their results to human dispersal and diversity have been criticized on the grounds of extreme variation in mtDNA substitution rates, indirect data estimation containing large standard errors and intentional bias. As Cann (2001) suggests, genetic research in human species focuses on genetic isolation and persistence of local human populations. In contrast, researchers working on nonhuman species incorporate a biogeographic ecological approach, recognizing that local continuity of populations is frequently caused by continual immigration and replacement. Some populations act as sources of species' genetic variation, while other populations are maintained only through immigration with allelic replacement of genes. However, there could be a genuine contrast between model populations, which are large, and in which relatively free gene flow is assumed, and small human populations that could migrate rapidly. Moreover, in small populations, isolated by rapid migration, chance events, genetic drift and inbreeding all affect the conventional predictions of population biology.

Variations in neutral alleles are crucial to an evolutionary geneticist who, in order to estimate the time of a dispersal event, sets a molecular clock 'on the basis of estimated rates of mutation and fixation' (Cann, 2001, p. 1743). Variations in the estimation of the rates of mtDNA mutation could result in the recalibration of the date of the original common ancestor of the anatomically 'modern' forms of *Homo* closer to one million years – the generally accepted date for the colonization of the northern latitudes and eastern Asia by *H. erectus* (for variations in mutation rates see Parsons *et al.*, 1997, and Strauss, 1999). In contrast, recent hominin finds from Chad, central Africa, dating between around 6 and 7 million years ago, are raising questions about the timing of divergence between humans and chimpanzees and its impact on the calibration of the molecular clock (Brunet *et al.*, 2002; Gibbons, 2002a; Vignaud *et al.*, 2002). If correct, these interpretations imply that the common ancestor of 'modern' humans may have lived almost anywhere in the Old World. On the other hand, if human emergence operated through unpredictable periods of static equilibrium punctuated by rapid change, and not through continuous evolution, it is difficult, if not impossible, to calibrate the molecular clock. A regular tick is the essence of a clock that keeps time. To average irregular ticks for the sake of estimating evolutionary points of origin may be wasting time.

2.2.6 Stone tools and art

In the 1950s, palaeoanthropologists considered toolmaking to be one of the defining characteristics of our *Homo* sp. ancestors (for an interesting perspective see Oakley, 1956). More recently, the diverse and prevalent use of tools by chimpanzees has led researchers to revise this hypothesis. In addition, the persistence and skill in tool use of West African chimpanzees suggests females rather than males may have played a leading role in the evolution of technology (Ambrose, 2001; Boesch and Boesch,

1984). Further, Darwin's theory that early humans developed tools to assist them in adapting to the savannah environment is being challenged after researchers observed forest-dwelling chimpanzees using tools (Boesch-Acherman and Boesch, 1994).

Irrespective of the uncertainty in the role of toolmaking in defining humanness, stone tools are one of the most prevalent artifacts that early hominins left behind for archeologists to ponder. As such, they remain an extremely important part of the database used to reconstruct our history. Stone tools provide a durable record; one that is far more complete and continuous than the skeletal or faunal record. Archeologists identify tool components in archeological assemblages and changes in those components as cultural transitions. The shape of large cutting tools is frequently assumed to relate to an associated cultural group and, thus, is used to identify regional cultural traditions (Ambrose, 2001; Wynn and Tierson, 1990). An assumption is made that specific tool kits represent 'stylistic microtraditions' transmitted by one generation to the next via an associated culture. Some researchers perceive transitions as *in situ*, implying continuity from one Palaeolithic assemblage to a subsequent one. Others argue that transitions in tool industries are an adaptive response; for instance, Neanderthals modified their Mousterian technologies when they came in contact with 'modern' humans, creating a mixed middle and upper Palaeolithic tool assemblage. Still others suggest that intermediate assemblages do not exist; contemporaneous late middle Palaeolithic and early upper Palaeolithic tool industries are evidence that 'modern' humans intruded into Neanderthal territory around 50 000 years ago, bringing with them their upper Palaeolithic tools (for a discussion of the rapid intrusion of modern humans into Neanderthal territory see Clark, 1999).

Although such behavioural interpretations are made from early tool kits, it is important to remember that other factors influence tool size and shape, including mechanical properties, abundance, access to and size and shape of raw materials, the primary form of the blank, the amount of resharpening, distribution of food resources, mobility strategies, anticipated tasks, group size and composition (larger populations have a cultural innovation advantage over smaller ones) and duration of site occupation. Such factors certainly influenced Palaeolithic toolmaking. Further, shifts in stone tool technology may have been more cyclical than directional, reflecting, for example, changes in animal exploitation: exploitation of mobile resources (caribou versus clams; terrestrial versus aquatic mammals) requires, and thus stimulates the development of, a relatively more complex toolkit (Mellars, 1996; Osborn, 1999; Oswalt, 1976).[2]

Alternatively, Torrence's risk-buffering hypothesis (Torrence, 1989, 2000) suggests that the more mobile the prey, the lower the probability of capture, and thus the more elaborate the tools necessary to reduce the risk of resource failure – failing to capture the prey. In other words, fear of failure stimulates complex toolmaking. However, a number of issues remain unresolved, including whether the variables used to ascertain the risk of resource failure – effective temperature and net above-ground productivity – are appropriate proxies. Both temperature and productivity relate to plant biomass, but some argue that herbivorous animal biomass is more important. However, as Collard *et al.* (2005) point out, numerous hunter–gatherer populations are heavily dependent on plant resources. Further, the size and composition of the hunter–gatherer sample populations needs to be considered but is as yet unresolved. Currently the hunter–gatherer database sample used in these studies contains a strong bias towards high-latitude, coastal populations. Thus, although this research highlights the need to consider alternative factors, other than advancements in cognition, influencing tool size and shape, more research is needed before anthropologists can confidently specify those variables.

The presence of art, symbolism and personal ornamentation were isolated occurrences in the middle Palaeolithic, but their appearance in the form of perforated animal teeth, marine shells, beads, the use of pigments, symbolic artifacts (including female vulva symbols and mammoth ivory animal statues) and burial practices became relatively frequent in the upper Palaeolithic. An interesting find is that made by Nicholas Conrad (2009) of perhaps the oldest sculpture of a human figure. It is a lifelike voluptuous female about 35 000 years old carved out of a broken mammoth tusk. The media pounced upon this find from Germany as an example of Palaeolithic pornography, but it could just as well have been a fertility charm. In any case, it shows not only remarkable artistic craftsmanship, but also that a capacity for vision and a highly developed mental concept were present in the earliest Palaeolithic.

The increased frequency of symbolism and personal ornamentation in the Upper Palaeolithic has led some researchers to suggest a coincident change in symbolic and cognitive requirements and, therefore, capabilities of 'modern' human over Neanderthal populations (for a good discussion on this subject see Mellars, 1996). The niggling problem with this idea is that it implies that Neanderthals were the sole creators of the middle Palaeolithic. But many middle Palaeolithic sites do not contain any skeletal evidence, thus their Neanderthal association is frequently based on the tool composition being middle Palaeolithic. Yet anatomically 'modern' humans were present in the middle Palaeolithic, and it was not until sometime after about 50 000 years ago, at least in Europe, that the shift from middle Palaeolithic to upper Palaeolithic occurred. Thus, the shift in the archeological record may not have been a species-correlated issue – or at least one not evident in the fossil skeletal record, but rather a timing and adaptability issue, perhaps in direct association with a rapidly changing climatic environment. For instance, some archeologists attribute cultural change to new conditions of fairly dense forests, which necessitated hunting smaller groups of solitary animals and the development of new hunting techniques and an altered social organization (Straus, 2000a).

Scientific evidence indicates that the climate during marine isotope stage 3 (MIS3; between 60 000 and 30 000 years ago) was highly unstable. We should note here that we will refer to marine isotope stages throughout this book. They refer to alternating cool (even-numbered) and warm (odd-numbered) periods in Earth's palaeoclimate based on variations in oxygen-isotope concentrations in the shells of tiny organisms (deep-ocean foraminifera) obtained from deep-sea cores. Although climate model simulations show greater needle-leaf tree coverage in Europe during MIS3 than in AD 1800, they also indicate that a deteriorating climate at this time reduced the extent and productivity of human habitat. In combination with climate variability typical of MIS3, this likely reduced available food resources, and provided an impetus to create technologies to increase the carrying capacity of human habitats. (For a more complete discussion of climate and productivity during MIS3 see Chapters 5 and 6.)

2.3 Conclusion

From the primordial fibre of all life in Erasmus Darwin's time to molecular genetics today; from the belief in the Judeo-Christian history of Creation at 4004 BC to the discovery of more than one species of *Homo*; from correlating brain size with intelligence to developing an understanding of brain structure and development, and despite continuing controversy for more than 200 years, we have developed a better understanding of our humanness. Much of our understanding of what it is to be human is based on fossils and stone tools discovered buried in the African and Eurasian dirt. From these scraps of our past we have generated an interpretation of our humanness in the modern sense of the term. Assumptions are made about the relationship between observed change in the fossil record and genetic change and speciation. As well, interpretations are made about behaviour from the fossil and artifactual evidence.

Inconsistencies in the interpretations of human evolution and adaptability in the early human record are in part the result of not identifying the difference between genetic adaptation and adaptability. We need to ascertain which responses are truly associated with slow and gradual genetic adaptation and which are rapid developmental, physiological and behavioural adaptabilities in response to rapid changes in environment. Our next chapter will focus on the complex issues of 'modern' human behaviour.

As we progress, two fundamental questions will continue to emerge: to what extent did climatic influences produce Darwinian adaptation in hominids? And to what extent was cognitive and social adaptability the emergent result of rapid climatic changes?

Notes

1. For a discussion of Neanderthals as a separate species see McBrearty and Brooks (2000). For a discussion of how Neanderthal stature helped to retain heat during the last glacial period see Wong (2003). For a discussion of similarities between Neanderthal body structure and modern humans see Trinkaus *et al.* (1998).
2. Recent research by Collard *et al.* (2005) failed to support mobility as a key factor in diversity or complexity of toolkits.

3

Human behavioural evolution

Even though the three living great apes are more closely related to us than
to any other living primates, and one or two of these great apes are
probably our closest living relatives, the actual, *closest* relatives of
H. sapiens, and of each great ape, as well, are now extinct. In contrast ...
the closest relatives of most living species are also living.

(Schwartz, 1999, pp. 18–19)

H. sapiens embodies something that is undeniably unusual and
is neatly captured by the fact that we are alone in the world today.
Whatever that something is, it is related to how we interact with
the external world: it is behavioral.

(Tattersall, 2003, p. 24)

3.1 Introduction

The human species has a number of uncommon features, one being that we are the
only surviving species of our genus *Homo*. We did have close human relatives such
as *H. erectus* and *H. neanderthalensis* at one time, and we probably coexisted with
both of them before they became extinct. As Ian Tattersall infers, explanations of
such an unusual condition might be found through a study of human behaviour and
its evolution. As a general principle, behaviour has been a primary factor in all of
animal evolution, since it presented organisms with novel conditions of life that
influenced their development and physiology. Once mammals acquired the adaptable physiological homeostasis typical of marsupials and placentals, exploratory
behaviour exposed them to extreme environments that demanded new behavioural
responses, and those ultimately included the ability to change the environment.
Particular principles that we will examine in this chapter are how uniquely different
human behaviour is from that of ancestral species and to what extent human
behaviour has changed from the emergence of *Homo sapiens* to the present.

Another particular is the influence of climate in causing behavioural, and possibly evolutionary, change in humans.

One issue makes us question the conventional wisdom assumed by some anthropologists and psychologists – that human behaviour is metaphorically 'selectable', on the grounds of being 'useful', 'energetically conservative' and 'competitive'. Consequently, particular behavioural patterns have and will be sustained and rewarded by an expansion of such behaviour in human populations. However, these assumptions have led 'evolutionary psychologists' and their anthropologist equivalents to further assume that behaviour is *literally* selectable, in neo-Darwinist terms, since it is *genetically determined*.

There is no evidential basis for such an assumption, though it has the allure of keeping psychology and anthropology within the Darwinist fold, and creates the illusion of a more 'scientific' treatment since it is 'theory-driven'. These issues are important for our discussion of human evolution and climate, since they conclude that human behaviour was naturally selected during the trials of the ice ages. In his 1997 book *How the Mind Works*, Stephen Pinker (1997) argues that there has been 'cognitive closure' since that time. In other words, human genes (for behaviour at least) have not faced the challenges of natural selection since the Pleistocene, which terminated ten thousand years ago. Hence there has been no cognitive evolution since then. Because we seriously challenge the assumption that there are 'genes for behaviour', we cannot ignore such arguments, and will return to them throughout this chapter. Meanwhile we will belabour the obvious – research that is 'theory-driven' is worthless if the theory is based on false premises.

We are also cautious about accepting the assumptions of the middle ground of the debate regarding the metaphorical selectability of human behaviour. Beneficial behaviour might indeed be perpetuated and expanded from one generation to the next. But the process has been more Lamarckian than Darwinian – i.e., acquired behaviours are inherited by subsequent generations through imitation and education, not through genetic inheritance. Moreover, there are too many examples of non-beneficial behaviours being perpetuated and expanded from one generation to the next for this to be axiomatic. The prevalence of 'cycles of abuse', whether physical, emotional, sexual, doctrinal, drug, alcohol etc., in our world provides ample evidence of how abusive behaviour may be instilled in the abused, facilitating a recurrence of the negative behaviour in subsequent generations. The institutionalization of religion is an example of how a monolithic belief system can dominate a culture, without competition and selective culling. Sometimes simple caprice is responsible; non-advantageous behaviours simply become fashionable and popular for novelty alone, rather than for cultural value. The modern fashion industry is a prime example.

The question of hominin behavioural evolution is potentially a simpler matter. Correlations between behaviour and survival and reproduction are easier to draw for societies living at the brink of survival, than for contemporary cultures and societies. But assuming that we want something better than a reductionistic analysis of the selection of genes for behaviour, there remains one important question. Is there a metaphorical natural selection of beneficial behaviours which slowly accumulate to define particular societies? And even if there were, is there a saltatory emergent element that brings about *rapid* human behavioural changes? Think of the social impact, for better or for worse, of innovators such as Confucius, Buddha, Hippocrates, Aristotle, Christ, Mohammed, Newton, Darwin, Pasteur, Marx, Mao and Einstein. Did such instant emergences occur during early hominin evolution as the consequences of the actions of unsung innovators and their new ideas and practical experiments? Anthropologists seek to answer that question by identifying the emergence of new anatomies in the fossil record and new kinds of tools in the archaeological record.

3.2 Interpreting behaviour from the archaeological record

Unlike morphological change, behavioural change is not immediately apparent in the fossil record. Some aspects of human behaviour can be ascertained from fossil bones. For example, increased bone density may imply increased physical activity, and worn teeth may indicate lack of cooking, coarse food ingestion or utilization of the teeth for other purposes including chewing of hides. Yet archaeologists base much of their interpretation about behaviour from changes in technology, artwork, ornamentation, burial practices, subsistence strategies, patterns of mobility, social and demographic organization and habitation site organization and complexity. Anthropologists make inferences about the relationship between behavioural change and biological change. For example, can specific behavioural tendencies or advances be used as hallmarks of one hominin type, or one population? Are behavioural changes associated with cognitive shifts? Are those behavioural, cognitive or biological shifts rapid or gradual? Are they mutually exclusive, mutually dependent or independent? Are they similar in all regions of the world? Are they associated with migrations, dispersals, interactions and/or extinctions of populations? Are they correlated with climate and shifts in the nature or availability of resources?

From a different perspective, are the interpretations made by palaeoanthropologists about behavioural change biased by the theory that underpins their research? Schwartz (1999) suggests that anthropologists are either consciously or subconsciously influenced by 'the traditional dogmas of Darwinism' including natural selection and a gradualistic evolutionary perspective (p. 33). *If so, the anthropological synthesis might benefit from a different theoretical approach.* As Schwartz might put it, the substitution of a saltatory emergence theory for a gradualistic Darwinist theory could be more productive.

In the Appendix we will expound on biological emergence theory. However, by first outlining it here we can better demonstrate the relevance of Schwartz's ideas, as well as providing the theoretical context of our own interpretations. Biological emergences can be defined as the sudden origination of novelties. They may be unpredictable, saltatory processes whose timescale ranges from microseconds (point mutations of DNA), through minutes (physiological changes) to more than ten years in the case of individual development to a reproductively viable adult. Emergence theory does not deal with the prolonged phases of stability that may last from centuries to millions of years. Nevertheless, just as a stone thrown into a pond creates waves and ripples that may agitate the water surface for some time, a biological emergence may have adaptational ripple effects that last for many years. Thus, a complete *biological* synthesis must not only include emergent evolution, and minor genetic adaptations that follow it over long periods, but also the prolonged states of equilibria punctuated by emergent evolution.

In addition to sudden saltatory emergences, there are threshhold effects resulting usually from developmental changes that have grown more prominent over many previous generations. Reid (2007) calls these 'critical-point emergences'. Although they appear to be sudden, they are the culmination of a prolonged series of processes. Furthermore, they are predictable, unlike most saltatory emergences. An example is the slow enlargement of wings prior to the emergence of flight. The animal finally reaches a threshold where the lift provided by the wings exceeds the drag of its body and the effect of gravity. Primate brain evolution may also illustrate a series of critical-point emergences in development, and this is clearly related to human behavioural evolution. Reid further categorizes 'physiological and behavioural multifunctionality and adaptability' as general properties of biological emergences. As an example of physiological multifunctionality in the early evolutionary stages of vertebrates, consider the multiple roles of gills, which not only function as respiratory organs, but also as ion exchangers and excretory organs. At the human behavioural level, we will shortly be discussing how the basic Acheulean handaxe could be made to serve additional functions as an effective projectile or as a hafted axe.

In order to understand what is meant by behavioural change in terms of the behaviour of *Homo sapiens*, as opposed to that of now extinct human species, palaeoanthropologists define what it is that makes us human. According to Potts (1996a), human uniqueness can be defined by 'two-legged walking that freed the hands, technology and fine manipulation of tools, hunting and eating meat, large brains, culture and language, the social complex of the pair bond, division of labor, and marriage, [which] all developed in tandem as a complexly woven package' (p. 9). This reference to multifunctionality places Potts's opinion under emergence rather than that of gradualistic Darwinism.

Pope (1995) suggests that the multiple functions of 'language, art, foresight, and complexity and standardization of artifacts' (p. 502) are the most accepted characteristics of what is usually referred to as 'modern' human behaviour. According to Ambrose (2001), additional hallmarks are tools of greater complexity, including those made with bone, ivory and antler. Art and symbolism, whether through ornamentation or ritual burial, sophisticated architecture and land-use planning, as well as resource exploitation and strategic social alliances, are also identified as trademarks of 'modern' human behaviour. Some archaeologists (see, for example, Ambrose, 2001, and Klein, 1999) propose that evidence, including artifacts representing this kind of 'modern' human behaviour, is extremely rare in middle Palaeolithic Europe and middle Stone Age African sites, but appears consistently in these regions after about 40 000 years before present (or about 45 000 calendar years ago).[1] However, others propose that evidence exists that 'modern' behavioural modifications appeared gradually in Africa during the late middle Stone Age – by at least 100 000 thousand years ago, with an onset perhaps as early as around 300 000 years ago (for a thorough discussion of this perspective, see McBrearty and Brooks, 2000). In their view, as we paraphrase it, 'modern' humans are characterized by abstract thinking, planning depth, and behavioural, economic and technological innovations. These generalizations could, however, apply not only to *H. sapiens*, but also to *H. erectus* and more primitive primate species.

The most common usage of the word 'modern' refers to what happens at the present time. Therefore to apply it to the behaviour of humans who lived 50 000 or 70 000 years ago or earlier creates a semantic problem for the general reader. Omitting our technical advances, some of what present-day humans do now may indeed have been done by early members of our species and by other *Homo* species as well. There is a consensus among many palaeoanthropologists that many of the behavioural features of present-day humans go as far back as a creative 'enlightenment' in Europe about 45 000 years ago. That is what they mean when they write about 'modern human behaviour'. Others push it back to 70 000 years ago and beyond in Africa. Up to this point we have restrained the word modern within inverted commas, to mitigate the semantic problems that arise for the general reader from its anthropological usage. However, this becomes tedious after multiple repetitions.

Thus we will use the expression *modern* human to signify those humans who had much of the behavioural repertoire of present-day humans. It does not deny that components of that repertoire existed long before. But we still have to solve the problems of where and when modern human behaviour can reliably be said to have begun, and what biological correlations might have led to modern human behaviour.

As mentioned above, Schwartz suspects that conventional wisdom may be influenced by the dogmas of Darwinism or selectionism. As a relevant behavioural example, there may be a neo-Darwinist 'theory-driven' reason why some people push 'modern human behaviour', as well as language, as far back in time as they can, in some cases as far back as the common ancestor of hominins and chimpanzees. This is where McBrearty and Brooks, for example, take us in their 2000 survey of human behavioural evolution. Darwinism demands a gradualistic rather than an emergentistic–saltatory view of evolution, and the longer and slower the process of behavioural evolution, the better it fits the driving theory. The gradualistic Darwinian theory has prevailed because it is virtually the only one that has been considered by anthropologists and archaeologists. This is despite evidence that suggests sudden emergences in behaviour, and the design of novel artifacts, at certain times during the evolution of *Homo sapiens*. These emergences might result from saltatory or critical-point changes in brain anatomy, but there is no evidence that such anatomical changes occurred since the origin of the species.

Before we leave our discussion of what 'modern' should imply, we must deal with the expression 'anatomically modern human'. As discussed in Chapter 2 this term applies to all members of *Homo sapiens sensu stricto* (i.e., excluding Neanderthals, which are regarded as a subspecies of *H. sapiens* by some palaeoanthropologists). *H. sapiens* differs from other *Homo* species in having a relatively large round cranium with no bony sagittal crest and with small orbital ridges. Perhaps the most significant difference between *H. sapiens* and *H. neanderthalensis* is the expanded frontal cortex region, which contains the functions of intelligence and language, in the brain of the former (see Jones *et al.*, 1992; Tattersall, 1998). It is assumed that the fusion and ossification of the cranial plates comes later in post-partum life than in other species, allowing for cerebral expansion. The 'modern' anatomy also includes a more basal position of the foramen magnum, allowing the head to be held upright on a vertical torso. *H. sapiens* also has a more pronounced chin, providing us with a flat, rather than a chinless, prognathous face.

There is no fossil evidence of a gradual series of aggregations of the individual features of *Homo sapiens*' anatomy, though some could have emerged singly or in simple combinations that were then exaggerated by cross-breeding. We incline to the view that the anatomy of our species is characterized by correlated emergent features that resulted from the combinations of some large epigenetic shifts (for a complete discussion on epigenetics see Appendix B Section 2 'Developmental evolution: Epigenesis and epigenetics), which are consequently responsible for the origin of our species. There have been no significant gross cerebral changes since that origin. For human behavioural evolution, this has major implications that we will shortly reveal.

Since all members of our species were anatomically as we are at the present, the use of 'anatomically modern' is not contentious – merely redundant. It is only necessary to refer to 'the anatomy of *Homo sapiens*' – or 'Sapiens" anatomy, if we may be allowed to give our species a trivial name semantically equivalent to 'Neanderthal'.

The remaining problem lies in what is meant by 'human', since biological and anthropological usage sometimes includes other *Homo* species, while popular usage refers only to our own species. And where to place Neanderthals is another issue. For the purposes of this book, we follow the scientific usage of 'human' – i.e., all members of the genus *Homo* – and pragmatically treat Neanderthals as a separate species. While we are in pedantic mode, we remind our readers that we have been using the recent expression 'hominin' for the bipedal lineage leading to humans; that is, we are including *Australopithecus* and other species of *Homo*.

Another unfortunate prejudice in historical anthropological studies is the Eurocentric bias in discussions about the origin and development of modern human behaviour. Although this may not be justified, the prevalence of research on this topic has focused on its European origin and comparisons between Neanderthals and modern humans. Mellars (2005) notes that there are two opposing theses that attempt to explain the origins of modern-human behaviour in Europe associated with the distinctive upper Palaeolithic tool assemblage.

[Either] these patterns of behavior and the implied levels of associated cognition emerged by a purely internal process of behavioral and cognitive evolution among the local European populations, extending directly through the European Neanderthal line; or, alternatively that at least the majority of the new behavioral patterns, as well as the cognitive hardware necessary to support these innovations, was due to a major influx of new populations into Europe deriving ultimately from either an African or Asian source.

(p. 12)

In other words, did the human brain found in present-day members of our species evolve progressively only in a European locale through Neanderthals? This hypothesis implies that the Neanderthals are ancestral to *Homo sapiens*; a premise that has gone out of favour with many anthropologists on grounds of molecular genetics. Alternatively, did it emerge sometime in Africa or Asia and contribute to the drive for a rapid and extensive emigration into Europe and other parts of the Old World, where it then flowered technically?

There are so many problems associated with where and when and how 'modern' human behaviour occurred, and what evolutionary changes in brain structure and function caused its emergence, that we propose the following working hypothesis. Leaving *Homo neanderthalensis* out of the equation for the moment, there is no evidence that the brain has changed anatomically since the origin of *H. sapiens* around 195 000 years ago. Let us therefore start with the parsimonious view that it has not. From the outset, those brains had the *potential* to express the same thoughts, ideas, communication, spirituality, artistry and technical complexities as our own brains. But it took a combination of environmental conditions, both favourable and stressful, and a social complexification to bring out that potential. As an example of a creative potential being realized after thousands of years, take the invention of writing. This emerged after the development of agriculture, when traders perceived a problem keeping track of what had been traded to whom and at what agreed price. This brought about a revolution of communication and learning that was only rivaled by the invention of the printing press 5000 years later. No gene mutation or heritable novelty was required for these revolutions, simply the application of intelligence to an identified obstacle. Thus, we argue that the behavioural evolution of *Homo sapiens* subsequent to the species' origin requires environmental, psychological and socio-logical explanation rather than biological explanation. This is in general agreement with Wynn (1991), who suggests that culture and learning play a role in toolmaking. In contrast, the idea that there was a biological brain reorganization is championed by Bickerton (1995, 1998), with particular reference to language.

We concede that brain reorganization is possible without leaving any evidence in the form of differences in endocranial casts. We certainly admit that some anatomi-cal and physiological changes occurred after the emergence of *Homo sapiens* – for example, in relation to water conservation and temperature adaptation (skin colour and sweat gland distribution) – and we are open to argument regarding the fine-tuning of cerebral evolution. However, for the time being our working hypothesis removes some of the anthropological contradictions and simplifies the where, when and how questions about the modern human brain and cognition. Incidentally, according to this premise, it should be hypothetically possible to adopt the children of the earliest members of our species and teach them to talk, read, write and operate computers.

Because anthropologists base so much of their interpretations of modern human behaviour on early stone tools that may or may not be associated with fossil bones, we will briefly describe these tool industries before going on to discuss in more detail the theses explaining the origins of modern human behaviour.

3.3 Early stone tool industries of the genus *Homo*

3.3.1 The early Stone Age of Africa including the Oldowan and the Acheulean

From the Ethiopian River Valley in Africa comes the first direct evidence of hominin technology at 2.5 million years ago (Figure 3.1) in the form of hammer stones, anvils and bones with hammer and cut marks. Although the Oldowan technology appears simple, its creation by *Homo habilis* required manual skills far exceeding those of chimpanzees (Ambrose, 2001; but see Boesch-Acherman and Boesch, 1994).

Oldowan techniques remained unchanged until 1.5 million years ago and subse-quently disappeared some time after 500 000 years ago (Clark *et al.*, 1994). The appearance of large cutting tools, including large, teardrop-shaped, two-sided handaxes and cleavers, marks the advent of the Acheulean Industrial Complex 1.5

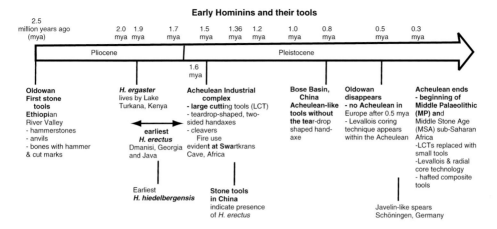

Figure 3.1 Early hominin timeline

million years ago. 'Handaxe' may be a misnomer, since the cutting edge went all the way round the tool. O'Brien (1984) proposes that the artifact was a projectile, having experimentally demonstrated that a 2-kg replica could be thrown like a discus a distance of 50 m, to land on a cutting edge, often point down. This could cause serious damage to back or shoulder muscles of a large prey animal in a closely packed herd. The large cutting tools (or projectiles) typical of the Acheulean and Acheulean-like tool industries were fashioned through the control of the primary form, size and mechanical properties of raw materials. According to Wynn (1991) and Gowlett (1984), their development implies preconceived designs and a high degree of standardization, reflecting higher conceptual abilities than those used for Oldowan toolmaking. The Acheulean toolkit was manufactured by *H. erectus* and the larger-brained *H. heidelbergensis*. Puzzling, however, is evidence indicating that the Acheulean tool industry disappeared in northern Europe 500 000 years ago, 200 000 years prior to its disappearance elsewhere (Ambrose, 2001; Asfaw *et al.*, 1992; Clark, 1994).

A contemporaneous toolkit dating to about 800 000 years ago, recently found in the Bose basin of South China (Hou *et al.*, 2000), demonstrates tool sophistication in China that is similar to that in Africa and western Eurasia, although the dating may be problematic (Langbroek, 2003). The Bose basin tools are found in association with tektites – glassy remnants of molten rocks displaced by the impact of a large meteorite that struck South East Asia approximately 800 000 years ago – and forest burning. The Bose basin is situated between the Loess plateau and the South China Sea, a region of significant environmental oscillation during the last 2 to 3 million years. Hou and colleagues suggest that the restriction of the Bose basin tools to the stratigraphic layer containing the tektites may be the result of a behavioural change associated with widespread forest destruction and the burning of woody plants caused by the meteor strike, which resulted in the exposure of cobble outcrops in the basin.

We can anticipate that early *Homo* used materials other than stone for tools. However, the poor preservation of nonlithic tools limits our understanding of this important dimension of palaeotechnology. However, at a human occupation site situated by a lake in Schöningen, Germany, wooden (specifically spruce, *Picea* sp.) javelin-like spears have been uncovered in association with the butchered remains of more than ten horses (Thieme, 1997). These are well-balanced, sophisticated hunting weapons. Each spear is made from the trunk of a 30-year-old spruce tree and has the proportions of a modern javelin – with the centre of gravity one-third of the way down the shaft. The tip of each spear is made from the base of the tree, where the wood is the hardest. Their creators were *Homo erectus*, or one of their close relatives who lived about 400 000 years ago. According to Thieme (1997) and Dennell (1997), the creation of such well-formed tools implies that early humans possessed a behavioural repertoire that included foresight, planning, the ability to choose appropriate technology and systematic hunting skills; their discovery requires revision of a previously limited interpretation of middle Pleistocene hunting and planning capacities and subsistence strategies. It also makes our working hypothesis regarding the evolution of the human brain in relation to behaviour more plausible.

During the later part of the Acheulean, between 300 000 and 500 000 years ago, tools were fashioned using a prepared Levallois core technique that generated large, well-formed flakes. Blades were struck from prismatic cores. Using this technique, it was the flakes early *Homo* chipped from the core, rather than the core itself, that became the new stone tool.

3.3.2 Middle Palaeolithic technologies

By at least 100 000 years ago, and perhaps as early as 300 000 years ago, technological and cultural evolution accelerated. The middle Palaeolithic tool industry appeared in Europe along with the coincident sub-Saharan African middle Stone Age (see Figure 3.1 above). The large cutting tools of the Acheulean were replaced by smaller tools made using Levallois and radial core technology. Hafted composite tools, including stone-tipped spears, knives and scrapers mounted on shafts and handles required sequential nonrepetitive fine motor control. Because speech also requires nonrepetitive fine motor skills, both are thought to be controlled by adjacent areas of the brain's inferior left frontal lobe (Greenfield, 1991; Kempler, 1993). These findings have led some researchers to suggest that composite tool industry and grammatical language coevolved around 300 000 years ago in the middle

Palaeolithic, and that the common ancestor of Neanderthals and our own species
could speak.

3.3.3 Upper Palaeolithic and late Stone Age technologies

Around 50 000 years ago, more sophisticated blade-based lithic industries developed in East Africa and the Levant (Ambrose, 1998a and 2001; Bar-Yosef and
Kuhn, 1999). Previously rare in the middle Palaeolithic and middle Stone Age,
ground, polished, drilled and perforated bone, antler, ivory and stone tools became
prevalent in upper Palaeolithic and later Stone Age sites after 40 000 years ago
(see Figure 3.2). The later Stone Age is further characterized by microlithic – small
stone tool – technology (for a nice review and discussion of the transition from the
middle Stone Age to the late Stone Age in Africa, see McBrearty and Brooks, 2000).
Also evident is the spear thrower (atlatl) that allowed hunters to extend the power
and velocity of the spear. This artifact is not likely to have been discovered by
accident, but rather made by design. Ample evidence exists of string and woven
fibres that may have been used to create rope, build nets and fashion clothing and
bags (Ambrose, 2001; Soffer *et al.*, 2000). An abrupt transition from the middle
Stone Age to the later Stone Age is evident in the Mediterranean zones along the
northern and southern margins of Africa, where archaeological evidence indicates
gaps in the settlement history. An abrupt transition from middle Palaeolithic to
upper Palaeolithic is also evident in Europe and is considered by some to reflect a
revolution of modern human behaviour. This is in contrast to some rock shelter sites
in tropical Africa, which contain relatively continuous settlement records that
indicate a gradual transition from middle Stone Age to later Stone Age

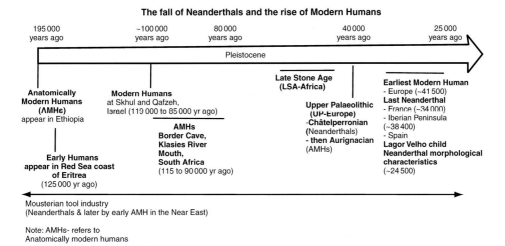

Figure 3.2 Neanderthal and modern human timeline

(McBrearty and Brooks, 2000; Mehlman, 1979, 1989). These contrasting findings may relate to what scientists identify as the greater impact of a changing climate on high- and mid-latitudes relative to tropical regions (for a further discussion on climate around the world during the last glacial cycle see Chapters 5 and 6; for the last glacial cycle's impacts on early humans see Chapter 7). It is also important to recognize that behavioural innovation need not take more than a single generation to be used by an extended family group, though the speed of its spread among the rest of the population might take as long as a century.

In an effort to associate the tools with their makers, archaeologists (see, for example, Mellars, 2005) had long considered the beads, bone carvings and upper Palaeolithic stone tools of the Aurignacian and those of the Châtelperronian, including the stone knife with a blunted diagonal edge, a Western European trade-mark of behaviourally modern humans. Alternatively, Neanderthals and early modern humans in the Near East were considered to have contributed the older middle Palaeolithic Mousterian industry (see, for instance, Mellars, 2005). In support of the interpretation that the Châtelperronian tools were made by modern humans, researchers explain that Neanderthals were cognitively inferior and, further, that they lacked language and could not have developed the more 'advanced' Châtelperronian stone tools without input from *H. sapiens* (see, for example, Mellars, 2005). However, the discovery of *c*.36 000-year-old Neanderthal skeletal remains alongside Châtelperronian stone tools that resemble both the immediately preceding middle Palaeolithic Mousterian and the succeeding upper Palaeolithic Aurignacian, at the site of Saint-Césaire, France, cast doubt on this exclusive archaeological categorization. Mercier *et al.* (1991) suggested that early upper Palaeolithic Châtelperronian industries may have been made by late Neanderthals and that the Saint-Césaire fossils indicate that some Neanderthals did survive several millennia after the arrival of *H. sapiens*. Evidence in support of this temporal overlap continues to build (see, for instance, Skinner *et al.*, 2005). Over a period of several thousand years, the temporal overlap between the Aurignacian, which extended across Western, central, and Eastern Europe, and the Châtelperronian, which was confined to France and Spain, may have permitted acculturation between *H. neanderthalensis* and *H. sapiens* (for a short review on the evidence of how, where and when Neanderthals may have coexisted with modern Cro-Magnons see Stringer and Davies, 2001).

Others have suggested that the apparent overlap between the Châtelperronian, which is now generally considered to be associated solely with Neanderthals, and the Aurignacian, which is associated with modern humans, is partially the result of the different nature of dating materials. According to d'Errico and Sánchez Goñi (2003), Châtelperronian dates have been made exclusively on bone and thus have produced potentially unjustifiably younger dates than the Aurignacian dates, which have been made on bone and charcoal. By disregarding the younger ages, the Châtelperronian is slightly older and therefore precedes the Aurignacian, thereby abolishing the overlap between Neanderthals and modern humans. In addition, d'Errico and Sánchez Goñi indicate that new accelerated mass spectrometry (AMS) dating of the French Mousterian tool industry places it immediately preceding or slightly overlapping the Châtelperronian. The Mousterian is typically considered to be associated with Neanderthals and early modern humans in the Near East. Their interpretation places the French Mousterian first, followed by the Châtelperronian, and then the Aurignacian. However, it is important to keep in mind that the Châtelperronian at Saint-Césaire, France, is the only upper Palaeolithic tool industry that can be directly associated with Neanderthal fossil bones (Stringer and Davies, 2001).

Despite the varied interpretations of the stone tool evidence, Mellars (1992) notes the vast expanse and uniformity of the Aurignacian, which extends from western France and Cantabria through central Europe and reaches south to the tip of Italy and east to northern Israel and Lebanon. It appears in stark contrast to the highly variable patterns of technology across the same region in the previous middle Palaeolithic interval. Such uniformity across such varied topographic and ecological zones has led some researchers (see, for example, Mellars, 2005) to imply a corresponding uniformity in culture, communication and language, which is used to support the hypothesis of a nonindigenous derivation of the Aurignacian, indicative of population intrusion. It has also been suggested that in some regions the appearance of an early Aurignacian was concurrent with an increase in human population densities as measured by the total number of occupation sites, and specialized faunal exploitation strategies that focused on single, migratory herd species (Mellers 1973, 1989, 1992).

Archaeologists are also beginning to correlate changes in cultural complexes with climate. According to d'Errico and Sánchez Goñi (2003), there exist two Mousterian concentrations – an older one, dating to between about 49 000 and 36 000 years ago, and a younger one, dated to between around 34 000 to 23 000 years ago – which are separated by a hiatus coincident with the Heinrich event 4 (H4), which d'Errico and Goni suggest took place between about 35 300 and 33 900 years ago. (However, see Chapter 5 for more information about H1–II9, and indications that H3 occurred between 30 000 and 33 000 and H4 between 40 000 and 43 000 years ago; a Heinrich event occurs when a massive amount of glacial ice, which has accumulated during an ice age, is discharged into the ocean.) This is also true of Mousterian sites all over Europe (see Sánchez Goñi *et al.*, 2001). Based on these findings, d'Errico and Sánchez Goñi suggest that a significant reduction in size of Neanderthal populations, if not a partial extinction, occurred during H4. Further, the apparent lack of Aurignacian sites in southern Iberia before 33 500 years ago may also imply a 'virtual absence' of modern populations due to low biomass in this region during the H4 event. Klein (2000) suggests that adverse climatic conditions restricted 'modern' human occupation until 30 000 years ago or later. However, archaeological evidence indicates a continued presence of Neanderthals in the southern Iberian Peninsula during and after the H4 event.

Finding tools in the company of fossils of humans and their prey is a good indication that humans used the tools. Although such finds suggest that the humans invented those tools, they might have borrowed the techniques of making and using them from others. Conversely, failure to find sophisticated tools or decorative artifacts with human fossils does not mean that the humans lacked the creativity to invent them. It may mean we have yet to find such sites. Alternatively, it may mean that whether their makers were generalists or specialists, their habits might have been stuck in a behavioural rut, or they might have had prohibitions against such novelties. Behavioural ruts do not matter as long as resources remain adequate. The discovery or non-discovery of tools provides a rule of thumb but does not close the arguments pertaining to the level that cognitive, behavioural and cultural evolution of humans might have reached.

3.3.4 Controlled use of fire

Fire has been used by humans as a source of warmth in cold climates, as a source of light and as a defense against animal and human intruders. Cooking detoxifies some plant poisons, makes plant starch stores and tough animal proteins more digestible and sterilizes microorganisms and parasites. The tactical use of grass fires could be used in hunting and firing hardened wooden spear tips (Ehrlich, 2000). The controlled use of fire may have been devised by *Homo erectus* between 1.5 and 1.0 million years ago (Brain and Sillen, 1988). According to Nicolas Rolland (2004), *H. erectus* had developed the ability to manipulate fire by between 400 000 and 350 000 years ago, which, along with the punctuated emergence of home-based settlement organization, facilitated environmental modification and expansion into an increasing range of habitats. However, other authors argue that full control was not accomplished until 200 000 years ago (Straus, 1989).

3.4 The origins of human behaviour

3.4.1 The European Neanderthals or an influx of non-local modern humans?

Did the behaviour that is taken to be characteristic of present-day humans derive from the European Neanderthal line, to be copied and extended by *Homo sapiens* who came later, or was it the result of an influx of *Homo sapiens* into Europe from either Africa or Asia? Mellars (2005) explains that the Neanderthal derivation thesis has a Darwinistic evolutionary context emphasizing the climatic and environmental changes that occurred in Europe during marine isotope stage 3 (MIS3, about 60 000 to 30 000 years ago). Rapid, repeated climatic fluctuations 'imposed strong selective pressures on almost all aspects of their [Neanderthals'] cultural and behavioral adaptations [sic] leading to a range of associated patterns of technological, economic, and social change' (p. 13; see also Mellars, 2004b; Shennan, 2002) and ultimately the upper Palaeolithic revolution. Mellars (2005) further explains,

This model, of course, carries with it the automatic implication that all of the necessary intellectual and neurological capacities for these behaviors were either present in the indigenous Neanderthal populations of Europe or that these capacities emerged, presumably as a result of one or more genetic mutations … as a further direct evolutionary consequence of the various environmental, demographic, or other selective pressures to which the European Neanderthals were subjected.

(p. 13)

The model also assumes that there are indeed genes for behaviour, and that behaviour changes when they mutate. It depends on the ability of a metaphorical 'selective pressure' to metamorphose into a literal creative force. It further implies that genetic changes occurred rapidly, coincident with MIS3, contradicting the Darwinistic assumption of slow and gradual evolution. However, these criticisms do not deny the possibility that there could indeed have been a unique and rapid cognitive emergence in Europe.

Mellars clearly supports the alternative thesis – that the European behavioural revolution was a consequence of a major influx of modern humans from Africa or Asia. He highlights the coincident timing of the middle-to-upper Palaeolithic transition and the dispersal of modern humans into the European continent. Evidence for the dispersal includes mitochondrial DNA, Y-chromosomal DNA, and the earliest presence of anatomically modern human fossil remains dated between 50 000 and 35 000 years ago in Europe and adjacent south-west Asia. Evidence for the behavioural transition is supported by the presence in the European archaeological record of bone, antler and ivory tools; innovative blade technology; notation systems on bone and antler artifacts; elaborate distribution systems; pendants and perforated marine shells; ivory and stone bead forms; and abstract and naturalistic art, which show a close correlation with the distribution of the Aurignacian stone tool technologies. Mellars further notes, in agreement with McBrearty and Brooks, the existence of an earlier appearance of upper Palaeolithic features evident in the South African archaeological record 'in close association with the biological emergence of our own species' (p. 20). He also agrees that the most striking behavioural innovations evident in Africa apparently appeared gradually between about 70 000 and 60 000 years ago, whereas those in Europe appeared suddenly about 20 000 to 30 000 years later. McBrearty and

Brooks (2000) date the early behavioural innovations to 250 000 to 300 000 years ago, when they believe the middle Stone Age first appeared. Although they associate this earlier onset of modern human behaviour with fossils attributed to the large-brained archaic human *H. helmei*, it also positions them at the time of *H. erectus*, which could more easily explain how Neanderthals came to have it.

McBrearty and Brooks have in fact presented a major challenge to the idea that the modern human 'revolution' of 40 000 to 50 000 years ago is a reality. They paint a plausible picture of the different elements of behaviour that many associate with the European 'revolution' arising individually and eventually coming together in the most striking way in communities of *H. sapiens* dwelling in Europe about 40 000 years ago. They suggest that the whole concept of the European revolution in human behaviour is Eurocentric and based on the myth of human uniqueness.

However, their antithesis has some inadequacies. Each of the components that McBrearty and Brooks say 'accreted' are revolutionary in themselves. The decision to use an ineffective handaxe as an effective projectile could have brought about an economic revolution in the group that thought it up; the same goes for the idea that a haft could be attached to a hand-held stone axe or hammer, or an atlatl to a spear. Moreover, these are all-or-nothing events that either worked or did not, and if they worked, they only needed minor adjustment to find the most efficient variation. Furthermore, an accretion of anything constructs a whole that is no more than the sum of its parts. Yet, the coming together of two or more novelties of general biology (such as symbiosis), or behaviour, is known to create a whole that is *greater* than the sum of its parts.

The generative conditions that led to a new social organization where all of the 'accreted' novelties could be expressed simultaneously produced nothing less than a revolution, even if the components had existed individually 70 000 or more years before. It constitutes an emergence in societal interaction analogous to the 'phase shift' to which Kauffman (1993) refers in *The Origins of Order*, which analyzes organizational change at the physicochemical and biological levels. In ignoring, if not denying, the emergent nature of behavioural change, McBrearty and Brooks, despite the evidential value of their survey, put themselves squarely in the gradua-listic Darwinistic camp.

3.4.2 The influence of the environment and natural selection

To explain the influence of the environment on the expression of early modern human culture, Mellars (2005) states: 'It goes without saying that these local contextual factors can be invoked only in contexts where both the innate cognitive capacities for particular patterns of behavior and the essential technological expertise for the behaviors in question are already present in the populations involved' (p. 23).

This means that the cognitive capacities and technological expertise were present to generate the conditions for modern behaviour for some time; it was a local environmental adjustment that triggered these capacities into full-blown 'modern' behaviour. We would qualify this conclusion by saying that, while we agree that the innate capacities have to be already present to interact with local contextual factors, technical expertise need not be present. It could emerge as a consequence of a general manipulative capacity interacting with environmental conditions and improved cognitive capacities, including learning.

To explain the rapid evolution of 'modern' behaviour in Europe, Klein (2000) emphasizes the natural selection of an advantageous genetic mutation in modern humans:

Behavioral and anatomical evolution were aspects of a single process driven by natural selection for advantageous genetic novelties. When the full sweep of human evolution is considered, it is surely reasonable to propose that the shift to a fully modern behavioral mode and the geographic expansion of modern humans were also co-products of a selectively advantageous genetic mutation. Arguably, this was the most significant mutation in the human evolutionary series, for it produced an organism that could alter its behavior radically without any change in its anatomy and that could cumulate and transmit the alterations at a speed that anatomical innovation could never match. As a result, the archeological record changed more in a few millennia after 40 ky [thousand years] ago than it had in the prior million years, and the long-standing pattern of coevolution between anatomy and behavior was ruptured. Before 50–40 ky ago, anatomy and behavior evolved relatively slowly and in parallel. Afterwards, gross anatomical change all but ceased, while behavioral (cultural) change accelerated dramatically.

(pp. 17–18)

Mellar's and Klein's interpretations and arguments in favour of a rapid onset of modern human behaviour are incongruously founded on Darwinistic natural selection mechanisms and genetic mutations, which are supposed to be gradualistic. In order to justify the sudden appearance of modern human behaviour, rapid 'selectively advantageous genetic mutations' are invoked, again prompting the question: is innovative human behaviour gene determined? Klein implies that there are indeed genes for components of complex behaviours, that such genes can mutate, thereby modifying behaviour, and if beneficial they will be naturally selected. In fact he refers to a 'most significant' mutation that pulled off multiple behavioural changes single-handedly (p. 18).

3.4.3 Cautionary note on the gene-determination of human behaviour

Despite the phyletic gulf between us and insects, the study of their heritable instincts has been used to justify conclusions about gene-determined human behaviour. However, since those studies were first conducted, molecular biology has moved us to a much deeper and more complex interpretation of even insect behaviour.

The expression of instinctive behaviour depends upon hormonal and neural triggers, sometimes cued by environmental stimuli. What *is* gene determined is the structure of protein hormones and of enzymes that participate in the synthesis of steroid hormones and neural transmitters. Even if those genes mutate, the production of 'better' enzymes and hormones is unlikely; dysfunction is more probable, though complex systems have some backups for molecular failures. Moreover, genes that code for proteins involved in behaviour are *not* synthesized *unless something above the gene level calls for them*. Behaviour cannot be comprehended unless all factors from environment to gene are taken into account and their coordination understood.

There is a more fundamental philosophical problem with the idea of gene-determination of behaviour in humans. Instinctive behaviour, in insects for example, may have evolved under stable environmental conditions where it remains appropriate from one generation to the next, i.e., it is adaptational. In principle, human individuals encounter a variety of conditions during their lifetimes and choose to behave according to intelligent assessment of the situation and the ability to foresee the consequences of their actions – at least, that is the ideal situation. Human behaviour has to be *adaptable*, i.e., capable of being rapidly modified, sometimes on a timescale of seconds. Klein appreciates this since he refers to 'an organism that could alter its behavior radically without any change in its anatomy and that could cumulate and transmit the alterations at a speed that anatomical innovation could never match'. However, this need not involve any genetic determination. Furthermore, mutations do not happen on demand. Also, there is no gene for adaptability, and a gene-determined behavioural component would be unable to cope with change. Even if a rare beneficial mutation occurred, it would take generations to establish itself in a population, and by that time the conditions of life might easily have been altered so that it was no longer relevant.

The concept of the slow, gradual evolution of behaviour, while it is driven by Darwinian theory, is entirely inconsistent with dramatic accelerated change. Such rapid change is more consistent with emergence theory. Even if one accepts the more gradual model of the evolution of modern human behaviours in Africa at or around 70 000 years ago, we are left with the real evolutionary question – what stimulated its initiation at that time? Alternatively if the advent of behaviour typical of present-day humans began in Africa even earlier, why is there such a large gap in the archaeological record before the sudden appearance of modern human behaviour in Europe? Despite the presence of humans who were anatomically like us, at Mughuret es Skhul and Jebel Qafzeh in northern Israel, between 119 000 and 85 000 years ago, why did it take another 50 000 years for modern human behaviour to emerge in Europe? An inconsistency exists between the evidence of sudden behavioural change and the explanation in terms of mutation and natural selection.

When it comes to biological behavioural ecology, which is fundamentally neo-Darwinistic, anthropologists adhere to one of two schools of thought: 'cultural ecology', which focuses on resources (energy), social homeostasis (including population regulation) and group selection; or 'evolutionary ecology', which emphasizes individual selection and deems fitness to be a function of lifetime reproductive success. Although cultural ecology is more holistic, both rely on Darwin for their foundation, assuming slow, gradual evolutionary progress. However this bias is countered by Schwartz (1999), who makes it clear that genetic adaptational changes in populations are of little consequence since natural selection strives to maintain stability and thus nothing of consequence happens until it is disrupted.

Although evolutionary biologists try to guess what the adaptive significance of a feature is or was, they can do so only in retrospect and with reference to what they think is the adaptive significance of similar features in other organisms. It might just be the case that, if a novelty doesn't kill you, you retain it, whether or not it will do you or your relatives any good. As Hugo de Vries saw it, if you need to invoke natural selection at all, it would be in the context of eliminating a feature by way of eliminating its bearers. Death is certainly a strong selective force. But if natural selection works as Darwin proposed, as opposed to just eliminating the unfit, it would be involved in fine-tuning some of the features or adaptations of those individuals that did survive the unexpected inheritance of a morphological, physiological, or behavioral novelty.

(p. 319)

According to Schwartz, it is saltatory emergences of novelties that constitute real evolution.

A more consistent explanation for rapid human behavioural change would be, for example, the emergence of an individual or individuals with superior logical analytical capacity and the ability to apply it. This would be a significant improvement in behavioural *adaptability*, as opposed to a naturally selected genetic behavioural *adaptation* that would be inflexible. Further, if rapid behavioural change did not require a genetic mutation, but was induced through physiological and developmental modifications, as well as intelligent analysis, those modifications could have been correlated with rapid changes in the environment, such as the effects of climate. It is possible that by limiting our perceptions and interpretations about the advent and development of human behaviour within unfounded neo-Darwinistic constraints, we have failed to make the environmental and climate connection that would otherwise elucidate rapid changes in human behaviour.

3.4.4 Contrasts between Darwinistic and emergentistic interpretations of human behaviour

In its primitive invertebrate and vertebrate stages, behaviour was genetically assimilated into stable configurations including conditional reflexes, fixed action patterns
and instincts. However, through time, some animal lineages moved beyond these
fixed adaptational limitations and acquired the capacity for freedom of behaviour,
thought and intelligent action. Even in groups such as insects and birds, which
retained complex instinctive behaviours, there is some degree of freedom of action.
Exposure to different environments created a feedback effect influencing physiology and development. Although many functions of the brain are epigenetically
hard-wired, many others are not, and could not be without losing their flexibility.
For example, cultural conditioning is an environmental input that modulates the
adaptability of the brain. There is a lot more to human behaviour than genes.[2]

The dissimilarity between hard-wired genetic *adaptations* and behavioural
adaptabilities has not been clearly distinguished in evolutionary biology.
Anthropology is even vaguer. Anthropologists label learned behaviour as 'adaptation' when they say 'We adapt to the cold by making shelters.' They are actually
referring to the application of behavioural *adaptability*. Calling it adaptation opens
the door to the implication that there must be a gene for such behaviour. It would be
simple enough to substitute words such as 'modification', 'adjustment' or 'accommodation', rather than 'adaptation', in the case of applications of adaptability. By
putting the emphasis on adaptability in human behavioural evolution, we can see
that the final result is *freedom of action*, rather than automata driven by an aggregate
of gene-determined behavioural adaptations.

Here is another example of neo-Darwinist bias in anthropology. A television
documentary on human evolution was recently aired that depicted hunter–gatherers
of the Fertile Crescent subsequent to the last big ice age. These people were so
'successful' that they had formed complex hierarchies based in large villages and
were accumulating symbolic wealth including gold ornaments. Then a sudden cold dry
climate – the Younger Dryas mini ice age – struck, killing forests, wiping out sylvan
food plants, and causing a retreat of grassland foods, including animal prey. The local
people switched from hunting and gathering to farming virtually overnight. (This is a
drastic oversimplification that we will clarify in a subsequent chapter on agriculture.)
To explain this sudden change in behaviour, the anthropologist in the documentary held
up a stone sickle that 'changed the selection pressures' when the cold snap came.[3]
However, there is no mutation for sickle-making and it is difficult to conceive of
anything determined by human genes that would have caused such a switch to farming.

Yet even in our present condition of adaptability that is grounded in intelligent
thought, humans, when faced with increased greenhouse gases, rising average

global temperatures and multiple hurricanes sweeping the Caribbean and the Gulf of Mexico – the red flags of catastrophic climate change – have difficulty overcoming the old social equilibria quickly enough to respond effectively. Many continue to deny global warming. There is still a need for improved analytical logical capacities and the willingness to apply them, as opposed to doing things 'because that's the way we've always done them.'

3.5 Language and foresight

Associated with improvements in logic and communication is the development and possession of language. This has frequently been referred to as a precursor to, or as a key innovation in, modern human behavioural evolution. Debate has focused on whether fully developed language occurred in a gradual step-by-step process, or instead through abrupt, saltatory change. Some suggest that an abrupt shift, which radically restructured language, would explain some of the significant transformations in human behavioural patterns associated with the transition from archaic to modern human populations. This catastrophic event is reflected in technology, subsistence and the social and symbolic behaviours identified in the middle to upper Palaeolithic transition in Europe explained above. Alternatively, McBrearty and Brooks (2000) argue for a long-term evolution of behaviour, implying that the first signs of what others take to be 'modern' behaviour were coincident with the appearance of the middle Stone Age (MSA) in Africa at 250 000 to 300 000 years ago and the appearance of fossils attributed to the large-brained archaic human *H. helmei*, which exhibit a mix of primitive (*H. ergaster* and *H. erectus*) and derived (*H. sapiens*) features. They also infer that language evolved gradualistically, beginning before the common ancestor of chimpanzees and humans.

From a purely biological perspective, Lieberman (1989) and others argue that the onset of language was a modern human phenomenon and that Neanderthals (especially from La Chapelle-aux-Saints) did not possess a fully developed vocal tract, restricting vocal communications. Others counter that the assumed inadequacy of vocal apparatus of non-modern *H. sapiens* is based on reconstructions of soft tissue and is incompatible with the anatomically modern Neanderthal hyoid found in Kebara Cave in Israel (see, for example, Frayer and Wolpoff, 1993; Pope, 1995). According to Pope, suggestions that 'early *H. sapiens* lacked the ability to reproduce the full range of modern human speech are especially inappropriate in large geographical areas of the Far East (i.e., China and Thailand) and in numerous North American languages (i.e., Amerind languages), where different tones of an identically produced syllable can carry as many as sixteen different meanings' (p. 503).

Some researchers have looked to the gross appearance of the brain as an indicator of language function, but this may be an erroneous approach (see Galaburda and Pandya,

1982; Gannon *et al.*, 1998; Wood and Collard, 1999. See also Gibbons, 2002b). There is no question that *Homo* experienced emergent neocortical expansions that were correlated with other anatomical adjustments. But recent research indicates that the language-related parts of the brain are not clearly localized. If, as has been widely believed, the evolutionary origin of human language is founded on the size predominance of the left hemisphere planum temporale (PT), within Wernicke's posterior receptive language area, then this region was already developed in the ancestors of humans and chimpanzees about 8 million years ago. Recent findings have generated several evolutionary hypotheses including: (1) the PT is unrelated to language or communicative functions, or did not evolve such a role in chimpanzee lineage but retains some other function; (2) the PT is involved in communication but developed different evolutionary trajectories in humans versus chimpanzees, whereby chimpanzees developed the neural substrate for 'chimpanzee language' involving 'gestural–visual' modes that we have yet to understand; (3) the PT is not, nor ever has been, related to language or communicative functions; (4) the PT became asymmetric in both lineages independently, but this is unlikely as the orangutan also has an asymmetric PT (Yeni-Komshian, 1976). Irrespective of the PT's role in communication or language, Gannon *et al.* (1998) imply a long and complex evolutionary history for the development of language-related portions of the brain in hominid primates. Further, they suggest that cognitive and communicative abilities may have co-evolved, with the PT evolving species-specific repertoires that currently distinguish those abilities in humans and chimpanzees.

True language must have emerged from numerous interconnected factors. Allometric growth of the neocortex would have simultaneously expanded the areas concerned with memory, logic, innate grammar and the coordination of sound reception and production. An increase in the total number of neurons improved the potential for an increase in the number of cell-to-cell connections. There is also a tie-in with vision, which is important for perceiving behaviour, body language and emotional context. Vision is also necessary for lip-reading and sign language, which obey the same rules of innate grammar. A conducive social environment is also necessary. It is difficult to conceive of language evolution as a naturally selective accumulation of alleles for all of these factors one at a time – if they did not all evolve simultaneously, the system would be inoperative. The 1.3% genetic difference between chimpanzees and humans does not begin to explain anatomical differences at the gene level, far less the linguistic dissimilarity. Most of the differences are above the gene and enzyme levels, operating at the level of intercellular communication, and at higher levels of neural integration.

The previous paragraph describes what we think was the potential of the brain of *H. sapiens* when the species originated. We do not rule out subsequent shifts in brain organization, but for the time being we take the most parsimonious hypothesis and ask the question: what released the existing potential of the original Sapiens' brain during the onset and rise of modern behaviour in Africa and subsequently Europe? We know that social interaction activates the potential of individuals' learning ability, by allowing them to listen to linguistic examples that can be mimicked and then understood. Though based on innate grammar and other neural furbishments, the actual language acquired is distinguished by characteristic word orders and distinctive tonal or phonemic inflections. These are learned. All of them can potentially be learned by the adaptable brain of any human. And for a language to be learned effectively requires the time that is afforded by a protected early childhood and its social interactions.

Some archaeologists have argued that the increased complexity of logistical planning interpreted from the upper Palaeolithic record demands the existence of a relatively advanced and structured language (Soffer, 1994; Whallon, 1989). An increased ability to share information and the distribution of economic resources could have been the critical factor allowing colonization of more extreme periglacial environments. But Pope (1995, p. 503) disagrees, citing the lack of convincing evidence that links the appearance of language with the increased complexity associated with the middle Palaeolithic in Europe or Africa. In fact, he suggests that complexity (at least as indicated by material culture) antedates the appearance of 'modern' humans everywhere. The well-balanced, sophisticated, wooden javelin-like spears found in Schöningen, Germany, that date to around 400 000 years ago imply sophisticated planning and hunting capacities, supporting this hypothesis.

Tattersall (2001) turns to children's games to explain the emergent linguistic and cultural differences between Neanderthals and modern humans. He rejects natural selection as a creative force acting as a physical 'buffing-up of the cognitive mechanism' (p. 58). Although we should be wary of committing ourselves to particular 'key innovations', group game-playing certainly involves communication through signing or spoken language, as well as memory, learning and parental care, none of which can operate in isolation from the others. Furthermore, goal-oriented group game-playing would have tended to enrich vocabularies. By putting ideas into words, a child innovator would more effectively invoke imitation than by simply showing by example. Dunbar (1996) suggests that the putatively universal human trait of gossiping stimulated language evolution. However, while gossiping transfers titillating information and alleviates boredom, it is not an activity that might develop new vocabularies or innovative techniques, far less new ideas. Dunbar's hypothesis also presupposes that the language of gossip was learned by the gossipers as children. Be that as it may, the ideas of both Tattersall and Dunbar raise the question of having the *leisure* to do these kinds of things, which we will discuss in the following section.

In summary, the structural foundations for language may have developed early on, as evidenced by the finds in Africa, the early Schöningen wooden spears, coeval with *H. erectus*, and an early enlargement of the planum temporale. But behavioural novelties emerged with the first members of our species, and something brought out its potential to permit a sudden flowering of culture sometime around 70 000 years ago in Africa, to be even more strongly expressed around 40 000 years ago in Europe.

3.6 General intelligence or cognitive capacities

The question remains, if 'anatomically modern' humans emerged over 100 000 years ago in the Middle East, and 195 000 years ago or earlier in Africa, with or without language, why did it take them until about 70 000 years ago in Africa and between 50 000 and 40 000 years ago in Europe to stop behaving archaically, and to suddenly begin to behave with the 'spark of creativity' that we identify with present-day humans (Tattersall, 2003)? The brain did not get bigger or more complex in its superficial anatomy. But there could have been a further emergent leap in neocortical evolution that was manifested as a multifunctional creativity. The human brain and all its functions, including linguistic ability, could be the result of innovative differentiation and reorganization at the cellular level. Maybe it was an emergent quality of neural adaptability relating to a differential increase in the number of neurons, and their more complex coordination. However, our working hypothesis proposes that all of the potential for sparks of creativity existed with the first *Homo sapiens*, and was realized as a consequence of suitable environmental, psychological and social developments.

The smart, modern human lineage that came out of those beneficial, or challenging, conditions could have rapidly expanded in numbers. Also, despite the risks of hybrid dilution (and over Tattersall's objections), it could have interbred with the biologically compatible larger population of the original, less creative *Homo sapiens*, or even with remaining representatives of *H. erectus* or *H. neanderthalensis*. As a general principle, it is not gene differences, but chromosome differences that make two species reproductively sterile, and we know nothing about the chromosomes of extinct human species, or those of early *H. sapiens*. Thus both epigenetic novelty and its cultural accoutrements could have been quickly spread. Despite George Bernard Shaw's reservations regarding actress Mrs Patrick Campbell's proposal that they mate to produce intelligent and attractive children, the desired combination of the brainy and the beautiful would have been likely in some of the offspring. But the other species of *Homo* seem to have paid a terminal price, probably, as Tattersall argues, because a cultural mean streak came along with the Sapiens' package.

The problem of understanding language evolution is paralleled in explanations of the evolution of intelligence. The standard neo-Darwinist approach to the latter is to identify a 'key innovation' that is then 'selected' and may or may not be accompanied by the other necessary functional and morphological correlations of intelligence. Unfortunately, they all have different pet novelties, which suggests groping in the dark rather than incisive logic. As we have seen, language is a strongly touted key innovation. However, William Calvin (2005) has argued that a key innovation in the emergence of Sapiens' intelligence is not linguistic but is, rather, related to our unique hand–eye coordination, combined with a primitive calculus necessary to estimate trajectories and allow for acceleration of both weapon and prey. But monkeys can throw accurately, llamas are spitting aces, fulmars are accurate projectile vomiters, archer fish can bring down a fly with a jet of water from several metres away and carnivorous clams can effectively aim their sucking siphons at passing zooplankton. None have the talents of language, logic, art and predicting the future that humans possess, and the clam is literally brainless. All such human characteristics are part of the same package of emergent properties resulting from a cerebral expansion and reorganization. Would a poker player with four aces spend any time after the game arguing about which of the aces had caused the win?

Byrne (1995) comprehensively marshals facts, anecdotes and shrewd conclusions to illustrate the difficulty of fitting an emergentist peg into a Darwinist hole. For Byrne, the early common ancestor of simians and humans demonstrates a '*quantitative* shift in cognitive ability'. The enlargement of the neocortex responsible was 'almost certainly an adaptive response to the increased social complexity of living in semi-permanent groups' (p. 230). Could it have been the other way round; the increased social complexity coming from the prior epigenetic enlargement and differentiation of the neocortex? Or was it a bootstrapping relationship between the two?

Shettleworth (2000) defines cognition as 'information processing in the broadest sense, from gathering information through the senses to making decisions and performing functionally appropriate actions, regardless of the complexity of any internal representational processes that behavior might imply' (p. 43). She breaks cognition down into 14 aspects, including 'spatial memory, circadian timing, interval timing, dead reckoning, landmark use, imprinting, song learning, motor imitation, associative learning (and components thereof), social intelligence, theory of mind (and components thereof), language, reasoning about social obligations, and consciousness' (cited in Heyes, 2000, p. 6). This adaptationist (i.e., neo-Darwinist) view of cognitive evolution contrasts with the more anthropocentric traditional view, 'encompassing perception, learning, and understanding' (Bitterman, 2000, p. 61), which implies a 'single hierarchical ordering of mechanisms … with capacities possessed only by humans at the "most complex" or "most advanced" end' (Shettleworth, 2000, p. 57).

The opposition of the adaptationist and anthropocentric points of view hint at rival schools of thought, rather than objective analysis. While clear definitions are desirable, they can also be prejudicial. Defining cognition in a way that would give the capacity to the birds and the bees as well as to humans serves the adaptationist agenda but obfuscates the understanding of human cognition. Therefore, 'anthropocentric' or not, we prefer the 'hierarchical ordering of mechanisms' approach to the evolution of cognition. We place it in the context of hierarchically emergent evolution in which innovations, including those of the nervous system, are either emergences that appear at critical points in allometric shifts, or are saltatory events that result in the acquisition of multiple functions. For example, cognition in humans circumscribes not only intelligence, but also logic and language. The latter depends upon an innate grammar correlated with logic, which allows complex propositions or information to be conferred by signing, speech and words that can be written and read, as well as spoken.

The successful acquisition of information requires interaction between the parts of the brain involved not only with logic but also with hearing, vision, olfaction and tactile sensitivity. Instead of thinking in terms of vocal chords, or the piecemeal expansion and natural selection of brain modules, we need a holistic conception of cognition.

Tomasello (2000) comes closer to such a holistic view. According to him, the evolution of human cognition is built on 'the uniquely primate cognitive adaptation for understanding external relations' (p. 171). Note that his use of 'adaptation' in this context is equivalent to how we have defined 'adaptability'. The one major difference between primate cognition and human cognition is the incorporation of the mediating forces of causes and intentions – 'understanding of others as intentional beings like the self' (p. 180). The evolution of human cognition allowed a 'radically new form of cultural inheritance' (*ibid.*) and probably happened with the emergence of 'anatomically modern humans' as they evolved ~150 000 to 200 000 years ago in Africa, allowing them to 'predict and explain the behaviour of conspecifics' (p. 173) and outcompete other hominins as they migrated beyond Africa and around the world. We infer that the same degree of intelligence would allow the prediction and explanation of the behaviour of other species, including the likely distribution of prey. Nevertheless, the spark of creativity associated with modern human behaviour took a lot longer to emerge.

What actually happened, and what explanations have been forthcoming? Above we noted Tomasello's opinion that an important advance made by *H. sapiens* was to understand that others of our species think the same way we do. The archaeological record of the lower and middle Palaeolithic demonstrates that tools were discarded quickly, resources procured locally and organization was small-scale. On these grounds, Binford suggests that degrees of 'time depth' and long-range planning, forethought and prediction were important advances (1985, 1987, 1989, 1992). Mellars (1996) advocates that upper Palaeolithic forethought greatly exceeded that of the middle Palaeolithic. The upper Palaeolithic people travelled greater distances, transported commodities including marine shells, possessed highly structured internal site organization (which almost certainly implies that they occupied sites for prolonged periods), deliberately stored food, developed specialized exploitation of reindeer and clustered their hunting sites along migration routes.

According to Mellars, these characteristics suggest either an increase in cognitive ability or more highly structured forms of language. However, all of these could relate to penetration into cold climates and adjustment to those conditions, with annual migrations from warmer climes by traders. Are we justified in relating these behavioural characteristics to increased cognitive ability or language? Is cooperation, and decreased aggression, which we associate with increased social organization and trade, related fundamentally to cognitive ability or language? Or is a particular kind of environment, or climate, a necessary correlation?

What was required to achieve the switch from the middle Palaeolithic to the greater creativity of the upper Palaeolithic? We have argued that the formation of more complex societies during times of rich environmental resources was important. But this runs counter to conventional wisdom that takes environmental, particularly climatic, *stress* to be the most significant component in so far as it creates selection pressures which then wait for appropriate adaptational mutations.

3.6.1 Stress and leisure

One of the factors pursued by evolutionists such as Alexei Severtsov, Ivan Schmalhausen, C. H. Waddington and Robert Reid is the disequilibration of stable systems that allows for innovation, or for the success of previously repressed innovations. We might ask what kind of disequilibrating force made the 'old ways' redundant and permitted the flowering of human creativity. In particular, was there a climate connection?

There is a consensus amongst those interested in palaeoanthropology that *stressful* climate is important (see, for example, d'Errico Sánchez Goñi, 2003; Klein, 2000; McBrearty and Brooks, 2000; Pope, 1995; Potts, 1996a). However, the causal interpretations are different, since neo-Darwinistic palaeoanthropologists believe that such stresses somehow changed 'selection pressures' that triggered 'adaptations' that could deal with the less favourable conditions. McBrearty and Brooks (2000) are of the opinion that 'social adaptation to scarcity may involve the maintenance of long-distance reciprocal exchange relationships, or the long-term retention of individual and group memories of distant resources or survival strategies, often aided by the use of symbol and ritual (de Garnie and Harrison, 1988). We argue that most, if not all, of this behavior was practised in the MSA (p. 484). Their thesis is that features we consider to be representative of *H. sapiens'* societies were already characteristic of

H. erectus. If accepted at face value, it still does not provide an explanation of how, and under what circumstances, abilities that are *brought out* by scarcity, for which read 'stress', originated. Moreover, we question their conclusion that stress leads to 'long-distance' cultural exchanges. However, their thesis can be reconciled with the idea that disquilibrating stress allows pre-existing adaptabilities, or hitherto unexploited novel behavioural approaches, to be expressed. And *short-distance* exchanges might be very effective if different cultures are brought together by migration into shrinking areas of adequate resources. Group memories might very well be significant under the latter conditions.

How can we apply our working hypothesis that most of the biological evolution of Sapiens' brains occurred with the original emergence of the species, and that we require sociological explanations for subsequent developments of that potential? The stress thesis can be synthesized with the 'time and leisure' antithesis as follows. Stress does not cause adaptations to evolve, but does have a significant effect. Rather than changing 'selection pressures', which then wait for appropriate adaptational mutations, stress calls upon existing *adaptabilities* to be applied in an intelligent way to new conditions. Stress can therefore disequilibrate change-resistant equilibria, and the novel application of existing adaptabilities might raise the affected human populations to a new way of dealing with their environment. Then, with improvements in climate that enrich resources, the potential for forming new and more adventurous if not more complex societies can be realized.

Such societies allow more time for children to be children, and for the divisions of labour that allow some adults to be creative. But the tendency of leisured societies to get into ruts, reinforced by the presence of elites characterized by wealth and political or religious power, requires that for there to be further progress there has to be disequilibration. War and catastrophic climatic or geologic events are the most likely disequilibrating forces. Cycles of stress, relaxation and stultification have probably existed throughout the history of the species.

3.6.2 Transmission of learned behaviour and the demise of the Neanderthals

The period of maturation for humans is twice as long as it is for the gorilla and the chimpanzee and is critical for ensuring not only game-playing among the young, but also the transmission of learned behaviour in young and old (Wood and Collard, 1999). Recent research on dental and femoral development indicates that Neanderthals had a growth development period more similar to ours than that of the living apes (Dean *et al.*, 1986; Smith, 1994; Tardieu, 1998), implying an extended childhood period of dependence.

To a large extent, the burials of Neanderthal people lacked the symbolic grave goods that are frequently associated with ritual and belief in an afterlife. This has led Tattersall (2003) to note that despite their ability to remain successful during the difficult environmental circumstances of the last ice age, they 'lacked the spark of creativity that, in the end, distinguished *H. sapiens*' (p. 25). Yet, as Tattersall points out, the brains of Neanderthals were sufficiently similar to *Homo sapiens* that gross anatomy does not altogether explain the cultural differences associated with each. Further, according to Smith (2003), since the Neanderthals remained in some of central Europe's most preferred real estate until about 28 000 years ago, and in Portugal until as late as around 24 500 years ago (Duarte *et al.*, 1999), they were not quickly relegated to the periphery but competed or even cooperated reasonably well with the invaders (Smith, 2003) – unless of course *H. sapiens* largely avoided them. Stringer (cited by Wong, 2003) thinks that 'modern' humans were just a little more innovative and had better social networks that allowed them to outcompete the Neanderthals in what was a rapidly changing environment. Wolpoff (cited by Wong, 2003) believes that Neanderthals were just outnumbered by the influx of 'moderns', and after thousands of years of interbreeding, Neanderthal genes and features became diluted and disappeared. He believes a similar result will be observed in Australia in a few thousand years, with the complete domination of European features over native Australians; simply the result of the overwhelming number of Europeans.

3.7 The bigger picture

A broader hypothesis of the evolution of intelligence is presented by Stewart and Cohen (1997):

Consciousness is emergent if anything is, so it may be difficult to find a convincing evolutionary story for the gradual evolution of consciousness. If it is a 'threshold' phenomenon, a sudden expansion of the phase space for brain function, then on an evolutionary timescale it would appear to happen virtually overnight. The underlying hardware – wetware – might evolve continuously, but an emergent feature like consciousness can appear only when that evolution reaches a 'trigger point'. So we should not expect to find any obvious evolutionary continuity of consciousness – but we should expect to find a general tendency towards more complex brains with more nerve cells, more memory, and faster transmission of nerve impulses. Which we do. For the same reason, we should not expect to find a series of organisms that fills the gap between non-conscious and conscious animals in a continuous manner.

(p. 221)

Stewart and Cohen also address related human emergent properties: i.e., multi-functional characteristics that appear suddenly.

One of the universal features of complicity is the emergence of new patterns, new rules, new structures, new processes that were not present, even in rudimentary form, in the separate components. For language and intelligence, this is abundantly the case. Here the most influential new possibility opened up by complicity is that language permits experience to be stored in the memories of older people and passed on to the young. The collective experiences of the tribe become a cultural lexicon stored in the people that surround each child.

(p. 245)

Thus, Stewart and Cohen reject the Darwinistic theory-driven stance and, like Schwartz and ourselves, prefer to see cognitive evolution as an emergent phenomenon. They use the word 'complicity' to mean the way in which a numerical complexity of components can be simplified by a new level of order. For example, if a maze of wires connecting all telephones with all other telephones becomes reconnected through an exchange, the wiring system is simplified, yet the exchange represents a new level of complexity in the form of increased organization. The evolution of the nervous system and its integration with the endocrine system is the best general biological example. For Stewart and Cohen, human emergent properties are characterized by novelties of mind that lead not to an identifiable key innovation that confers high fitness, but to a very general adaptability. That then bootstraps itself to greater cultural heights by feedback between language, memory, education and tradition, accompanied by adaptational improvements in vocalization and cerebral coordinative conditions, once mind has emerged.

An aspect of human adaptability that could also be taken as a key emergent innovation is identified by Chris Sinha (1987).

Thus, the evolution of infancy was the biological mechanism through which the potential for inter-generational cultural transmission created by tool use was optimized, and appropriation was the specific social and ontogenetic process in response to which the niche of infancy evolved. In this respect, the evolution of infancy was not so much a terminal point of biological evolution, but the crucial inaugural moment in the socialization of biology (Riley, 1978). Such an evolutionary process, I suggest, would lead to a rapidly expanding endogenous spiral of change and adaptation.

(p. 282)

Although Sinha wraps together adaptability and ontogenic modifiability under 'adaptation', he anticipates Tattersall's recognition of the evolutionary importance of the behaviour of children. Sinha adds that while tool use long antedated the emergence of 'modern' humans, the extension of infancy potentiated 'the emergence of complex categorization and social-technological capacities' (p. 283). A longer childhood provided a greater opportunity to learn more complex methods of tool manufacture and use, as well as language and art, and in so doing changed the human environment in a way that affected the behavior of subsequent generations. Sinha does not exclude the possibility that hard-wired linguistic and manipulative skills were involved, but merely notes that his hypothesis does not imply them.

Thus we can add to the constellation of particular emergent qualities of a more organized neocortex, which characterize *Homo sapiens*, the adaptable quality of *learning how to use them effectively, combined with the opportunity to do so.* We have noted that a protected period of childhood and group play, as well as an increase in educational opportunities, was important in the enrichment of language that would carry on into adulthood. Children have leisure if adults supply food, shelter and protection from hostile animals including other humans.

The same principle applies to adult activities. Much has been made of cave-painting, religious rituals, manufacturing and playing musical instruments, making artistic ornaments and so forth. The popular literature on this sometimes conveys an image of groups of renaissance Cro-Magnons communing with nature. But what members of groups of people living close to the edge of survival could really indulge in such activities? First of all, those who wanted to and had the brain capacity to carry out their wishes. The artistic urge can surpass the pangs of hunger and lack of social approval. However, they might have been the children of those who had acquired power as clan leaders, or who had accumulated some kind of wealth. It seems more likely that the most creative humans were products of a society organized in such a way that it could afford some specialists in religion and the arts as well as in tool-making, i.e., division of labour.

Social organization for the purposes of hunting, finding or making shelter and defending the clan goes back to the earliest human species. But the creative spark attributed uniquely to 'modern' humans required a *hierarchical* structure that included priests and artists, either two types, or a single type with both functions. Shamans, or priests, who were involved in the passage of souls and who were believed to wield power over health, life, death and the afterlife, might have been the first artists as well. However, research into the shamanism of present-day humans reveals that the shaman's primary responsibilities, which might have existed from the earliest emergence of the human species, were confined to spiritual journeys and the interpretation of dreams, in the service of the tribe, and that the addition of healing, ceremonial and artistic functions is very recent (G. Rhyon-Berry, personal communication, 2005).

The first artists and musicians were probably self-chosen in the first place and given time to practise their arts because of their spiritual and aesthetic appeal to the rest of their society. This implies that while the capacity to develop the creative spark might have been possessed by the earliest Sapiens, if not to some extent in their antecedents, it only emerged under special circumstances. Favourable climates, producing favourable environments and resources, might have allowed the emergence of the social hierarchy under discussion. Prior to those rare moments when creativity sparked (prior to 70 000 years ago in Africa; ~50 000 years ago in Europe), humans were organized into extended family groups or clans. But were they universally engaged with the challenges of mere survival? Do studies of present-day isolated human groups suggest an answer? Before we get to details, we must remind ourselves that all contemporary humans are believed to have gone through the Pleistocene enlightenment, since they all have language, religion, art, ornaments, music etc. Isolated modern human groups, unaffected by 'civilized' humans, seem to have had a lot of leisure. They made no special effort to formally educate their children but afforded them more liberty than urban children are allowed. There is evidence that their cognitive development was faster than is evident in children from less 'advanced' social systems. As to the shamans' roles, in dealing with the troubles of individuals they maintained the equilibrium of the community, and in some cases that equilibrium may have been maintained for tens of thousands of years. The social equilibrium must have weathered minor disasters such as cooling periods, forest fires, earthquakes, tsunamis and volcanic eruptions, but it did not rise to a higher level of innovation and experiment in the subsequent restoration period. Any change in behavioural patterns was aimed at restoring the status quo (G. Rhyon-Berry, personal communication, 2005). This correlates with the apparent inertia of groups of *H. sapiens* for most of the history of the species. Yet, clearly, there were occasionally rare, revolutionary innovations. What caused them? Can we infer anything from disasters that *have* affected isolated non-civilized human groups permanently? These disasters all come down to contact with 'civilized' humans from the northern hemisphere. But it is possible that cross-cultural exchanges, hinted at by McBrearty and Brooks (2000), could have been benign and have led to innovative changes.

There are at least four possible causal explanations of human revolutionary innovation.

1. The African enlightenment of *c.*70 000 years ago and the European enlightenment of *c.*50 000 years ago were caused by complexifications of the human brain. This is a popular explanation, but it creates more problems than it solves. The first brain change should have provided evidence in the form of new artifacts that should have spread, and whose manufacture should have been continuously maintained. New artifacts there were, but their use petered out, and further improvements, which might have been expected if there had been a permanent brain change, did not occur.

   A second brain change triggered the European enlightenment, bringing more complex language and increased cognition, resulting in experiments with new artifacts. This is tempting, but evidence is lacking, and the hypothesis implies that there were two independent brain changes.
2. The second explanation is the McBrearty and Brooks' model of accretion of different behaviours and artifacts because of stressful conditions that promoted group interaction. Their interpretation does not require the anatomical and functional evolution of the brain.

We hasten to repeat that any change in behaviour that produces innovative artifacts or novel interpretations of phenomena are likely to be revolutionary in their impact. Furthermore, such changes do not require the assistance of natural selection – imitation and learning are all that are required for useful novelties to spread.

3. Our working hypothesis has been that the Sapiens' brain had adequate potential to produce creativity under certain circumstances. Those circumstances included a period of stress that disequilibrated normal behaviour and released repressed behavioural adaptabilities that allowed the affected humans to accommodate to the stress. This was followed by a time of improving resources when new behaviours continued to be applied prior to settlement into stasis once again.

    However, the evidence from studies of isolated present-day human groups suggests that our working hypothesis is inadequate, that unchanging stability is the rule, that natural disasters are accommodated without permanent behavioural change, and that only invasion by technologically modern humans was sufficient for disequilibration and, in many cases, demise.

4. The last sentence of the third explanation suggests its synthesis with the McBrearty and Brooks' model, which would emphasize behavioural hybrid vigour implicit in the latter. (Hybrid vigour is the genetics' term for new phenotypic characteristics that emerge when two previously isolated strains are allowed to interbreed.) Such a behavioural vigour, marked by flashes of creativity, might have arisen when two or more previously isolated stable groups congregated in the same area due to climate change. For example, a cooling and drying period might have desertified former hunting and gathering grounds. In the case of the European enlightenment of *c.*50 000 years ago, several groups of *H. sapiens* may have congregated due to glaciation of formerly benign environments. Closer contact with Neanderthals may have been a significant part of the mix, even though it was finally deadly for the Neanderthals.

This synthesis must also allow for a period of improvement in living conditions that allowed new behaviours to be consolidated, providing in particular the leisure for children to innovate with play, language and perhaps new artifacts such as toys. Also, enlargement of social groups in restricted areas can lead to greater stratification and an associated division of labour that makes more efficient use of limited resources and time (Ehrlich, 2000, p. 531).

3.8 Corollary on social stratification

We here are dealing with a 40 000- to 50 000-year-old stage of social evolution where stratification may have been just sufficient for modern humans to specialize as priests, healers, artists and musicians. This process is more obvious in later agrarian societies with complex social structure, and even later in industrial and global 'hi-tech' cultures. As Ehrlich (2000) conjectures, 'Human natures have been strongly shaped by multilevel human societies – families, bands, tribes, trade networks, and so on – and by population pressures, all interacting in a positive feedback system' (p. 113). However, he then argues that the more complex societies became, the more social problems arose, and this stimulated brain evolution. In contrast, we argue that *biological* evolution of the human brain had largely occurred with the emergence of *Homo sapiens*, and that what subsequently evolved in response to social complexification was the *behavioural* realization of the brain's potential.

Hierarchical stratification is equivalent to Stewart and Cohen's 'complicity', that is a simplification of a chaotic, i.e., disorganized and ineffective system, through reorganization. Instead of independent, extended family groups, in competition with each other despite a richness of resources, a society emerges with an ordered political complexity that must provide leisure not only for its children but also for its priests, healers and artists. The emergentist principle of re-equilibration following disequilibration continues to apply. Climate volatility results in emergent behavioural innovation, followed by periods of climate stasis when that innovation can be transferred to the larger population, while hierarchical societies are in the simplest stages of construction. Eventually, hierarchical societies are bound to be caught in behavioural binds that resist innovation. Institutionalized religions and political systems are good examples of such change-resistant equilibria.

According to McBrearty and Brooks (2000), there is little evidence for behaviour or artifacts that were formerly designated as 'modern' prior to the African renaissance of 70 000 years ago. If the *potential* to develop the creative spark was an emergent property of the anatomically modern humans prior to that time, if not to the first members of *Homo sapiens*, it was probably realized in significant ways such as improved hunting behaviour and tool manufacture and the development of a simple language associated with such activities. Neurologist William Calvin (draft chapter for book in preparation, http://williamcalvin.com/2005/CreativeExplosion. htm, 2005) believes that early language consisted of short sentences based on a limited vocabulary, presumably for presenting the orders of the day and passing practical information. There may have been environmental conditions that periodically realized stronger expressions of the original creative potential, but the evidence of thousands of years of dynamic equilibria is in favour of a return to the old norms in isolated groups. Archaeological clues of innovation finally appeared in Africa 70 000 years ago, and, like McBrearty and Brooks, we attribute them to stress and cultural exchange. At this time there probably was an advance in language beyond simple domestic and hunting-and-gathering applications to a more complex form that was able to use abstractions and make plans as well as convey information.

At some periods during the existence of our ancestral *H. erectus*, conditions were probably favourable for social progress that might have left an archaeological record. The existence of balanced wooden javelins hints at such a possibility. So too does the evidence of *H. floresiensis*, a dwarf relative of *H. erectus* that might have had the ability to both raft or boat to Flores Island and hunt the formidable komodo dragon, *Varanus komodensis*. Thus, a small brain does not exclude certain behaviours that might once have been regarded as unique to 'modern' humans. On the other hand, those small brains probably lacked the potential to express the full range of creativity found in *H. sapiens*.

As to *H. neanderthalensis*, its brain size was superior not only to that of *H. erectus*, but also to that of our own species. Its increase in brain size alone may have been enough to account for its apparent behavioural superiority over *H. erectus* – the bigger the brain, the more neurons and the greater the complexity of axonic and dendritic connections – even without further cerebral differentiation. But despite their occupation of desirable real estate, Neanderthals seem to have lacked the behavioural potential that might have led to the division of labour and greater creativity when social conditions were suitable. It might have been a combination of environmental factors and behavioural stagnation that held them back. We must also remember that their frontal lobes were relatively smaller than those of *H. sapiens*, which may suggest less thinking power and less linguistic capacity.

In short, what we propose here is that progress in creative human activities can be largely explained in environmental, behavioural and sociological terms. The emergent biological properties of brain evolution were associated with the origin of our species. These, in turn, made the new type more adaptable in terms of brain function and behavioural innovation. Thus there was an immediate *potential* for progressive creativity; however, that needed the appropriate environmental and social conditions to be strongly expressed.

Nevertheless we admit that it would be oversimplifying human evolution to exclude further biological brain evolution after the emergence of the first members of *H. sapiens* and after the emergence of anatomically modern humans. Self-generating allometric shifts in brain growth might have continued after the primary appearance of the species. That is, some parts of the brain could have become larger or more complex relative to other parts (see Appendix B – Developmental evolution). Such changes might have been related as critical-point emergences to the sudden appearance of novel behaviours.

All of this is a far cry from the antithetical worldview of 'evolutionary psychology' and its anthropological equivalent. For example, Morton (1997) puts it this way in his review of Pinker's 'evolutionary psychology' book *How the Mind Works*:

The human's mind is the way it is because big complicated minds have, in the past, helped human beings get their genes into the next generation. From this it follows that the mind is not the all-purpose thinking machine philosophers and psychologists have often taken it for. The mind has been developed to think in specific useful ways, and its structure reflects this.

(p. 102)

The implication of 'evolutionary psychology' is that intelligence and related mental qualities are aggregates of gene-determined behaviours that appeared one at a time and were naturally selected and accumulated before and during the Pleistocene. The aggregates, they argue, create the illusion of 'all-purpose' thinking abilities.

We take the opposite view that many features of the modern human brain and behaviour emerged simultaneously, that natural selection had no creative role, and that the brain is indeed potentially an all-purpose thinking organ that can be stimulated to be creative, especially under favourable environmental conditions that allow the society to expand sufficiently, to progress educationally and to develop specialized roles for individuals. It was certainly capable of dealing with the stresses of the Pleistocene, but both before and after that period it had the all-purpose adaptabilities that can deal with new situations. The 'cognitive closure' of the Pleistocene is an 'evolutionary psychological' fantasy.

Behaviours that are genetically fixed are *non-adaptable*, incapable of being integrated into complex adaptable systems. The evolutionary psychologists' point of view makes it impossible for them to explain the major characteristic of the phyletic line that gave rise to humans; namely a progressively increasing physiological and behavioural adaptability.

To understand the evolution of mind it is also instructive to detail the contrast between the emergentist stance taken by Stewart and Cohen (1997) and the dour ultra-Darwinism of the 'evolutionary psychologists' in greater detail. The gene components of the brain cells are the same as those of liver cells and are almost identical in all human brains and livers. At the level of neural and hormonal signals in the brain, there are some differences arising from differential gene expression. Brain anatomy is also gene based but ontogenically variable. That is, its embryonic development can vary somewhat, and after birth its cerebral development depends upon the experiences, environmental and educational, of the individual. It is *experience* that stimulates the complex differentiation of axonic and dendritic connections between neurons, including memory storage. Hormonal conditioning has also been imposed along the way, and all together they produce personalities further modified through education and indoctrination. We do need an evolutionary psychology to understand the evolution of human behaviour, but not one that is founded in a simplistic genocentrism.

To further our argument it is necessary to backtrack through neo-Darwinism to Darwinism. It would be difficult for firm adherents to those principles to accept that a cerebral complexity could emerge with only a *potential* to give rise to the behavioural characteristics that appear thousands of years after their origin. This is because they believe that there has to be a 'selection pressure' to produce the complexity in the first place, and hence, a 'selective advantage' that would make those individuals with the novel characteristic immediately superior to competitors. But as Schwartz (1999) infers, emergent novelty need not be adaptive; provided that it is not harmful, it will be retained. To this we add that if it is undetrimental, it means that the novelty has also integrated effectively into the pre-existing ancestral physiological and developmental systems. Furthermore, the emergentist sees such changes as having multifunctional potentials that might not be expressed until appropriate conditions arise.

In the Appendix we will discuss relevant physiological and developmental factors that contributed to the generative conditions for brain evolution in the placental mammals that were involved in the run-up to the emergence of *H. sapiens*. For example, the placenta was a feature that allowed extended foetal brain development within the mother. The corpus callosum was an innovation that integrated the two expanding sides of the cerebrum. Proportional changes in brain areas were also involved. Here are some of the generally accepted points about hominin brain evolution, for the moment without interpretive bias:

1. If the species of *Australopithecus* and *Homo* are examined in chronological sequence, saltatory brain enlargement is found to be a prominent aspect of their evolution. (Some anthropologists challenge the legitimacy of such a phyletic line. The antithesis is that an even greater saltation produced hominins from an early anthropoid close to the chimpanzee line, and that the australopithecines are a separate but parallel clade.)
2. Brain enlargement is a hypermorphic process correlated with enlargement of the cranium and, in females of the species, the birth canal. Stereoscopic vision, an upright stance, manipulative skills and complex vocalization are other correlates.
3. Brain enlargement has involved convolution of the neocortex and its lamination into layers of neurons.
4. Neocortical packaging allows for compactness within the limited capacity of the cranium, as well as greater computing power.
5. The neocortex integrates with, but can override, some lower homeostatic functions.
6. During the development of individuals, learning is correlated with the differential establishment of dendritic connections between neurons in the central nervous system and probably with synaptic facilitation through modification of cell surface receptors for neurotransmitters. This means that the microstructures of mature brains differ according to experience as well as to the underlying adaptability of neural functional anatomy – even in genetically identical twins.
7. Unless children learn how to use the potential of their higher brain functions at the appropriate threshold stages of their development, they may not be able to acquire full potential later on.
8. The brain is an adaptable organ in two ways. Even in mature adults, brain damage can be repaired or circumvented. The adaptability of the brain, in the sense of having multiple, modifiable, organized functions, allows behavioural adaptability.

Removing the conception that natural selection is the universal cause of an evolution that must be seen as gradual in order to fit the Darwinistic driving theory would free the creative imaginations of those who delve into the complexities of evolving intelligence from logical obstacles. For example, Ehrlich (2000) goes to some pains to demonstrate that many aspects of modern human behaviour that characterize the 'great leap forward' of 50 000 years ago are detached from gene determination. Yet his more adventurous conclusions are qualified with the observation that 'there might be a gene for it', and although he is one of the more enlightened neo-Darwinist population geneticists, he unthinkingly fills in significant gaps in our knowledge of evolutionary causation with natural selection as an active force. Byrne (1995), Calvin (unpublished, 2005), and Tattersall (2001, 2003), along with Stewart and Cohen (1997), are almost selection-free, and their very disagreement over the 'key adaptive quality' of intelligence strengthens the case for emergent multifunctionality. Schwartz (1999) seems to have arrived already. 'Evolutionary psychologists' ought to be there, without their restraining inverted commas and without compromising the validity of their analytical studies, as opposed to their metaphorical causes and metaphysical conclusions.

So why are humans so unreal a species in nature with all our closest living relatives extinct? Granted, our behaviour is unlike any other on the planet, but why then the demise of our closest relatives? Perhaps it is an *H. sapiens*' mean streak that led to the extinction of *H. erectus* and *H. neanderthalensis*; perhaps the Sapiens' behaviour we interpret as 'modern' is no more than a fundamental adaptability stimulated by our cultural and social interactions and experience in different environments. Perhaps the Neanderthals, despite the general abilities that they had developed to deal with a wide variety of resources, were fundamentally more set in their ways than *H. sapiens*, and less capable of modifying their behaviour in the face of climatic change. Perhaps, despite having larger brains than most present-day humans, their relatively smaller frontal lobe regions reflect lower intelligence and linguistic ability. During the course of human evolution, climatic change per se, and behavioural responses to it, probably had a feedback effect on physiology and development. Perhaps *H. sapiens* had learned during their rapid migrations to respect the value of new ideas. Maybe the specialized skills of their hunters were linked with an opportunism that gave them a strong competitive edge at a time of scarce resources due to climatic change. Take the example of hunting down large cold-hardy mammals over long distances. This illustrates a link between specialized hunting (behavioural), taking a risk to reach a distant goal (planning or foresight) and the disappearance of normal local resources due to temperature decrease or overpopulation.

However, our own adaptabilities also have their limitations, especially if we lose our connection to other species or deny that climate change matters. In the past we have always 'won'; in the future we may not be so lucky. During the course of human evolution, climatic change per se, and behavioural responses to it, probably had a feedback effect on physiology and development. Those behavioural responses had to be adaptable, rather than culturally fixed. Flexibility was necessary in an environment whose climate, and hence resources, might change drastically from generation to generation. In climatically and economically stable environments, social hierarchies and a division of labour can arise. There is more time for the education of children, and for children to play and thereby expand their vocabularies and the complexity of their vocal communications. Then, when they are adults, some might opt to become teachers, storytellers and repositories of the group's history. Thus the potential for creativity can be better expressed. However, it is under the same conditions that cultures get into ruts, with members who demand that 'we do things the way we've always done 'em.' We do not argue simplistically that climatic change causes evolution. But the environment is so strongly correlated with behaviour, physiology and development, that there is an evolutionary boot-strapping effect when the environment changes.

With this grounding in human behavioural evolution in mind, we now move on to the last chapter in this human history section, which reviews the migration and diaspora of *Homo*.

For those seeking a more in-depth understanding of the biological background to evolution, please read Appendix A, where we review evolutionary theory, which provides the foundation necessary to take a closer look at developmental evolution and human physiological adaptability; the mechanisms that allow mammals, including humans, to respond quickly to changes in their environment. These are in contrast to slow, gradual genetic adaptations operating under the constraints of natural selection, which preserves stasis during periods of climate stability.

Notes

1. Throughout this book many archaelogical and some geological dates are given in radiocarbon years because radiocarbon dates provide the vast majority of ages used by researchers when dating artifacts and fossils that are up to about 50 000 years old. However, radiocarbon dating does not provide the calender age; it measures the concentration of 14C in a specimen relative to the concentration of 14C present in the atmosphere in the year 1950 (year 'zero' for radiocarbon dating). Conversions from radiocarbon age to calendar age are provided as original authors published them or by us based on the 'Fairbanks0805' calibration curve (Fairbanks *et al.*, 2005).
2. For some interesting insights into environmental impacts on the brain and addictive behaviour see Maté, 2008.
3. The documentary, Planet Discovery: *Human Evolution*, aired 20 November 1997 on the Discovery Channel. The series is commendably entertaining and quite adequately researched but is an unwitting and rich source of unexamined selectionism.

4

The migrations and diaspora of *Homo*

4.1 Introduction

As we discussed in the previous chapter, human physiological and behavioural adaptability provides for immediate responses to rapid climate change. During the history of hominins one typical behavioural response to changing climate has been dispersal or migration. Hominins first dispersed out of Africa nearly 2 million years ago, or earlier (e.g., Dennell and Roebroeks, 2005), but it was during the highly variable climatic period of the last glacial cycle (LGC), 135 000 to 11 650 years ago, that a new species of hominin – *Homo sapiens* – embarked on a series of rapid and extensive migrations that resulted in their populating both the Old and the New World.

During the LGC, flora, fauna and the geographic territories they inhabited expanded and contracted in response to climate change. Hominin populations also shifted their territories, at times following the migrations of large mammalian fauna. Movement within the same ecological niche did not require the same large behavioural adjustments that movement beyond would require. A true dispersal event, one in which hominins expanded beyond their known territory, necessitated an adjustment to new habitats and their associated subsistence resources and may have required the invention and adoption of new technologies. Therefore, if we want to understand whether hominin movements were associated with habitat expansions or were true dispersal events, we need to understand changing climate and its impact on human habitats through our history. Geographic barriers, sea level change and climatic factors, including glaciation, desertification, temperature, precipitation and related changes in vegetation and edible biomass, influenced hominin expansion and dispersal. By combining archaeological and environmental data with climate model simulations at critical intervals in human history we are able to begin to develop a better regional understanding of these changing conditions during hominin expansion and dispersal events.

Hominins' reproductive life spans and life histories are similar to those of their closest extant relatives, the great apes *Pan*, *Pongo* and *Gorilla*. During the Miocene, apes had distributions from southern to northern Africa, from Western Europe to China, and in Europe from the Mediterranean to Poland, but more recently their distribution has been more restricted (Fleagle, 1998), whereas hominins rapidly spread to inhabit the farthest corners of the Earth. Early hominin expansions appear to correlate with changes in the size of home ranges and diet quality, which are associated with ecological changes in vegetation productivity and extent; geographic barriers, including expansion or contraction of deserts such as the Sahara (Antón *et al.*, 2002); as well as faunal interchange (Kurten, 1968; Opdyke, 1995; Tchernov, 1987 and 1992b; Vrba, 1995). Intervals of significant climate change had major impacts on both the development and behaviour of early *Homo*. For instance, Antón *et al.* (2002) suggest that it was during a period of increased aridity and coincident expansion of African grasslands and reduced primary productivity that changes in hominin diet and foraging behaviour potentially increased *H. erectus*'s reliance on meat and fat and led to increased brain and body size. During periods of less extreme climate change, dispersal barriers for early hominins were reduced, facilitating their ability to follow migrating animals. Although temperature reductions were not as significant during glacial intervals in the tropics as in high-latitude regions, the climate of Africa was still impacted. For example, during the last glacial maximum (LGM), increased desertification appears to have reduced productivity in some regions of interior Africa. Flora, fauna and human populations likely inhabited the potentially more productive and proximally located African continental shelf exposed by lowered sea levels (Hetherington *et al.*, 2008). Earlier migrations may have been similarly influenced. Some researchers suggest that earlier modern human dispersals out of Africa may have been related to human exploitation of coastal shellfish resources that provided extensive and valuable nutritional advantages (Crawford *et al.*, 1999; Foley, 2002; R.G. Klein, 1999).

These examples emphasize the importance in changing climate to human migration and dispersal. This chapter reviews the history of hominin migration throughout the world and provides a brief synopsis of the prevailing climatic conditions at a number of archaeologically relevant locations. A more detailed analysis of regional climatic conditions and changing landscapes during the last glacial cycle will follow in Chapters 5 and 6, respectively, followed by an interpretation of the impacts of climate on humans in Chapters 7 and 8.

4.2 Out of Africa – population expansions and bottlenecks

Prior to 2.0 million years ago, the African climate was temperate and humid (O'Brien and Petters, 1999), and mammal diversity was evident (Behrensmeyer and Bobe, 1999). But soon after 1.9 million years ago the climate became colder and more arid, sea level lowered and faunal turnover increased (Vrba, 1995). However, while climate was arid elsewhere in Africa, particularly in North Africa, East Africa experienced two major high lake level intervals between 1.9 and 1.7 million years ago and between 1.1 and 0.9 million years ago (Trauth *et al.*, 2005). Changes in lake levels are believed to correlate in the first instance with a shift and intensification in the East–West zonal atmospheric circulation, known as the Walker circulation (Ravelo *et al.*, 2004), and in the second to the beginning of the mid-Pleistocene revolution, when glacial/inter-glacial cycles shifted from every 41 000 years to every ~100 000 years (Berger and Jansen, 1994). It is believed that the high relief in East Africa affected the dominant climate-forcing factors and may explain the anticorrelation between high lake levels in East Africa and the dust records from ocean sediment records off West Africa and Arabia (deMenocal, 2004). These disparate regional responses to climate imply rapid variation between humid and arid intervals and are thought to be major influencing factors in human evolution and dispersal (Trauth *et al.*, 2005).

It was coincidently about 2.0 million years ago that the first hominin dispersals likely occurred out of Africa – the out-of-Africa 1 hypothesis. An initial dispersal of *Homo erectus* (*sensu stricto*), bearing core-chopper industries, migrated to Pakistan by 2.0 million years ago (Dennell *et al.*, 1988b; Rendell *et al.*, 1987) and to Java, Indonesia, before 1.5 million years ago and perhaps as early as 1.9 million years ago (Aguirre and Carbonell, 2001; Larick *et al.*, 2001; Swisher *et al.*, 1994; although see Bar-Yosef and Belfer-Cohen, 2001; Langbroek and Roebroeks, 2000; Sémah *et al.*, 2000). A small-brained hominin possessing progressive 'modern' traits and classified as *H. erectus/ergaster* came to Dmanisi, Georgia, by ~1.75 million years ago if not before (Aguirre and Carbonell, 2001; Dzaparidze *et al.*, 1989; Gabunia *et al.*, 2001; Gabunia, Vekua, Lordkipanidze, Swisher *et al.*, 2000; Vekua *et al.*, 2002) (see Figure 4.1). During this time Africa was experiencing increased aridity and open vegetation (Aguirre and Carbonell, 2001). A pronounced shift to an arid-adapted C_4 grass dominated ecosystem had occurred in East Africa (Wynn, 2004). C_4 grass signifies a particular kind of plant metabolism that reduces water loss, and uses carbon dioxide more efficiently than C_3 grass. C_3 grass is more common in humid environments, is easier for most browsers to digest and is more nutritive.

The small-brained, tiny, early *Homo* spp. at Dmanisi, Georgia, used the Mode 1 stone tool industry, which was similar to the Oldowan tool industry of East Africa (Vekua *et al.*, 2002). African hominins are believed to have been separated into two culturally and technologically distinct populations possessing two different tool industries – Mode 1, Oldowan, and the more advanced Mode 2, Acheulean (Carbonell *et al.*, 1999). Although the initial Mode 1 hominin dispersal out of Africa may have been coincident with faunal dispersals, Mode 2 technology is thought to have left Africa later, when African areas had become saturated with Mode 2 producers. Competition for resources in Africa is believed to have driven the initial Mode 1 producers out of Africa. By around 1.5 million years ago, when climate was warming, sea level was rising, and woodlands

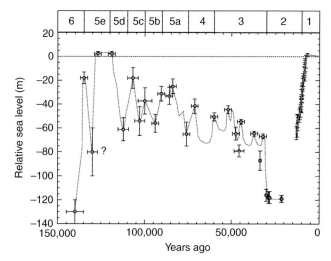

Figure 4.1 The relative sea-level curve for the last glacial cycle. Error bars define the upper and lower limits. The main oxygen isotope stages MIS6 to MIS1, including substages, are identified. Reprinted by permission from Macmillan Publishers Ltd: *Nature* (Lambeck *et al.*, 2002), copyright 2002.

in Africa were expanding (Aguirre and Carbonell, 2001), its hominins found their way to Ubeidiya, in the Jordan Valley, Israel. At this site, hominin remains have been found associated with the early Acheulean toolkit and dated to *c*.1.4 to *c*.1.5 million years ago (Bar-Yosef, 1994; Belmaker *et al.*, 2002; Tchernov, 1992c).

However, Tattersall (2000) argues that technological innovations, such as the shift from Mode 1 to Mode 2, are not explained by, nor coincident with, the advent of new hominin species. Species evolution, including that of hominins, may have been 'driven by shifts in climatic instability and habitat variability' (Wynn, 2004, p. 116). Habitat heterogeneity opens new opportunities for species to utilize or develop a variety of habitat and dietary specializations, whereas during intervals of habitat homogeneity, those species well-adapted to the 'stable refugia' or 'resource exploitation strategies' associated with those habitats have a better chance of survival. Thus, during periods of climatic instability, hominins were stimulated or forced to 'exploit and conquer new and varied habitats' to facilitate their survival (*ibid.*).

By 1.3 million years ago, climate again deteriorated; cooler temperatures resulted in increased grassland coverage in Africa (deMenocal, 1995), and reduced flora in Western Europe (Zagwijn, 1992) and midlatitude Eurasia (Grichuk, 1997). A large dispersal of mammals out of Africa and into Europe occurred at 1.2 million years ago (Maglio and Cooke, 1978; van der Made *et al.*, 2003).

Hominins appeared in Europe around 900 000 years ago (Bermúdez de Castro *et al.*, 2004), around the time of marine isotope stage (MIS) 25, a very warm, humid interval, when sea levels were raised. It was during MIS25 that mammalian faunas changed in East Africa and a major faunal dispersal occurred out of Africa and into Europe (Maglio and Cooke, 1978; van der Made, 2001; Vrba, 1995). At the Atapuerca-Gran Dolina site in north-central Spain, human remains have been dated to this interval and attributed to the species *Homo antecessor*, a putative intermediate between *Homo erectus/ergaster* and modern humans. *Homo antecessor* has been suggested as the last common ancestor of Neanderthals and behaviourally modern humans (Carbonell *et al.*, 1999). After MIS25, climate continued to vary. While climate in the temperate regions was very cold and dry during MIS22, around 830 000 years ago, it became very warm and humid shortly thereafter in MIS21, and woodland was generated in East Africa (Aguirre and Carbonell, 2001).

During the interglacial period correlated most closely with MIS19 (~750 000 years ago), hominins arrived in northern Europe and inhabited the Pakefield, England, site (52° N) on the exposed continental shelf, which then connected what is currently England with Europe. By MIS16, around 615 000 years ago, climate in temperate regions was extremely cold and arid. Climate became once again very warm in MIS15 before cooling again during MIS14, between 565 000 and 515 000 years ago (Aguirre and Carbonell, 2001). The earliest archaeological sites containing the Acheulean tool industry, with their large cutting tools including the 'teardrop' shaped handaxe, are found in Africa and dated to 1.5 million years ago, but their later appearance in Europe at Cagny, France; Boxgrove, England; Fontana Ranuccio and Notarchirico, Italy; Kärlich-E, Germany; and Laguna de Medina, Transfesa, Ambrona (lower unit), and Atapuerca-Galería (Beds TG 6–8, or GIIa), Spain, date to later than 600 000 years ago. These sites, combined with the multitude of sites dating to around 500 000 years ago in the Old World, imply a demographic explosion, the result of either *in situ* expansons or another dispersal 'out of Africa'. There is also evidence that a concurrent major mammalian dispersal out of Africa and into Europe occurred between 500 000 and 450 000 years ago (Tchernov, 1992a).

Homo erectus's survival was different in disparate geographic locations (Rightmire, 2001). *H. erectus* disappeared relatively early in the West, but in China, at Zhoukoudian, *H. erectus* persisted until as late as 230 000 years ago (Etler, 1996; Wu and Poirier, 1995), and at Ngandong in Java until after 50 000 years ago (Swisher *et al.*, 1996), although the Ngandong dates are contested (Grün and Thorne, 1997). Early humans reached the temperate grasslands surrounding the Tibetan massif, tropical India, and South East Asia (Gamble, 2001), setting the stage for the later and far more pervasive *Homo sapiens*' dispersal.

Based on scattered fossils and artifacts, two competing hypotheses have arisen and been given the greatest consideration in the debate about the origin and subsequent migration of behaviourally modern humans. As stated in Chapter 2, they are the out-of-Africa hypothesis (Stringer and Andrews, 1988) and the multiregional hypothesis (Wolpoff, 1989).

4.2.1 Out-of-Africa hypothesis

Fossil evidence and genetic research indicates *Homo sapiens* originated in Africa between *c.*195 000 and *c.*200 000 years ago, prior to the onset of the last glacial cycle, which began about 135 000 years ago, and subsequently migrated around the world (see Quaternary timeline Figure 1.4). Geneticists generally support the out-of-Africa hypothesis.[1] Using mutation rates to set the human 'molecular clock', they suggest that modern humans are the descendants of an ancestral African population living between 100 000 and 200 000 years ago (Balter, 2002; Bräuer, 1989; Stringer 1989, 1990). *H. sapiens*-like crania found on the ancient margin of a freshwater lake alongside butchered hippopotamus carcasses in Middle Awash, Ethiopia, have been dated to between 160 000 and 154 000 years ago (Clark *et al.*, 2003; T.D. White *et al.*, 2003). The Aduma crania attributed to *H. sapiens*, also from Middle Awash, has been dated to between 105 000 and 70 000 years ago (Haile-Selassie *et al.*, 2004). These findings combined with the *H. sapiens'* fossil remains found at Omo Kibish, Ethiopia, and dated to about 195 000 years ago (McDougall *et al.*, 2005) and the 118 000-year-old fossils from Klasies River Mouth, South Africa (McBrearty and Brooks, 2000), support the out-of-Africa model for the origin of behaviourally modern humans. Archaeological evidence suggests a rapid migration of *H. sapiens* out of Africa into the Middle East and Asia at around 100 000 years ago followed by an extended period of *in-situ* diversification and restricted migration to the north (Endicott *et al.*, 2004). According to the out-of-Africa theory, *Homo sapiens* then replaced pre-existing populations of *H. erectus* or archaic *H. sapiens*. The partial *H. sapiens'* skeletons found at Skhul and Qafzeh, Israel, which date to between 119 000 and 85 000 years ago, support this interpretation (Valladas *et al.*, 1998). Studies of mtDNA variation and X-chromosome haplotype variation indicate that Africans possess the greatest number of population-specific alleles relative to the paucity in genetic diversity evident in populations outside of Africa. This implies that the African populations of *H. sapiens* are the oldest in the world, and supports the out-of-Africa hypothesis. Furthermore, a genetic bottleneck may have restricted gene flow of behaviourally modern humans during their migration out of Africa (Tishkoff and Kidd, 2004). A genetic bottleneck is usually caused by a geographical bottleneck, but the settlement of the Middle East by the first behaviourally modern humans may also have been a barrier to free through-migration by later emigrants from the south.

4.2.2 Multi-regional evolution hypothesis

Not all researchers agree. Some still support the multi-regional evolution hypothesis, proposing that humans evolved simultaneously around the globe beginning around 2 million years ago (see, for example, Wolpoff, 1989; Thorne and Wolpoff,

1992). They argue that recent African origin theories that suggest dispersal out of Africa at around 90 000 to 100 000 years ago are inconsistent with the timing of the emergence of the upper Palaeolithic 'modern' behaviours, including art and water transport, which they say were evident only around 50 000 years ago. Further contested are suggestions that modern human behaviours evolved simultaneously and that mass migration was the only mechanism of behavioural change through time. Instead multi-regional supporters suggest that such behaviours had a gradual origin. Suggestions of recent dispersals at 50 000 years ago are criticized due to their inconsistency with the existence of 'anatomically modern' human fossils outside Africa that are dated to 90 000 years ago (Hawks and Wolpoff, 2001). Yet, the unexplained morphological continuity observed in fossil hominins in Australasia, one of the most serious difficulties with the out-of-Africa replacement model, supports the multi-regional hypothesis (Ayala, 1995).

However, the multi-regional hypothesis contravenes both Darwinian and neo-Darwinist interpretations of speciation. According to conventional evolutionary theory, isolated populations should produce different varieties, through adaptation to different environments. These might then get to the stage where the different populations are mutually sterile. However, human evolution, by whatever route it took, left all *H sapiens* capable of interbreeding and producing viable offspring. The only agreement between the multi-regional hypothesis and Darwinism is that evolution is a gradual process. How then does the multi-regional hypothesis stand up against emergence theory? The only possible explanation for multi-regional emergences of independent fringe populations that would remain reproductively compatible relates to common allometric trends (i.e., shifts in anatomical proportions during early development) and common epigenetic constraints that would prevent the differentiation of isolated groups. This combination could hypothetically lead to multiple emergences that would in effect result in the appearance of *H. sapiens* in several different parts of the world. As discussed in more detail in Appendix B: Developmental evolution, this is equivalent to orthogenesis, a concept that is totally unacceptable to the modern synthesis of evolution. Yet several molecular biological mechanisms clearly fall within the definition of orthogenesis (Reid, 2007). Although we are certainly not averse in principle to this kind of emergent evolution, we believe that the originating molecular mechanisms appear at random. Therefore there is a vanishing probability that *identical* orthogenetic trends would appear in numerous isolated populations. Thus, we conclude that the multi-regional hypothesis is invalid by *any* biological evolutionary theory.

4.2.3 Hybrid theories

Since the induction of the two major competing theories, new fossils have been uncovered, new analytical techniques devised and new hybrid theories have been advanced that argue that out of Africa is too simplistic. They propose various degrees of hybridization – reproductive intercourse – between different populations that had previously been isolated. Furthermore, bottleneck restriction of gene variety might have occurred in the isolated populations where unusual gene re-ordering might be possible – genetic revolution, as Ernst Mayr (1954) called it. If the gene pools of the previously isolated groups were considerably different, interbreeding would affect hybrid vigour. Supporters of the Weak Garden of Eden Model (Ambrose, 1998; Harpending *et al.*, 1998; Sherry *et al.*, 1994) propose that after the initial out-of-Africa expansion, population did not increase but dispersed widely and suffered genetic bottlenecks. It was not until between about 70 000 and 40 000 years ago that development of the new upper Palaeolithic (UP)–late Stone Age (LSA) technology provided humans an opportunity to increase the environmental carrying capacity of their regions and facilitated a major population expansion (Finlayson, 2004, p. 72).

Dennell and Roebroeks (2005) suggest that dispersals occurred both into and out of Africa. Sparse Asian hominin evidence may be explained by regional imbalances in paleoanthropological investments and taphonomic preservation. Further, they highlight the weakness in attributing stone tool assemblages to specific *Homo* species when associated hominin fossil evidence is lacking. Instead, hominin dispersal and adaptability research needs to focus on local populations identified by hominin fossil remains and their association with comparable grassland paleoenvironments, in which all early hominins apparently resided.

The Multiple Dispersals, Bottlenecks and Replacement Model outlines a series of climate-related population expansions and bottlenecks that characterized human dispersal (Foley and Lahr, 1997; Lahr and Foley, 1994, 1998). According to this theory, the ancestors of the Neanderthals dispersed out of Africa during oxygen isotope stage (OIS) 8 or 7 (~250 000 to 185 000 years ago), followed by dispersal into the Middle East during OIS5 (~135 000 to 75 000 years ago) (Lambeck *et al.*, 2002), with a subsequent retreat due to climatic cooling. During OIS4 (~75 000 to 60 000 years ago) or early OIS3 (~60 000 to 30 000 years ago), a second dispersal occurred into Asia. A final dispersal into the Middle East happened coincident with the middle–upper Palaeolithic transition around 45 000 years ago, and from which behaviourally modern humans colonized Europe, replacing Neanderthals. Finlayson (2004) advocates that the driving force behind dispersal and survival was natural selection based on a changing ecology and geography of Pleistocene Earth. Those hominin populations most able to adapt to change were most successful. Finlayson includes behavioural change in his definition of 'adapt' (p. 73), with rapid behavioural adjustments facilitating increased survival. According to Braüer (1984; 1989, p. 124) and Clark (1989), who support the out-of-Africa thesis, the impetus for the original expansion of modern humans out of Africa and into Asia and Europe may have been primarily climate change-induced, related to increased desertification of the region now known as the Sahara.

Irrespective of which side of the debate one resides on, as stated by S. L. Smith and Harrold (1997), 'both share a commitment to Darwinian evolutionary theory' (p. 113). Lost in the long-standing anthropological conflict has been the need to recognize the difference between slow, gradual, genetically influenced morphological change and saltatory emergence of behavioural adjustments that depend upon a pre-existent adaptability. The persistence of novel behaviours does not qualify as a process of natural selection by any formal definition, whether Darwinist or neo-Darwinist. As neo-Darwinist Richard Dawkins (1999) has observed, the process would be better described as Lamarckist. Further, although population expansion has been frequently cited as an explanation for dispersals of *H. erectus* and *H. sapiens* out of Africa (Bar-Yosef and Belfer-Cohen, 2001), justifications for such increases remain unclear. Critical in understanding these relationships is the role rapidly changing environmental or social factors played in stimulating saltatory human behavioural changes and associated dispersals or *in situ* population expansions.

4.2.4 The role of environment and climate in early hominin dispersals

Environment and climate are critical influencing factors in some human dispersal models. During the early Pleistocene, glaciations were generally low amplitude and high frequency, occurring about every 41 000 years. In the middle Pleistocene, after ~800 000 years ago, glaciations were of higher amplitude and reduced frequency, occurring about every 100 000 years (see Chapter 5 for a complete discussion of these 'Milankovic cycle oscillations'). Transitions between glacial and interglacial conditions were more pronounced, resulting in greater latitudinal shifts in fauna and flora (Dennell, 2004). Beginning 1.2 million years ago, a series of mainly mammalian dispersals began in Asia. By about 800 000 years ago, more than 25 species of mammals had left Asia and relocated in central and Western Europe. These dispersals are believed to be related to climate change (van der Made, 2001; van der Made *et al.*, 2003). It is possible that early hominins were part of this major mammalian dispersal out of Asia.

During much of the middle Pleistocene (780 000 to 126 000 years ago), ice volume changes combined with solar insolation oscillations resulted in increased aridity in south-western Asia, central and northern Asia, and a weakening of the monsoon in South and East Asia. These climatic changes resulted in the development of a desert barrier, still currently evident between East Africa and South Asia (Dennell, 2004). (See Chapter 6 for a more thorough discussion on desert formation during the LGC.)

In general, researchers agree that hominin expansion into the tropics occurred during glacial intervals or intermediate phases between glacial and interglacial intervals when sea levels were lowered. In contrast, expansion into the extra-tropical or more northern latitudes occurred during warmer interglacial intervals or during transitions from glacial to interglacial intervals (Overpeck *et al.*, 1996; Walker and Shipman, 1996). Changing precipitation levels and desert extent likely played a role in the capacity of humans to remain *in situ* and/or to disperse. If this was the case, then the Sinai Peninsula and the Levant potentially acted as an environmental 'corridor' or refugium and the primary migration route for hominins out of Africa (Bar-Yosef, 1992; R.G. Klein, 1999). On the other hand, during glacial intervals the region may have behaved as an environmental barrier as desert regions expanded and habitable range contracted (Foley and Lahr, 1997; Lahr and Foley, 1998; Rolland, 1998). (See Chapter 7 for a discussion of desert formation and its impact on modern human migration.)

Population expansions may result in the depletion of resources in the home territory and, as a consequence, in a shifting of the population into proximal territories possessing similar habitats. However, although this may explain the colonization of temperate Europe, it may not apply to the Mediterranean (A. Turner, 1999). Further, although it may describe how hominins followed shifting habitats, it does not explain the true dispersal of hominins into new and diverse habitats. Using 'survival of the fittest' as an explanation, some researchers imply that because *H. erectus* was a successful species, it goes without saying that it expanded its geographic distribution irrespective of habitat changes (Bar-Yosef and Belfer-Cohen, 2001, p. 25). Such question-begging is superfluous if we were simply to argue that the success of *H. erectus* was owing to its pre-existing adaptability. Alternatively, Bar-Yosef and Belfer-Cohen (2001) suggest that *Homo* may have left Africa to escape zoonotic diseases – those transmitted by plants or other animals. Africa is home to most of the world's zoonotic diseases, and *Homo* may have been in search of cooler, drier, less-diseased environments. However, this would require knowledge that the diseases were zoonotic and the prediction that there were other places less affected by such diseases. Although some of these theories are dependent for their causality and validation on neo-Darwinism, substantive support awaits scientific evidence.

Dennell (2003) suggests that prior to about 735 000 years ago, hominin habitation of North Africa, Europe and Asia was temporary and spatially discontinuous, and occupation of latitudes north of 40° was generally limited to nonglaciated intervals. Around this time the Earth made the transition from low-amplitude–low-frequency changes to high-amplitude–low-frequency changes associated with a shift to the 100 000-year Milankovic cycle. Resident populations of *H. erectus* were evident in the Levant and Java, and *H. erectus/ergaster* were present in East Africa when conditions were hot and dry, implying that the earliest Eurasians occupied grasslands and open scrub- and wood-lands and avoided forests. However, absence of grassland habitats does not appear to have been the sole limiting factor in the generation of successful *H. erectus* occupation sites outside of Africa. Although discussion remains as to whether *H. erectus* was a hunter or a scavenger, there is general agreement that it consumed meat as opposed to eating a vegetarian diet. Dennell further suggests that hominin expansion into northern latitudes was limited by reduced biomass availability. As hominin survival skills improved over time, they were able to 'adapt' to a greater variety of environments and climates.

Recently discovered 125 000-year-old artifacts found on the Red Sea coast of Eritrea imply early human occupation of coastal areas and exploitation of near-shore marine food resources in East Africa (Walter *et al.*, 2000) and may imply an early expansion out of Africa and into South East Asia along the subaerially exposed continental shelf. Interestingly, the Red Sea find dates to ~125 000 years ago, and although sea levels at that time were very similar to those today, the region had just experienced a period of rapid sea-level change (see Figure 4.1). Global eustatic sea level rose rapidly (~120 m) between 140 000 and 135 000 years ago and then subsequently fell rapidly (~60 m) between about 135 000 and about 130 000 years ago (Lambeck *et al.*, 2002). After 130 000 years ago, rapidly rising sea levels inundated subaerial continental shelves. By 125 000 years ago, shelves previously exposed to a depth of as much as 100 m were inundated as sea levels reached ~6 m above present levels.

4.2.5 Genetic evidence of behaviourally modern human dispersals

Mitochondrial DNA evidence suggests dispersal of *H. sapiens* out of Africa between 80 000 and 60 000 years ago followed by rapid expansion into Asia and a more restricted migration into colder northern environments (Forster, 2004; Oppenheimer 2003a) (see Figure 4.2). Behaviourally modern humans dispersed out of Africa around 65 000 years ago, using the southern coastal route to reach the coast of the Indian Ocean, South East Asia and finally Australia (Macaulay *et al.*, 2005). This may explain their presence in South East Asia so much earlier than in Europe, where behaviourally modern humans did not appear until between 50 000 and 40 000 years ago. Based on global mtDNA, diversity in humans of diverse origins provides a lower bound for the 'modern' human exodus out of Africa at 52 000 ± 27 500 years ago, with the most recent (female) ancestor present at 171 500 years ago (Ingman *et al.*, 2000). Climate evidence supports these interpretations and indicates that Egypt and the Middle East possessed an arid desert climate, restricting this northern route until ~50 000 years ago (Hopkins, 2005; see Chapters 5 and 6).

Mitochondrial DNA evidence indicates that the earliest maternal lineages of all living humans, known as L1 lineages, are restricted to Africa. Between 60 000 and 80 000 years ago, hominins expanded in Africa generating the L2 and L3 lineages.

Figure 4.2 Hypothetical routes out of Africa highlighting a southern coastal route along the Indian Ocean coastline that may have been taken by humans migrating out of Africa. (Based largely on Forster [2004], Forster and Matsumara [2005] and Oppenheimer [2003a].) Also shown is the theoretical 'Movius line' first proposed by archaeologist H. L. Movius.

L3 lineage gave rise to M and N, from which all non-Africans now derive. It is unclear where M and N originally derived. However, a high diversity of M in India, and its restriction to eastern Africa and Afro-Asiatic-speaking areas, may imply its origin in the east and subsequent reverse migration back to Africa some time in the last 20 000 years. Alternatively, M haplotype may have originated in eastern Africa approximately 60 000 years ago and quickly spread along the South-Arabian coast to South East Asia, then to Australia and the Pacific Islands. The remaining M haplogroup in eastern Africa after 60 000 years ago did not spread until about 10 000 to 20 000 years ago. It is interesting to note that the mtDNA M-haplotype is not present in Europeans today (Forster, 2004). However, recent nonrecombining Y-chromosome analyses shows an early out-of-Africa range expansion followed by subsequent migrations out of Asia, with Europe a primary receiver. Differences in male versus female migration rates and/or effective population sizes are also evident within sub-Saharan Africa. Alternatively, Y-chromosome sequence variation indicates that the descendents of most patrilineal lineages of anatomically modern humans currently reside in East and South Africa and that dispersal of the most recent common ancestor out of Africa occurred between 35 000 and 89 000 years ago. More recently, analysis of human Y-chromosome haplotypes shows the presence of haplogroup IX Y chromosomes in northern Cameroon that may be related to a back migration of males from Asia to Africa, perhaps associated with an upper Palaeolithic expansion from Asia into Europe ~30 000 years ago (Cruciani *et al.*, 2002; Semino *et al.*, 2000; Underhill, Passarino *et al.*, 2001; Underhill, Shen *et al.*, 2000; Wells *et al.*, 2001). Human migratory movements between Eurasia and Africa, particularly those subsequent to 45 000 years ago, are thought to have occurred primarily via the Levantine corridor (Luis *et al.*, 2004). Genetic research supports a non-African period of population growth estimated at 1925 generations ago, or about 38 500 years ago, just subsequent to a period of cultural change that is evident between 50 000 and 40 000 years ago (Ingman *et al.*, 2000; R. G. Klein, 1999).

However, it still remains for those supporting theories that assume an early behaviourally modern human expansion out of Africa to explain why modern behaviour did not appear in Europe until much later. Alternatively, theories assuming late population expansion are limited in their ability to explain the expansive behaviourally modern human dispersals throughout Asia and into Australia by at least 50 000 years ago (Hawks and Wolpoff, 2001). If the populations that recently migrated out of Africa were small, an explanation in terms of rapid population expansion is required to explain subsequent global dispersals.

These interpretations draw attention to the need to ascertain climate-related shifts in habitat throughout Africa and Eurasia during the period of early human history and to understand developmental, physiological and behavioural adaptabilities that may have allowed hominins to respond rapidly to sudden climate change. Further, there exist numerous perspectives on the meaning of 'modern' behaviour and how it correlates with the genetic, archaeological and anatomical evidence of *H. sapiens* and their dispersal. Each region has its unique archaeological and climatological history that must be understood if we are to gain a clearer picture of the history of our global dispersal and the demise of our cousin *Homo* species.

4.3 The Middle East

In the past, much as it does today, the Middle East served as a crossroads between Africa, Asia and Europe. Pleistocene glaciations combined with reduced precipitation and increased desertification in Africa probably resulted in population dispersals into the the Middle East (Kramer *et al.*, 2001). Two typical routes cited for hominin dispersals out of Africa and into the Arabian Peninsula (presently including Saudi Arabia, Kuwait, Bahrain, Qatar, United Arab Emirates, Oman and Yemen) include: up the Nile Valley, across the Sinai, and northward into the Israel/Levant region; through the Afar Depression at the Horn of Africa and across the Strait of Bab el Mandeb [Petraglia, 2003]. A third, less-studied option is across the Tunisia – Sicily straits, although this does not appear to be the route of entry into Italy (Villa, 2001). The route across the Red Sea via the Strait of Bab el Mandeb may have been used during intervals of lowered sea level or may have involved the use of rafts (Whalen *et al.*, 1989).

Lower Pleistocene Turkey and the Arabian Peninsula possessed a more open grassland environment and moister conditions than today, and although numerous Acheulean and Oldowan tool assemblages have been discovered in the Arabian Peninsula, none have been firmly dated, and as yet, no hominin fossils have been recovered (Petraglia, 2003). However, the presence of lower Palaeolithic sites along coast margins of the Arabian and Red Seas would imply that more than one route was taken by early humans as they dispersed out of Africa. These provide persuasive evidence of coastal occupation in the Arabian Peninsula, as is the case in Africa, Europe, the Levant and South Asia (for example, see de Lumley, 1969; Fleisch and Sanlaville, 1974; R. G. Klein *et al.*, 1999; Korisettar and Rajaguru, 1998; Petraglia, 2003; M. Roberts and Parfitt, 1999).

Once early hominins migrated into the Levant, it is believed that they migrated along the margin of the Zagros Mountains before traversing the Iranian plateau and then dispersing farther eastward. In north-east Iran, Oldowan-type stone artifacts have been discovered along the shoreline of a vast paleolake in the Kashafrud basin at 36° N (Ariai and Thibault, 1975; P. E. L. Smith, 1986). The Levant region is one of the few areas outside Africa containing evidence of repeated and consistent habitation, implying persistent *in situ* habitation and/or, as some researchers suggest, waves of dispersals out of Africa (Dennell, 2003). There is also evidence that extensive changes in faunal distribution occurred farther to the north-east, through the region now known as Georgia, during the Plio-Pleistocene (Gabunia, Vekua, and Lordkipanidze, 2000). The region is also cited as the centre of the 'Neolithic revolution', where domestication of animals and systematic cultivation originally occurred (Bar-Yosef, 1992).

The earliest hominin evidence that includes Oldowan-type artifacts has been found in the archaeological site of Erq el-Ahmar, Israel, and dated to between 1.77 and 1.95 Ma (Ron and Levi, 2001). To the north-east, in Dmanisi, Georgia, bones from four hominins have been found that include evidence of progressive 'modern' traits. The crania are very small. The most recently discovered has a cranial capacity of ~600 cm^3, which is substantially smaller than that expected for *H. erectus* and is the smallest *H. erectus* crania ever found outside of Africa. This find has led to serious reconsideration of the theory that early dispersals from Africa were the result of expanding brain size (see, for example, Dennell, 2003). The fossil hominins have characteristics similar to both early African *H. erectus* and Asian *H. erectus* and show close affinities to *H. ergaster*. They have been found together with a core-chopper industry differing little from the basic African Oldowan industry and dated to approximately 1.7 to 1.8 Ma (Gabunia *et al.*, 2001). The site of Dmanisi appears to be the most northerly latitude location (42° N) inhabited by early Pleistocene hominins. At the time of its occupation, the Dmanisi site was located near a lake, surrounded by a semi-arid to Mediterranean-like open steppe and a gallery forest environment with abundant fresh water and edible resources (Dennell, 2003; Gabunia, Vekua, and Lordkipanidze, 2000; Gabunia, Vekua, Lordkipanidze, Swisher *et al.*, 2000; see also Antón and Swisher, 2004). Fauna included *Stephanorhinus etruscus etruscus* (rhinoceros), *Cervus perrieri* (deer), *Dama nesti* (deer), *Canis etruscus* (wolf), *Equus stenonis* (Steno's horse), *Megantereon* (tiger), *Allophaiomys-Microtus, Mimomys* (extinct vole), *Tcharinomys (Pusillomimus)* (vole), *Parameriones* sp. (jird), *Cricetus* sp. (hamster), and *Allocricetus bursae* (rodent). The concentration of resources, combined with a climate warmer than today, made this part of Georgia favourable for early hominin occupation (Gabunia, Vekua, and Lordkipanidze, 2000). These findings support the theory that early hominins migrated through the 'Levantine corridor' en route to the colonization of Eurasia. Further, the region may have acted as a refugium during periods of climatic stress in Eurasia. In the latter part of the occupation of this site, climate change resulted in increased aridity, reduction of forested areas and a reduction in number and diversity of faunal species observed. This shift is coincident with an increase in abundance of stone tools. Gabunia, Vekua and Lordkipanidze (2000) suggest that the diverse environments and resources encountered by hominins expanding into Eurasia may have resulted in 'rapid evolutionary changes' in successful Eurasian colonizers. This may provide an example of saltatory behavioural evolution occuring during rapid climate change. In support of this theory they cite the advanced traits seen in the Dmanisi fossil bones. However, although these advanced traits are evident at the Dmanisi site 1.8 million years ago, *Homo* did not appear in Europe until between about 1.0 and 0.8 million years later. The mixture of morphological characteristics is also thought to reflect the admixture of local and immigrant populations (Bar-Yosef, 1992).

The site of Ubeidiya in the Jordan Valley, Israel, dated to between 1.4 and 1.5 million years ago, is the most well-known of numerous early Palaeolithic sites in the Levant. It contains flake and core tools attributed to a Developed Oldowan and several hominin teeth that have been tentatively attributed to *H. ergaster* (Belmaker *et al.*, 2002). The upper levels of the site contain tools with Acheulean influences including bifaces, handaxes and picks, which are the oldest known Acheulean tools outside of Africa. This lakeside site was inhabited during a semi-arid interval. Fauna are indicative of a woodland environment and include numerous cervids, a few other grazers and East African species including *Oryx* (antelope), *Kolpochoerus olduvaiensis* (extinct bush pig), *Hippopotamus gorgops* (extinct hippopotamus), *Pelorovis* (extinct African bovid), and *Crocuta crocuta* (spotted hyena) (Belmaker *et al.*, 2002; Tchernov, 1992c).

It is thought that the nearby site of Gesher Benot Ya'aqov, Israel, which contains cleavers made with large flakes and is dated to *c.*780 000 years ago (Verosub *et al.*, 1998), was created by a new group of African foragers exploiting lava material for their tools (Goren-Inbar and Saragusti, 1996). The introduction of a new biface tool component into the site is considered evidence of a diffusion of African ideas or the presence of African immigrants around 750 000 years ago (Saragusti and Goren-Inbar, 2001). Also in Israel, the Tabun Cave site yielded a hominin mandible whose taxonomy is still being debated, an almost complete skeleton of a Neanderthal (McCown and Keith, 1939), and one of the longest lower and middle Palaeolithic sequences in the Near East. Recent redating of this site stretches its occupation over 275 000 years, from nearly 400 000 years ago to 125 000 years ago (Grün and Stringer, 2000; Mercier and Valladas, 2003; Rink *et al.*, 2004). This site is often raised in discussions about the origin of modern humans and their coexistence with Neanderthals.

Partial *H. sapiens* skeletons have been uncovered at the sites of Skhul and Qafzeh, Israel and have been dated to between 119 000 and 85 000 years ago (Valladas *et al.*, 1998). Their elongated limbs and linear physique combined with a low estimated body mass resembles that of recent humans from hot, dry climates and contrasts with the heavier, shorter-limbed Neanderthals that more closely resemble recent humans from the Arctic (Holliday, 1997a, 1997b; Pearson, 2001; Ruff, 1994; Trinkaus, 1981). The presence of these hominins near those of a 60 000-year-old Neanderthal skeleton at Kebara 2, Israel, has led to suggestions of coexistence of Neanderthals and behaviourally modern humans.

It is evident from the dating and distribution of early archaeological sites in the Middle East that this region played a key role in the dispersal of hominins. Paleoenvironmental reconstructions of early Middle Eastern sites indicate that many were located near water. Thus, as is evident in Africa, early hominins were attracted to waterside locations and their associated faunal and floral resources (R.G. Klein, 2000; Petraglia, 2003). These factors, combined with similarities in genetics, raw material preference and manufacturing strategies between sites in East Africa, the Levant and Arabia, and the warm-adapted body proportions of behaviourally modern humans supports a hominin dispersal link between these regions. Y-chromosome analyses support this interpretation for Neolithic expansions into Europe from the Middle East, out of the southern Balkans and from North Africa into the Iberian Peninsula (Semino *et al.*, 2004). They also suggest behaviourally modern humans moved from the Middle East back toward East Africa, and more recently toward North Africa and Europe.

4.4 Europe

Although *Homo* was present at the 'gates' of Europe by 1.7 million years ago or earlier (Gabunia, Vekua, Lordkipanidze, and Swisher, 2000), there is no clear evidence that they dispersed into Europe at this time, although expansion into Asia has been inferred. The Iberian Peninsula appears to have been inhabited by *Homo* about 300 000 years earlier than central and northern Europe. Paleomagnetic age dating suggests two sites in southern Spain – Barranco Leon and Fuente Nueva-3 – may date to at least 1.0 Ma. Artifacts were deposited adjacent to a lake coincident with faunal material including those listed in Tables 4.1 and 4.2 (Oms *et al.*, 2000).

The earliest fossil hominin evidence in southern Europe was discovered at the TD6 horizon at Gran Dolina site on Atapuerca Hill, Spain, dated to 780 000 years ago. It includes 86 hominin remains from at least six individuals, including juveniles. Also at Ceprano, Italy, an adult skullcap, intermediate between *Homo ergaster/erectus* and the later middle Pleistocene *Homo heidelbergensis* ancestor to the original behaviourally modern humans, has been uncovered with an estimated age between 900 000 and 800 000 years ago (Bermúdez de Castro *et al.*, 2004; Falguères *et al.*, 1999; Manzi *et al.*, 2001). These early hominins lived in a Mediterranean forest environment amongst oak and cypress trees, mammals and birds very similar to those inhabiting the region today (Rodríguez, 1997; Sánchez-Marco, 1999). The TD6 hominin remains, which possess some Neanderthal characteristics combined with an anatomically modern human face, appear to have been cannibalized. Hominin bones were apparently butchered in order to extract meat and marrow – bones have frequent cut marks and have been smashed with a stone hammer. This is

Table 4.1. *Faunal material from Fuenta Neuva-3*

Fuente Neuva-3	
Mammuthus meridionalis	Southern mammoth
Hippopotamus antiquus	Hippopotamus
Stephanorhinus etruscus	Rhinoceros
Equus altidens	Horse
Megaloceros solihacus	Giant deer
Cervus sp.	Deer
Bovini	Bovines
Hermitragus sp.	Caprines
Megantereon whitei	Saber-toothed tiger
Ursus sp.	Bear
Hystrix major	Porcupine
Allophaiomys bourgondiae	Vole
Allophaiomys chalinei	Vole
Mimomys	Extinct vole

Table 4.2. *Faunal material from Baranco Leon*

Baranco Leon	
Castillomys crusafonti spp.	Rodent
Apodemus mystacinus	Broad-toothed field mouse
Apodemus aff. sylvaticus	Wood mouse
Eliomys intermedius	Dormouse
Mimomys sp.	Extinct mole
Allophaiomys pliocaenicus	Vole
Oryctolagus cf. lacosti	Rabbit
cf. *Homotherium* sp.	Saber-toothed cat
Equus sp.	Horse
Megaloceros sp.	Giant deer
Cervidae gen. et sp. indet.	Cervid
Bovini gen. et sp. indet.	Bovine
Hemitragus alba	Goat
Soergelia minor	Ovibovine = sheep
Hippopotamus antiquus	Hippopotamus
Mammuthus meridionalis	Southern mammoth
Bovini indet.	Bovines

the earliest evidence of cannibalism recorded in hominin evolution (Bermúdez de Castro *et al.*, 2004). Nearby, at the site of Sima de los Huesos, 4000 hominin fossil remains, making up more than 80% of the worldwide *Homo* middle Pleistocene record, have been uncovered in what is probably a burial ground associated with possible symbolic rituals and dated to between 400 000 and 500 000 years ago (Bischoff *et al.*, 2003). The temperate climate believed to have existed at the time of occupation has led researchers to suggest that hominins were not practicing ritual cannibalism. Instead, despite the temperate climate that existed at the time of occupation, hominins were consumed as part of the regular dietary regime (Bermúdez de Castro *et al.*, 2004).

Initially it was believed that prior to 500 000 years ago, hominins were unable to settle in northern latitudes (Dennell and Roebroeks, 1996; Roebroeks, 2001). The oldest fossil hominin sites in north-western Europe include the site at Boxgrove, England, and Mauer, Germany (Bosinski, 1995), and date to ~500 000 years ago. These hominin fossils have alternatively been attributed to *H. heidelbergensis* and 'archaic *Homo sapiens*', suggesting a species intermediate between *H. erectus* and behaviourally modern *Homo sapiens*. Some researchers believe that the reason *Homo* did not enter the European regions north of 45° N until the middle Pleistocene was because *H. erectus* was limited in its ability to adjust to northern latitudes. This limitation is thought to be due to restricted cranial capacity and low behavioural adaptability; 'a quantum leap in their behaviour' would have been required to 'take them north of 35° and across the many inland barriers of central

Spain, the Pyrenees, Massif Central, Alps, Dolomites, Transylvanian Alps and so on into northern and even central Europe' (Dennell and Roebroeks, 1996, p. 540; see also Dennell, 1997).

However, flint artifacts, provisionally interpreted as Mode 1 Oldowan, recently uncovered in the Cromer Forest-bed Formation at Pakefield (52° N), Suffolk, UK, and dated to about 700 000 years ago by Parfitt *et al.* (2005), suggest that hominins were evident at least 200 000 years earlier. During this period, exposure of the continental shelf between England and Europe connected what is now England with the European mainland. The artifacts were discovered in river sediments located on the floodplain on the north-east edge of the shelf, below the English Midlands, a region that subsequent to habitation was covered in glacial ice. The oldest artifact was found in a sediment layer containing brackish-water ostracods, foraminifera, dolphin and walrus. The fossil flora and fauna indicate a climate warmer and drier than today, with mean temperatures ranging between 18 and 23 °C in the summer and between −6 and +4 °C in the winter. The more typically Mediterranean climate, which supported large grazing and browsing mammals including *Mammuthus trogontherii* (steppe mammoth), *Stephanorhinus hundsheimensis* (rhinoceros), *Megaloceros savini* (giant deer), *M. dawkinsi* (giant deer), and *Bison* cf. *schoetensacki* (bison), and predators including *Homotherium* sp. (saber-toothed cat), *Panthera leo* (lion), *Canis lupus* (small wolf) and *Crocuta crocuta* (spotted hyena), implies early hominins may have dispersed northwards into familiar climates necessitating few adjustments in their behaviour (*ibid.*). Further, the presence of the lion, *Panthera leo*, and the larger saber-tooth felid *Homotherium* sp. suggests that, contrary to earlier interpretations (Dennell, 2003; Hemmer, 2000; A. Turner, 1992), these large carnivores did not preclude hominins from accessing carcasses, if this was their habit, and in fact hominins were able to coexist with large carnivores in northern latitudes at least 700 000 years ago.

The Neanderthals – a morphologically distinct population of human relatives whose identity was first discovered when a skullcap and partial skeleton were uncovered in Feldhofer Cave in Germany's Neander Valley in 1856 – remain an enigma. The Neanderthal lineage emerges in Spain before 350 000 years ago (Quam *et al.*, 2001; Rosas, 1995, 1997). They roamed Europe and western Asia, according to some, alone and unchallenged until ~40 000 years ago (Mercier and Valladas, 2003; Tattersall and Schwartz, 1999; Wong, 2003), when the Cro-Magnon behaviourally modern humans appeared in Eastern Europe (Kozlowski, 1982) and the Iberian Peninsula (Bischoff *et al.*, 1989). Early reconstructions portrayed the Neanderthal as hunched-over, shambling, apelike brutes, in contrast to the much more gracile humans. But later re-evaluation of Neanderthal fossils indicated a posture and movement similar to ours (Wong, 2003). However, the elongated limbs, linear physique and low estimated body mass of behaviourally modern humans resembles that of recent humans from hot, dry climates and contrasts with the heavier, shorter-limbed Neanderthals that more closely resemble recent humans from the Arctic and the South American Alto Plano (Pearson, 2001).

The last known Neanderthals date to ~33 000 BP (38 400 years ago) at Zafarraya on the Iberian Peninsula (Hublin *et al.*, 1995), ~34 000 years ago in France (Hublin *et al.*, 1996), between ~28 000 and ~29 000 years ago for two specimens from Vindija, in the Hrvatsko Zagorje of Croatia (F. H. Smith *et al.*, 1999) and as late as ~24 500 years ago in Portugal (Duarte *et al.*, 1999). The Portugal find is a skeleton of a four-year-old child that possesses a mix of Neanderthal and behaviourally modern human morphological characteristics. When buried, it was wrapped, stained with red ochre, and emplaced with specifically oriented animal bones and a pierced shell, typical of European behaviourally modern human burial practices.

The earliest generally accepted *Homo sapiens* fossils in Europe are from Hahnöfersand, Germany, and date to 36 300 years BP (41 500 ka ago) (Bräuer, 1980), and from south-western Romania, where a robust human mandible was found and dated to between ~34 000 and 36 000 years BP (39 200 and 41 400 years ago) (Trinkaus and Duarte, 2003; Trinkaus *et al.*, 2003). They are contemporaneous with Neanderthal sites in Europe including Saint-Césaire, in western France, dated to 36 000 years BP (41 400 years ago) (Lévêque and Vandermeersch, 1980; Mercier *et al.*, 1991). Stone tools and a mammoth tusk exhibiting incised grooves inflicted by humans using a sharp stone tool have been found in the Eurasian Arctic at the Mamontovaya Kurya Palaeolithic archaeological site. Mammoth, horse, reindeer and wolf bones were collected and dated to between 34 400 and 37 400 years BP (39 800 and 42 500 years ago) and were associated with a treeless steppe environment dominated by herbs and grasses. It is unknown whether the individual who inflicted the marks on the mammoth tusk was a Neanderthal or a behaviourally modern human (Pavlov *et al.*, 2001). If a Neanderthal, then this site would indicate that Neanderthals had expanded much farther north than previously assumed; if a behaviourally modern human, then the Russian Arctic was occupied by *Homo sapiens* shortly after their first appearance in Europe.

One theory for the migration of behaviourally modern humans into Europe suggests that Neanderthals disappeared and were subsequently replaced by behaviourally modern humans (such as Cro-Magnon) through conflict or economic competition. This 'replacement' theory supports a separate species status, whereas a theory of rapid evolution or a swamping of Neanderthal genes by behaviourally modern human genes entails some degree of interbreeding. Mousterian technology, thought to be contributed by Neanderthals, as well as Neanderthal anatomy, persisted in southern Iberia some 10 000 years after the appearance of Aurignacian technology in nearby Cantabria and Catalonia (Straus, 1997). These findings may suggest that behaviourally modern humans arrived in Western Europe via the Middle East and Eastern Europe. However, as discussed in the previous chapter, they also raise questions about the behavioural qualities of 'modern' humans relative to Neanderthals. Why, for instance, did it take about 50 millennia after the appearance of anatomically modern humans at Qafzeh and Skhul in Israel before behaviourally modern human behaviour appeared in Europe? Further, what generated the proposed shift in mental and behavioural capacity, and what resulted in the expansion of *H. sapiens* and the demise of the Neanderthals? We concluded in the previous chapter that the mental capacity and potential for culturally advanced behaviour existed in the originating members of the species. This is borne out to some extent by the discovery of shell beads at Skhul and Algeria that may date as far back as 135 000 years, close to the origin of anatomically modern *H. sapiens*. The authors who report this finding conclude that the kind of mind normally associated with behaviourally modern humans goes back much earlier than previously thought, as we have already argued. Nevertheless, it would stretch credulity to claim that a few beads confirm the complete repertoire of modern human behaviour. Our own interpretation is that the complete realization of the potential to become a behaviourally modern human required additional environmental and social occurrences. *Crisis* (especially climatic), *communication* between hitherto isolated groups, and their subsequent *collaboration* to produce a new social whole, could all have been involved in a revolutionary change. Migration was clearly an important part of such progress, as well as a precursor to the putatively 'modern' human condition in the Levant and North Africa.

Based on non-recombining Y-chromosome analyses, M89/M213 lineage dispersed out of East Africa into the Levant and expanded west, north and east ~40 000 years ago (Underhill *et al.*, 2001). This dispersal into Europe represents the earliest upper Palaeolithic occupation, whose Y-chromosomes later became nearly extinct. Subsequent dispersals into Europe occurred between 30 000 and 20 000 years ago, followed by significant population contraction coincident with the LGM. Neolithic dispersal of farmers out of the Levant and into Europe is also evident, although the impact is relatively localized.

The two most recognized potential routes for dispersal of *H. sapiens* into Europe from the Middle East and Eastern Europe include the northern route along the Danube River Valley and the southern Mediterranean route. The southern route is exemplified by 'proto-Aurignacian' bladelet technologies that are smaller than the traditional Aurignacian technology, that appear quite suddenly in the European record, and that are believed to derive from the Near East (Mellars, 2004a). The northern route hypothesis implies behaviourally modern humans rapidly entered Europe through the 'Danube corridor' (Conard and Bolus, 2003), bringing with them their 'classic' Aurignacian technologies.

However, ascertaining the route of behaviourally modern humans into Europe depends to a significant degree on reliable dating of upper Palaeolithic deposits. Recent dating of skeletal remains from Vogelherd Cave in south-western Germany, which were previously believed to be the best evidence that behaviourally modern Cro-Magnons produced the Aurignacian artifacts, places them between 3900 and 5000 years BP (4300 and 5700 years ago). Such a relatively recent age implies instead that they are intrusive Neolithic burials. These new data remove one of the previously most convincing associations between behaviourally modern humans and early Aurignacian stone tool assemblages and weakens the arguments for the Danube corridor hypothesis (Conard *et al.*, 2004).

An alternative theory to the Danube corridor is the *Kulturpumpe* theory, which sets out to explain the sudden appearance of behaviourally modern human behaviour and cultural innovation of the Aurignacian and Gravettian tool complexes in south-western Germany, which antedate similar innovations in most if not all other European locales. Several quasi-Darwinist *Kulturpumpe* hypotheses exist to explain this cultural florescence including: (1) competition between archaic and behaviourally modern humans that resulted in a dramatic increase in symbolic expression and technological advancement in the upper Danube Valley around 40 000 years BP (44 700 years ago); (2) increased competition in the foothills of the Alps as a consequence of climatic stress associated with climate change; and (3) cultural innovations resulting from social–cultural and demographic changes that had little to do with Neanderthal and behaviourally modern human competition or climate-related stress. Although we concur with Conard and Bolus (2003) that cultural modernity and technical innovations may have been driven by social–cultural dynamics, we suggest that such innovations, instead of being independent of environmental factors, were strongly influenced by a rapidly changing climate. This stimulated migration, cultural mixing and behavioural and technological innovations in Europe. Subsequently, intervals of environmental stability allowed behavioural and technological changes to be reinforced and transferred among groups in the broader populace.

An improved understanding of the rapidly changing European climate over the past 100 000 years has led Finlayson (2004) to propose that the robust Neanderthals became extinct because the severity of climatic change around 30 000 years ago negatively impacted their small heterogeneous territories, whereas the more gracile, behaviourally modern humans were better 'adapted' to exploit the expanding open plains – an environment that stimulated inventive behaviour. However, Finlayson's assumptions that Neanderthals were warm-adapted omnivores, whereas behaviourally modern humans were cold-adapted meat eaters, are not supported by current evidence, which instead shows that Palaeolithic Europeans and coastal Neanderthals maintained a varied diet, whereas the high- and mid-latitude

Neanderthals possessed a restricted carnivorous diet. However, climate reappears as a key factor in other interpretations about Neanderthal extinction. Stringer and Davies (2001) advocate that their extinction was a result of competition for resources and climatic instability.

4.5 Asia

4.5.1 Introduction

Archaeological sites dating to pre-2.0 million years ago in Asia suggest a very early Asian dimension to toolmaking, one which may predate the initial dispersal of *H. erectus* proposed in the out-of-Africa 1 hypothesis. Further, the absence of a true middle Palaeolithic in China is indicative of a very different behavioural development pattern in East and South Asia relative to Europe or Africa, with a transition directly from early to late Palaeolithic occurring around 30 000 years ago, coincident with the presence of other archaeological indicators of modern human behaviour (Gao and Norton, 2002). South Asia, represented today by Bangladesh, India, Nepal, Pakistan, Sri Lanka and the Maldives, is located at the juxtaposition between the west, where prepared core technology is everywhere apparent, and the east, where it is absent. Thus, South Asia may provide important insights into the relationship between environment and hominin behaviour.

The ongoing debate between the out-of-Africa and the multi-regional hypotheses is based on whether *H. erectus* populations in Java and China evolved independently of others elsewhere in the world, ultimately giving rise to distinct, yet genetically compatible, populations of *H. sapiens* in Asia and Australia (Bird *et al.*, 2004; Stringer and Andrews, 1988; Wolpoff, 1989). As discussed above, debate also persists as to whether there was a single or multiple *H. sapiens* migration out of Africa.

Climate, tectonic and geological events likely played a prominent role in human dispersal and behaviour throughout South Asia. By the middle Pleistocene, the Qinghai–Tibet plateau had reached an altitude of 4000 m, strongly influencing the monsoonal weather pattern, particularly in the African and Indian monsoon region (Tong and Shao, 1991). Tectonic changes in the Himalayas–Karakorum and the evolution of Indonesia were also dramatic during the Pleistocene (1.8 million years to 11 700 years ago), with reduced geological impacts felt in the Red Sea, the Caucasus Mountains and Caspian Sea, and the Sea of Japan. These tectonic changes and the subsequent development of major river systems likely impacted the biogeography of plants and animals, and the dispersion of hominins (Dennell, 2004). During warm intervals between 240 000 and 180 000 years ago and between about 130 000 and 100 000 years ago, vegetation resembled that found today in the subtropical rainforest (Shen, 1993). During cold intervals, falling sea level resulted in seasonal expansion of forests (Heaney, 1991) and a drop in treeline of up to 1500 m (Bellwood, 1992), resulting in the southward movement of warm-adapted species including primates (Jablonski, 1997). The fauna in mainland South East Asia and southern China show fewer changes than that in northern China, as most persist throughout the Pleistocene, indicating the continued presence of forested tropical refugia (Schepartz *et al.*, 2000). However, by the Holocene, fauna including pandas, orangutans, *Stegodon* (a primitive elephant) and tapirs no longer appear associated with archaeological sites (Hoáng, 1991).

The availability of fresh water, raw materials and high food resource biomass, combined with the geological and geotectonic framework, were factors critical for early hominin survival in the Gondwana and Purana basins in South Asia. During the last 80 000 years the Ganga Plain of northern India remained tectonically active (Srivastava *et al.*, 2003). The Toba, Sumatra, volcanic eruption 71 000 years ago influenced the paleoenvironment and putatively resulted in population reductions and extinctions (Ambrose, 1998b; Zielinski *et al.*, 1996b). The extended volcanic winter resulting from Toba's eruption may have severely reduced behaviourally modern human populations, reducing genetic diversity in regions outside the tropical refugia, including that of equatorial Africa. Ambrose argues that this resulted in an abrupt differentiation in the surviving groups of behaviourally modern humans that may explain the differences currently apparent in modern humans. According to Ambrose, subsequent to the volcanic winter associated with the Mt. Toba eruption, a climatic amelioration resulted in the expansion of isolated populations.

Hominin settlement in Asia is believed to have been confined to warm grasslands and open woodlands, 'generally south of 40° N in West Asia and 30° N in South and Southeast Asia', with habitation in the colder regions to the north restricted prior to the early middle Pleistocene (Dennell, 2004, p. 207). During the late Pleistocene, increased monsoonal activity across the peninsula augmented fresh water resources, and this, combined with mineral resources suitable for tool making, resulted in increased habitation in adjacent regions (Korisettar, 2005). The ability of early hominins to exploit upland environments facilitated their expansion into South East and East Asia, and these upland regions provided a relatively rich, stable and versatile source of resources rich in animal biomass and non-lithic resources (Schepartz *et al.*, 2000). Settlement of lower river floodplains occurred later, during agricultural periods. During the LGM, increased aridity in the Thar Dessert in the north-west and a weakening in the south-west monsoon resulted in monsoonal shifts (Deotare *et al.*, 2004); these, combined with glaciations, are believed to have also influenced hominin settlement behaviours (Korisettar and Rajaguru, 1998; Korisettar and Ramesh, 2002; see also Paddayya, 1982). (Please see Chapter 7 for further discussion on the impact of shifting monsoons and increased desertification on human settlement and behaviour.)

South Asia is also critically featured in colonization models of Australia (James and Petraglia, 2005), particularly those suggesting expansion via a coastal route along the South Asian coastline (for example, see Oppenheimer, 2003b). These models are supported by the apparently long-time depth of mitochondrial DNA found in the Malaysian and Andamanese populations (Endicott *et al.*, 2003). Although this evidence was initially contested (Cordaux and Stoneking, 2003), it has subsequently been further substantiated (Macaulay *et al.*, 2005; Thangaraj *et al.*, 2005).

4.5.2 India and Pakistan

If *Homo* dispersed out of Africa according to the out-of-Africa 1 hypothesis, it was not long in finding the land bordered to the north by the Himalayas and to the west and east by the Arabian Sea and the Bay of Bengal, respectively. Numerous crude, flaked stone tools dated to 2.0 million years ago (Dennell *et al.*, 1988b; Rendell *et al.*, 1987), and a stone tool core-chopper dated to 2.4 million years ago, have been found in Riwat, Pakistan, although the dating has been the subject of some criticism (Bar-Yosef and Belfer-Cohen, 2001). Stone artifacts have also been found on the erosional surfaces of fossil-bearing deposits in the nearby Pabbi Hills and dated to between *c.*2.2 to *c.*1.8 million years ago and 1.4 to 1.2 million years ago (Hurcombe and Dennell, 1992). The vegetation in this region during the lower Pleistocene was predominantly C_4 grass, indicative of open grassland (Quade *et al.*, 1993). Fauna included *Equus, Rhinoceros, Elephas, Stegodon*, ostrich, numerous bovids and the giraffid *Sivatherium giganteum* found in the older horizons, dated to between *c.*1.8 and *c.*2.2 million years ago, as well as carnivores *Pachycrocuta, Megantereon* (tiger), *Panthera* and *Canis cautleyi*. The browsing herbivores disappeared after 1.8 million years ago, potentially as a result of increased aridity and contracting woodland (Dennell, 2003).

Fossils of the hominins who manufactured the early stone tools in Pakistan have not yet been uncovered. However, fossil bones have been discovered associated with the more advanced stone tools found in the younger Acheulean sites in India (Korisettar and Rajaguru, 1998; Petraglia, 1998; Petraglia *et al.*, 1999), including the Narmada Valley assemblage, and indicate the presence of *Homo* in India during the middle Pleistocene (Rightmire, 2001, p. 83). A robust fossilized skullcap – attributed to *H. heidelbergensis* (Cameron *et al.*, 2004) or, alternatively, 'archaic' *H. sapiens* by Kennedy (2000) and others (see also Kennedy *et al.*, 1991) and dated to *c.*250 000 to *c.*300 000 years ago – was uncovered in a river bed along with quartzite handaxes, cleavers, flakes, choppers and flint and chert scrapers resembling an Acheulean tool industry. Mammalian bones associated with the skullcap, including *Bos namadicus, Stegodon* sp., *Hippopotamus namadicus, Equus namdicus* and *Cervus* sp., support a middle Pleistocene age (Sonakia and Kennedy, 1985).

No hominin remains have yet been uncovered associated with later middle Palaeolithic stone tool assemblages that date to between more than 390 000 years at 16R Dune in Rajasthan and as young as the 56 000 years in the Hiran valley (Baskaran *et al.*, 1986; James and Petraglia, 2005). Elsewhere, such technology has been found associated with both behaviourally modern human and archaic human populations, preventing clear identification of its creators. South Asian stone tools were generally produced on flakes struck from prepared cores using methods developed from preparation of the preceding Acheulean tool assemblage, implying local evolution of prepared-core technology. This suggests to us that a refugium of sorts may have existed within South Asia that was sufficiently amenable to continued habitation despite being proximal to regions that were volatilely impacted by changing climates, and in which *in-situ* evolution occurred.

Industries from Sri Lanka and Nepal differ from those in other parts of South Asia (James and Petraglia, 2005; Petraglia *et al.*, 2003). The use of points is much less common in the north-west and north-central regions of the Indian subcontinent, where there exists a preponderance of scrapers.

The increased variability in stone tool composition and use of blades marks the beginning of the late Palaeolithic in the South Asian archaeological record. It is different from both the late Stone Age in Africa and the upper Palaeolithic in Europe. The appearance of a microlithic blade industry predates that in Europe and postdates that in Africa. James and Petraglia (2005) anticipated that the replacement of anatomically modern *Homo* sp. with behaviourally modern humans migrating into this region would be marked by a distinct archaeological signal, yet there is no apparent 'symbolic revolution' like that evidenced in Europe. The earliest behaviourally modern human evidence, including *H. sapiens* fossils found with microlithic industries, is dated to *c.*31 000 years ago at Fa Hien Cave and *c.*28 500 years ago at Batadomba-lena (Deraniyagala, 1992).

Although evidence of functionally relevant modern behaviour, including structured site use, is visible earlier, the first appearance of symbolic behaviour in the form of beads and 'art' on the Indian subcontinent is dated to between 30 000 and 20 000 years ago, including engraved ostrich-eggshell pieces at the site of Patne (Misra, 2005), and, according to James and Petraglia (2005), is coincident with proposed population expansion in the region (Kivisild *et al.*, 1999). James and Petraglia suggest that the gradual appearance of modern behaviour in the Indian subcontinent is not the result of a late arrival of behaviourally modern humans, but instead implies that populations arriving as early as *c.*70 000 years ago were anatomically and cognitively modern but, for reasons not yet clear, did not express that modernity in an expected manner. They assume that 'the components of the modern human behavioral package … would be advantageous under certain conditions but … that even if they were "adaptive" at a given point in time they may not necessarily all have spread within a given population's cultural repertoire' (p. S22). They suggest that increases in populations, combined with dispersals, contractions and isolations, explain the appearance of these behaviourally modern human char- acteristics and are subject to the influences of natural selection and cultural drift. However, the very limited fossil evidence combined with a dearth of well-supported chronometric dates currently limits the ability of researchers to ascertain clear dispersal routes and the timing and impetus of modern behaviour (Lukacs, 2005). Furthermore, we dispute the assumption that innovative human behaviour is gene determined and hence subject to selectionist theory.

4.5.3 *China*

In 1941, Teilhard de Chardin suggested that eastern Asia was geographically isolated for most of the Pleistocene and thus 'closed to any major human migratory wave' (pp. 87–8). This idea was further supported by other researchers emphasizing the significance of mountain and desert barriers in China to dispersal. The multi- regional hypothesis emphasized the isolation of populations on different continents, implying that variation between continents should be greater than variation among groups within a continent. However, as additional *H. erectus* specimens were uncovered in China, there was much more variation among the various groups than initially observed (Etler, 1996, p. 288), and that variation was much more evident than the variation among present-day populations of different continents (S. Wu, 2004). By 2004, Keates showed that the faunal dispersal evidence does not support geographic isolation of eastern Asia during the Pleistocene. However, climate evidence suggests that at times during the last glacial cycle, expansive deserts stretched across the continent extending into China almost to the Pacific Ocean (for a more in-depth discussion, see Chapter 6; also see Yu *et al.*, 2000).

The earliest Chinese archaeological sites date to between 3.0 and 1.0 Ma (Bar- Yosef and Belfer-Cohen, 2001; Huang, Ciochon *et al.*, 1995; Wei, 1999) and include a hominid incisor, a partial mandible and stone tools attributed to *Homo erectus* at Longgupo Cave (30.4° N, 109.1° E), 20 km south of the Yangtze River in Chongqing province, in the Three Gorge area of southern China. Huang, Ciochon *et al.*'s dating of the site to between 1.8 and 1.9 Ma is similar to the original dating by C. Liu *et al.* (1991) to the Reunion normal subchron (2.13–2.15 Ma) (Baksi and Hoffman, 2000). Interestingly the Longgupo dates predate, or are coincident with, the northward expansion of *H. erectus* into temperate Eurasia, evidenced by early hominin remains at Dmanisi, Georgia, and Erq el-Ahmar, Israel. However, Schwartz and Tattersall (1996) and X. Wu (2000) question the assignment of the Longgupo site to hominins, suggesting instead that it should be attributed to an ape.

The earliest evidence of hominin presence at high northern latitudes is evident at Xiaochangliang (40.2° N, 114.65° E) in the Nihewan basin in northern China. Magnetic polarity stratigraphy has resulted in a determined age of 1.36 million years for stone tools associated with animal processing (Zhu *et al.*, 2003), only slightly younger than the evidence of human presence in western Asia. The presence of *H. erectus* at Dmanisi, Georgia, in western Eurasia by ~1.7 million years ago and in the Nihewan basin, China, by 1.36 million years ago suggests hominins inhabited northern latitudes over a prolonged period and were required to cross the Tibetan plateau and traverse around the Himalayan Mountains. Zhu *et al.* (2001, 2004) propose that early hominins rapidly migrated out of Africa, potentially during a warm climate phase. Alternatively, Wood and Turner (1995) propose that *H. erectus* instead evolved in Asia and subsequently dispersed back to Africa. Late Pliocene to early Pleistocene faunal remains found at the site, including *Allophaiomys* cf., *A. pliocaenicus, Mimomys chinensis, Hyaena (Pachycrocuta) licenti, Palaeoloxodon* sp., *Hipparion* sp., as well as *Proboscidipparian sinensis, Equus sanmeniensis* from the early to middle Pleistocene, *Coelodonta antiquitatis* (woolly rhino), *Martes* sp., and *Cervus* and *Gazella* (Qiu, 2000; Tang *et al.*, 1995; Wei, 1997; You *et al.*, 1980), imply conditions were not cold (Dennell, 2003). Also in the Nihewan basin is the prolific Donggutuo locality (40.2° N, 114.67° E), where stone cores, choppers and flake scrappers have been unearthed. The age of this site was originally estimated by Li and Wang (1982) to be 1.1 million years ago, a date which was subsequently supported by Schick and Dong (1993), although Jia and Wei (1987) determined an age of 900 000 years ago for the Donggutuo site.

The earliest substantial mainland hominin specimen in China is a damaged *Homo erectus* cranium uncovered at 34° N from Gongwangling, Lantian country, on the southern Loess plateau in north-central China. It dates to 1.15 Ma (An and Ho, 1989; An *et al.*, 1990; Woo, 1966).[2] Wang *et al.* (1997) suggest the presence of *H. erectus* at this site may represent an 'evolutionary milestone in human adaptability' (p. 228), representing the first habitation of a non-tropical environment. However, the primary utilization of warm-adapted species suggests that habitation was limited to warm seasons, with continuous presence indicated only after 650 000 years ago at Chenjiawo, some 150 000 years earlier than in Europe. Nearby, in the Xihoudu locality (34.7° N, 110.7° E), stone artifacts associated with a rich late Pliocene to early Pleistocene fauna – including *Proboscidipparion sinense, Elaphurus bifurcatus, Euctenoceros boulei, Bison palaeosinensis, E. sanmeniensis, C. antiquitatis* and *Elasmotherium inexpectatum* – have been dated to about 1.27 Ma (Jia, 1989; Zhu *et al.*, 2003).

The dates of these early *H. erectus* sites substantiate the theory that the middle region of the Yellow River has been inhabited by humans since about 1.2 Ma (An and Ho, 1989). Further, the dispersion of sites between the middle Yellow River and the Nihewan basin, a distance of about 800 to 900 km, indicates an ability to 'occupy or shift their range…during a time of enhanced global and regional climatic variability that included intermittent aridification of north China' (An *et al.*, 2001; T. S. Liu and Ding, 1999; Zhu *et al.*, 2003).

Two crania found at Quyan River Mouth, Yunxian, dated to between ~581 000 ± 93 000 years ago (T.-M. Chen *et al.*, 1997), lack many diagnostic features typical of Western archaics and later elaborated by Neanderthals, and instead possess features seen universally in fossil and modern Asians (Etler, 1996). South of the Changjiang (Yangtze) River, in the Hulu Cave on Tangshan hill, east of Nanjing, Jiangsu province, a large *H. erectus* craniofacial fragment was uncovered and initially dated to between 207 000 and 417 000 years ago and subsequently dated to greater than 500 ka (Q. Chen *et al.*, 1998; Zhou *et al.*, 1999). Fauna associated with this site are similar to those found at the famous Zhoukoudian site in Bejing, China, and likely represent the southward dispersal of Paleoarctic fauna during a cold climate interval (Etler, 1996).

Six skullcaps, 12 cranial fragments, 157 teeth and 15 mandibular remains were recovered between 1921 and 1966 from Zhoukoudian, China (Black, 1931; Etler, 1996; Weidenreich, 1936a, 1936b, 1939, 1941, 1943). Although excellent castes of the original material still remain, nearly all the fossils were lost during World War II. The morphological features of the Zhoukoudian fossils have become the standard benchmark for *Homo erectus*, who intermittently occupied the site from ~580 000 to 440 000 years ago until perhaps as late as 240 000 to 230 000 years ago (Etler, 1996; Wu and Poirier, 1995). In 1943, based on his analysis of the skulls of 'Peking Man' from Zhoukoudian, Weidenreich noted 12 morphological features indicative of the continuity of human evolution in China, providing support for the multi-regional evolution hypothesis. According to S. Wu (2004), since that time additional Pleistocene hominin fossils found in China provide supplementary evidence of continuity and transition between primitive hominins and modern *H. sapiens* in China. Wu suggests that evidence indicates intermittent gene flow between the West and China.

More recently discovered middle and late Pleistocene sites, including Dali and Jinniushan, have yielded 'archaic' crania more advanced than *Homo erectus* (Rightmire, 2001, p. 83). At what is thought to be a stone knapping site in the mountainous Guizhou karstland, stone tools uncovered in the Guanyingong Cave associated with *Stegodon-Ailuropoda* fauna have been dated to 230 000 years ago (Leng, 1992). Nearby, the site of Panxian Dadong disclosed evidence of four human teeth with stone tools and fauna. Uranium-series dating of a rhinoceros tooth places the mean age of the upper levels of the site between 156 000 ± 19 000 and 138 000 ± 16 000 years ago and the lower cave deposits between 261 000 ± 31 000 and 214 000 ± 24 000 years ago (Rink *et al.*, 2003), or older than 300 000 years ago

(Shen *et al.*, 1997). At the Jinniushan site in the Liaoning province, a female skeleton identified as a premodern *H. sapiens* dates to between 200 000 and 300 000 years ago (T. M. Chen *et al.*, 1994; Etler, 1996; Wu and Poirier, 1995). These premodern finds are in addition to a partial cranium at Maba, Quijiang county, Guangdong, dated to between 135 000 and 129 000 years ago (Yuan *et al.*, 1986); numerous fossil bones, including those from a six-year-old child, at Xujiayao in Yanggao county, Shanxi, dated to 125 000 years ago (T. M. Chen *et al.*, 1984), and cranial remains from Yanshan, Chaoxian county, Anhui. The relatively recent age of the *H. erectus* at the Zhoukoudian site suggests an overlap with the more derived forms, including the Jinniushan hominin species, and indicates potential coexistence of *H. erectus* and *H. sapiens* (T. M. Chen and Zhang, 1991).

Although difficulties exist with uranium-series dating, the timing of premodern *H. sapiens* in China at about 300 000 years ago appears to lag slightly their initial appearance in Africa and the appearance of Neanderthals in Spain. Archaic *H. sapiens* persisted in China until about 100 000 years ago, when they experienced an abrupt decline. Facial and cranial vault morphology of early Chinese *H. sapiens*, although transitional between *H. erectus* and *H. sapiens*, show clear affinities with modern Asian populations, including the presence of central incisor shoveling, a key characteristic of the East Asian 'sinodont dental pattern' (C. G. Turner, 1990; see also Etler, 1996). This characteristic is absent in other modern people, leading some researchers to argue against an event in which archaic *H. sapiens* simply replaced *H. erectus*, instead suggesting that Asia played a key role in the development of behaviourally modern humans. However, if this were the case, we would expect that many other behaviourally modern humans would have the sindodont dental pattern.

However, the absence of any securely established sites between 100 000 and 30 000 years ago in east and South East Asia, during a time when numerous hominin localities exist in Europe, makes it difficult to ascertain when behaviourally modern humans arrived or evolved (Brown, 2001). The earliest date for behaviourally modern *H. sapiens* in China comes from the Liujiang Cave site and is dated to 67 000 years ago (Yuan *et al.*, 1986), although its association has been questioned. The oldest generally accepted behaviourally modern human remains date to between 34 000 and 29 000 years ago from the Upper Cave, China, and include three complete skulls and bone fragments. Based on Y-chromosome evidence, behaviourally modern humans are believed to have migrated into Asia from Africa at about 60 000 years ago (Su *et al.*, 1999). However, according to S. Wu (2004), if this were the case, the cultural record should show a sudden shift from Mode 1 Oldowan to Mode 3 Mousterian technology. However, no such interruption is observed in the Chinese Palaeolithic record. Instead, eastern Asia possesses a unique stone tool tradition.

A variation in the stone tool cultural evidence is apparent between eastern Asia and elsewhere in the Old World. Unlike in the Western World, where there is a transition from Mode 1 Oldowan through Mode 2 Acheulean and then Mode 3 Mousterian before the emergence of behaviourally modern humans and their Mode 4 technology, in China the Mode 1 Oldowan tradition persisted throughout the entire Palaeolithic age, except for the last 40 000 years. Even then, the Oldowan technique still formed the dominant mode for artifact creation, despite the capability for advanced toolmaking techniques that may have been developed locally or resulted from intermittent introductions of Western Palaeolithic techniques (S. Wu, 2004).

A limited number of Acheulean Industrial Complex lithic assemblages have been found in China (see Huang, Si *et al.*, 1995; Huang and Wang, 1995). However, if the dates of the 800 000 year old Acheulean-like bifacial handaxes found in the Bose basin in China associated with tektites are substantiated, their discovery suggests that Asian *H. erectus* were no less intelligent and adaptable than their African counterparts. In fact, it may represent an atypical eastward dispersal event and associated behavioural change related to widespread forest burning (Hou *et al.*, 2000; Yamei *et al.*, 2000). Yamei *et al.* (2000) argue that although the Bose basin handaxes do not exactly replicate the classic Mode 2 Acheulean technology found in Africa, the Middle East and Europe, their manufacturing technique is similar enough to suggest a sharing of technological traditions between east and west. However, despite the Bose finds, China lacks the teardrop-shaped handaxe – the signature tool of the middle Palaeolithic Acheulean – and the Asian Acheulean sites are restricted to south-west Asia and the Indian subcontinent (Gao and Norton, 2002). Yet the Bose basin finds do imply that instead of a cultural backwater in Asia, east of what is known as the 'Movius line' (shown on Figure 4.2), the forested regions of South East Asia may have provided a unique environment where bamboo or other woods replaced or provided an alternative resource to stone (Pope, 1989). Alternatively, there may have been a greater reliance on plant foods or distinct technological traditions may have existed, making the Acheulean bifaces unnecessary (Watanabe, 1985; S. Wu, 2004).

4.5.4 Thailand, Burma, Laos, Vietnam, Korea

The earliest hominin evidence from Thailand appears at the Mae Tha and Ban Don sites in the Lampang Phrae provinces of northern Thailand, where quartzite river cobbles have been dated to between 800 000 and 600 000 years ago (Pope *et al.*, 1986; Pope and Keates, 1994; Reynolds, 1990). Several large pebble-tool localities uncovered on terraces and adjacent upland areas along the Irrawaddy River, Upper Burma, may be of middle Pleistocene age (Movius, 1944; contested by Bartstra, 1982). A fragmented maxilla and a partial mandible have been found at New Gwe Hill and Letpan-Chibaw, Burma, respectively, and have been attributed to *H. erectus* (Ba, 1995; Ba *et al.*, 1998). Evidence from Laos is restricted to the Tam Hang Cave locality, where isolated molars, a temporal bone, and a partial subadult skull have been found (Olsen and Ciochon, 1990; Schepartz *et al.*, 2000). The subadult skull has been alternatively attributed to *H. erectus* and *H. sapiens* (Schepartz *et al.*, 2000). Numerous caves in Vietnam have generated hominin teeth, although as in Laos, no stone tools have been associated with the remains. Attribution of the teeth to hominin species as opposed to fossil orangutans, as well as mixing and reworking at the sites, has led to difficulty dating, although a middle Pleistocene age seems likely. Ciochon *et al.* (1996) suggest an age of 475 000 years ago. In the eastern Thanh Hoa province, Vietnam, surface-collected stone tools may be as old as 750 000 years ago (Olsen and Ciochon, 1990), although opinion is still divided, with some researchers considering the finds to be geofacts, whereas others identify them as Bronze Age axes and adzes (Bui, 1998; Schepartz *et al.*, 2000).

4.5.5 *Java, Indonesia, Malaysia, Philippines*

Most researchers agree that the first hominins in Indonesia were *H. erectus*. Unlike in Europe or western Asia, where there appears to have been a split in the evolution of the genus *Homo* around 1.0 million years ago, the general consensus is that within the balance of Asia, *Homo erectus* continued their existence from the early Pleistocene through to the end of the middle Pleistocene or later (Larick *et al.*, 2001; Swisher *et al.*, 1994).

Much of the evidence for the early arrival of *H. erectus* has come from Java. The Java *H. erectus* are regarded by some, although not all, as different from *H. ergaster* of East Africa. Fossil hominins in Java show similarities with early western Asian characteristics. However Antón and Indriati (2002) suggest this may be due to varying source populations and/or endemism. Associated artifacts indicate a core-chopper, Oldowan-like tool kit (Antón and Swisher, 2004; Bar-Yosef and Belfer-Cohen, 2001).

In 1891, near Trinil, Java, archaeologist Eugène Dubois discovered 'Java Man'. Hominin fossils included a skullcap, a femur and three teeth, which he attributed to the species *Pithecanthropus erectus*. In 1950 the fossils were reclassified as *H. erectus* by Ernst Mayr. A second 'Java Man' was found by Andoyo, a geological assistant with the Geological Survey of the Netherlands Indies in the Perning district of Java (recently relocated at 7° 22' 36.1" S; 112° 29' 01.5" E) (Huffman and Zaim, 2003), and was later identified as an early hominin by paleontologist G. H. R. von Koenigswald in 1936. This infant's skull – frequently referred to as the 'Mojokerto child' – has been dated to 1.81 million years ago (Swisher *et al.*, 1994). *H. erectus* fossils subsequently appear at Sangiran, Java, where nearly 80 fossils have been dated to 1.6 million years ago (*ibid.*; Aguirre and Carbonell, 2001; Antón and Swisher, 2004; Larick *et al.*, 2001), although this has been contested by

Langbroek and Roebroeks (2000), who suggest that the Javan hominin sites are younger than 1.0 million years. However ^{40}Ar/^{39}Ar dates throughout the Sangiran section support the initial interpretations (Sémah *et al.*, 2000). Artifacts indicate a core-chopper, Oldowan-like tool kit (Antón and Swisher, 2004; Bar-Yosef and Belfer-Cohen, 2001). Teeth and mandibles of the older hominin fossils are equally or more primitive than the early *H. erectus* from Africa dated to *c.*2.0 million years ago (Kaifu *et al.*, 2005).

The Mojokerto *H. erectus* lived on a seacoast in an environment that was inhabited by 'deer, muntjak, bovids, pig, hippopotamus, rhinoceros, *Stegodon*, and large cat' (Huffman and Zaim, 2003, p. 1) as well as turtle, crocodile and edible molluscs. They inhabited this seacoast during a glacio-eustatic lowstand when conditions were dry. Preliminary carbon isotope analyses of herbivore tooth enamel indicates a primarily C_4 grass diet. Analyses of the plant material from the site suggests *H. erectus* lived near a coastal delta that supported mangroves and was surrounded by open grassland with scattered trees and bamboo thickets, swamps and distant montane forests predominated by mountain *Casuarina*. Volcanos were likely active in the distant highlands of East Java (Huffman and Zaim, 2003).

There appears to be a relatively continuous presence of *H. erectus* up until the Brunhes-Matuyama boundary (0.78 million years ago), with dates as young as 53 000 to 27 000 years ago suggested for the youngest *H. erectus* remains at Ngandong, central Java, implying their contemporaneity with *H. sapiens* in South East Asia (Larick *et al.*, 2001; Swisher *et al.*, 1994, 1996). Grün and Thorne (1997) have questioned these dates, but if they are correct, they compare with the youngest in Africa, a cranium obtained from Bed IV in the Olduvai Gorge, dated to 1.0 million years ago, and evidence from mainland Asia dated to between 420 000 and 290 000 years ago from Zhoukoudian, China (Swisher *et al.*, 1996). Thus, as stated by Rightmire (1998), the Ngandong fossils may record the last appearance of *H. erectus*.

However, difficulties with the Javanese evidence have resulted in significant debate about the arrival and subsequent development of *H. erectus* (O'Sullivan *et al.*, 2001). Thus, it has been with increased interest that researchers are reviewing the fossil evidence of early hominids within the 1000 km^2 Soa basin, which is surrounded by mountains and active volcanoes on Flores Island.

The region of Wallacea, including the islands of Kalimantan, Bali, Java and Sulawesi, were intermittently connected to South East Asia or Sunda and Australia or Sahul during intervals of lowered sea level since the beginning of the Pleistocene, 1.8 million years ago. This contact resulted in a full range of continental Asian mammals inhabiting Wallacea, whereas the Indonesian islands to the east remained isolated by water. This 25 to 30 km wide deep-water gap, known as Wallace's line, is recognized as a major biogeographical barrier (Wallace, 1890). Although the main island of Java lies to the west of Wallace's line, the island of Flores lies to the east.

Volcanic rock and chert artifacts have been discovered deposited in a grassland savannah that was created by the draining of basin lakes in the Soa basin, Flores Island, Indonesia. These artifacts are amongst volcanic tuff deposits containing the remains of large *Stegodon*, crocodile, giant rat, as well as fresh-water molluscs. The volcanic tuffs have been dated and indicate that hominins, presumably *H. erectus*, reached Flores by 840 000 years ago and occupied the island until at least 700 000 years ago (Morwood *et al.*, 1998; O'Sullivan *et al.*, 2001). The date of the artifacts suggests they were made by *H. erectus* at a time that roughly coincides with the arrival of *H. erectus* in central Europe. Their arrival at Flores Island, in the Wallacean Islands belt, means that even during periods of maximum sea-level regression, *H. erectus* would have been required to make a water crossing of 19 km (Morwood *et al.*, 1998; Straus, 2001), a talent previously attributed only to modern humans. This feat has made researchers rethink the technological and intellectual capacity of *H. erectus* (Bird *et al.*, 2004; Morwood, 2001; Morwood *et al.*, 1998; O'Sullivan *et al.*, 2001).

More recently, finds of a new *Homo* species, *Homo floresiensis*, have been discovered on Flores Island. This very small-brained hominin was only 1 m tall, walked on two legs, and was found associated with a vertebrate faunal assemblage (Brown *et al.*, 2004; Morwood *et al.*, 2004). These small humans possessed a formal stone tool kit and hunted the dwarf elephant *Stegodon*, which was previously thought to have gone extinct sometime around 840 000 years ago (Rolland and Crockford, 2005; van den Bergh *et al.*, 2001). *Homo floresiensis* survived on the island from before 38 000 years ago to at least 18 000 years ago, a time that overlapped with the presence of *H. sapiens* in the region. Their demise, coincident with that of the *Stegodon*, occurred at the same time that a volcanic eruption destabilized what was probably a fragile environmental equilibrium.

In the Great Cave at Niah, on the coastal plain of Sarawak in Borneo, a modern human skull was discovered in the 1950s by Tom and Barbara Harrison and subsequently dated to at least 40 000 years BP (44 400 years ago) and perhaps as early as 45 000 years BP (48 900 years ago) (Harrison, 1958; see also Barker, 2005). Only a few hundred metres away, the Harrisons unearthed humans buried in wooden boats in Kain Hitam Cave, where cave walls are painted with pictures of boats and dancing figures. Pleistocene sediments indicate drier, more varied, seasonal and open conditions than today, with fauna associated with 'a mosaic closed forest alternating with scrub, bush, or parkland, and including extensive areas of swamp, lakes, or large rivers' (Cranbrook, 2000, p. 83). The first Pleistocene people were hunters, fishers, and shellfish and plant gatherers who possessed a remarkable knowledge of rainforest plant distribution and extraction and processing technologies that allowed them to eat and digest carbohydrates from roots, tubers and palm pith (Barker, 2005).

By the LGM the Sunda Shelf was fully exposed, joining Java, Sumatra, Palawan, and Borneo to the Asian mainland; a narrow strait separated the Philippines from Sunda. Shelf exposure likely facilitated the migration of peoples to these more distant regions. However, Thiel (1987) theorizes that dispersion of *H. sapiens* to the eastern islands that have always been separated from Sundaland by deep straits would have occurred only during intervals of higher sea level, when reduced land area resulted in increased land-use pressure, stimulating dispersal in search for food. Yet the earliest evidence of *H. sapiens* in Thailand appears at Lang Rongrien Cave and dates to 37 000 years BP (42 000 years ago) (Anderson, 1990). The earliest archaeological evidence in the Philippines is found at Tabon Cave in Palawan and dates to 30 000 years BP (35 000 years ago), although modern human fossil evidence dates to between 20 000 and 22 000 years BP (24 500 and 24 000 years ago) (Fox, 1970). The timing of these finds coincides with intervals of lowered sea level. At around 42 000 years ago, global eustatic sea level was about 73 m below present. Between 35 000 and 30 000 years ago, sea level fell to 120 m below today and remained there until about 21 000 years ago (Lambeck *et al.*, 2002), implying that lowered, as opposed to raised, sea levels facilitated habitation of these distant lands.

4.5.6 Japan

Palaeolithic archaeologists claim *H. erectus* was in Japan ~500 000 years ago, presumably arriving via Sakahalin to the north when sea levels were lowered. It is not unreasonable to assume that, during intervals of lowered sea levels, when Japan, Taiwan and Hainan were joined to mainland Asia, early hominins already present in China would have crossed to Japan (Bird *et al.*, 2004). However, the validity of evidence supporting such claims is suspect in light of the recent discovery of fraud, including the planting of artifacts at various 'palaeolithic' archaeological sites in Japan, and the suicide of a prominent senior figure in the Japanese Palaeolithic community (Bhattacharjee, 2004).

Irrespective of these difficulties, what does remain secure is the existence of the Neolithic Jomon period, *c.*12 000 to *c.*2300 years BP (~13 900 to 2300 years ago), with evidence of a distinct pottery culture (Meggers, 1998c; Omoto and Saitou, 1997). Initial genetic evidence partially sustained the hypothesis that the modern Ainu and Ryukyuan (Okinawa) populations are directly descended from the Jomon people who originated in South East Asia, whereas the Hondo – Main Island – Japanese are descended from migrants from mainland north-east Asia (Hanihara, 1991). However, more recent research shows that although all three populations belong to a north-east Asia group, the Ainu and Ryukyuan share a group that differs from the Hondo-Japanese group, which is more closely aligned with the Korean (Omoto and Saitou, 1997; see also Horai *et al.*, 1996). Also demonstrated are genetic similarities between Koreans and northern Chinese, which are clearly different from southern Chinese populations (Derenko *et al.*, 2004). Omoto and Saitou propose that the Jomon people arose in north-east Asia and gave rise to the Ainu and Ryukyuan. Mitochondrial DNA reconstructions identify direct mtDNA connections between Tibet and these Palaeolithic Japanese (Tanaka *et al.*, 2004). A subsequent migration from north-east Asia, via Korea, to Japan likely brought the ancestors of the Hondo-Japanese whose genetic influence predominates in the majority of modern Japanese.

4.5.7 Siberia and the Russian Far East

Lower Palaeolithic artifacts, including stone choppers, scrapers and cores, recovered from the highest terrace of the Lena River at the Diring Yuriakh site, 140 km south of Yakutsk in central Siberia (61° 12' N, 128° 28' E), first discovered and dated by Mochanov in 1983 to 1.5 to 2.0 million years ago, have been subsequently dated to older than 260 000 years ago (Waters *et al.*, 1997). Based on these findings, it is believed that early people migrated into the area at the end of the glacial MIS8 or during the interglacial stage 7. However, irrespective of the exact timing, the location indicates climatic conditions would have been severe, requiring clothing, the controlled use of fire and shelter for hominin survival. The duration of the occupation is unknown. Other lower Palaeolithic sites have been found north-west of Lake Baikal in the Cis-Angara region as well as in the Lower Amur, containing Levallois cores, choppers, scrapers and handaxes (Larichev *et al.*, 1987).

Electron spin resonance dating of *Homo neanderthalensis* remains found in seven middle Palaeolithic (Mousterian) levels and three upper Palaeolithic levels in Mezmaiskaya Cave in the northern Caucasus in western Russia have been dated to between 73 000 and 36 200 ±5000 years ago. Skinner *et al.* (2005) suggest that the prolonged presence of *H. neanderthalensis* in this region may indicate an overlap with *H. sapiens*. The largest group of middle Palaeolithic surface sites has been found in the Sagly Valley, Tuva, in addition to the Rivers Tuecta in the Altai, and includes river pebble cores and Levallois flakes. Redeposited material in T'umechin I on a terrace above the river in the Altai contains protoprismatic, Levallois, and radial cores, retouched tools, points, blades, denticulates, burins and scrapers. The principal typology of the group is Mousterian. T'umechin II, located 2 km east of T'umechin I, is near the entrance to a canyon. It contains redeposited material of an unstandardized nature and is notable for the absence of the Levallois technique, the low frequency of blades and the high incidence of crude flakes (Larichev *et al.*, 1987). The stratified cave deposit of Ust'-Kanskaya contains faunal and bird evidence including the extinct species *Crocuta spelaea* (spotted hyena), *Rhinoceros tichorhinus* (extinct arctic woolly rhino), and *Spiroceros kjakhtensis* (antelope), along with species presently found to the east or south including *Poephagus grunniens* (yak), *Equus caballus fossilis* (horse), *Equus hemionus* (Asian wild ass), and *Gazella* cf. *gutturose*, as well as extant species (Rudenko, 1960). It was initially dated to 'the warm phase before the last glaciation of the Altai, the Upper Pleistocene' (*ibid.*, p. 125), but is more recently believed to correspond to the Kargin interglacial (40 000 to 20 000 years ago) (Zeitlin, 1979).

One of the richest faunal assemblages has been unearthed at the Strashnaya Cave, in the Charish River basin, south-west of the Altai, where 37 species have been identified, including *Coelodonta antiquitatis* (woolly rhino), *Mamonteus primigenius* (mammoth), *Equus hemionus* (Asian wild ass), *Bison priscus* (steppe bison), *Alces alces* (moose), *Saiga tatarica* (saiga antelope), *Ovis ammon* (argali), *Crocuta spelaea* (spotted hyena), *Canis lupus* (small wolf) and *Ursus* (*spelaearctos*) (cave bear) (Okladnikov *et al.*, 1973, pp. 49–50). Pollen analysis of this site indicates a predominance of grass with some fir, pine, spruce and birch (Zeitlin, 1979, p. 88). Stone tools are primarily Levallois cores made on river pebbles, with blade tools and points. Radiocarbon dates from the lower cultural layers from Ush'-Kanskaya and Strashnaya suggest an age of between 45 000 and 40 000 years ago (Larichev *et al.*, 1987). Okladnikov Cave, on the outskirts of the village of Sibir'achikha, contains bones and a tooth from two archaic *Homo sapiens* (Derev'anko, 1986), a few Lavallois flakes and several scrapers. Not far to the north-east lies the Dvuglazka Cave site where Levallois Mousterian tools have been unearthed (Abramova, 1985), along with fossils of *Equus* cf. *hemionus* (Mongolian wild ass), *Equus caballus* (horse), *Coelodonta* cf. *antiquitatis* (woolly rhino), *Bison priscus* (steppe bison), *Ovis ammon* (argali), *Cervus elaphus* (wapiti or elk), *Saiga* sp.(antelope), and carnivores *Crocuta* sp. (hyena), *Panthera* sp. (lion), *Ursus* sp. (bear), *Canis lupus* (small wolf), *Vulpes* sp. (fox) and *Gulo* sp. (wolverine). The Mousterian stone tool facies show considerable variety in all the Siberian sites, and their presence in southern Siberia could mean that they may be found elsewhere in northern Asia (Larichev *et al.*, 1987). It is thought that climate change resulted in reductions and expansions of Palaeolithic populations. It is also clear that early peoples must have been required to adjust to arctic desert conditions.

As in southern and eastern Asia, early upper Palaeolithic cultures in Siberia lack the European characteristics seen in the Aurignacian and Gravettian industries (Foley and Lahr, 1997; G. K. Klein, 1992). Underhill *et al.* (2001) suggest this represents a rapid expansion of *H. sapiens* out of Africa and their potential extinction and replacement by behaviourally modern *H. sapiens* in subsequent expansions out of the Levant 45 000 to 30 000 years ago and out of central Asia at 30 000 to 20 000 years ago. While Larichev *et al.* (1992, pp. 441–6) consider the D'uktai culture archaeological sites north and east of Lake Baikal to date to 32 000 years ago, others place these sites no older than 18 000 to 15 000 years ago (Bird *et al.*,

2004). However, it is from north-east Asia that researchers believe that humans migrated into the Americas during and subsequent to the last glacial maximum (about 21 500 years ago).

The distinctly different genetic diversity among lineages in Europe and Asia is also apparent in Siberia. Native north-eastern Siberians show genetic similarities to Asian specific mtDNA groups, whereas the Saami aboriginals from north-western Siberia are thought to be genetically similar to the upper Palaeolithic Europeans. Puzzling to researchers is how this geographic separation in lineages arose. Was it the result of different migrations out of Africa or due to subsequent genetic diversity? An ancient group of fishers and hunters living amongst the pine-birch forests on the eastern slope of the northern Ural Mountains in northern Siberia provides some insights (Derbeneva *et al.*, 2002). Their mtDNA contains both ancient European haplogroups, including the rare U7 subhaplogroup present in the Middle East and A, C and D, which are included in the migration from Siberia to the Americas. This may represent a recent blending of eastern and western populations, or alternatively may reflect a much earlier distribution of mtDNA that was more general and subsequently became geographically diversified. The rare U7 haplogroup may have come from a Palaeolithic dispersal to the Ural Mountains from the Middle East.

Ancient DNA from a bone sample from a 3600 years BP (3900 years ago) woman's skeleton unearthed in north-eastern Yakutia in east Siberia has recently been analyzed. Her nuclear DNA links her to both east Siberian/Asian and Native American populations; she belongs to haplogroup C, which is one of the four main founding groups in Native American populations (Ricaut *et al.*, 2005). Analysis of the woman's allelic frequencies showed greatest affinity with populations in native Alaska, native western Canada and East Asia.

4.6 Australia and New Guinea

R. G. Roberts *et al.* (1990, 1994) have suggested that behaviourally modern human occupation of Australia occurred by about 55 000 years ago. New optical ages for human occupation at Lake Mungo III, 2700 km from the current north-western Australian coast, by Bowler *et al.* (2003), which was originally dated by Thorne *et al.* (1999) to 62 000 ± 6000 years ago, has resulted in dates between 50 000 and 45 000 years ago for the initial occupation and 40 000 ± 2000 years ago for two human burials. A review of recent research by O'Connell and Allen (2004) resulted in dating the colonization of Pleistocene Australia and New Guinea more conserva-tively at between 45 000 and 42 000 years ago. At this time sea level was between about 55 and 73 m lower than today, respectively (see Figure 4.1 above) (Lambeck *et al.*, 2002), exposing portions of the Sunda and Australian shelves. Even during periods of maximum glaciation, when sea-level regression reached its greatest, the water crossing required to reach Australia, although reduced, would still be as much as 100 km. Hydrological evidence at Lake Mungo indicates that initial lake filling began 60 000 years ago, followed by a period of intermittent changes in lake levels of limited duration between 50 000 and 40 000 years ago, coincident with the arrival of people, and a subsequent major hydrological change at about 42 000 years ago reflecting increased aridity in the continental interior (Bowler *et al.*, 2003).

By 21 500 years ago, during the LGM, lowered sea levels connected Australia to New Guinea, Tasmania and other islands of Melanesia, forming the continent of Sahul. Large freshwater lakes developed in northern Western Australia in the Bonaparte Gulf (Yokoyama, De Dekker *et al.*, 2001; Yokoyama, Purcell *et al.*, 2001), between Australia and New Guinea in the Gulf of Carpentaria (Torgerson *et al.*, 1988), and between Tasmania and Australia in the Bass Strait (Lambeck and Chappell, 2001). Isolation by the Bass Strait, which separated Tasmania from Australia until 43 000 years BP (47 300 years ago), likely prevented colonization of Tasmania until about 35 000 years BP (40 350 years ago) (Porch and Allen, 1995), when falling sea levels resulted in intermittent connection resulting in the exposure of the entire Bass Strait by 32 000 years BP (37 400 years ago) (Lambeck and Chappell, 2001).

Despite the probability that early colonizers of Australia would have been adept at utilizing marine resources, key coastal locations, including the north-west Cape in Western Australia, are at present bereft of habitation evidence – although Golson (1971) notes the barrenness of the north-west corner of the continent and its potential inhospitality to late Pleistocene migrants. Thus, it may be that after the initial waterborne colonization into northern Australia, rapid southward expansion occurred along the periodically wetter than present savanna–woodland of the continental interior and was followed by radiation into less favourable regions (Bird *et al.*, 2004). O'Connor and Veth (2000) hypothesize that increased aridity associated with the onset of the LGM may have initiated a peripheral movement back to the coastal environments. Alternatively, Bowdler (1990) suggests that a coastal occupation was followed by migration up major river systems from the south, facilitating inland habitation. Climate simulations and other geologic evidence discussed more fully in Chapters 6 and 7 indicate greater precipitation and improved habitat in Australia during the LGM. These improved conditions may have stimulated a population expansion and encouraged the reoccupation of coastal environments.

While many of the early occupation dates for the colonization of Australia have been questioned, the conservative later colonization date for Australia coincides with the habitation dates of the earliest upper Palaeolithic sites in the Levant. This has led some researchers to suggest that behaviourally modern humans did not use an interior northern route out of Africa but used a southern coastal dispersal route. Early people are thought to have left the Horn of Africa and crossed to the southern Arabian Peninsula before traversing the coasts of India and South East Asia, and then eventually colonizing Australia and New Guinea (Field and Lahr, 2006; Lahr and Foley, 1994). The timing of Australia's colonization remains the strongest evidence supporting a southern Asian dispersal route. Additional climatological evidence is further discussed in Chapters 6 and 7. However, additional evidence includes the morphological dissimilarities between Australo–Melanesian and some southern and south-eastern Asian populations and their purported upper Palaeolithic ancestors. The Australo–Melanesian and identified Asian groups appear to more closely resemble African groups. Evidence includes the M haplotype mtDNA lineages found in Indian and East African groups as discussed above. Further, according to non-recombining Y-chromosome loci, the M168 African populations dispersed out of Africa 50 000 to 45 000 years ago, reaching southern Asia where the RSP4Y/M216 mutations likely originated, and then colonized South East Asia, Australia and New Guinea (Underhill *et al.*, 2001). Yet, although sea level fell briefly between 50 000 and 45 000 years ago, reaching a low of about −68 m, it was not until after 45 000 years ago that sea level began rapidly falling, before reaching a low of −120 m 30 000 years ago where it stayed until at least 21 600 years ago (Lambeck *et al.*, 2002). Previous intervals of falling sea level occurred between 85 000 and 76 000 years ago (reaching a low of about −65 m), 72 000 and 63 000 years ago (reaching a low of about −73 m), and 60 000 and 55 000 years ago (reaching a low of about −62 m) (refer to Figure 4.1). Thus, it is not clear that falling sea levels per se were the sole motivating impetus behind a potential migration out of Africa at between 50 000 and 45 000 years ago.

Geneticists suggest that a subsequent mutation of the RSP4Y/M216 southern Asian group, the M217 mutation, reached central and eastern Asia, Japan, and, later, North America. Further, they suggest that the southern route from Africa into southern Asia may have been used more than once by a virtually contemporaneous YAP/M145/M203/M174 lineage expansion from the Horn of Africa into southern Asia (Underhill *et al.*, 2001). However, they limit the usage of the southern route to only those climatic and geographic windows of opportunity when lowered sea level and light monsoonal regimes facilitated human dispersal. More recent population expansion events have subsequenlty largely replaced the colonizing lineages in Asia.

Interestingly, according to Field and Lahr (2006), Australians seem to be unrelated to highland New Guinea populations, suggesting multiple dispersal events into the region between 60 000 and 30 000 years ago. New Guinea, which lies east of Wallace's line, was populated between 45 000 and 40 000 years BP (49 000 and 44 700 years ago) (Groube *et al.*, 1986; O'Connell and Allen, 2004; R. G. Roberts *et al.*, 1994). Plant and animal populations on each side of Wallace's line had dispersed and redistributed themselves in their own continental and island geographies and probably were inhibited from attempts at migration into what for a long time was a void. It was a behavioural barrier rather than a physical barrier. So, the question is, did migrating humans even notice it as *any* kind of barrier, apart from recognizing the different flora and fauna on either side of the line?

Occupation of the interior highland forested regions of New Guinea seems to have lagged coastal habitation, with the first occupation appearing at Kosipe dated to around 30 000 years BP (35 100 years ago) (Hope and Golson, 1995; J. P. White *et al.*, 1970). Colonization of the Aru Islands, located on the exposed continental shelf south-west of New Guinea during intervals of lowered sea level, occurred by 32 000 years BP (36 900 years ago), and farther to the west, the eastern region of Timor appears to have been colonized by at least 32 000 years ago (O'Connor *et al.*, 2001). East of New Guinea, New Britain and New Ireland were populated by 35 000 years BP (40 600 years ago), requiring water crossings of about 40 km (Pavlides and Gosden, 1994). Farther to the east, colonization of Oceania by modern humans, in this early phase, apparently stopped at the Solomons' archipelago about 30 000 years BP (35 100 years ago).

Archaeologists remain puzzled over the demise of megafauna in Australia and New Guinea. Although poorly dated, extinction of several forest-dwelling marsupials in New Guinea may have occurred after 40 000 years BP (44 700 years ago), leading some to suggest a relationship with early human colonization (Flannery, 1995). Megafauna extinction in Australia began between about 50 000 and 45 000 years ago, coincident with, or just subsequent to, the proposed first arrival of behaviourally modern humans (Miller *et al.*, 1999; R. G. Roberts *et al.*, 2001). All 19 Australian marsupial species exceeding 100 kg, and 22 of the 38 species between 10 and 100 kg, in addition to 3 large reptiles and 1 large flightless bird, became extinct. Two other large flightless birds, including the emu, survived. Many other species experienced range reduction and dwarfing (Miller *et al.*, 1999). One theory for their demise suggests that species such as the 130-kilogram marsupial lion, *Diprotodon*, which resembled a three-tonne wombat, giant kangaroos and huge lizards, were destroyed by the recently arrived over-zealous human hunting society within a few generations of its arrival. However, archaeological sites indicating predation by humans are extremely rare. Consequently, an alternative theory suggests that megafuanal extinction was the indirect result of human activity including burning practices that resulted in fundamental changes in ecosystems (Jones, 1973). A third theory suggests instead that the megafauna lived side by side with humans for perhaps as long as 15 000 years or longer and that climate change, including increased aridity, may have been responsible for their demise (Price and Sobbe, 2005; Trueman *et al.*, 2005). Although the flightless bird *Genyornis newtoni*, which fed primarily or exclusively on C_3 grasses, went extinct, the flightless emu,

Dromaius novaehollandiae, which possessed a broader diet including C$_4$ grasses, survived because of its dietary adaptability to changing ecological conditions (Miller *et al.*, 1999).

Although systematic agriculture is not evident until the Holocene, the widespread use of fire by peoples in Australia dates to 45 000 years BP (49 000 years ago) and may have significantly impacted vegetation and climate (Turney *et al.*, 2001; see also Bird, 1995). This is substantiated by the New Guinean clearing of forested valley areas by burning for selective promotion of food plants (Haberle, 1998; Haberle *et al.*, 1991; Hope and Golson, 1995; J. P. White *et al.*, 1970).

4.7 The Americas

We suggest that climate has played an important role in the history of human migration across all regions of the world. In Chapters 5 and 6 we will investigate how climate and vegetation has changed in all the regions discussed over the last glacial cycle. In Chapters 7 and 8 we will delve more deeply into how climate has influenced human behaviour. For now, however, because of our greater research background in the Americas, we have chosen to use this section to explore climate as a key consideration in modern human migration.

In 1926, near Folsom, New Mexico, archaeologists uncovered beautifully fashioned biface spearpoints embedded in the skeletons of extinct bison. Within ten years, similar bifaced points with fluted bases were found near Clovis, New Mexico, buried underneath layers of dirt containing Folsom points (Cotter, 1937; Sellards, 1952). Clovis points and other stemmed fishtail points, which some researchers consider to be Clovis derived, have been subsequently discovered across North America, in Central America, and as far south as the southern tip of South America. These tools were evidence that early peoples inhabited the vast wilderness of the New World contemporaneously with extinct megafauna including mammoth, mastodons, horses, short-faced bear, dire wolf and American lion (Dyke, 2004a; Dyke *et al.*, 2004). By the late 1950s, radiocarbon dating of the Clovis points suggested that they were as old as ~11 300 years BP (~13 200 years ago) (Haury *et al.*, 1959; Haynes, 1969, 1992), although more recently the earliest site – the Aubrey site in Texas – has been dated to 11 500 years BP (~13 400 years ago) (Ferring, 1994; Meltzer, 2002). This led Haynes (1969) to suggest a theory for the peopling of the Americas – now termed the 'Clovis first hypothesis' – in which the first inhabitants of the continent were big-game hunters who crossed the Beringian landbridge no earlier than ~12 000 years BP, when global sea level was drawn down by massive glaciers during the last ice age, resulting in the connection of north-east Asia to North America. Later, C. G. Turner (1983) and Greenburg *et al.* (1986) suggested a three-migration hypothesis from Siberia with the oldest migration being the

Paleoindians (Clovis), the second being the Beringian Paleo-Arctic people, the ancestors of the Na-Dené, and the third the south Beringian coastal proto Eskimo-Aleuts. Anthropological studies of fossil teeth found that both prehistoric and living Native Americans possess the sinodont dental pattern, which is elsewhere evident only in north-east Asia. According to Turner, the first Americans originated in China about 20 000 years ago and subsequently migrated from Siberia. Those from the east Siberian coast became the Eskimo-Aleut speakers, those from a forest-adapted culture in Siberia were the Na-Dené speakers, and a steppe-adapted group migrated south to escape the forests with their Clovis culture and were the ancestors to all other Native Americans. According to Turner, the three migrations were stimulated by shifts in vegetation. However, Matthews (1982) found that the timing of the shift from open to taiga in interior Alaska occurred at 10 000 years ago, not between 16 000 and 12 000 years ago as suggested by Turner, limiting acceptance of the three-migration hypothesis. Further, the dental pattern of the 8400 years BP (9500 years ago) Kennewick Man, found along the Columbia River in Washington is sundadont (Chatters, 2000; Powell and Rose, 1999). Sundadonty occurs in Polynesia, South East Asia, Micronesia and among the Jomon-Ainu people of Japan and is more similar to the European dental pattern than is sinodonty. In addition, the cranial morphological features most closely resemble crania from individuals of South Pacific, Polynesian and Ainu/Jomon origin (Powell and Rose, 1999; Steele and Powell, 2002; C. G. Turner, 2002).

Any theory about the migration of early peoples into the Americas must take into account the times when potentially inhabited environments could viably support biotic communities, particularly human life. By the LGM, about 18 000 years BP (~ 21 500 years ago), immense ice sheets had expanded to their greatest extent across Russia, Europe and the Americas. In South America, glaciation was limited to high-altitude and high-latitude areas of the Andes. In North America, however, the Laurentide ice sheet covered the Canadian shield and spread west to the Rocky Mountains (Figure 4.3), where it coalesced with the Cordilleran ice sheet (Bobrowsky and Rutter, 1992; Jackson and Duk-Rodkin, 1996), which covered virtually all of British Columbia west of the Rocky Mountains. Ice stretched onto the Pacific and Atlantic continental shelves, south into the northern United States, and north into the southern half of the Yukon (Clague, 1989). Significant archaeological controversy exists as to the existence and timing of a passage – referred to as the 'ice-free corridor', between the two ice sheets: the Cordilleran to the west, and the Laurentide to the east – through which early 'Clovis first' hunters potentially migrated following the megafauna, and subsequently dispersed south of the ice sheets (Easton, 1992; Fladmark, 1979; Gruhn, 1994). Recent geological evidence implies that the corridor was closed between from 11 500 to 12 000 and 20 000 years BP (~13 500 and 24 000 years ago) (Dyke, 1996, 2004b; Jackson *et al.*, 1997;

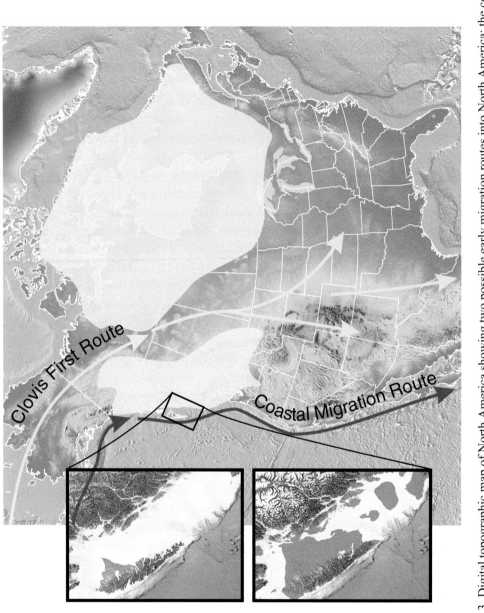

Figure 4.3 Digital topographic map of North America showing two possible early migration routes into North America: the coastal migration route (red) and the Clovis first route (yellow). The terrain and bathymetry inset maps show the Queen Charlotte Islands region off the coast of British Columbia, Canada as it appears today and as it likely appeared around 13 500 years ago after the last glacial maximum. Lowered eustatic sea level, combined with a forebulge that developed when ice loaded the mainland, formed a coastal plain that linked the islands to the mainland. This landbridge probably facilitated the migration of plants, animals and humans between the mainland and the islands. (See colour version in colour plate section.)

Lemmen *et al.*, 1994; J. M. White *et al.*, 1985). If people lived in North America prior to the LGM and occupied regions that were subsequently glaciated, evidence of their habitation, if it existed, probably would have been destroyed as glaciers advanced and then retreated.

Restricted glaciation in lower Central America, South America, the south and south-eastern United States, as well as the uplifted Beringian subcontinent likely facilitated the development of a relatively predictable resource base for early peoples (Dillehay, 1999). Beringia extended from north-eastern Siberia to the Mackenzie River in the Yukon of Canada. During the LGM, central and southern Beringia were milder than west or east Beringia and provided expanded and productive shrub and grassland habitat as vegetation coverage and productivity diminished in east and west Beringia. The exposed Beringian shelf probably served as a major refugium for plant and animal species (Guthrie, 2001; Hetherington *et al.*, 2008). Another glacial refugium likely existed on the uplifted coastal plains of the Queen Charlotte Islands region, British Columbia, Canada (Figure 4.3) (Hetherington, Barrie, Reid *et al.*, 2004; Hetherington and Reid, 2003; Hetherington *et al.*, 2003; Josenhans *et al.*, 1995; Mathewes, 1989). The unglaciated subaerial shelf may have facilitated the migration of faunal and floral species between the islands and the mainland until the early Holocene (Chinnappa, 1997; Hebda, 1997; Hetherington and Reid, 2003; Hetherington *et al.*, 2003; Nagorsen and Keddie, 2002; Ward *et al.*, 2003; Warner *et al.*, 1982). The unglaciated areas of the exposed Grand Banks of Newfoundland provided a glacial refugium on the Atlantic coast of Canada during the LGM (Shaw, 2005; Shaw *et al.*, 2006).

When temperatures began to warm, glaciers started to recede, and sea level began to rise. Previously exposed continental shelves, including the Bering landbridge, as well as uplifted regions along the glaciated north-west Pacific coast, and the expansive shelves along the Atlantic seaboard of both North and South America were flooded. Early people inhabiting formerly exposed continental shelves would have been forced to migrate as rising sea levels shifted coastlines more than 100 kilometres within the span of a few human lifetimes (Hetherington, Barrie, MacLeod *et al.*, 2004; Hetherington, Barrie, Reid *et al.*, 2004). The submergence of coastlines would have had a significant impact on the preservation of any early archaeological sites in these regions. However, due to the complex response of sea-level change to glacio-isostacy, some early coastal archaeological sites may yet be preserved (for example, see Hetherington *et al.*, 2003). Coastal submergence would have resulted in the migration of coastal peoples inland; recession of the giant ice sheets would have opened new regions for habitation in the interior of the continents and the changing climate would have induced shifts in vegetation biomes and faunal paleofaunistic zones.

However, the timing of habitation by biotic communities did not always coincide with deglaciation. In some areas, vegetation colonized stagnant ice; in other areas, extensive proglacial lakes precluded the establishment of biotic communities after glaciers melted (Driver, 1998). For example, the melting of the Laurentide ice sheet generated glacial Lake Agassiz, which extended over thousands of square kilometers flooding much of what is now Ontario, Manitoba, and parts of Saskatchewan in Canada, and south of the Great Lakes into the United States, making colonization of these areas difficult (Driver, 1998; Fenton *et al.*, 1983). The presence of ice and glacial Lake McConnell would have prevented humans and animals from moving east of the Mackenzie River between 11 500 and 9000 years BP, and long glacial lakes precluded colonization of many valley systems in British Columbia (Sawicki and Smith, 1992; D. G. Smith, 1995). Glacial Lake Missoula, an enormous meltwater catchment in a region now containing parts of Montana, Idaho and Washington, formed the Channeled Scablands every time it burst, estimated at more than 30 times.

According to Martin's blitzkrieg theory (1973), when early peoples finally dispersed south of the ice sheets, they advanced rapidly at a speed of 16 km/year and caused a massive extinction of megafauna around 10 000 years BP (~11 700 years ago). But paleoparasitalogical findings and palaeoclimate modelling simulations indicate that in order to reach archaeological sites where pre-Columbia American hookworms have been found, migrants carrying human intestinal hookworms would have had to migrate through the interior of Beringia at rates far exceeding those surmised by archaeologists (Montenegro, Araujo *et al.*, 2006). Further, the megafaunal extinction, which is evident in the fossil record, is coincident with the change from Clovis to the non-fluted Folsom and side- and corner-notched stone tool technology (Ellis *et al.*, 1998). Clovis points disappeared sometime after 10 900 years BP (~12 800 years ago) in western and central North America and by 10 600 years BP (~12 600 years ago) in eastern North America (Meltzer, 2002). Puzzling however, is why so few species (mammoth, mastodon, horse and possibly camel) appear in Palaeoindian kill sites, to the exclusion of the balance of the 30 or so genera that went extinct. Also perplexing is why some species that were hunted did not go extinct (e.g., *Bison* sp., caribou). MacPhee and Marx (1997) propose another theory for the prehistoric American megafaunal extinctions. They suggest it was caused by deadly microbes carried by unwitting early people who had developed genetic immunity. Alternatively other scientists argue for a climatic explanation. The sudden late Pleistocene extinction is coincident with a Younger Dryas cooling (Graham, 1998). However, according to Dyke *et al.* (2004), the faunal record appears to be equally abundant in the Younger Dryas interval between 10 900 and 10 000 years BP (~12 800 and 11 700 years ago) than it is in the preceding interval between 11 900 and 11 000 years BP (~13 900 and 12 900 years ago) (see also Dyke, 2004a). Further, the fossils are not clustered early in the cool interval near 10 900 years BP (~12 800 years ago). In addition, vegetation history and climate simulation (see Chapters 6 and 7) reconstructions do not clearly show a cooling event on the grasslands that resulted in shrinking the available grazing habitat for the megafauna. What is more apparent is the coincidence of the megafauna extinction with the end of the Younger Dryas interval – the rapid onset of climate warming. But in southern South America, at the location of the Monte Verde archaeological site, early inhabitants butchered mastodons at 12 500 years BP (~14 700 years ago), even though deglacial warming began around 14 000 years BP (~16 900 years ago). In Patagonia, hunters were killing now extinct horses at 12 800 years ago (Fiedel, 2000). The contrariety of evidence has led other scientists to consider whether it was the more extreme seasonal variations as opposed to the overall warming or cooling of temperatures experienced during the Holocene that induced the rapid extinction of the megafauna. We are likely to discover that it was a combined interaction between human impacts and climate change that affected faunal survival. Thus continued interdisciplinary research may prove the most elucidating avenue of research in the future.

Since the induction of the 'Clovis first' hypothesis, new evidence from early archaeological sites has revised early opinions on the peopling of the Americas. The critical Ushki Lake site in Siberia originally dated to 16 800 years BP (19 900 years ago) (Dikov, 1979) was considered to be the Clovis antecedent and the earliest archaeological site in Beringia. But subsequent redating of the site indicates that the average age of its earliest component is only 11 000 years BP (12 900 years ago) (Goebel *et al.*, 2003), making it centuries younger than the oldest Clovis points uncovered in the Americas and eliminating the only potential Clovis progenitor in Siberia. However, the Yana River site, 500 km above the Arctic Circle in northern Siberia, dates to 27 000 years BP (Pitulko *et al.*, 2004). And although it is not likely an ancestral Clovis site, its existence indicates that humans could survive in the cold Siberian climate and were at least available to migrate east to the Americas. The redating of the Ushki Lake site makes it younger than sites in the Americas including Cactus Hill in Virginia (McAvoy *et al.*, 2000; McAvoy and McAvoy, 1997), Meadowcroft in Pennsylvania (Adovasio *et al.*, 1998), Topper in South Carolina (Goodyear, 1999), Taima-Taima in Venezuela (Oschenius and Gruhn, 1979), Alero Tres Arroyos in Tierra del Fuego (Massone, 1996), Toca do Boqueirão da Pedra Furada in north-eastern Brazil (Guidon, 1986; Guidon and Delibras, 1986; Parenti, 1993, 1996, 2001; Pessis, 1996) and others that have associated dates equal to or older than Clovis.

Although many of these pre-Clovis sites lack precise dates, one in particular – Monte Verde, Chile – has received much interest and acceptance. Non-Clovis stone tools, animal skin dwellings, and a child's footprints were uncovered at this site dated to 12 500 years BP (~14 700 years go), and seaweed samples dated to about 12 300 years BP (~14 200 years ago) (Dillehay, 1989, 1997; Dillehay *et al.*, 2008).

Another astonishing find was the discovery of footprints, preserved in volcanic ash, of perhaps two adults and two to four children who walked barefoot along the shoreline of a lake that is now a reservoir called Lake Valsequillo near Pueblo, Mexico. However, their dating to about 40 000 years BP (44 700 years ago) is controversial (González *et al.*, 2006). Renne *et al.* (2005) have suggested a date of 1.3 Ma and question their categorization as human. Although the dating of some of these early sites may in the end prove erroneous, there are others dated to earlier than 11 500 years BP (~13 500 years ago) found south of glaciated regions, including the Monte Verde site, which are difficult to refute. Their existence implies that people had reached these unglaciated regions prior to or coincident with the opening of the 'ice-free corridor'.

Based on these findings, archaeologists propose a number of potential alternatives regarding the peopling of the Americas: (1) problems exist with the stratigraphic integrity and dating of pre-Clovis sites; (2) early North American and Beringian sites exist, but we have yet to find them or accept their early antiquity; and (3) initial migration into the Americas occurred earlier or via a different route(s) than terrestrial migration across Beringia and via the 'ice-free' corridor (for a discussion of coastal migration in the Americas see Surovell, 2003). These scenarios are in addition to numerous aboriginal oral traditions that depict an origin story in the New World (Echo-Hawk, 2000). However, few archaeologists, even multiregional hypothesis advocates, have delved into this arena. Even within the domain of recovered archaeological finds in the Americas, there is disagreement over the age of pre-Clovis archaeological sites. However, should more early sites be uncovered and validated with repeated independent dating, it will become increasingly difficult to dismiss them.

There is no question that there is a severe lack of early North American and Beringian sites. One archaeological find particularly relevant to this discussion are the Bluefish Caves in northern Yukon where, according to Cinq-Mars and others, pre-Holocene evidence exists of human occupation of eastern Beringia, dating to 13 500 years BP (~16 100 years ago) and earlier (Cinq-Mars, 1979; Cinq-Mars and Morlan, 1999; Burke and Cinq-Mars, 1998). Late Pleistocene horse bones and teeth, mammoth, caribou and other mammal bones, as well as birds and fish, have been found in three caves along with stone and bone tools. If the dates on the Bluefish Caves site and other early sites north of the ice sheets are valid, then, in order for the hypothesis of a terrestrial migration across Beringia to be maintained, people must have migrated across Beringia earlier than the Clovis first model would suggest (see Meltzer, 2002). Yet even if they are refuted and the dates obtained from sites south of the ice sheets are valid (e.g., Monte Verde in South America), then people migrating across the Bering landbridge must have come before the Laurentide and Cordilleran ice sheets had coalesced at 20 000 years BP (~24 000 years ago) (Dyke,

1996; Jackson *et al.*, 1997; Lemmen *et al.*, 1994; J. M. White *et al.*, 1985), or they arrived in the Americas a different way. For example, early people may have travelled along a coastal route from north-east Asia moving from north to south along the Pacific coast of the Americas, or directly to South America or other regions south of the ice sheets via transoceanic voyages.

The coastal migration route from north-east Asia and down the west coast of North and South America was first suggested by Heusser (1960), Kreiger (1961) and Macgowan and Hester (1962). Fladmark (1979) later did extensive analysis of the palaeoenvironmental and archaeological data pertaining to this route lending it further support. Fladmark attributed the lack of archaeological evidence for early settlement along the coast to inundation of the coastline caused by rising sea level subsequent to the LGM. However, if recent reconstructions are accurate, there exist some late Pleistocene–early Holocene coastlines that are not drowned and which may provide evidence of early archaeological sites (Hetherington *et al.*, 2003). Subsequent to Fladmark's work, R. A. Rogers (1985a, 1985b; Rogers *et al.*, 1990) found that glacial ice acted as an isolating agent during the last ice age and that both glacial ice and coincident biogeographic zones were important in influencing linguistic distribution and divergence. Rogers found that unglaciated areas to the south of the ice sheets have substantially greater aboriginal language diversity – indicated by highly ramified phyla[3] and concentration of language isolates – than do unglaciated regions to the north of the ice sheets. Further, the greatest aboriginal linguistic diversity is found along the Pacific north-west coast of North America. Elaborating on Roger's work, Gruhn (1988, 1994) discovered that the greatest diversity of aboriginal language distributions occur along the Pacific north-west coast of North America, where six of seven language groups are considered language isolates; California, where 64 languages exist, of which 18 are isolates; Middle America, where the number of languages ranges from about 200 to 350; South America, where approximately 1500 native languages exist; the northern coast of the Gulf of Mexico, where there are seven language isolates on the northern coastal plain and five farther east. These findings compare with the eastern and central interior of North America where only one language isolate is evident. Based on the assumption that language diversification is positively correlated with the time depth of occupation of an area, Gruhn suggests that humans have occupied most of the Americas for at least 35 000 years, with a longer history in South and Central America and along the Gulf and Pacific coasts. This evidence, particularly the greater time depth of west-coast populations compared with interior populations, supports a Pacific coastal-entry model.

Although no early archaeological evidence exists conclusively supporting either the Clovis first or the coastal migration hypothesis, some discoveries along the Americas' Pacific coast are intriguing. A human mandible unearthed from a cave on Prince of

Wales Island, Alaska, has been reported by Dixon (2001) and dated to 9800 years
BP (11 200 years ago). Human bones discovered by Orr in 1962 at Arlington
Springs, Santa Rosa Island off the southern coast of California have been subsequently dated to 11 500 years BP (13 400 years ago) (cited in Dixon, 1999, p. 129).
These are now considered to be some of the oldest reliably dated human remains in
the Americas. The finds at the Arlington Springs' site also imply the earliest use of
watercraft in North America; the island was not connected to the mainland even
during lowered sea levels during the last ice age (Dixon, 1999; Erlandson, 1994;
Moratto, 1984). Also evident at the Prince of Wales Island, Alaska site are numerous
mammals including a 41 000 years BP (45 600 years ago) black bear found by
Heaton *et al.* (1996). To the south, in a cave on Vancouver Island, Canada, 12 000
years BP (13 800 years ago) mountain goat fossils imply sufficient resources were
available for their survival on adjacent unglaciated exposed coastal plains
(Nagorsen and Keddie, 2002). The discovery of parkland terrestrial fauna dating
to between 18 000 and 16 000 years BP (21 500 to 19 200 years ago) at Port Eliza
Cave on Vancouver Island further supports the presence of a glacial refugium in the
region during the last ice age (Ward *et al.*, 2003). Research along Canada's Pacific
margin indicates the availability of food-mollusc species from at least 13 200 years
BP (15 700 years ago), ice-free terrain from 13 800 years BP (16 600 years ago) and
a landbridge connecting the mainland with the islands (Hetherington and Reid,
2003; Hetherington, Barrie, MacLeod *et al.*, 2004; Hetherington, Barrie, Reid *et al.*,
2004). Much farther south, the coastal archaeological site at Quebrada Tacahuay,
Peru, contains abundant processed marine fauna, including birds and fish, and dates
to between 10 800 and 10 500 years BP (12 700 and 12 500 years ago) (Keefer *et al.*,
1998). These exciting finds provide added support for the feasibility of a coastal
migration route, yet they still do not prove its existence.

4.7.1 Other routes

Other potential migration routes into the Americas include the Atlantic route from
Europe to eastern North America by boat, either south of the polar ice that covered
Iceland, Greenland, and all but the most southerly parts of Ireland and England, or a
direct route across the Atlantic from the Iberian Peninsula. Archaeologists Stanford
and Bradley (2000, 2002; see also Holden, 1999) suggest that similarities between
the Clovis complex artifacts in North America and those of the Solutrean period
dating to between 22 000 and 16 500 years BP (~26 500 and 19 500 years ago) of
prehistoric Europe maritime hunters and fishers support the North Atlantic migration route. Anatomical similarities between some human skeletons found in
America and modern Europeans (e.g., Kennewick Man) have increased the profile
of this potential route. Other archaeologists, including Straus (2000b; see also

Straus *et al.*, 2005) disagree, citing the lack of fluting on Solutrean points and the fact that Solutrean sites contain many stone and bone tools not found in the Americas. Further, the 5000-radio carbon-year gap between the end of the Solutrean period ~16 500 years BP (~19 500 years ago) and the earliest dated Clovis site ~11 500 years BP (~13 500 years ago) has led naysayers to suggest that the artifactual similarity between Clovis and the Solutrean is due to cultural convergence. An experiment was done by Montenegro, Hetherington *et al.* (2006) simulating prehistoric transoceanic crossings into the Americas from disparate locations including central Europe. They found that the probability of a successful crossing was at or below 1% when paddling from central Europe in under 180 days. Crossings from Europe via Iceland and Greenland had greater probabilites of occurrence and were much faster.

New DNA, craniometric, climate modelling and other research findings suggest possible contact or colonization of the Americas from Asia, Australia, Polynesia, Europe and Africa. Palaeoanthropologists typically use craniometrics to make interpretations about shared genetics, where similar craniofacial characteristics imply a shared common ancestry and genetic relationship. Comparisons of crania indicate a wide variation in craniofacial features between the Palaeoindian – the Americas' first settlers, and the late and modern Native Americans – populations dating from about 7000 years ago to modern times. For example, based on the pattern of human craniofacial affinities, Neves *et al.* (1999a) first suggested a double migratory event: the first by South East Asians during preglacial times, the second by Mongoloids who originated in north-east Asia and from whom Amerindians derived. More recently these findings were supplemented with additional evidence from Lagoa Santa, Brazil, suggesting that late and modern Native Americans resemble north-eastern Asians, whereas the earlier Palaeoindian crania resemble Australo-Melanesians and Africans (Neves *et al.*, 2003; see also Lahr, 1995).[4] The Hominid-1 adult female from Lapa Vermelha IV, the earliest known female skeleton from South America, considered to be about 12 000 years old, is strikingly similar to the Dogo, Yeit and Zulu of Africa (Neves *et al.*, 1999c; Powell and Neves, 1999; Steele and Powell, 2002). But others disagree, pointing to the limitations of the research sample – most were based on single skulls or poorly dated material (Dillehay, 1989, 1997; Roosevelt *et al.*, 1996; Van Vark *et al.*, 2003; but for a response see Neves *et al.*, 2003). However, more recently evidence of an 8860 years BP (~10 000 years ago) skeleton found in Capelinha, Brazil, shows a clear associa-tion with the Palaeoindians and non-Mongoloid groups and possesses a typical Australo-Melanesian-like morphology (Neves *et al.*, 2005). According to Neves and others, these findings are not restricted to Brazil, but are evident in southern Patagonia (Neves *et al.*, 1999b), the Columbia Highlands (Munford *et al.*, 1995; Neves and Pucciarelli, 1989), Mexico (González-José *et al.*, 2002) and Florida

(Powell *et al.*, 1999). Using craniometric evidence González-José *et al.* (2003) show an affinity exists between a modern Amerindian group situated on the Baja California Peninsula in Mexico and early Holocene human fossil remains. The isolated Baja California inhabitants potentially suffered genetic isolation, resulting in the temporal continuity of their craniofacial morphology which is different from later Amerindian groups inhabiting mainland areas. The authors suggest the best explanation for this divergence in morphology is that Baja California Amerindians are the descendants of the original Palaeoamericans, who came from South Asia and the Pacific Rim and subsequently became isolated due to increased aridity in the region. More recently, comparative analyses of late Pleistocene–early Holocene human skulls from the Americas correlate prehistoric, recent and present Native Americans with northern Asians, while South American Palaeoindians are correlated more closely with Australo-Melanesians' groups (Neves and Hubbe, 2005). Lahr (1995) found that when comparing Eskimo populations near the Arctic Circle with Tierra del Fuego hunter–gatherers living near the Antarctic Circle, the Tierra del Fuegans resembled more closely southern Pacific Islanders, leading her to suggest that any morphological changes resulting from environmentally determined natural selection were of a superficial nature. Yet there are others who feel that the divergence in craniofacial morphology observed in early North and South American skeletal samples is a result of genetic drift and gene flow (see discussion in Steele and Powell, 2002). It is apparent that these differences in opinion have yet to be resolved. However, as more thoroughly addressed in Appendix B (Developmental evolution), because of the limitations of morphological analyses and interpretations, it would benefit us to review research findings in related fields.

4.7.2 Genetic evidence

According to a recent review by Schurr (2004), mtDNA and Y-chromosome research supports an initial entry into the Americas between 20 000 and 15 000 years ago, during or subsequent to the LGM. The early colonizing mothers of the Americas brought with them mtDNA haplogroups A, B – which is evident elsewhere only in East and South East Asia – C, D and potentially X – which is Eurasian, or possibly European and is absent in virtually all aboriginal Siberian populations with the possible exception of Altaian groups (Derenko *et al.*, 2000, 2001; Schurr, 2003; Schurr *et al.*, 2000, 2004; Schurr and Wallace, 2003; Starikovskaya *et al.*, 1998; Torroni *et al.*, 1993b cited in Schurr, 2004, p. 563). However, not all aboriginals possess the same mtDNA haplotypes. Amerindians possess a different set of haplogroups than do Na-Dené and Eskimo-Aleut who possess mostly haplogroups A and D, potentially lacking B and C (Rubicz *et al.*, 2003; Saillard *et al.*, 2000; Shields *et al.*, 1993; Starikovskaya *et al.*, 1998; Torroni *et al.*, 1992, 1993a; Ward *et al.*, 2003 cited in Schurr, 2004, pp. 556–7), and thus may represent a different migration into the Americas than that of the Palaeoindians. In the New World, haplogroup A decreases in frequency from north to south (Schurr, 2004). Haplogroup B is virtually absent in northern North America and is not clearly evident in the Andean Indians (Bonatto *et al.*, 1996 cited in Schurr, 2004 p. 565; Lorenz and Smith, 1996, 1997; Schurr *et al.*, 1990; Torroni *et al.*, 1992, 1993a, 1994a, 1994b cited in Schurr, 2004, p. 556). Haplogroups C and D increase in frequency from north to south (Schurr, 2004), and haplogroup X is virtually exclusively found in North America amongst only North American Amerindian populations (Bolnick, 2004; Bolnick and Smith, 2003; Brown *et al.*, 1998; Huoponen *et al.*, 1997; Malhi *et al.*, 2001, 2003; Scozzari *et al.*, 1997; D. G. Smith *et al.*, 1999; Torroni *et al.*, 1992, 1993a cited in Schurr, 2004, p. 565). Central American populations possess virtually only mtDNA haplogroups A and B (Batista *et al.*, 1995; Gonzalez-Oliver *et al.*, 2004; Kolman *et al.*, 1996; Kolman and Bermingham, 1997; Santos *et al.*, 1994; Torroni *et al.*, 1993a, 1994b cited in Schurr, 2004, p. 556).

The early colonizing fathers of the Americas brought with them their non-recombining Y-chromosomes including Q-M3, R1a1-M17, P-M45, F-M89 and C-M130. Q-M3 and P-M45 are widely distributed in the Native American populations and represent the majority of Y-chromosomes. Q-M3 increases in frequency from north to south (Bianchi *et al.*, 1998; Karafet *et al.*, 1997, 1999; Lell *et al.*, 1997, 2002; Santos *et al.*, 1999; Underhill *et al.*, 1996, 1997 cited in Schurr, 2004, p. 557). P-M45a is the ancestral haplogroup to Q-M3 and is found in populations from central Siberia to South America (Bortolini *et al.*, 2003; Lell *et al.*, 2002 cited in Schurr, 2004, p. 558). A subsequent P-M45b haplogroup is evident in eastern Siberia and in North and Central America, but is absent in South America and central Siberia and may represent a second later expansion from Beringia (Lell *et al.*, 2002 cited in Schurr, 2004, p. 558; Schurr, 2004). The remaining non-recombining Y-chromosome haplotypes comprise 5% of the Native American Y-chromosomes and appear in isolated regions. For example, C-M130 is apparent in the Tanana, Navajo, Chipewayans and the Cheyenne (Bergen *et al.*, 1999; Bortolini *et al.*, 2003; Karafet *et al.*, 1999; Lell *et al.*, 2002 cited in Schurr, 2004, p. 558). The R1a1-M17 haplotype is found only in the Guayami tribe from Costa Rica. This haplotype is not common in Siberia, emerging there only towards the end of the LGM around 13 800 years ago, after the Americas had been settled, suggesting its presence in the New World was a consequence of a subsequent expansion out of Asia (Lell *et al.*, 2002 cited in Schurr, 2004, p. 558).

It is interesting to observe that mtDNA haplogroup C and D and Y-chromosome haplotype Q-M3 increase in frequency from north to south, and mtDNA haplogroup B is absent in the north. Alternatively, mtDNA haplogroup A and Y-chromosome

P-M45 are more predominant in the north and X is found only in North America. These disparate distributions are interpreted to reflect different patterns of original settlement as well as subsequent genetic differentiation. In 1998 Brown *et al.* (cited in Schurr, 2004, p. 558) estimated the age of mtDNA haplogroups and used their findings to suggest that haplogroup B and probably X were brought to the Americas after haplogroups A, C and D. But subsequent research suggested this was not the case. More recently, D. G. Smith *et al.* (2000 cited in Schurr, 2004, p. 560) observed that only haplotypes B, C, D and X were found in the DNA of ancient skeletal material from North America, suggesting that haplogroup A arrived later than the other mtDNA haplogroups. Although not all scientists agree with this interpretation, these findings lend support to the possibility of separate migrations to the Americas. According to Y-chromosome research, the New World Adam lived roughly 22 500 years ago, and his ancestors lived in the Yenissey River basin and Altai Mountains of Siberia (Hurtado de Mendoza and Braginski, 1999). He was followed to the Americas by a second, less-common, rival Adam whose Y-chromosomes are found more commonly in North as opposed to South American natives, suggesting more than one migration wave to the Americas.

According to Hey (2005), it took a group of fewer than 80 individuals to found the New World. One of those individuals may be related to the ancient human fossil found on Prince of Wales Island, Alaska. DNA was recently extracted from teeth obtained from the 10 300 years BP (12 100 years ago; marine reservoir corrected to 9800 years BP) human mandible (Dixon, 2001). Analysis of the mtDNA and Y-chromosomes indicates the individual likely originated in Asia and belongs to haplogroup D and haplogroup Q-M3, respectively. Both of these haplogroups are found to increase in frequency from north to south. Kemp *et al.* (2007) and others (Dalton, 2005) suggest the specimen's haplotype is a Native American founding lineage and that it is linked with populations along the western edge of North and South America, including the Cayapa coastal tribe in Equador and the Chumash tribe in California, as well as populations in Brazil, Mexico and the south-eastern United States. Their findings also indicate faster mutation rates, almost four times higher than previous estimates, and point out the need for a calibration correction for molecular rates of evolution. Bianchi *et al.* (1998 cited in Kemp *et al.*, 2007, p.618) imply that previous estimates of the emergence of Y-chromosome haplogroups predating 20 000 years BP (about 23 900 years ago) overestimate the timing of arrival of the first peoples in the Americas. Their findings also support a coastal migration route. The Chumash, who inhabited the southern California coast, are also the people who possessed the tomol or sewn-plank canoe, which is only found on the coast of Chile and among the Pacific Islanders. Questions remain as to whether the tomol was invented by the Chumash or was a Polynesian influence, implying a potential Polynesian–Californian connection (Edgar, 2005).

Arnold (1995 cited in Davis, 2001, p. 61) argued that large canoes facilitated transportation between the Chumash people from the Channel Islands, California and the Nootka of coastal Canada and facilitated the development of trade, social, political and economic complexity. The Zuni people of New Mexico, who possess morphological and cultural similarities with the Japanese, are linked to the people of southern California by a system of trails, trade items and religious ideas. According to Davis (2001), the combination of cultural finds with an Asian influence discovered along the coasts of the Americas continues to accumulate, leading to speculation of an Asian–American transoceanic connection. For example, Chinese coins have been found along the coast of the Americas. One, more than 2000 years old, was found in a pottery jar by the Tlingit Indians near the village of Wrangell, Alaska. Chinese coins are also frequently found on shaman's masks, further suggesting prehistoric trade or contact (Fitzhugh, 1988 cited in Davis, 2001, pp. 76–7). Iron chisels and tropical bamboo have been found at the Ozette site in Washington, where five longhouses were buried by a sudden mudslide 500 years ago. Pottery evidence in South America antedates that found in North America. Pottery that dates to pre-Columbian contact has been found near the Columbia River. Ceramic figurines appear to incorporate Japanese manufacturing methods including kiln firing, yet the Native Americans now living in the area do not have a tradition of pottery making (Davis, 2001, p. 77; Stenger, 1991). According to Meggers (1971, 1998a, 1998b) and her colleagues (Meggers *et al.*, 1965; Estrada and Meggers, 1961; Imamura, 1996; see also discussion in Davis, 2001, pp. 92–3), ceramic pottery excavated at Valdivia, Ecuador, is related to the oldest known ceramics in the world, the Jomon – 'cord-marked' – pottery of Japan, which dates to 12 000 years ago, providing evidence of trans-Pacific voyages from Asia to America since 6000 years BP (6800 years ago). Iron, which was sought after as a highly prized trade item, was already present in North America when the Europeans first arrived on the west coast (Davis, 2001, p. 79). Divers off the south coast of California have discovered large stones, weighing as much as one thousand pounds (454 kilograms), with holes drilled through their centres. Some people think they are Chinese anchors, with the original stone having been quarried from a source in southern China (Pierson, 1979; Pierson and Moriarty, 1980).

Dispersion patterns of plants and animals to and from the Americas may also provide insights into human migration. American loggerhead turtles (*Caretta caretta*) possess the same mitochondrial DNA as do those in Japan (Bowen *et al.*, 1995 cited in Davis, 2001, pp. 65–6). Further, their young hatch on Japanese shores and then swim 10 000 km across the North Pacific before heading down to the coast of Baja California where they grow up. They then make the return journey to Japan to breed and hatch the next generation of turtles. Human skeletons from La Jolla, California, an archaeological site in the same coastal vicinity as the loggerhead turtles, possess remarkable similarities to prehistoric Japanese skeletons, and differ from other North American aboriginal skeletons (Davis, 2001, pp. 113–4; S. L. Rogers, 1963; Ushijima, 1954), making us ponder whether it was not just the turtles that made the cross-oceanic passage.

The bottle gourd, coconut, cotton, sweet potato and corn are plant species that were potentially distributed in pre-Columbian times. The bottle gourd and coconut may have floated by natural means from Africa and the southern Pacific to the Americas. But the American sweet potato and corn generate more lively debate. Although sweet potato seed capsules may have floated on the water, the transportation of corn (maize) would require human assistance. Maize was domesticated in Meso-America and introduced into Europe, probably by the Spanish, in the sixteenth century. However, its appearance on eleventh-century sculptures in India is far more puzzling (Davis, 2001, pp. 69–70; Johannessen, 1998 cited in Davis, 2001, p. 70).

4.7.3 A different perspective

In summary, we know that high-frequency human genetic haplotypes were present south of the ice sheets prior to deglaciation of the ice-free corridor between about 13 500 and 13 000 years ago and subsequently dispersed northward. Thus, it is possible that an early human migration reached the southern regions of the Americas prior to or during the LGM. These people may then have migrated northward subsequent to the LGM. They may have followed the dispersion of mammals as vegetation shifted with the changing climate during and after the last ice age. Subsequent to the LGM, mammal dispersion in North America occurred from areas in the south, where climate was conducive to their survival, into more northerly areas as ice melted and warming regenerated conditions suitable to their survival (see Dyke, 2004a; Dyke *et al.*, 2004). Coincidentally, it is interesting to note that the oldest Clovis lithic culture occurs in the south-east and progressively gets younger to the north and west (Meltzer, 2002). It is probable that as mammal populations moved from the south and south-east to the north as climate ameliorated subsequent to the LGM at 21 500 years ago, so too did humans.

This early migration may have been followed by a subsequent migration flowing from the north to the south, culminating in the last population expansion into the Americas of the Eskimo-Aleuts and the Na-Dené people subsequent to the LGM. These findings imply that earlier and/or alternative migration routes other than the Clovis first route across Beringia and down through the ice-free corridor require further consideration.

The lack of early archaeological sites in the north, the variable presence of recoverable DNA, and the existence of potential pre-Columbian transoceanic connections, lead the authors to consider some interesting migration scenarios other than the Clovis first hypothesis. One possibility is that early people migrated into northern North America across the Bering landbridge during a period of lowered sea levels prior to the LGM and moved south between the ice sheets before the ice sheets coalesced *c.*18 000 radiocarbon years BP (Dyke 2004b) (21 500 years ago), settling in South America and in some cases never returning north. According to this scenario, all potential early northern migrants would have to have moved south or died, and evidence of their presence is well hidden or obliterated. To support the DNA findings, any subsequent northward movement of these people would have been slow, and was overwhelmed by subsequent incoming migrants across the Bering landbridge just before its inundation around 11 000 radiocarbon years BP (~12 900 years ago).

A second potential migration scenario would include initial colonization of southern areas prior to northern areas. This could be accomplished via waterborne migration, either rapidly along the coast, for instance from north-east Asia and down the North American Pacific coast to South America during the last glacial period, or alternatively, at any time prior to the LGM from other islands or continents. This last alternative would require a transoceanic crossing as opposed to coastal migration.

Climate modelling has recently been used to ascertain the feasibility of potential prehistoric transoceanic routes to the Americas by providing temporal and spatial climatic constraints (Montenegro, Hetherington *et al.*, 2006). Potential crossings include those from Japan to North America in 83 days or from northern Africa to South America in 91 days. The probability of success on the Japan route is up to 8%, while that on some African routes exceeds 13%. Other potential routes occurred from northern Africa to northern South America and the Caribbean, from Scandinavia to Iceland and Greenland, and in the Pacific from the Kamchatka Peninsula to North America and Alaska, as well as from Australia to New Zealand, and New Zealand and Australia to South America. Of lesser potential success were routes from Iberia to the Caribbean, South America, and southern North America, and from central Europe to the Americas. Modelling did not provide any evidence of potential direct transoceanic Pacific equatorial crossings from Asia to the Americas. Further, it is obvious that any waterborne route is only feasible if watercraft building technology and materials were present in the migrating populations, and the migrating populations were there to migrate. The authors note that the Pacific Islands and New Zealand were some of the last locations on Earth to be inhabited by people. Further, the failure of initial Australian immigrants to occupy New Zealand until recently may indicate that these people lost their boating technology after occupying Australia and were subsequently unable to migrate either to New Zealand or South America. As well, the failure of Africans to occupy Madagascar until only very recently raises questions about the pervasiveness of boating technology among East Africans. Thus, although researchers are aware of the use of boating technology by some early peoples, we are far from developing a clear understanding of its origin, diffusion, and use as a mechanism of dispersion for early peoples into the Americas. Despite these limitations, modelling of prehistoric transoceanic crossings into the Americas by Montenegro, Hetherington and others identifies potential transoceanic crossings from north-east Asia and Africa to the Americas, and provides parameters and provocative insights into the feasibility of coastal and transoceanic migration.

4.8 Islands of the Pacific

The Polynesian islands were one of the last regions on Earth peopled, having been colonized only in the last 3000 years. New Caledonia was colonized by 3100 years ago, Samoa by between 3000 and 2800 years ago, the Marquesas by 1200 years ago, Hawaii and Easter Island by 1100 years ago, Pitcairn Island by between 1000 and 950 years ago, New Zealand by 800 years ago and the Chatham Islands by 700 years ago (Bird *et al.*, 2004). Our understanding of the routes and impetuses stimulating this final migration into the islands of the Pacific remains limited and disjunct. Bellwood (1989) concludes from the archaeological record that Polynesian ancestors originated in Taiwan and South East Asia but arrived via New Guinea. Non-recombining Y-chromosome evidence shows Polynesian lineages derive from New Guinean populations, supporting genetic as well as linguistic evidence suggesting at least a stopover in Melanesia, where migrants spent sufficient time to accumulate a genetic fingerprint (Kayser *et al.*, 2000; Underhill *et al.*, 2001). This rapid expansion may have been propogated by a burgeoning population of agriculturalists at the expense of hunter–gatherers (Bellwood, 1996; Bird *et al.*, 2004). Alternatively, rising sea levels during the Holocene flooded Sundaland, forcing people to disperse both east and west prior to 6000 years ago (Oppenheimer, 1998). Irrespective of which hypothesis is more valid, each illustrates a close connection between a changing environment and human societies and the importance of understanding their interrelationship.

Most archaeologists agree that early Oceanic settlers were competent mariners and navigators who purposefully explored the Pacific (Irwin, 1992). Using a computer simulation, University of Victoria graduate student Chris Avis and others (Avis *et al.*, 2007) showed that both drift and downwind sailing could account for all the major crossings in the Lapita region, which extends from Near Oceania to Samoa. During El Niño Southern Oscillation (ENSO) conditions, surface easterlies are reversed, and this would have facilitated the rapid discovery of islands by drifting or downwind sailing. These anomalous winds would allow explorers to proceed against the typically prevailing trade winds. This means that changing El Niño conditions would have had a major influence on colonization of the islands in the Pacific.

4.9 Concluding thoughts

Although some animal populations increase their territory at the fringes simply through increasing numbers, in this chapter we have stressed mobility and migration as a feature of human social behaviour – characteristics compatible with the out-of-Africa hypothesis. An intelligent species with itchy feet need not remain *in situ*, simply relying upon its physiological adaptability to deal with climatic change or increasing competition for resources. It can go looking for better living conditions, make choices about trade-offs between eating and freezing, and contemplate the value of tolerating discomfort in the short term for overall gain in the longer term.

In conclusion, there exist a number of key environmental factors that have influenced hominin dispersal:

1. Mountains, rivers, deserts, glaciers and glacial lakes can act as geographic barriers.
2. Lowered sea level can generate expansive coastal plains rich in resources during glacial intervals.
3. Hominin dispersal was likely influenced by climate change and faunal and floral dispersal associated with shifts in terrestrial biogeography. Such dispersals would have required less behavioural and potential morphological adjustment than would have hominin dispersals beyond accustomed habitats.
4. Faunal and potentially hominin dispersal in Eurasia was probably from south to north and from east to west during warm (interglacial) intervals, and from north to south during cold (glacial) intervals. This movement potentially resulted in greater population dispersal during warm intervals than cold intervals when populations likely concentrated in glacial refugia and warmer southern regions.
5. Genetic exchange between populations probably increased during glacial intervals when populations likely congregated in productive refugia and especially when landbridges linked regions unconnected during intervals of higher sea level (Antón, 2002; Keates, 2004).
6. The genetic makeup of separate populations is probably the result of genetic drift, inbreeding, genetic bottlenecks and geographical isolation. A particular genotype is not necessarily more adapted to a particular environment than any other, with some exceptions relating to physiology. On the other hand, social intercourse that results from migration and communication, and consequent interbreeding, combine to make a whole greater than the sum of its parts and provide the potential for behavioural and technical innovation.
7. Physiological change or behavioural change, particularly migration or dispersal, is a more reasonable response to rapid climate change than slow gradual genetic change associated with natural selection. As stated in Pease *et al.* (1989), 'migration may often be more important than selection in allowing a population to respond to changing environmental conditions' (p. 1662).
8. Individuals living in regions where the quality of resources is patchy, for example in high-latitude regions, are more likely to disperse over greater distances and be somewhat

less affected by barriers to migration (Chown and Gaston, 2000). Further, species occupying large territories frequently exhibit migratory behaviour and exhibit high levels of gene flow (Schwartz *et al.*, 2002, p. 521).

9. Fresh water was likely a key influencing factor in hominins' ability to adapt to particular locations and in dispersal routes taken. The location of fresh water and the level of precipitation would have changed through time in response to a changing climate (Keates, 2004).

We now look more closely at the changing climate of the last glacial cycle, which we suggest has played a key role in early human history and in the development of human adaptability.

Notes

1. Much of the early genetic work was done by Allan C. Wilson as well as R. L. Cann and M. Stoneking. See, for example, Cann *et al.* (1987).
2. For a list of hominid fossil remains from China see Etler (1996).
3. This is a specialized use of the biological term phylum to indicate, in this case, a major group of related languages.
4. For morphological comparisons of Kennewick Man with the South Pacific Islanders, Polynesians, and the Ainu of Japan see Powell and Rose (1999); Steele and Powell (1992, 1994); Jantz and Owsley (1997).

Part II

Climate during the last glacial cycle

5

Climate change over the last 135 000 years

5.1 Introduction

Current archaeological evidence indicates that modern humans, *Homo sapiens*, appeared in Africa by 200 000 years ago (McDougall *et al.*, 2005; Shea *et al.*, 2004). Their arrival occurred just prior to a very cold interval known as marine isotope stage 6 (MIS6), when ice sheets expanded across much of Europe, northern North America, and parts of Asia and southern South America. *H. sapiens'* migration out of Africa occurred during the last glacial cycle (LGC), an oscillating climate interval that saw the Earth move from the warmth of interstadial environments to the extreme cold of the last ice age. The highly variable climate of the LGC was followed by the onset of the relatively warm and stable Holocene period, beginning 11 650 years ago (see Chapter 1, Figure 1.2). About 10 000 years ago, at a number of disparate locations around the world, humans began to ensure a more stable source of food through the development of agriculture and the domestication of animals. The appearance of agriculture and advances in its techniques usually coincided with major climatic changes (see Chapter 8). Since the onset of agriculture we have continued to increase our capacity to manipulate Earth's resources and as a consequence have spawned an exponential increase in global population and consumption. These changes have also begun to seriously impact Earth's climate.

We will delve more deeply into Earth's changing climate and landscape, and our ongoing relationship with climate, in the following chapters, but before doing so it is important to clarify the key forcing mechanisms that influence climate, how they have affected climate through the last glacial cycle, and how scientists use these data to understand past and present climate and make predictions about future climate.

5.2 Climate change forcing mechanisms

Earth's climate changes when the amount of energy stored in the climate system is varied. The most significant changes in the climate system occur as a result of

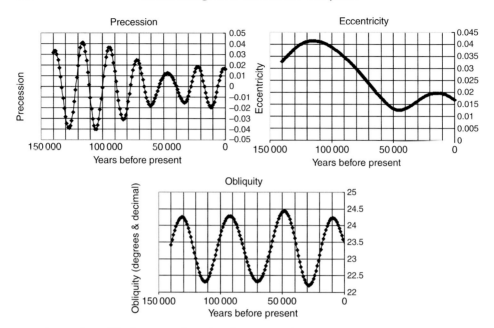

Figure 5.1 Milankovic cycles showing precession, eccentricity and obliquity cycles over the last 140 000 years.

variations in the amount of solar radiation coming in from the Sun and/or the amount of outgoing energy from the Earth. These variations occur as a result of a number of important natural factors. They include changes in the Earth's orbit, variations in ocean circulation and changes in the composition of Earth's atmosphere – for example, changing levels of greenhouse gases, which have recently been heavily impacted by human-induced emissions that are now also forcing climate change.

5.2.1 Milankovic cycles

Of the natural climate-forcing mechanisms, cyclical changes in the Earth's circumnavigation of the Sun are critical. The Serbian mathematician and planetary physicist Milutin Milankovic (Milankovic, 1998) is generally credited with calculating the magnitude of three dominant cycles (Figure 5.1). The first Milankovic cycle reflects Earth's precession, which has a periodicity of about 21 700 years and occurs because Earth wobbles as it spins around its axis. As a result, there are times when Earth's northern hemisphere is closest to the Sun during the winter, and other times when it is closest to the Sun during the summer. It is when the northern hemisphere is closest to the Sun during the summer and farthest from the Sun during the winter that seasonal contrasts in the northern hemisphere are exacerbated. Because the greatest amount of landmass on Earth is located in the northern hemisphere and, therefore, its climate is less influenced by the modifying effect of the vast oceans in the southern hemisphere, the amount and timing of solar insolation reaching the northern hemisphere plays a key role in glacial onset and termination. For example, during intervals of reduced seasonality, warmer northern hemisphere winters typically generate increased precipitation in the form of snow, which, combined with cooler summers, facilitate the retention of the winter's snowfall and stimulate the expansion of glacial ice.

The second Milankovic cycle is Earth's eccentric orbit that changes from more to less elliptical. This changing orbit alters the distance between the Earth and the Sun, influencing the amount of solar radiation that reaches Earth's surface during different seasons. The eccentricity cycle, with a periodicity of 98 500 years, has a significant impact on Earth's climate and the amount of glacial ice that persists from one year to the next. The larger the eccentricity the less insolation reaches the northern hemisphere in the summer and the more insolation reaches the northern hemisphere in the winter, reducing seasonality.

The third Milankovic cycle is Earth's obliquity, or the inclination of the Earth's axis in relation to its plane of orbit as it rotates around the Sun. This ranges from 21.5 to 24.5 degrees and operates on a periodicity of 41 000 years. During intervals when the Earth has less tilt, solar radiation is more evenly distributed over the planet, thereby reducing the difference between seasons. However, reduced tilt also increases the difference in solar radiation received between polar and equatorial regions and amplifies the equator-to-pole temperature gradient. As a result, the hydrological cycle is intensified by the warming of tropical oceans. For example, in the Pacific, a higher incidence of El Niño events would be observed relative to La Niña events (Kukla and Gavin, 2004). Decreased obliquity thus contributes to the growth of land-based ice in the high latitudes.

5.2.2 Impacts of variations in solar radiation

The solar insolation from all three of these forcing mechanisms combines to generate the total solar insolation reaching Earth in watts per metre squared (W/m^2). Figures 5.2a and 5.2b show the insolation parameters from a 122 000-year time-series climate simulation performed on the UVic Earth system climate model (UVic ESCM), indicating simulated insolation over the last 120 000 years for January and July, respectively, at 65° N. In contrast to the small variations seen in winter solar insolation levels, summer insolation levels have varied by as much as 90 W/m^2 over the last 120 000 years. The January insolation curve shows extended intervals where little to no solar insolation reached this northern latitude. Today, this is the case in high northern latitudes where, during the winter solstice, winter 'days' consist of 24 hours

(a)

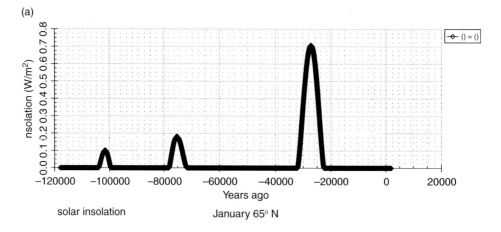

solar insolation   January 65° N

(b)

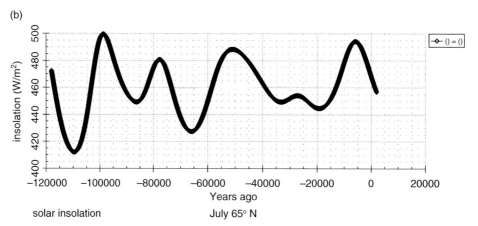

solar insolation   July 65° N

Figures 5.2a and 5.2b Solar insolation over the last 120 000 years for (a) January and (b) July at 65° N, respectively in watts per metre^2. These solar insolation parametres were used in a 122 000-year time-series climate simulation performed on the UVic Earth system climate model (UVic ESCM).

of darkness. The greatest amount of solar insolation in over 120 000 years to have reached 65° N during the winter (just under 1 W/m^2) did so about 29 000 years ago. It occurred coincident with an extended period of relatively low solar insolation (approximately 450 W/m^2) during the summers and contributed to rising winter snowfalls that persisted from one year to the next. This period of relatively low summer insolation (cold summer) and high winter insolation (warm winter) in the northern hemisphere was coincident with the onset of the last glacial interval – MIS2.

Despite these correlations between changing seasonal solar insolation and glacial events, solar insolation forcing alone is insufficient to explain the temperature variations associated with glacial cycles. The combined changes in radiative forcing generated by the three Milankovic orbital changes must be amplified to promote the ice-sheet growth that is evident during glacial phases. The single biggest amplifier of the orbital signal is the feedback associated with the albedo of sea ice, land ice and snow. Generally, conditions with warmer winters are associated with greater high-latitude precipitation in the form of snow because the warmer atmosphere holds more moisture. With cooler summers come extended intervals of snow-covered landscapes and sea-ice coverage. As a result, such conditions are associated with a brightening of the surface of the Earth, which reflects a higher proportion of the incident solar radiation back to space, and further intensifies the cooling (for a discussion on the causes of glacial inception at 116 000 cal. yr BP see Yoshimori *et al.*, 2002).

Solar insolation forcing and the albedo feedbacks that amplify its effects on climate are also strongly influenced and driven by other natural climate-forcing mechanisms, including those associated with the atmosphere, oceans, biological productivity and the hydrological cycle.

5.2.3 Climate forcing by changes in atmospheric composition

Changes in the Earth's atmosphere also play an important role in climate forcing and are of particular interest with recent concerns about rising greenhouse gases. Earth's atmosphere consists of 78% nitrogen, 21% oxygen and several greenhouse gases, including water vapour, carbon dioxide (CO_2), methane (CH_4), nitrous oxide and others. Greenhouse gases act like the glass in a greenhouse, allowing visible light, the shorter wavelength radiation from the Sun, to reach Earth's surface, but preventing longer wavelength, infrared energy emanating from the Earth from escaping to space. The effect of greenhouse gases in the atmosphere is to raise the overall temperature of the Earth; the greater the concentration of greenhouse gases, the more heat is trapped, and the higher Earth's temperature rises and vice versa. This natural process is what keeps the Earth warm enough for us to live on. In 1896, Arrhenius suggested that CO_2 had a role in amplifying glacial cycles. Further, he correctly estimated the climate effects of doubling atmospheric CO_2. This has since been substantiated by climate model simulations.

Ice cores taken from the Greenland and Antarctic ice sheets provide scientists with a long history of Earth's climate (see Chapter 1 Figure 1.3). The Vostok and Greenland ice-core records indicate that temperature change has led to atmospheric changes in greenhouse gases (CO_2 and CH_4) over at least the last 650 000 years, suggesting that concentrations of these greenhouse gases respond to changes in global temperature. It is also clear that they act in unison; temperature does not increase without an associated increase in CO_2 and CH_4 and vice versa. Shifts in greenhouse gases, combined with associated changes in water vapour content, act to amplify global warming and cooling forced by Milankovic forcing. These natural cycles are evident over the last 650 000 years as temperature, CO_2 and CH_4 levels oscillate between the now predictable limits associated with ice ages and warm intervals.

5.2.4 *The oceans' role in a changing climate*

The oceans have not always been as we observe them today. Their extent and dimensions have changed over millions of years as tectonics shifted the plates on which continents reside. The shifting of continents and the resultant opening or closing of straits that link oceans has a profound effect on ocean circulation patterns and thus climate. Over much shorter timescales, global sea level has also varied. During intervals of glaciation, water evaporates from the oceans and is trapped in ice, causing global glacio-eustatic sea level to drop; during the last ice age, global sea level dropped by about 120 m. Alternatively, during warm interglacial periods when relatively little water is locked up in ice sheets, global eustatic sea level rises. Changes in global sea level can also have an impact on ambient surface-air temperature. A drop in sea level would cause a very slight warming over the ocean, associated with higher air pressure as the air over the oceans thickens, and a slight cooling over land, associated with lower air pressure as the air over the land thins. If sea level were to fall by 100 m, we might see a rise in the ambient air temperature over the oceans of perhaps 0.3 to 0.5 °C and a fall in ambient air temperature over land of about 0.5 to 0.7 °C. Theoretically, the air over land where ice sheets were growing would push away the warming atmosphere above non-ice-sheet locations slightly, cooling ice-sheet locations.

Falling sea level also exposes large expanses of the continental shelf and thereby restricts connections between ocean basins (see Figure 5.3). In recent geological history, such changes have influenced ocean currents and impacted climate to some degree. But perhaps more critically, from a human perspective, the relative exposure of continental shelves and their productivity has played a key role in the capacity of humans to inhabit coastal regions and to migrate to new territories. For example, during the last glacial maximum, when mainland North America was covered with extensive ice sheets, the exposed Beringian and Pacific eastern continental shelves probably provided productive refugia for plants, animals and potentially humans as they emigrated from diminishing mainland habitats (Hetherington *et al.*, 2003, 2004, 2008).

The oceans are an important component of the natural climate system. Oceans play a large role in absorbing and storing vast amounts of CO_2 and thus perform a critical role in influencing and maintaining climate. However, because of their size and large heat capacity, oceans operate on a timescale far longer than that of the atmosphere. As a result, the effects of perturbations in the oceans take much longer to become apparent in the global climate. For example, it takes the deep ocean many millennia to respond to instantaneous changes in atmospheric levels of greenhouse

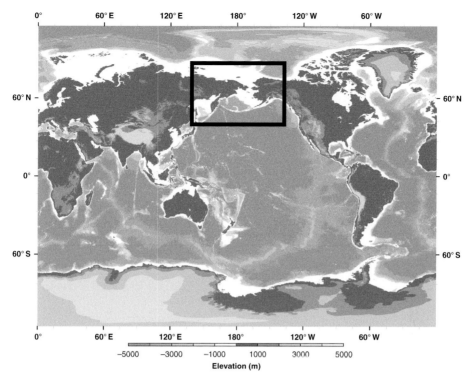

Figure 5.3 Land elevation and ocean depths showing continental shelves of the world and indicating the location of the Bering Strait continental shelf (Beringia).

gases and solar insolation, and hence global climate responds on the same longer timescale.

The oceans' role as an important natural climate-system component has much to do with the fact that the oceans store a vast amount of Earth's heat energy, far more than the atmosphere. Oceans transfer heat into the atmosphere through evaporation and transfer heat vertically from the warmer surface waters to the colder deep ocean water. Oceans also assist in the transfer of heat from warmer low latitudes to colder high latitudes. In the Atlantic Ocean this heat transfer occurs through a mechanism called the North Atlantic meridional overturning (AMO; see Figure 5.4), which operates much like a conveyor belt, bringing warm water from the equatorial Atlantic up to the North Atlantic, giving up heat to the atmosphere, and drawing cold, dense water which sinks in the Greenland and Labrador seas, south to the Indian and Pacific Oceans as deep currents. Water movement is driven through global-scale differences in water density – generated at the sea surface through solar heating, rain and clouds, and eroded by the mixing of tides and winds (for a discussion of the dynamics of global ocean circulation see Visbeck, 2007). In the past, during intervals when sea water in the North Atlantic was cooler and less saline, either as a consequence of

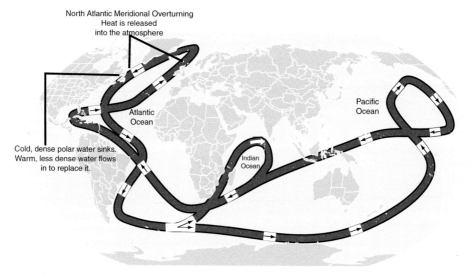

North Atlantic Meridional Overturning
Heat is released
into the atmosphere

Atlantic
Ocean

Pacific
Ocean

Cold, dense polar water sinks.
Warm, less dense water flows
in to replace it.

Indian
Ocean

Figure 5.4 Schematic diagram of the ocean conveyor belt with the AMO noted.

reduced evaporation or freshwater influx from melting ice sheets, the northern extent
of sea ice shifted southward and was associated with the slowing and, on rare
occasions, the shutting down of the AMO, which considerably cooled the currently
mild European climate. These slowdowns or shutdowns, which are not completely
understood, have occurred over very short time spans – decades to centuries – and are
thought to significantly impact climate. However, Seager *et al.* (2002) argue that the
role of the AMO for European climate is exaggerated. They suggest that European
climate warms by about 3 °C and the principal cause of these milder winters is
advection by the south-westerly winds that bring warm maritime air into Europe.

The AMO has also experienced intervals when its strength has increased; these
intervals result in amplified warming. During the transition from the depths of the
last ice age to the Holocene, around 14 600 years ago, there was a large increase in
the strength of the AMO. This event is called the Bølling–Allerød interstadial and
was responsible for a 150% decrease in atmospheric radiocarbon and a 3 °C increase
in annual mean sea-surface temperature over a period of about 500 years (Weaver
et al., 2003). Interestingly, this event was concurrent with an exceptionally large
melting event, meltwater pulse 1A (mwp-1A), which resulted from the melting of
continental ice sheets that had accumulated during the previous glacial period (Clark
et al., 1996, 2002; Weaver *et al.*, 2003). The source of the meltwater has been the
subject of controversy, and although the Laurentide ice sheet in North America has
been commonly cited as its source, more recent research indicates that AMO
strengthening may have been due to the partial collapse of the Antarctic ice sheet
(Clark *et al.*, 2002; Weaver *et al.*, 2003). Research has shown that future increases in

atmospheric greenhouse gas levels will probably be accompanied by declining AMO circulation of between 15% and 31%. It appears that the dominant cause of this decrease is changing surface-heat fluxes associated with rising levels of CO_2 in the atmosphere rather than changes in freshwater fluxes, which appear to reinforce the declining AMO (Weaver *et al.*, 2007).

A quasi-cyclic phenomenon associated with ocean–atmosphere coupling, the El Niño–Southern Oscillation (ENSO), also occurs in the Pacific Ocean and strongly influences the distribution of rainfall in the tropics. El Niño is defined as an anomalous warming of approximately 2 °C in sea-surface temperatures in the central and eastern Pacific Ocean (Glantz, 1996). Research into isotope ratios taken from corals on the Huon Peninsula, Papua New Guinea, indicates that ENSO interannual variance around 85 000 and 40 000 years ago was less than that observed in modern corals (Tudhope *et al.*, 2001 cited in Burroughs, 2005, pp. 54–5). Further, Timmermann and An (2005) have linked El Niño suppression with a reduced meriodional overturning in the Atlantic associated with a large freshwater influx from glacial meltwater lakes. When the Atlantic thermohaline circulation shuts down, the thermocline deepens, which leads to a considerable damping of ENSO amplitude. When the amplitude of El Niño is reduced, colder drier conditions are apparent, particularly in coastal Peru. There is evidence that ENSO activity was virtually absent in the early Holocene and that ENSO frequency became significant around 7000 years ago. ENSO variability increased, peaking around 4900 years ago, and remained high until 1200 years ago, when it suddenly declined (Burroughs, 2005, pp. 234–5). These findings highlight the importance of recognizing the variability of changing climatic conditions in disparate regions. During glacial intervals temperatures drop more in high latitudes than in low latitudes. Temperatures also rise more during warm intervals in high latitudes than they do in low latitudes. This variation between latitudes is linked to greater climatic variability in high and middle latitudes.

5.2.5 Additional natural climate feedback mechanisms

In addition to the feedbacks associated with solar radiation, the atmosphere and the oceans, other natural feedbacks are also fundamental in influencing the major forcing mechanisms affecting climate. They include the albedo of sea ice, land ice and snow, as well as dust and some high-latitude temperature-related feedbacks. Sea ice, land ice and snow have a high albedo; they reflect large amounts of solar radiation back into the atmosphere, creating a positive cooling feedback. Densely vegetated regions absorb solar radiation, contributing to warming.

Another feedback is associated with the amount of dust in the atmosphere. During glacial intervals, as the climate cools, the total amount of precipitation decreases at middle latitudes. Dust from lower-latitude regions is then able to travel farther poleward before precipitating out. According to evidence obtained from the ice-core records, dust levels increase in high-latitude regions during glacial intervals (Kumar *et al.*, 1995; Petit *et al.*, 1999). Today, the biological productivity of the southern ocean is iron-limited, and adding iron directly increases carbon uptake. It is thought that enhanced biological carbon uptake, due to enhanced southern ocean iron fertilization resulting from higher dust levels in a drier mean glacial climate, is a key feedback mechanism during interglacial-to-glacial transitions (Kumar *et al.*, 1995; Watson *et al.*, 2000). This potentially explains, at least to some degree, glacial–interglacial CO_2 differences. Coincidently, as the climate cools in the northern hemisphere, the generation of permafrost and freezing of the wet lowlands of the periglacial region shuts down natural microbial processes involved in the emission of CO_2 and CH_4, further contributing to the decline of the atmospheric concentration of these greenhouse gases.

In the northern hemisphere, vegetation feedbacks amplify the cooling that results from reduced solar insolation and higher albedo. This occurs when the northern treeline shifts southward and broadleaf trees increasingly convert to shrubs and C_4 grasses in tropical regions, creating a positive feedback (Meissner, Weaver *et al.*, 2003). The shift in treeline is due to the higher albedo of snow on open ground than on forest. The conversion to shrubs and C_4 grasses in the tropics is the result of reductions in precipitation and soil moisture as well as to the physiological response of photosynthesizing plants to lower atmospheric CO_2 concentrations. In addition, northern sea-ice extent shifts southward and is associated with lowered sea-surface temperatures and a reduced Atlantic meridional overturning (Meissner, Schmittner *et al.*, 2003; Meissner, Weaver *et al.*, 2003; Schmittner *et al.*, 2002).

These feedbacks operate in a similar but opposite direction when orbital cycles shift to those associated with the warmer summers and cooler winters that are conducive to interglacial intervals. Warmer high-latitude surface temperatures are coincident with, and amplified by, a shifting northward of the treeline, a conversion of C_4 grasses to broadleaf trees in the tropics, a northward shift in the northern hemisphere sea-ice edge and higher sea-surface temperatures. The warmer climate also results in changes to the hydrological cycle, enhancing annual mean runoff into, and precipitation over, the Arctic (Weaver, 2004). It also slightly increases the strength of the AMO (Weaver *et al.*, 1998). The amplitude of the temperature shifts between cold glacial and warm interglacial intervals is higher in the middle and high latitudes than in the tropical and equatorial regions due to the stronger albedo and vegetation feedbacks (McManus *et al.*, 1999; Rostek *et al.*, 1993; Schneider *et al.*, 1996).

Another observed phenomenon is what appears to be an out-of-phase climate response between the northern and southern hemispheres – a bipolar see-saw effect. During cold stadial phases, when the northern Atlantic Ocean was cool, the southern Atlantic Ocean was warm, and during warm interstadial phases, when the northern Atlantic Ocean was warm, the southern Atlantic Ocean was cool (Crowley, 1992; Schmittner *et al.*, 2003; Steig, 2006; Stocker, 1998).

5.2.6 Volcanism

Explosive volcanic eruptions that spew large amounts of sulphur-rich gases into the stratosphere can cause the global climate to cool by about 0.2 to 0.3 °C for several years (Zielinski, 2000). This is because explosive volcanic activity injects sulphates into the stratosphere, which can have a significant although relatively short-term cooling impact on climate. Based on the sulphate levels recovered from GISP 2 ice-core data, scientists have a good representation of equatorial and mid- to high-latitude eruptions over the last 110 000 years (Zielinski *et al.*, 1996a). A mega eruption between about 73 000 and 71 000 years ago, likely that of Mount Toba (Zielinski *et al.*, 1996b), on what is now the island of Sumatra, Indonesia, briefly raised the GISP 2 volcanic sulphate levels to about 466 ppb (Zielinski *et al.*, 1996a). Sunlight would have been reduced, and climate was potentially impacted over centuries as a consequence of positive climate feedback mechanisms and the cooling of global surface temperatures by 1 °C or more (Rampino *et al.*, 1988; Zielinski, 2000; see also Oppenheimer, 2002). However, climate is impacted differently depending on the location of the eruption. Eruptions that occur in equatorial regions appear to impact global climate more forcefully than do midlatitude eruptions, which can cool the climate in the hemisphere of origin. The Z2-ash eruption in Iceland around 52 000 years ago appears slightly larger in the GISP 2 ice-core record than the Toba eruption, yet, because of its location in the midlatitudes of the northern hemisphere, its global climatic impact was less. This may also just be an issue of midlatitude eruptions looking bigger than they are because they are closer to the ice cores.

There is some evidence to suggest that rapid fluctuations in climate can actually force volcanism. By comparing an updated sea-level curve with the number of eruptions over the last 110 000 years, scientists have ascertained that the number of volcanic eruptions was greatest when sea level was changing quickly. This is believed to be a result of glacial loading and unloading of the crust when large ice sheets grew and melted (Nakada and Yokose, 1992; Zielinski, 2000; Zielinski *et al.*, 1996a).

Based on the GISP 2 ice-core records, climatologist Zielinski (2000) suggests that volcanic signals were most concentrated between about 13 000 and 7000 years ago. Multiple volcanic eruptions that occur closely spaced in time appear to have the potential to force climate over decadal to multi-decadal time frames, particularly when climate is already cool. It is clear that volcanic eruptions have an impact on climate and that the impact is large enough to affect humans. Although large eruptions can drop global or hemispheric climate by 0.2 to 0.3 °C for a number of years after the eruption, multiple, closely spaced eruptions and mega eruptions likely affected human social and economic systems.

5.2.7 Human-induced climate forcing

Since the Industrial Revolution, humans have been impacting natural climate-forcing mechanisms. Through the burning of fossil fuels, cement production, changing land-use patterns and agricultural processes, humans are emitting record levels of greenhouse gases (particularly carbon dioxide and methane) into Earth's atmosphere. As a consequence, we have increased atmospheric greenhouse gas levels well beyond the traditional predictable oscillating limits observed from ice-core records extending over the last 650 000 years. Scientists around the world now agree that rising atmospheric CO_2 levels far exceed any level observed in the atmosphere over the last 650 000 years and predict global warming over the next few decades that is unprecedented in human history.

This is a frightening prospect and highlights the importance of understanding our ongoing relationship with our planet and its climate – the objective of this text. Through greater understanding of human history, Earth's climate history and our changing interaction with that climate, we seek to gain and provide greater wisdom for dealing with future climate-related issues, particularly as they influence our ability to adjust our behaviour to anticipated rapid change.

Unfortunately, a lack of understanding about the slow nature of thermodynamics in the oceans relative to the virtually instantaneous changes apparent in the atmosphere with changing greenhouse gas levels has resulted in conflicting stories in the public media about climate change. Thus, it is important to recognize that Earth's climate system takes time to equilibrate. The very high and rising greenhouse gas levels in the atmosphere today are very real. The recent rapid rise in the amount of CO_2 that humans have emitted and continue to emit into our atmosphere will take time to impact climate. The climate we are experiencing today will most likely not be the climate that our children and our grandchildren will experience. They will feel the consequences of the greenhouse gases that we have already dumped and have yet to dump into the atmosphere as their effects work their way through the climate system.

5.3 Identifying climate change and its impacts

5.3.1 Proxy indicators

It is by gathering and analyzing proxy (observational) indicators that scientists obtain information about how climate has changed over long timescales and how those changes have impacted the environment. Over shorter timescales we can use direct observations. Proxy evidence can be found in ice, ocean and lake cores, as well as in glacial and loess deposits. Frequently used proxy indicators include oxygen isotopes and pollen and plant macrofossils. For example, oxygen isotope ratios in microfossil shells cored from the ocean floor have been used to derive and date the varying global ice-volume record through time (Imbrie *et al.*, 1984; Linsley, 1996; Shackleton, 1987); changes in glacio-eustatic sea level have been derived through the dating of fossil coral reef terraces, which thrive only in a narrow sea-water depth range, but are found today both far below and high above current sea level (Esat *et al.*, 1999). Through the identification of glacial striations left in rocks, glacial deposits including erratics deposited far from their host rocks, and drop stones released from the bottom of melting glaciers and icebergs and left lying on the ocean floor, geologists have identified the timing and extent of many of the world's ice sheets as they have expanded and retreated through time (Ehlers and Gibbard, 2004a, 2004b, 2004c). By coring through glaciers in Greenland and Antarctica, scientists have been able to measure the levels of the greenhouse gases carbon dioxide (CO_2) and methane (CH_4) in ancient air trapped in the ice and also to use oxygen isotope data to ascertain the temperature above the ice sheets and also where water evaporated in Greenland over the last 100 000 years (Dansgaard *et al.*, 1984) and above Antarctica over the last 650 000 years (Petit *et al.*, 1999; Spahni *et al.*, 2005) (see Chapter 1 Figs. 1.2 and 1.3). These remarkable records show that climate warming is not maintained without an excess of green-house gases, including CO_2 and CH_4, nor are cold climates maintained without a depletion of these gases. By combining the Vostok ice core and the Dome C extension from Antarctica, Spahni *et al.*'s (2005) findings imply that the current level of atmospheric CO_2 (approximately 390 ppm in 2009) is more than 30% larger than any level observed in the ice cores for the last 650 000 years. Current levels of atmospheric CH_4 (approximately 1790 ppb in 2009) are about two and a half times that observed at any time during the last 650 000 years. Furthermore, the increase in atmospheric CO_2 observed between the depth of the last ice age, about 21 400 years ago, and AD 1850 (from 190 ppm to 280 ppm) is less than that observed between AD 1850 and today (from 280 ppm to 390 ppm), a period of just over 150 years.

It is by analyzing and interpreting this growing body of proxy indicators and combining it with our understanding of climate change forcing mechanisms that scientists have been able to gain great insights into how Earth's climate has changed. However, proxy evidence is limited because it provides data only for points in time at particular locations. We cannot assume that proxy findings are universal and so there are great gaps in our data. Thus, scientists make climate models to simulate Earth's climate and help us discover what may be in those gaps and to identify changes in Earth's climate.

5.3.2 Coupled atmosphere–ocean general circulation models (GCMs)

'The weather is not so bad. It is pretty nice where I live,' climate-change dissenters claim. And they may be correct. But weather is not climate. Weather is what you see outside your window today; climate is the probability of that weather continuing. Climate models have been developed to predict the probability of long-term changes in climate. However, many climate models do simulate weather, and indeed, many climate models are based on the same models as those used for weather prediction (e.g., the Hadley Centre models).

Climate modelling efforts aimed at understanding the ocean's and atmosphere's role in climate variability have evolved substantially over the last decade. Initially, stand-alone component models, such as oceanic or atmospheric general circulation models (GCMs) with specified surface boundary conditions – including, for example, changing solar insolation and CO_2 levels – were used almost exclusively. However, most climate models now consist of an atmospheric component that is coupled with, or linked to, ocean and sea ice models. All general circulation models (GCMs) include a land-surface scheme, and some now allow predictions of how terrestrial vegetation will respond to changes in climate.

Today a new breed of intermediate complexity models have evolved that incorporate many interactive subcomponent models, some or all of which are of reduced complexity. Intermediate models are used to understand and evaluate only those processes and parameterizations that are deemed important to answer a particular set of scientific questions. Intermediate complexity models utilize far less computer time and thus are very useful in performing long-timescale climate simulations or multiple iterations that can be used to test the impact of changing climate variables. Over the next decade, as computer technology is enhanced, these intermediate complexity models will likely be surpassed by Earth system GCMs capable of efficiently undertaking ensembles of millennial timescale integrations.

Climate models are built using the physical and mathematical principles that govern the climate system. Before their results are accepted as useful for future climate predictions, they are tested, and results must satisfactorily compare with the present-day and transient twentieth-century climate. Model simulations of past climate are also compared with palaeoreconstructions and proxy data to evaluate the model's performance. The combined efforts of many scientists, using both proxy data and climate models, have resulted in fundamental advances in our understanding of the processes involved in climate variability over the last glacial cycle and beyond. This knowledge provides a foundation on which to gain greater understanding about the historical relationship between humans and climate.

5.4 Modelling with the UVic Earth system climate model

The UVic Earth system climate model (ESCM) is an intermediate complexity model consisting of a three-dimensional general circulation model coupled to a dynamic–thermodynamic sea-ice model, an ocean carbon-cycle model, a dynamic energy–moisture balance atmosphere model, a land-surface model, and a terrestrial vegetation and carbon-cycle model. The model is driven by time-dependent changes in atmospheric CO_2 and orbital forcing. It has been extensively and successfully evaluated against both contemporary observations (Weaver *et al.*, 2001) and palaeoproxy records (Meissner, Schmittner *et al.*, 2003; Meissner, Weaver *et al.*, 2003; Schmittner *et al.*, 2002; Weaver *et al.*, 1998).

We used the UVic ESCM in a recent project to understand the world's changing climate since MIS6. Because of the important influence that changing ice sheets can have on climate, land-ice extent and thickness was included in the model along with atmospheric CO_2 and orbital forcing. Land ice-sheet extent for MIS6 was interpolated based on Ehlers and Gibbard's (2004a, 2004b, 2004c) ice-sheet distributions for the world obtained from geological data. The last glacial cycle land-ice was interpolated by Sean Marshall at 1000-year intervals from prior ice-sheet simulations.[1] Using the UVic ESCM, the equilibrium climate and vegetation of MIS6 was then simulated. A much longer time series was then performed for the last 122 000 years, with output generated every 500 years. The model simulations provide climate variable output including land, sea and surface-air temperatures; precipitation; meridional overturning; changing vegetation coverage and productivity; as well as many other variables. Results were compared with ice and ocean core, pollen and other proxy evidence and with climate-related research. The findings were then analyzed from an anthropological perspective to ascertain interactions between climate and human evolution, migration and behavioural change.

Figure 5.5 shows global average surface-air temperature for the MIS6 equilibrium and 122 000-year time-series simulations. The simulations clearly capture glacial MIS6 temperatures, followed by a warm period beginning in MIS5e, progressive cooling into the LGM and then warming after the LGM. Figure 5.6 shows maximum meridional overturning for MIS6 and the 122 000-year time series. The simulations do a remarkable job of capturing the Younger Dryas cool interval and Heinrich events 1 through 9. These findings provide confidence in the more detailed analysis that follows.

The balance of this chapter is spent chronologically reviewing our best estimate of the climate during MIS6 and the LGC. The following chapter reviews the impact that changing climate had on vegetation. Then the book deals with our interpretation about what these changing conditions meant for early modern human evolution, migration and development. However, a caveat to these

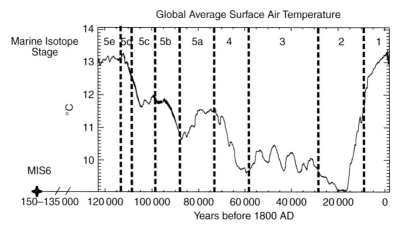

Figure 5.5 Global average surface-air temperature (°C) simulation results from a 122 000-year time-series climate simulation performed on the UVic ESCM and the MIS6 equilibrium simulation result (9 °C). Marine isotope stages (MIS5e through MIS1) are indicated by broken vertical lines.

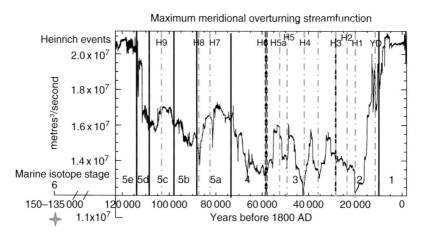

Figure 5.6 Maximum meridional overturning streamfunction simulation results from a 122 000-year time-series climate simulation performed on the UVic ESCM and the MIS6 equilibrium simulation result (1.1 x 10[7] metres/second). Marine isotope stages (MIS5e through MIS1) are indicated by solid vertical lines. Heinrich events (H9 through H1) and the Younger Dryas event (YD) are indicated with broken vertical lines.

findings and their interpretation must now be made. The limitations of this work are many. Huge gaps remain. Proxy evidence, fossil bones, stone tools and modern computer climate simulations cannot and do not tell the entire story. Further, we can neither understand nor discuss all the climatic changes that occurred during this period nor begin to comprehend the many ways climate change impacted early modern human development and behaviour. We urge readers to be cognizant of

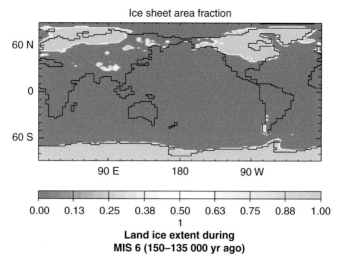

Figure 5.7 Land-ice extent during MIS6 based on recent proxy and field data-based global digital glacial ice-extent compilations of Ehlers and Gibbard (2004a, 2004b, 2004c). Land-ice extent used in the MIS6 equilibrium climate simulation. (See colour version in colour plate section.)

these limitations, while also recognizing that this work and other related research provide important contributions on which future research can improve.

5.5 Climate during the origin and dispersal of *Homo sapiens*

5.5.1 Marine isotope stage 6 glaciation (150 000 to 135 000 years ago)

Homo sapiens appeared in Africa by 200 000 years ago (McDougall *et al.*, 2005; Shea *et al.*, 2004). Their arrival occurred just prior to a very cold interval known as marine isotope stage 6 (MIS6). This major glacial interval lasted for about 15 000 years from about 150 000 to 135 000 years ago and caused ice sheets to expand across much of Europe, northern North America and parts of Asia and southern South America.

Proxy evidence indicates that extensive MIS6 glaciation (Figure 5.7) (based on Ehlers and Gibbard, 2004a, 2004b, 2004c; and climate simulations done for this publication) caused global eustatic sea level to fall to about 130 m below present; a level below which sea level has not fallen since, even during the last ice age (see Figure 4.1). Ice-core records indicate that global average surface temperatures over Vostok, Antarctica were almost 10 °C colder than during the Industrial Revolution of the 1800s and have not subsequently fallen below this level (see Figure 1.2).

Although in many ways MIS6 appears to have had very similar conditions to those of the last ice age (MIS2) a few differences are evident. Based on the evidence from the Vostok ice-core record, MIS6 appears to have been slightly colder than MIS2 (Petit *et al.*, 1999). Sea level dropped somewhat more during MIS6, to 130 m below present versus about 120 m below present for MIS2 (Lambeck *et al.*, 2002). The geological evidence indicates that ice sheets extended over a slightly greater expanse during MIS6 than they did during MIS2, although according to our climate models' simulations, sea ice was less extensive in both the northern and southern hemispheres during MIS6 than it was during MIS2. Model simulations also indicate a significant slowing of the AMO, which likely acted as a feedback mechanism, further cooling the European continent. Climate simulations also indicate that global average net primary productivity appears to have been less during MIS6 than present, but still higher than during the last glacial maximum (LGM; ~21 500 years ago) and during another interval of extremely low productivity 60 000 years ago.

5.5.2 The last glacial cycle

At the close of the MIS6 ice age, the last glacial cycle (LGC) began. The LGC evolved progressively through a series of warm and cold intervals – MIS5e through MIS5a to MIS2, which spanned the period from about 135 000 years ago to about 11 650 years ago. The overall evolution of global temperature saw warming with the onset of MIS5e, followed by progressive cooling, then warming after the LGM (see Figure 5.5). Climate during this time was highly variable, with rapid transitions between cool climate intervals (stadials) and warm intervals (interstadials) lasting hundreds to thousands of years (Dansgaard *et al.*, 1984; Oeschger *et al.*, 1984). Global eustatic sea level ranged from about 6 m above present to 120 m below present levels (Lambeck *et al.*, 2002).

The strength of the thermohaline circulation, which directly influenced northern hemisphere climate, was also highly variable during the LGC. During warm phases the maximum meridional overturning streamfunction reached levels of 20 Sv and higher, whereas during cold phases it fell to below 14 Sv (see Figure 5.6). Global average sea-surface temperatures oscillated between a high of about 20 °C and a low of 14 °C (Imbrie *et al.*, 1984). Surface-air temperatures also fluctuated. Based on interpretations from ice cores, surface-air temperatures over Vostok, Antarctica, fluctuated from about 2 °C warmer than today to about 9 °C colder than today. Northern-latitude landscapes saw the advance and retreat of massive continental ice sheets. Continental shelves emerged and subsided in response to the changing ocean levels as well as local changes in water, ice and sediment loading.

High-amplitude millennial-timescale variability between warm (interstadial) and cold (stadial) states was a prominent feature of the LGC (Bond *et al.*, 1997; Dansgaard *et al.*, 1993; Schulz *et al.*, 2002). Each abrupt warm interstadial event coupled with the cooling that immediately followed it are referred to as Dansgaard–Oeschger (D–O) oscillations; 24 of these short-lived climatic oscillations transpired during the LGC between about 110 000 and 14 000 years ago (for a discussion see Lowe, 2001). D–O oscillation transitions were associated with abrupt local mean surface-air temperature changes that included a warming over Greenland of up to 10 °C. These occurred on decadal timescales (for example, see Lang *et al.*, 1999) and were followed by a series of gradual steps that reverted conditions to the glacial mode. However, the effects of D–O oscillations were not local (Justino, 2004); evidence suggesting a global impact has been found in the North Pacific (Kotilainen and Shackleton, 1995), the Santa Barbara basin (Behl and Kennett, 1996; Kennett and Ingram, 1995), South America (Lowell *et al.*, 1995), the Arabian Sea (Schulz *et al.*, 1998) and the South China Sea (Wang, 1999). Furthermore, scientists have observed that conditions were cooler in the southern hemisphere when they were warmer in the northern hemisphere. Thus, during stadial phases of D–O oscillations, when the northern Atlantic Ocean was cool, the southern Atlantic Ocean was warm, and during interstadial phases when the northern Atlantic Ocean was warm, the southern Atlantic Ocean was cool (for example, see Crowley, 1992; Schmittner *et al.*, 2003; Stocker, 1998).

Less frequent occurrences, known as Heinrich events, are also evident in the form of layers of coarse debris in marine sediments in the North Atlantic. At least 11 (H11 to H1) Heinrich events occurred during the LGC (based on Figure 5 in Rasmussen and Thomsen, 2004, p. 105) (see, for example, Figure 5.6) and are associated with the deposition of ice-rafted debris, believed to be the result of the partial collapse of the massive Laurentide ice sheet in North America and the Fennoscandian-Greenland ice sheet. Further, an ocean core drilled in the sediments of Brazil's continental margin contains large terrigenous sediment pulses for each Heinrich event, suggesting humid conditions in north-east Brazil (Arz *et al.*, 1998). Trade winds may have altered the north Brazil current, affecting the growth and decay of glacial ice sheets. These Heinrich events occurred only once every four to five D–O cycles, suggesting that some critical threshold was crossed that facilitated the purging of massive icebergs (Cortijo *et al.*, 1997, 2000 Grousset *et al.*, 1993; Sarnthein *et al.*, 1995). They represent perhaps the most extreme example of rapid global climate change.

Heinrich events are thought to have induced rapid changes in the Earth's ocean and atmospheric circulation that were linked with abrupt climate change in tropical/subtropical areas. A Chinese loess record indicates intensified winter monsoons correlate with the timing of Dansgaard–Oeschger and Heinrich events 1 through 5 (Porter and An, 1995). The abundance of planktonic foraminifera assemblages from ocean core sediments taken from the bottom of the Alboran Sea in the western Mediterranean links colder sea- and land-surface temperatures with Heinrich events

1 through 4 (Cacho *et al.*, 1999). Heinrich events are also correlated with dry episodes in Africa through sediment cores taken from the equatorial Atlantic beneath Africa's bulge (Broecker and Hemming, 2001). Heinrich events also appear to have influenced vegetation. A mid-Pleistocene marine record showing the long-term variations in southern African C_4 grass vegetation indicates variations are coincident with tropical sea-surface change, including a decrease in sea-surface temperatures of up to 5 °C during glacial intervals (Schneider *et al.*, 1995). Others have interpreted this pattern as indicating that tropical sea-surface temperatures directly controlled continental aridity by reducing tropical evaporation and the atmospheric moisture content, and thus African precipitation (Schefuß *et al.*, 2003).

The climatic consequences resulting from Heinrich events likely influenced human dispersal and behaviour. The last six main Heinrich events are well dated and studied (H1-H6: H1: 16–22 000 years ago, H2: 24–26 000 years ago, H3: 30–33 000 years ago [H3a: 37–38 000 years ago], H4: 40–43 000 years ago [H4a: 46–47 000 years ago], H5: 49–54 000 years ago [H5a: 56–58 000 years ago] and H6: 60 000 years ago) (Broecker and Hemming, 2001; Chapman and Shackleton, 1998; Hemming, 2004; Lambeck *et al.*, 2002; McManus *et al.*, 1994). However, only H7 (85 000 years ago) and H9 (105 000 years ago) were associated with cooling of the subtropical Atlantic during MIS5. H9 is believed to have led to abrupt changes in climate that may have caused North, East and West Africa to be unsuitable for human occupation and thus associated with an early migration of *H. sapiens* out of Africa (Carto *et al.*, 2008).

It was during the highly variable climate of the last glacial cycle that *H. sapiens* began their pervasive expansion out of Africa and throughout the world. Thus, let us now look in more detail at each stage of the last glacial cycle.

5.5.2.1 Marine Isotope Stage 5e – the Eemian interglacial (135 000 to 116 000 years ago)

Proxy evidence Earth's annual mean surface-air temperatures during the Eemian rose to as much as 2 °C warmer than pre-industrial times (White, 1993). Greenhouse gas levels rose rapidly in association with this rise in surface-air temperature; atmospheric CO_2 reached about 280 ppm and CH_4 levels reached just over 700 ppb. Coral proxies indicate that sea level rose, reaching up to 6 m higher than today, during a time when ice volumes were at their minimum (Ku *et al.*, 1974; Lambeck *et al.*, 2002). Global mean sea-surface temperatures rapidly escalated during the interglacial reaching about 3.5 °C warmer than latest Holocene values in the Angola basin along the West African margin (Imbrie *et al.*, 1984; Schneider *et al.*, 1995).

Climate simulations Proxy evidence suggesting significant warming in MIS5e is supported by the authors' recent time-series simulation showing climate over the last 122 000 years. Simulations are consistent with proxy evidence and indicate that global average surface-air temperatures and global average sea-surface temperatures reached their highest levels (13.3 °C and 18.1 °C, respectively) between 122 000 and 115 000 years ago. Temperatures this warm returned only recently within the last two thousand years. In Greenland climate simulations show 122 000 years ago air temperatures warmed to more than 4 °C greater than AD 1800. This was followed by warming up to 0.5 °C higher than AD 1800 in subtropical and equatorial regions by 118 000 years ago. However, simulations show land-surface temperatures lagged behind air temperatures and warming was not as dispersed. Land-surface warming was initially absent from Asia, Antarctica, Canada, north-east Africa and the Middle East. Although warming expanded across Africa and southern Asia by 118 000 years ago it never clearly developed poleward of 50° N and 50° S.

Climate simulation results for MIS5e also show a strong maximum global meriodonal overturning (MOT) of between 20 and 22 Sv, which is higher than today's 19 Sv (Cunningham *et al.*, 2007). Model simulations indicate that the Greenland ice sheet was thinner than today, in some places by up to 1000 m. Southern hemisphere sea-ice area and sea-ice volume reached their lowest modelled levels during MIS5e (levels not seen again until only very recently), whereas northern hemisphere sea-ice extent remained higher than that modelled for the Holocene (last 11 650 years). Simulated atmospheric humidity levels were greater during MIS5e than AD 1800, particularly in Europe and Africa. A belt of raised humidity gradually developed between 20° N and 50° S which intensified through 118 000 years ago. Alternatively, northern hemisphere and Antarctic humidity levels fell below AD 1800 levels by 122 000 and 120 000 years ago, respectively, and remained so through 118 000 years ago (Figure 5.8). Increases in surface specific humidity are associated with global warming and are related to increases in heavy precipitation events (Weaver, 2008, p. 175).

5.5.2.2 *Marine isotope stage 5d (116 000 to 110 000 years ago)*

Proxy evidence By 116 000, glacial inception had begun (Imbrie *et al.*, 1984). Between 116 000 and 110 000, rapid cooling resulted in a sudden drop of about 65 m in global eustatic sea level. Sea level fell from the LGC high experienced during MIS5e of 6 m above present to about 60 m below present (Yoshimori *et al.*, 2002; see also Cutler *et al.*, 2003; Lambeck *et al.*, 2002). Coincidently global average sea-surface temperatures dropped from 20 °C to below 18 °C (Imbrie *et al.*, 1984). A strong reduction in surface-air temperature is evident from the Vostok ice-core records (see Chapter 1 Figure 1.2 above). This appears coincident with a rapid reduction in atmospheric methane concentrations. However, for reasons not yet clear to researchers, the extent of the coincident rapid lowering of atmospheric CO_2 is not nearly as great as that observed in subsequent cold intervals.

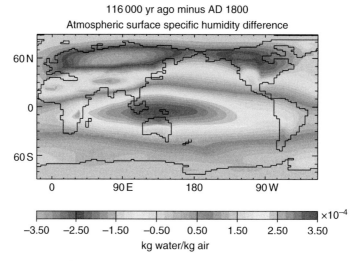

Figure 5.8 Atmospheric surface specific humidity anomaly for 116 000 years ago relative to AD 1800. Figure shows simulated surface specific humidity was higher 116 000 years ago than AD 1800 in the equatorial regions and lower across the northern hemisphere. (See colour version in colour plate section.)

Climate simulations Consistent with proxy evidence, our climate simulations show global average surface-air temperatures declining steadily after about 115 000 years ago, dropping from a high of 13.3 °C at 115 000 years ago to 11.6 °C at 107 000 years ago. This drop was coincident with a sudden drop in the maximum global MOT from about 22 Sv to about 17 Sv. Although global surface-air temperatures remained higher than AD 1800 until 112 000 years ago, model output indicates that regions poleward of 60° N and 60° S always remained colder than AD 1800 throughout the entire interval. Global average precipitation, surface specific humidity and global average sea-surface temperatures peaked at 114 000 years ago before suddenly and continuously dropping until 107 000 years ago (Figure 5.9 and Figure 5.10). Total global vegetation carbon, soil carbon, net primary productivity, and gross primary productivity all reached their second-highest levels modelled at 114 000 years ago, after which they rapidly dropped until 107 000 years ago. By 114 000 years ago, southern hemisphere sea-ice volume had reached its lowest levels modelled, while northern hemisphere sea-ice volume was gradually building from the extremely low levels it had experienced during MIS5e. While the Greenland ice sheet thinned during this interval, the Antarctic ice sheets thickened.

5.5.2.3 Marine isotope stage 5c (110 000 to 100 000 years ago)

Proxy evidence This brief, highly variable interval began about 110 000 years ago and lasted until about 100 000 years ago. It was a very important interval in human

Global average precipitation
122 000 years ago to present

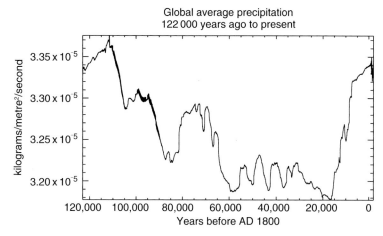

Figure 5.9 Global average precipitation for 122 000-year time-series climate simulation performed on the UVic Earth system climate model (UVic ESCM).

history. It was clear that by this time *Homo sapiens* were leaving Africa for the Middle East. Although surface-air temperatures were warmer than the previous MIS5d interval, they were still 2 °C colder than today at Vostok, Antarctica. Global sea levels rose from the low levels evident during MIS5d to about 20 m lower than today before retreating back down to about 55 m below present around 103 000 years ago and then rising back up to about 38 m below present by 100 000 years ago (Lambeck *et al.*, 2002). Global average sea-surface temperatures also warmed slightly, but still remained cooler than present (Imbrie *et al.*, 1984). However, off the west coast of Africa, sea-surface temperatures in the Angola basin dropped by about 4 °C by around 110 000 years ago (Schneider *et al.*, 1995).

Climate simulations Consistent but slightly lagging proxy evidence, our time-series climate simulations indicate a variable, but relatively warm interval persisted from about 107 000 to 97 000 years ago. During this time, global average surface-air temperatures oscillated between 11.6 and 12 °C, or between about 1.1 and 1.4 °C colder than AD 1800, and maximum global MOT remained between 16 and 17 Sv. Higher-latitude regions experienced greater cooling relative to AD 1800 than lower-latitude regions. Global average sea-surface temperature, surface specific humidity levels, and precipitation levels oscillated around levels reached 107 000 years ago. However, northern hemisphere precipitation developed a downward trend, whereas southern hemisphere precipitation increased. Global average net primary productivity generally improved over a low reached 107 000 years ago, as did total global vegetation carbon and soil carbon. Northern hemisphere ice stabilized around 1.1×10^{13} m^2, and southern-hemisphere sea ice rapidly increased from the low experienced prior to this interval. Ice developed in north-eastern North

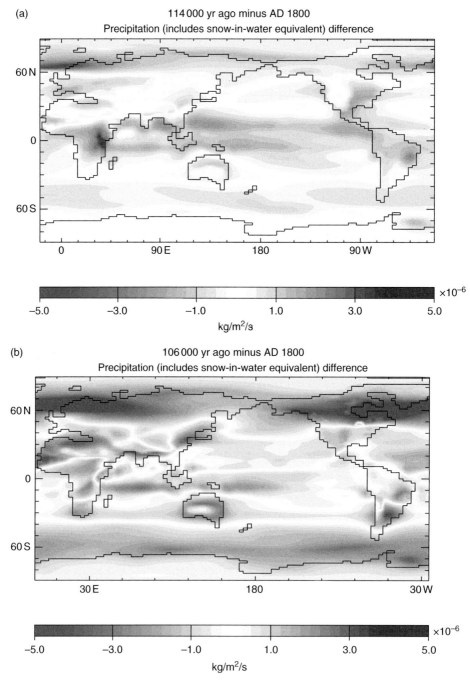

Figure 5.10 Global average precipitation anomaly for (a) 114 000 years ago and (b) 106 000 years ago relative to AD 1800. Global average precipitation derived from 122 000-year time-series climate simulation performed on the UVic Earth system climate model (UVic ESCM). (See colour version in colour plate section. Red indicates lower precipitation levels relative to AD 1800 and blue represents higher precipitation levels relative to AD 1800.)

America and northern Europe reaching thicknesses of over 2000 m, while ice still remained thinner in Greenland than it was in AD 1800. Ice building persisted until about 102 000 years ago, when ice began to retreat, although not as extensively as observed by AD 1800.

5.5.2.4 Marine isotope stage 5b (100 000 to 90 000 years ago)

Proxy evidence The brief relative warming of MIS5c was followed by a mild glacial interval that lasted from 100 000 years ago until about 90 000 years ago. Proxy evidence suggests that surface-air temperatures over Vostok, Antarctica, were about 6 °C colder than today. Atmospheric CO_2 fell to just above 200 ppm, and CH_4 to about 440 ppb. At the onset of MIS5b, global eustatic sea level was about 38 m below present but quickly fell to about 55 m below present before rising to 30 m below present by the end of the interval.

Climate simulations Climate simulations indicate that global average surface-air temperatures rapidly dropped by more than 1 °C from nearly 12 °C at 97 000 years to about 10.5 °C at 89 000 years ago resulting in temperatures about 2.5 °C colder than AD 1800. Temperatures remained below 11 °C until about 85 000 years ago. At the onset of MIS5b maximum global MOT fell from 17 Sv to 16 Sv where it oscillated until 97 000 years ago before falling to 15 Sv 95 000 years ago and then suddenly dropping to less than 14 Sv by 89 000 years ago. Global average precipitation and net primary productivity levels began to plummet after 97 000 years ago and by 89 000 years ago had reached their lowest levels since the onset of the LGC. Global sea-ice area and volume rapidly oscillated, particularly in the southern hemisphere, throughout most of MIS5b before peaking at 89 000 years ago. Global snow volume rapidly escalated for the first time since the onset of the LGC reaching a peak around 89 000 years ago.

5.5.2.5 Marine isotope stage 5a (90 000 to 75 000 years ago)

Proxy evidence MIS5b was followed by another brief relative warming with temperatures reaching about 2 °C colder than today during an interval that spanned from 90 000 to 75 000 years ago. Fossils of *Homo sapiens* uncovered in the Middle East have been dated to this time. Sea levels oscillated between a high of about 25 m below present before falling to over 65 m below present levels. Global average sea-surface temperatures warmed slightly.

Climate simulations Our climate simulations show a brief warming occurred 89 000 years ago. However, consistent warming did begin until 87 000 years ago. Simulations also indicate that global average surface-air temperatures rose to about 11.5 °C or about 1.5 °C colder than AD 1800 by 82 000 years ago and remained around this level until 75 000 years ago when they began to drop. Global MOT steadily rose with the onset of MIS5a from the low of 14 Sv seen 89 000 years ago to over 16.5 Sv by 85 000 years ago when a brief, but sudden drop to less than 15 Sv was sustained. As quickly as global MOT dropped, it spiked, reaching more than 17 Sv by 82 000 years ago. Global MOT remained between 16 and 17 Sv until 75 000 years ago when it again suddenly fell. Climate simulations support proxy evidence and show that sea-surface temperatures warmed slightly from a low of about 16.5 °C experienced at the end of MIS5b to about 17 °C by the end of MIS5a. Global average precipitation levels remained low until about 84 000 years ago when they rapidly rose, almost but not quite reaching levels seen in MIS5c where they remained until about 75 000 years ago. Global average surface specific humidity levels reached their lowest yet observed during the LGC at 90 000 years ago before gradually climbing almost to levels seen during MIS5c by 84 000 years ago where they remained until 75 000 years ago. Global total net primary productivity reached its lowest level yet achieved during the LGC at 90 000 years ago and again at 87 000 years ago. Global total net primary productivity then increased until the end of MIS5a 75 000 years ago, despite brief drops at 85 000 years ago and 80 000 years ago.

Global sea-ice area fell after 89 000 years ago reaching an interval low by 85 000 years ago before expanding only slightly towards the end of MIS5a. Global sea-ice volume on the other hand began to gradually fall after 89 000 years ago reaching an interval low at 82 000 years ago and then rapidly increasing for the balance of MIS5a. Sea-ice response varied between hemispheres. Northern hemisphere sea-ice area and volume suddenly decreased 89 000 years ago; sea-ice area reached an interval low at 85 000 years ago where it remained until 83 000 years ago when it quickly expanded. Northern hemisphere sea-ice volume did not reach an interval low until 82 000 years ago after which it rapidly expanded. Alternatively, southern hemisphere sea-ice area and volume continued to decrease after 87 000 years ago until the end of MIS5a.

Global snow volume reached its highest levels yet during the LGC at 90 000 years ago and then 84 000 years ago rapidly dropped hitting an interval low at 80 000 years ago before rapidly expanding through the end of MIS5a. Ice sheets were significantly more extensive than today 90 000 years ago extending across much of what is now eastern, central and parts of western Canada as well as northern Europe. Although ice sheets reduced their extent somewhat by 86 000 years ago, by 84 000 years ago ice had once again extended virtually across all of Canada, Iceland, northern Europe and parts of northern Asia. By 80 000 years ago ice had retreated from most of Canada and northern Europe. However, by 76 000 years ago ice was once again expanding its extent. Although ice sheets in Antarctica were thicker than AD 1800 for the entire duration of MIS5a, particularly between 0 and 180° W, the

Greenland ice sheet was at times more than 460 m thinner than AD 1800, particularly between 80 000 and 77 000 years ago.

5.5.2.6 Marine isotope stage 4 (75 000 to 60 000 years ago)

Proxy evidence By 75 000 years ago the last glaciation had evolved into a full glacial, known as the lower Pleniglacial, which lasted until 60 000 years ago. MIS4 does not appear to have been as intensive a glacial interval as either MIS6 or MIS2 with global eustatic sea level falling to less than 70 m below today relative to beyond 120 m below present levels in MIS6 and MIS2 (Lambeck *et al.*, 2002). However, temperatures did plummet to about 9 °C colder than today over Vostok, Antarctica (Petit *et al.*, 1999). Ice sheets were more extensive than today, extending over much of Canada, northern Europe, north-central Asia, as well as the southern tip of South America. Sea-surface temperatures dropped to a low of around 15.5 °C by around 65 000 years ago before beginning to warm slightly through the end of MIS4 (Imbrie *et al.*, 1984).

Climate simulations Consistent with proxy evidence, our climate simulations show global average surface-air temperatures fell from a high of about 11.5 °C at the onset of MIS4 to a low of less than 10 °C 62 000 years ago. Surface-air temperatures then climbed to 10.5 °C by 58 000 years ago. Maximum global MOT fell from a high of 17 Sv at the beginning of MIS4 to a low of less than 13 Sv 68 000 years ago. MOT remained at or below 14 Sv until about 58 000 years ago when it suddenly jettisoned to nearly 16 Sv. Consistent with proxy evidence, simulations show global average sea-surface temperatures fell to a low around 16 °C about 65 000 years ago, although warming did not begin until 60 000 years ago. Simulated global average precipitation levels also declined, reaching their lowest levels by 62 000 years ago and remaining there until just after 58 000 years ago. At about 61 000 years ago global total net primary productivity hit a low not seen in the simulations since the onset of the LGC and not observed again until MIS2.

Ice sheets were gradually building throughout MIS4 despite a slight retreat 68 000 years ago. Maximum ice extent was not reached until 58 000 years ago when ice covered northern Europe, Iceland, Greenland, north-central Asia and most of northern North America with the exception of what is now Alaska and parts of the Yukon Territory. Simulations suggest that the Antarctic ice sheet thickened particularly between 0 and 180° W, while the central area of the Greenland ice sheet remained thinner than AD 1800 throughout MIS4. Global snow volume peaked between 60 000 and 58 000 years ago. Global sea-ice area reached the greatest extent seen during the entire LGC at 66 000 years ago before rapidly dropping after 61 000 years ago. However, maximum sea-ice extent was reached between about 69 000 and 68 000 years ago in the northern hemisphere before rapidly dropping.

Alternatively maximum sea-ice extent was not reached in the southern hemisphere until between about 64 000 years ago and 61 000 years ago.

5.5.2.7 *Marine isotope stage 3 (60 000 to 30 000 years ago)*

Proxy evidence It was after the cold, dry MIS4 stage that climate became highly unstable. It was generally cooler and drier than present for the duration of MIS3, which lasted from about 60 000 until 30 000 years ago. Ice-core records show that temperatures oscillated between about 4 and 8 °C colder than today over Vostok, Antarctica (Petit *et al.*, 1999). Low-latitude regions also experienced significant cooling. Temperature-dependent amino acid racemization of emu eggshell fragments indicate that beginning at least 45 000 years ago, when early modern humans first arrived in Australia, until about 16 000 years ago, average subtropical air temperatures in the continental interior of Australia were at least 6 °C cooler than present (Miller *et al.*, 1997). Sea level oscillated around a high of about 45 m below present at 52 000 years ago before falling to about 67 m below present just before 30 000 years ago when it plunged to 115 m below present (Lambeck *et al.*, 2002). Global average sea-surface temperatures cooled about 1 °C throughout this interval (Imbrie *et al.*, 1984).

Climate simulations Our climate simulations support proxy evidence indicating a highly unstable MIS3. Global average surface-air temperatures oscillated around 10 °C, or about 3 °C cooler than today for the duration of MIS3. Highs of about 10.5 °C were reached between 57 000 and 55 000 years ago, between 43 000 and 42 000 years ago, and again between about 40 000 and 38 000 years ago. Global average surface-air temperatures dropped to about 9.5 °C 45 000 years ago and again 38 000 years ago. Simulated global average precipitation levels were also highly variable oscillating between LGC lows reached between about 61 000 and 58 000 years ago, around 50 000 years ago, 45 000 years ago, and again around 38 000 years ago and intermittent increases between 57 000 and 55 000 years ago, 49 000 years ago, 44 000 and 41 000 years ago, 35 000 years ago and 32 000 years ago (see Figure 5.9).

   Model simulations indicate that maximum global MOT was also highly variable during MIS3. It peaked at about 16 Sv four times during MIS3, between 57 000 and 54 000 years ago, about 50 000 years ago, again at 41 000 years ago, and finally between 35 000 and 33 000 years ago. These were interceded by drops, the greatest of which occurred at around 44 000 years ago when maximum global MOT reached the lowest level seen during the LGC, and was matched only by that subsequently achieved during the LGM (see Figure 5.6). Simulations show global average sea-surface temperatures oscillated between about 16.5 °C and 16 °C throughout this interval.

Simulations indicate that there appeared to be little change in the extent of ice sheets subsequent to the end of MIS4 with ice sheets still extending over much of what is now Canada, northern Europe, north-central Asia and the southern tip of South America. These conditions persisted throughout MIS3, although there were times when ice did retreat from western Canada, for example between 50 000 and 48 000 years ago, and again between 44 000 and 42 000 years ago. Ice sheets progressively thickened until about 38 000 years ago, in places reaching over 4200 m thicker than AD 1800 and then thinned slightly for the duration of MIS3. Simulations show global average snow volumes oscillated around the peak they reached during MIS4 at 58 000 years ago before rapidly escalating around 30 000 years ago. Global sea-ice area fell to levels attained during MIS5a where it oscillated before expanding about 40 000 years ago and again about 34 000 years ago. These increases appear to have been stimulated to a large extent from expansions of sea ice in the southern hemisphere.

Although global total net primary productivity made intermittent gains between 58 000 and 55 000 years ago, and again at 49 000 years ago, 44 000 and 41 000 years ago, it also dropped between 55 000 and 52 000 years ago, 45 000 years ago, and finally at 38 000 years ago after which it did not recover again until after the last glacial maximum around 17 000 years ago.

5.5.2.8 *Marine isotope stage 2 and the last glacial maximum (30 000 to 11 650 years ago)*

Proxy evidence MIS2, commonly referred to as the last ice age, began around 30 000 years ago. Glacial ice was most extensive around 21 400 years ago during what is termed the last glacial maximum (LGM) as glacial ice extended over Europe, most of the northern half of North America, and parts of Asia and South America. Global eustatic sea level reached 120 m below present levels. Continental shelves around the world were exposed including the huge subcontinent of Beringia in the Bering Strait that linked north-east Asia with North America and blocked the Bering and Artic Seas from the Pacific Ocean. It is the exposed Beringian continental shelf that is believed to have been critical in the peopling of the Americas.

At 14 600 years ago the Bølling–Allerød interstadial event occurred. A large increase in the strength of the AMO was responsible for a 150 per mil decrease in atmospheric radiocarbon and a 3 °C increase in annual mean sea-surface temperature over about 500 years (Weaver *et al.*, 2003). Global eustatic sea level rose by about 20 m during this same interval. Concurrent with the Bølling–Allerød interstadial event was an exceptionally large melting event – meltwater pulse 1A (mwp-1A) – resulting from the melting of continental ice sheets that had accumulated during the last glacial period (Clark *et al.*, 1996, 2002; Weaver *et al.*, 2003). The source of the meltwater has been the subject of some controversy, and although the Laurentide ice sheet in North America is commonly cited as its source, other researchers suggest that partial collapse of the Antarctic ice sheet led to a strengthening of AMO, warming the North Atlantic (Clark *et al.*, 2002; Weaver *et al.*, 2003). This also provides an explanation for the onset of the Bølling–Allerød warm interval and a coincident cooling – the Antarctic cold reversal – in the southern hemisphere, which was associated with the warm interval in the north (Weaver *et al.*, 2003). Global sea level rose by 20 meters in less than 500 years (Fairbanks, 1989), air temperature warmed to near-interglacial levels, particularly in Europe, and continental ice sheets retreated significantly.

Within a few decades, North Atlantic warming was interrupted by the last known Dansgaard–Oeschger event (after the pioneering work of Dansgaard *et al.*, 1984; Oeschger *et al.*, 1984) – the Younger Dryas (sometimes also called Heinrich event 0 [H0]), an abrupt, brief cold interval between 12 700 and 11 650 years ago, during which sea-surface temperatures in the Atlantic dropped by several degrees (Dansgaard *et al.*, 1989). According to geochemist W. S. Broecker (1995), the Younger Dryas cooling event was triggered by the sudden discharge of glacial Lake Agassiz, a very large proglacial lake at the head of the Laurentide ice sheet in North America. At its largest, Lake Agassiz reached nearly 1 million km^2 (Leverington *et al.*, 2000). During discharge, huge amounts of glacial meltwater are thought to have suddenly flushed into the St. Lawrence River, resulting in a likely slowdown of the AMO. Coincident cooling was also evident in the Pacific Ocean.[2] Within a few decades the Younger Dryas abruptly terminated, setting the stage for the onset of the Holocene, a period of remarkable relative climate stability (Weaver, 2004).

Climate simulations Our simulations indicate global average surface-air temperatures dropped steadily with the onset of MIS2, remaining at about 9 °C, or about 3 °C cooler than today until around 18 000 years ago. Temperatures then rapidly increased by more than 2 °C until about 12 500 years ago, when a brief cooling, coincident with the Younger Dryas, occurred. Consistent with proxy evidence, global average sea-surface temperatures cooled between 30 000 and 20 000 years ago, and around 18 000 years ago sea-surface temperatures warmed rapidly, although the rise was not as large as indicated by proxy evidence. Subsequent to the Younger Dryas event, temperatures rapidly rose into the Holocene. With few minor interruptions, global average precipitation levels continued to fall from the onset of MIS2 until about 18 000 years ago, when they spiked upward with only a brief fall again during the Younger Dryas cool interval at about 12 000 years ago. Maximum global MOT dropped to 13 Sv at the LGM and then quickly rose with numerous large perturbations between about 16 000 years ago and the end of MIS2 when it reached to close to 19 Sv. Global total net primary productivity steadily declined reaching its lowest level during the LGC at 18 000 years ago before rapidly rising through the end of MIS2.

Our simulations show ice sheets progressively extended their coverage until the LGM, when ice covered much of Canada, northern Europe, north-central Asia, Greenland, Antarctica and the southern tip of South America. In locations of maximum thickness, ice sheets reached nearly 4 km thicker than AD 1800 before steadily thinning. By 14 000 years ago, ice was retreating from northern Europe and western Canada, and by 10 000 years ago, ice remained only in the northern reaches of Eurasia, north-eastern Canada, Greenland and Antarctica. Climate simulations indicate that during LGM, global average snow volume reached levels just higher than those seen during MIS6. With the exception of brief delays at around 16 000 years ago and again at 12 000 years ago, which coincide with Heinrich events 1 and 0, simulations show snow volume rapidly and consistently dropping before reaching AD 1800 levels at about 9000 years ago. Global sea-ice volume reached its highest levels modelled during the LGC at LGM, and then rapidly dropped after 20 000 years ago reaching slightly below AD 1800 levels around 10 000 years ago. This pattern was mimicked by northern sea-ice volume, whereas southern sea-ice volume did not begin to drop until about 17 000 years ago.

5.5.2.9 The Holocene (11 650 to present or AD 1800)

Proxy evidence The Younger Dryas cooling event ended abruptly. Its culmination was coincident with the onset of the warm and relatively stable Holocene epoch (for a discussion of the exceptional nature of the Holocene's extreme stability see Dansgaard *et al.*, 1993), which began around 11 650 years ago, and with the extinction of Pleistocene mammals in North America (Dyke, 2005). This interval also witnessed the rise of human civilizations and concentrated population centres, the domestication of plants and animals and the Industrial Revolution.

The warming of the early Holocene persisted until a relatively mild cold climate event occurred 8200 years ago. It was triggered, scientists say, by the collapse of an ice dam holding back an immense amount of freshwater from two large glacial lakes in Canada's Hudson Bay region (Barber *et al.*, 1999). Temperatures over Greenland dropped by about 4 °C, whereas temperatures over Germany dropped by slightly more than 0.5 °C (Von Grafenstein *et al.*, 1998). Temperatures rapidly recovered so that by 8000 years ago they were once again warmer than today.

The mid-Holocene, an interval spanning between about 7000 and 5000 years ago, was generally warmer than today. However, warmer temperatures were limited to the summer in the northern hemisphere. Temperature changes were much less significant in the lower latitudes. With the commencement of the Industrial Revolution, temperatures again warmed, and by the late twentieth and early twenty-first centuries, temperatures had exceeded those observed during at least the previous 1200 years.

Climate simulations Simulated global average surface-air temperatures continued their rapid rise after the brief and minor cooling coincident with the Younger Dryas, reaching more than 13 °C at their height in the recent Holocene (see Figure 5.5). After the brief fall in precipitation during the Younger Dryas, global average precipitation levels rapidly rose, peaking just within the last few hundred years at levels slightly below the LGC high modelled at 114 000 years ago (see Figure 5.9). Maximum global MOT continued its dramatic rise after a drop of about 1 Sv coincident with the Younger Dryas cooling event (see Figure 5.6). Another drop of about the same magnitude occurred around 8200 years ago after which MOT again spiked reaching over 21 Sv by about 8000 years ago. Climate simulations indicated that MOT then stabilized around 20.5 Sv from about 7000 years ago until the last few hundred years when oscillations of up to 1 Sv began.

With the onset of the Holocene our simulations show ice sheets rapidly retreating. By 10 000 years ago, the Cordilleran ice sheet in western Canada had virtually disappeared and conditions in the southern tip of South America were much as they are today. By 7000 years ago ice extent was virtually the same as today, although ice remained thicker than AD 1800 in Greenland and Antarctica, particularly western Antarctica, until about 6000 years ago. Simulations indicate that snow volumes dropped precipitously with the onset of the Holocene and remained very low right through to the present. Northern sea-ice volume rose nominally after about 10 000 years ago, whereas southern sea-ice volume maintained the fall that initiated around 17 000 years ago.

5.5.2.10 The Anthropocene

A new geological epoch has been suggested for the recent period of Earth's history, beginning in the early 1800s, when human activities began to have a noticeable impact on Earth's landforms. Impacts include increased erosion and denudation of continents through agriculture, construction and damming of major rivers; a large increase in carbon dioxide and methane levels since pre-industrial times; rising global average temperatures; accelerated animal and plant extinctions and biotic population declines particularly on coral reefs and oceans; global species' migrations; replacement of natural vegetation with agricultural monocultures; rising sea levels and acidification of surface ocean waters (Zalasiewicz *et al.*, 2008). As we progress through this epoch, temperatures are expected to continue to rise and result in habitat changes that are beyond the environmental tolerances for many taxa. The consequences are expected to be more dire than previous climatic transitions because many migration alternatives that were previously available to species have been eliminated (Thomas *et al.*, 2004).

5.6 Conclusion

Earth's climate changes when the energy stored in the climate system is varied, due to changes in incoming solar radiation or outgoing energy from the Earth, for example Milankovic cycles, which alter the amount, timing and location of incoming solar radiation. However, Milankovic cycles must be amplified to promote large climate changes, including the ice-sheet growth that is generated during glacial phases. The largest amplifier is the feedback associated with sea ice, land ice and snow. Other natural climate-system components include the atmosphere, the oceans, biological productivity and the hydrological cycle. Both climate modelling and proxy evidence show a natural characteristic pattern of gradual cooling followed by abrupt warming (see, for example, Martrat *et al.*, 2004). The last ice age was followed by one of these rapid warming events that culminated in the more stable Holocene epoch. With the onset of the Industrial Revolution, humans also joined the list of climate-forcing agents. Through fossil fuel burning, cement production, changing land-use patterns and agricultural processes, humans now emit large amounts of greenhouse gases into the atmosphere, which have raised atmospheric CO_2 levels beyond any observed in the 650 000-year ice-core record.

Scientists use proxy indicators from ice, ocean and lake cores, and glacial and loess deposits to obtain information about how climate has changed and the resultant impacts on the environment. Fossil corals from reef terraces have been used to determine changing glacio-eustatic sea level. Using glacial striations, drop stones and erratics geologists can determine when and where ice advanced and retreated in past ice ages. Climate models are used in conjunction with proxy data to predict long-term changes in climate both in the past and into the future.

These data have provided a greater understanding of the complexity and variability of the climate over the last 135 000 years of modern human history. The next chapter reviews the impact that changing climate had on the landscape. Of particular interest are those regions in which early humans emerged and into which they subsequently migrated. The connection between changing vegetation and climate and humans is the subject of the following section on the interaction between climate and humans.

Notes

1. Ice-sheet simulations were performed by Shawn Marshall, University of Calgary with input from Michael Eby, University of Victoria. Ice-sheet simulations for North America and Eurasia were based on Marshall *et al.* (2000); Greenland was based on Marshall and Cuffey (2000); application to the northern hemisphere is outlined in Marshall and Koutnik (2006). Note: in all cases, climate forcing is based on a perturbation of mean monthly present-day climate fields (precipitation and temperature). For Greenland, the perturbation that is applied comes directly from the Summit ice cores. For the rest of the northern hemisphere the forcing applied is based on a linear interpolation between present-day and LGM climate fields from the CCCma model (Vettoretti *et al.*, 2000). The

interpolation is based on 100-year delta ^{18}O values from the Summit, Greenland, ice cores. We assign LGM a 'glacial index' of 1 and present day a glacial index of 0, and intermediate δ^{18}O values give a glacial index between 0 and 1. A value of 0.5 would mean a 50–50 blend of LGM and modern climatology, with a compensation for altitude based on the contemporary ice topography. This is a crude climate forcing that drives the ice sheets through a glacial cycle but should not be mistaken for 'realistic' or freely modelled climate.

2. For example, for cooling evident in the Santa Barbara Basin see Kennett and Ingram (1995) and Behl and Kennett (1996). For cooling in the north-east Pacific see Lund and Mix (1998).

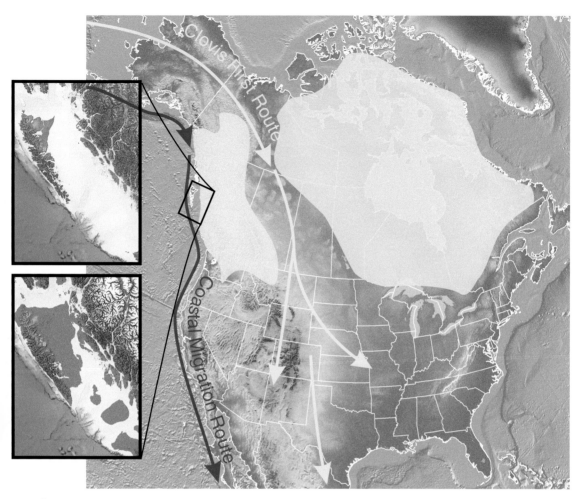

Figure 4.3

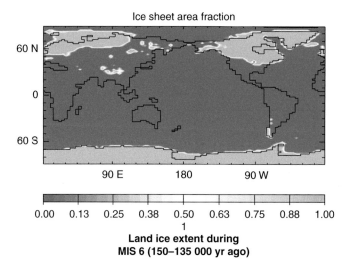

Ice sheet area fraction

0.00 0.13 0.25 0.38 0.50 0.63 0.75 0.88 1.00
                        1

**Land ice extent during
MIS 6 (150–135 000 yr ago)**

Figure 5.7

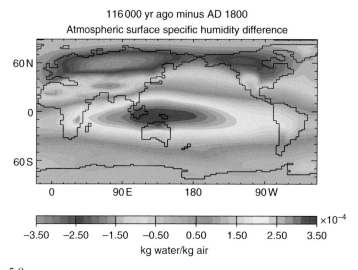

116 000 yr ago minus AD 1800
Atmospheric surface specific humidity difference

−3.50 −2.50 −1.50 −0.50 0.50 1.50 2.50 3.50 ×10⁻⁴

kg water/kg air

Figure 5.8

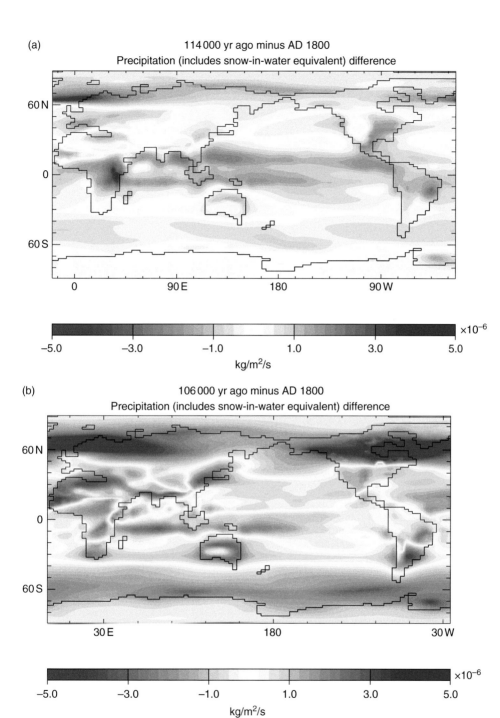

(a) 114 000 yr ago minus AD 1800
Precipitation (includes snow-in-water equivalent) difference

×10⁻⁶

−5.0   −3.0   −1.0   1.0   3.0   5.0
kg/m²/s

(b) 106 000 yr ago minus AD 1800
Precipitation (includes snow-in-water equivalent) difference

×10⁻⁶

−5.0   −3.0   −1.0   1.0   3.0   5.0
kg/m²/s

Figure 5.10

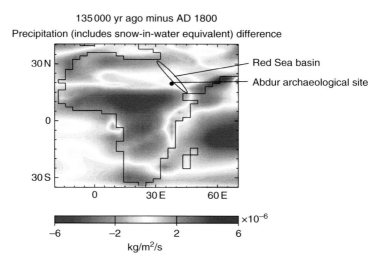

135 000 yr ago minus AD 1800

Precipitation (includes snow-in-water equivalent) difference

Figure 6.1

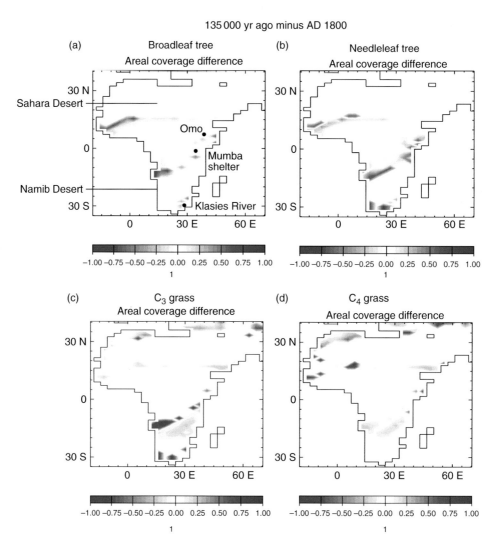

135 000 yr ago minus AD 1800

Figure 6.2

122 000 yr ago minus AD 1800

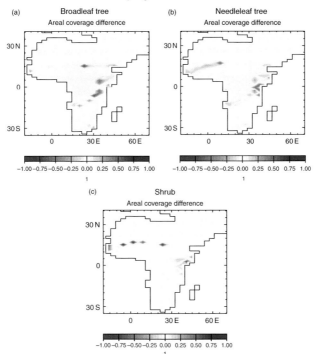

Figure 6.3

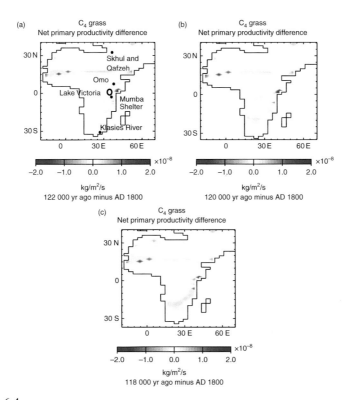

Figure 6.4

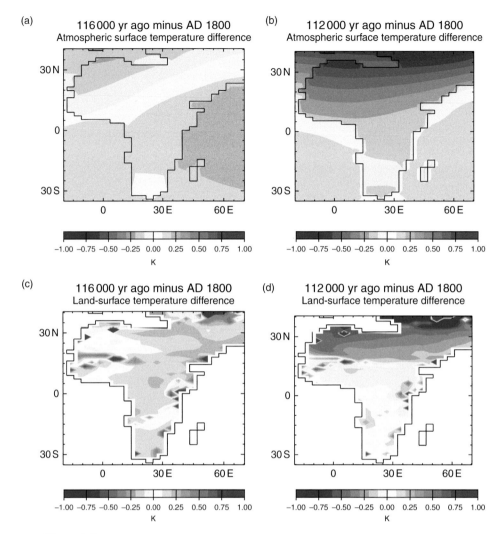

Figure 6.5

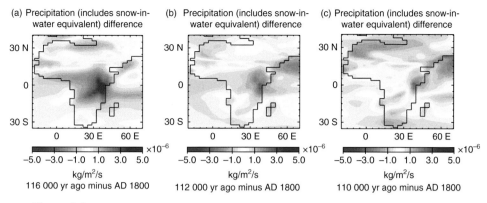

(a) Precipitation (includes snow-in-water equivalent) difference

(b) Precipitation (includes snow-in-water equivalent) difference

(c) Precipitation (includes snow-in-water equivalent) difference

$\times 10^{-6}$
−5.0 −3.0 −1.0 1.0 3.0 5.0
kg/m²/s
116 000 yr ago minus AD 1800

$\times 10^{-6}$
−5.0 −3.0 −1.0 1.0 3.0 5.0
kg/m²/s
112 000 yr ago minus AD 1800

$\times 10^{-6}$
−5.0 −3.0 −1.0 1.0 3.0 5.0
kg/m²/s
110 000 yr ago minus AD 1800

Figure 6.6

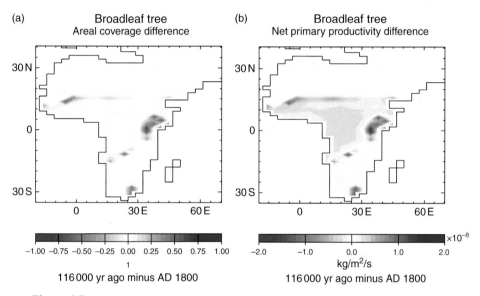

(a) Broadleaf tree
Areal coverage difference

(b) Broadleaf tree
Net primary productivity difference

−1.00 −0.75 −0.50 −0.25 0.00 0.25 0.50 0.75 1.00
1
116 000 yr ago minus AD 1800

$\times 10^{-8}$
−2.0 −1.0 0.0 1.0 2.0
kg/m²/s
116 000 yr ago minus AD 1800

Figure 6.7

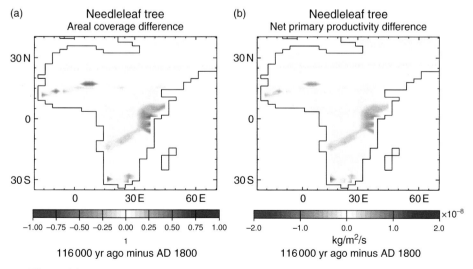

(a) Needleleaf tree
Areal coverage difference

(b) Needleleaf tree
Net primary productivity difference

−1.00 −0.75 −0.50 −0.25 0.00 0.25 0.50 0.75 1.00
1
116 000 yr ago minus AD 1800

$\times 10^{-8}$
−2.0 −1.0 0.0 1.0 2.0
kg/m²/s
116 000 yr ago minus AD 1800

Figure 6.8

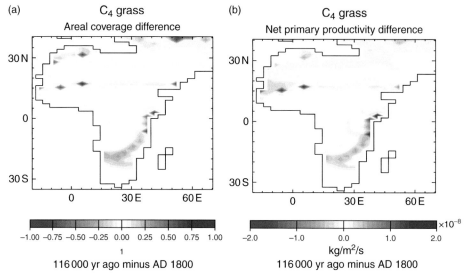

(a) C$_4$ grass
Areal coverage difference

(b) C$_4$ grass
Net primary productivity difference

30 N

0

30 S

0   30 E   60 E

−1.00 −0.75 −0.50 −0.25 0.00 0.25 0.50 0.75 1.00
1
116 000 yr ago minus AD 1800

−2.0   −1.0   0.0   1.0   2.0
×10^{-8}
kg/m^2/s
116 000 yr ago minus AD 1800

Figure 6.9

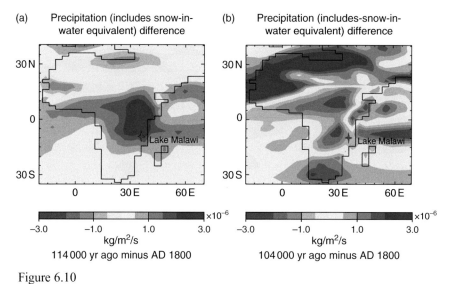

(a) Precipitation (includes snow-in-water equivalent) difference

(b) Precipitation (includes-snow-in-water equivalent) difference

30 N

0

Lake Malawi

30 S

0   30 E   60 E

−3.0   −1.0   1.0   3.0
×10^{-6}
kg/m^2/s
114 000 yr ago minus AD 1800

−3.0   −1.0   1.0   3.0
×10^{-6}
kg/m^2/s
104 000 yr ago minus AD 1800

Figure 6.10

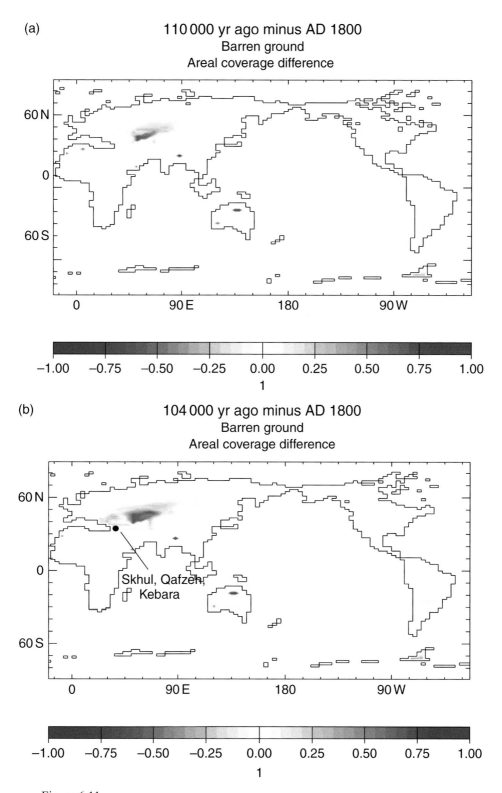

(a)

110 000 yr ago minus AD 1800
Barren ground
Areal coverage difference

60 N

0

60 S

0   90 E   180   90 W

−1.00   −0.75   −0.50   −0.25   0.00   0.25   0.50   0.75   1.00

1

(b)

104 000 yr ago minus AD 1800
Barren ground
Areal coverage difference

60 N

0

Skhul, Qafzeh,
Kebara

60 S

0   90 E   180   90 W

−1.00   −0.75   −0.50   −0.25   0.00   0.25   0.50   0.75   1.00

1

Figure 6.11

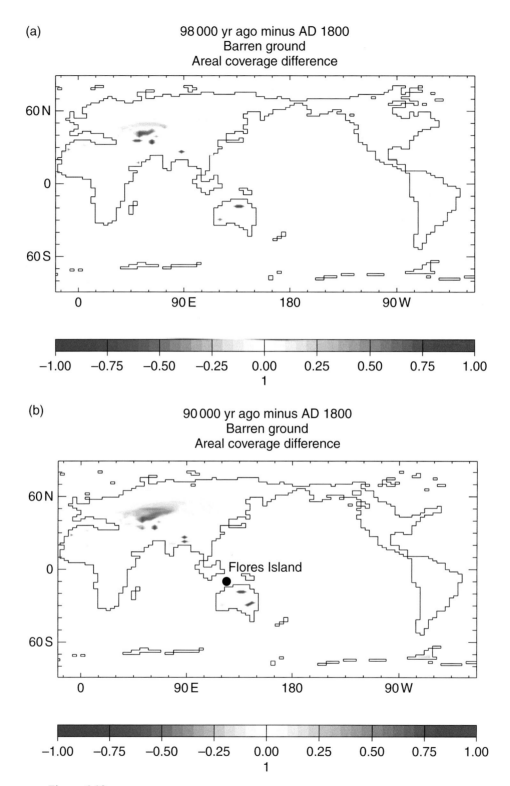

Figure 6.12

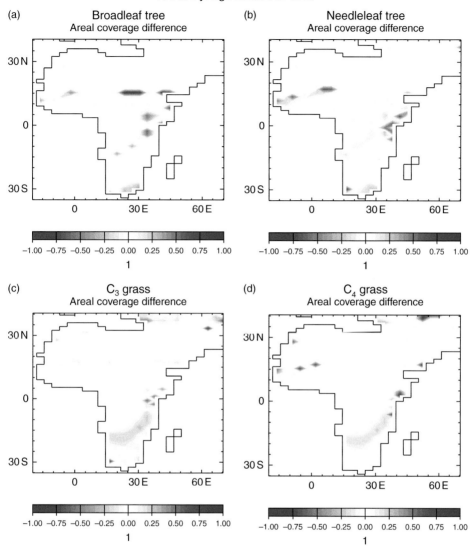

Figure 6.13

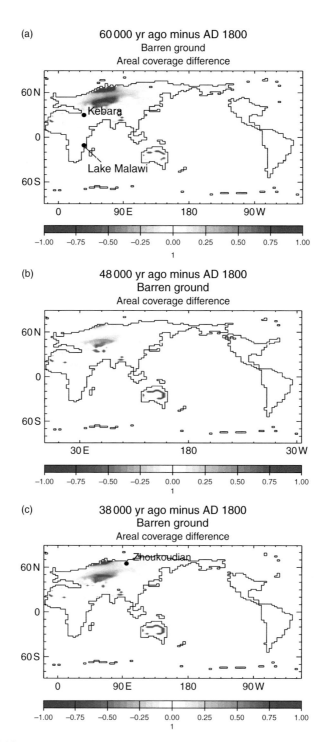

Figure 6.14

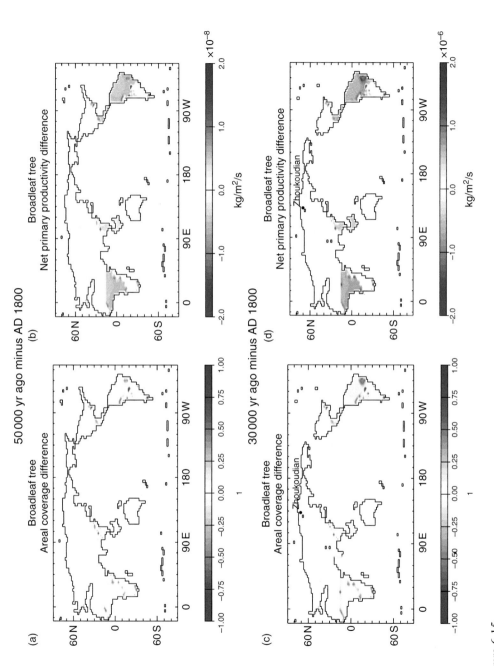

Figure 6.15

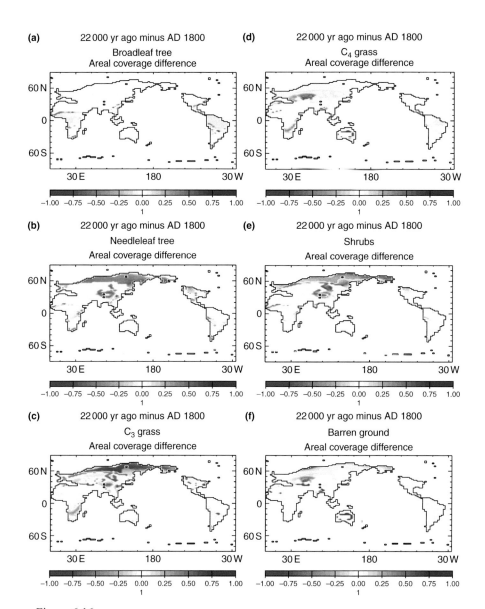

Figure 6.16

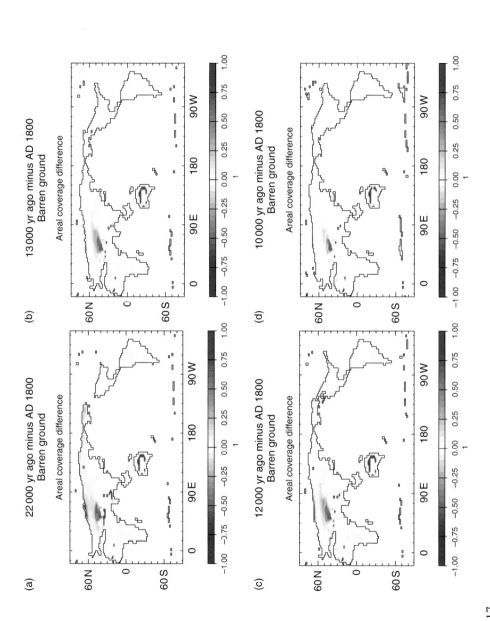

Figure 6.17

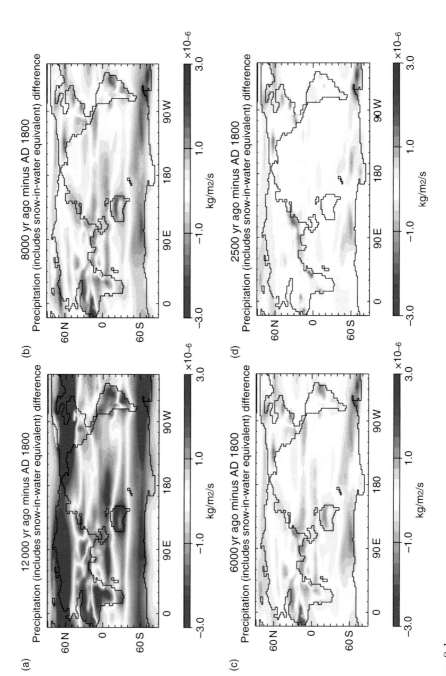

Figure 8.1

6

The effect of 135 000 years of changing climate on the global landscape

6.1 Introduction

The previous chapter provided a general review of the changing climate on Earth during, and subsequent to, the last but one glacial maximum. We reviewed proxy evidence and climate model simulations, particularly focusing on a recently completed 122 000-year transient or time-series simulation prepared for this book. In this chapter we examine the impact of that changing climate on the landscapes in which early humans lived and subsequently migrated. Of particular interest are the effects on vegetation. Vegetation provides fodder that promotes game populations and various vegetable, fruit and cereal crops for humans. Its absence implies deserts that can act as barriers to habitation and migration.

In this chapter we again use proxy evidence combined with UVic Earth system climate model (UVic ESCM) simulation results, particularly the 122 000-year time series, to ascertain changing vegetation over the LGC. The UVic ESCM consists of a three-dimensional (3D) ocean general circulation model coupled to a dynamic–thermodynamic sea-ice model, an ocean carbon-cycle model, a dynamic energy-moisture balance atmosphere model, a land-surface model and a terrestrial vegetation and carbon-cycle model (Ewen *et al.*, 2003; Matthews, Weaver, Eby *et al.*, 2003; Matthews, Weaver, Meissner *et al.*, 2003; Meissner *et al.*, 2003; Weaver *et al.*, 2001). As stated in the previous chapter, we also included changing land-ice extent and thickness in model simulations.

The coupled vegetation component of the model defines the terrestrial biosphere in terms of soil carbon, five plant functional types (PFTs) and barren ground. The PFTs are broadleaf trees, needleleaf trees, C_3 grass, C_4 grass and shrubs. Vegetation change is driven by net carbon fluxes. Carbon fluxes combined with changes in the land-atmosphere fluxes and water runoff are calculated by the land-surface model, time-averaged, and then passed on to the atmospheric, ocean and terrestrial components of the model. Changes in these fluxes influence the net

175

primary productivity of each plant type and its capacity to expand its vegetated area or coverage. It is the changes in these two model outputs – net primary productivity and coverage – that are discussed below in light of their influence on early humans.

Changes in extent and productivity of each plant type also act as a feedback further influencing climate. For example, as noted in Chapter 5, during glacial intervals, snow-covered landscapes amplify cooling in high latitudes, facilitate the retention of the winter's snowfall and stimulate the expansion of glacial and sea ice. This in turn results in further cooling and partly reduces meridional overturning. Studies validating UVic model vegetation against observed vegetation for the present and/or the past have been performed by Matthews, Weaver, Eby *et al.* (2003), Matthews, Weaver, Meissner *et al.* (2003) and Meissner *et al.* (2003).[1] Potential limitations of the simulated vegetation do exist, particularly when a shortage of palaeovegetation proxy data does not allow us to assess model results. This is particularly evident the further we go back in time. Our ability to assess the accuracy of vegetation modeling results would clearly benefit from more long sequences of pollen data, particularly in tropical monsoon regions.

We begin our review of the changing landscape in Africa, the birthplace of *Homo sapiens*, during MIS6. We then move forward in time through the last glacial cycle, expanding our analysis to those regions into which humans initially migrated and subsequently inhabited.

6.2 Marine isotope stage 6 – the changing environment of Africa, the birthplace of *Homo sapiens*

In Africa, where the first *Homo sapiens* had appeared by about 195 000 years ago, the cool, dry climate of the MIS6 glacial interval (150 000–135 000 years ago) was having its impact on the landscape. The African atmosphere was about 3.75 °C cooler than it was in AD 1800, and African land surfaces were nearly 4 °C cooler than in AD 1800. Climate simulations suggest precipitation was lower than in AD 1800 over most of Africa with the exception of North Africa and the southern tip of Africa (Figure 6.1). Pollen taken from deep-sea cores from the Gulf of Guinea concur with these simulations, indicating drier conditions generated reductions in rainforest and mangrove swamp extent. In West Africa, along the northern coast of the Gulf of Guinea, savanna grassland and open dry forest developed (Dupont *et al.*, 2000). Figure 6.2, showing simulated vegetation coverage during MIS6, supports these proxy findings. The Sahara extended southward to 14° N and the Namib Desert reached far to the north (*ibid.*). In West Africa, vegetation appears to have been similar to that typical of the last ice age (Willoughby, 2007).

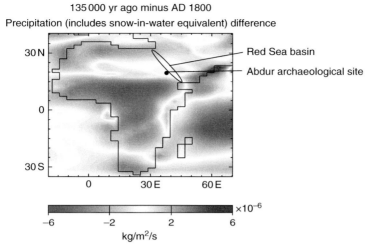

Figure 6.1   Average precipitation anomaly for Africa at 135 000 years ago relative to AD 1800. Global average precipitation derived from MIS6 equilibrium climate simulation performed on the UVic Earth system climate model (UVic ESCM). (See colour version in colour plate section. Red indicates lower precipitation levels relative to AD 1800 and blue represents higher precipitation levels relative to AD 1800.) Locations of Red Sea basin and Abdur archaeological site noted.

6.3 Marine isotope stage 5e – the Eemian interglacial

Just prior to the onset of the Eemian interglacial (MIS5e: 135 000–116 000 years ago), global eustatic sea level rose by about 120 m in a span of 5000 years between about 140 000 and 135 000 years ago. Then over the next 5000 years global eustatic sea level fell about 60 m (Lambeck *et al.*, 2002). The warm and mild MIS5e caused global sea level to rapidly rise, inundating exposed continental shelves as sea level reached about 6 m higher than today by 125 000 years ago. These, and coincident changes in climate and vegetation, had a substantial impact on the habitat of early *H. sapiens*.

6.3.1 The changing climate of Africa

With the onset of MSI5e, warm, wetter conditions may have initially improved living circumstances for *H. sapiens*. Consistent with proxy evidence suggesting wetter conditions in Africa during the Eemian, climate simulations show that by 122 000 years ago, all of Africa with the exception of central Africa was experiencing greater precipitation than today. The time-series climate simulation over the period of the last glacial cycle also shows that by 122 000 years ago, surface-air temperatures in South and central Africa had warmed to about the same as AD 1800,

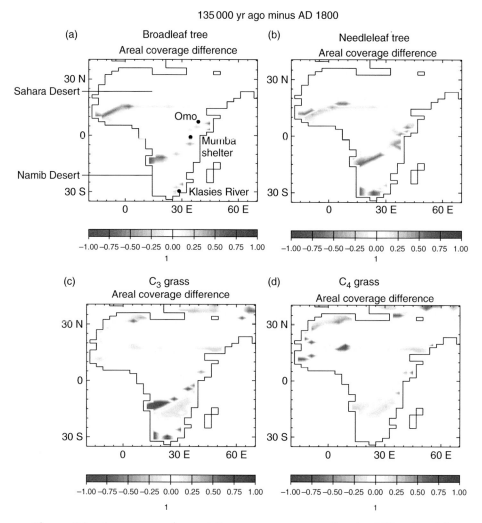

Figure 6.2   Average areal vegetation coverage anomalies in Africa for (a) broadleaf trees, (b) needleleaf trees, (c) C$_3$ grass, and (d) C$_4$ grass at 135 000 years ago relative to AD 1800. Average areal vegetation coverage derived from MIS6 equilibrium climate simulation performed on the UVic Earth system climate model (UVic ESCM). (See colour version in colour plate section. Red indicates lesser vegetation coverage levels relative to AD 1800 and blue represents greater vegetation coverage levels relative to AD 1800.) Locations of archaeological sites Omo and Klasies River noted.

temperatures in north-west Africa had become slightly warmer than AD 1800, and temperatures in north-east Africa reached levels just slightly cooler than AD 1800. Within 2000 years, South Africa remained the only region in Africa with surface-air temperatures warmer than today. However, the African land surface warmed more slowly than did the atmosphere. During MIS6 African land surfaces averaged nearly

4 °C cooler than AD 1800, but subsequent warming to about 0.3 °C warmer than AD 1800 was delayed until about 118 000 years ago. Land-surface temperature usually responds relatively quickly to changes in climate if the land cover remains the same. This delay in land-surface warming in Africa during MIS5e was thus likely the result of changes in vegetation or land cover.

6.3.2 The changing vegetation of Africa

The changing MIS5e climate caused the African rainforest and mangrove swamps to expand and dry forests and savanna to shrink (Willoughby, 2007, p. 87). Warmer, wetter conditions encouraged extensive forest development. The montane pine tree *Podocarpus* is thought to have made a large westward expansion (Dupont *et al.*, 2000, p. 116). Desert limits shifted to higher latitudes, and Africa experienced a reduction in barren land during the warm Eemian relative to the cold MIS6 glacial interval (*ibid.*, p. 113). Thus, at the outset, the amelioration of Africa's climate during MIS5e may have expanded the extent of *H. sapiens'* habitable terrain in Africa. Vegetation is believed to have been extensive covering much of today's desert areas.

However, although climate simulations show an overall reduction in barren land in Africa, more specifically they show an increase in barren ground coverage in North Africa and South Africa and a decrease in barren ground coverage in central Africa that persisted between 122 000 and 118 000 years ago. 122 000 years ago a reduction in C_3 grass productivity and coverage to less than the AD 1800 level was offset by greater coverage and productivity of C_4 grass. C_4 grass possesses a particular kind of plant metabolism that reduces water loss and uses carbon dioxide more efficiently than C_3 grass. C_3 grass is more common in humid environments, is easier for most browsers to digest and is more nutritive. African broadleaf trees expanded their coverage and experienced increased productivity. This was evident in North and South Africa, where broadleaf trees expanded their extent and productivity to levels greater than AD 1800 and along a band between central and South Africa where broadleaf productivity intensified (Figure 6.3). By 122 000 years ago, the expansion of needleleaf trees that had occurred during MIS6 along the northern border of central Africa had receded and, as with broadleaf trees, needleleaf tree productivity and extent was generally lower in central Africa than the present, and higher in North and South Africa.

Whereas the cold temperatures during MIS6 restricted the productivity and coverage of shrubs, with the onset of the warmer temperatures during MIS5e, shrubs in Africa proliferated. Between 122 000 and 118 000 years ago, shrub coverage and productivity expanded, particularly along the southern periphery of what is now the Sahara desert (see Figure 6.3 above). By 120 000 years ago, although C_3 grass productivity was still

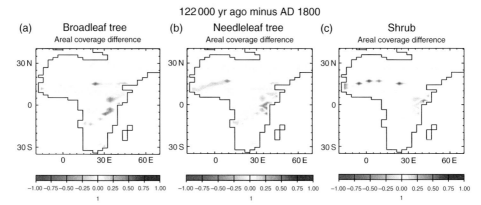

Figure 6.3   Average areal vegetation coverage anomalies in Africa for (a) broadleaf trees, (b) needleleaf trees, and (c) shrub at 122 000 years ago relative to AD 1800. Average vegetation coverage derived from 122 000-year time-series climate simulation performed on the UVic Earth system climate model (UVic ESCM). (See colour version in colour plate section. Red indicates lesser vegetation coverage levels relative to AD 1800 and blue represents greater vegetation coverage levels relative to AD 1800.)

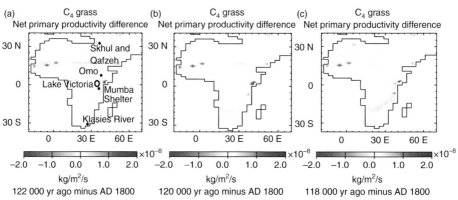

Figure 6.4   Net primary productivity anomalies in Africa for C_4 grass at (a) 122 000 years ago, (b) 120 000 years ago, and (c) 118 000 years ago relative to AD 1800. Average C_4 grass net primary productivity derived from 122 000-year time-series climate simulation performed on the UVic Earth system climate model (UVic ESCM). (See colour version in colour plate section. Red indicates lower net primary productivity levels relative to AD 1800 and blue represents higher net primary productivity levels relative to AD 1800.)

less than the present, C_3 grass coverage had expanded and C_4 grass productivity continued to be higher than today, particularly in South Africa (Figure 6.4).

However, these improved conditions were short-lived. By 118 000 years ago, things had rapidly changed again. Most of Africa was warmer than today, with the exception of north-west Africa, which remained slightly cooler than today. Most of

North Africa, including the Sahara, experienced less precipitation than AD 1800. C_3 grass coverage had again retreated to less than today. C_4 grass coverage and productivity fell in North Africa and rose in South Africa. Needleleaf tree coverage remained higher in North and South Africa than AD 1800 through to 116 000 years ago, whereas net productivity remained higher only in North Africa.

6.4 Marine isotope stage 5d

The cooler temperatures of MIS5d (116 000 to 110 000 years ago) brought the onset of another glaciation and a 65-m drop in global eustatic sea level. By the termination of MIS5e, a limited number of *H. sapiens* had dispersed into the Levant, which was biogeographically part of Africa (Willoughby, 2007, p. 94).

6.4.1 The changing climate of Africa

In Africa, the home of early *H. sapiens*, our climate simulations indicate that the warming observed at the end of the MIS5e remained in all but north-west Africa until suddenly, at 112 000 years ago, atmospheric temperatures across all of Africa became colder (Figure 6.5). This was particularly the case in North Africa, where temperatures were nearly 1 °C colder than AD 1800. Atmospheric temperatures continued to cool through 110 000 years ago. Land-surface temperatures followed a similar pattern, except South Africa remained warmer than AD 1800 until about 110 000 years ago. Simulations show that lower precipitation levels in North Africa and higher precipitation levels in South and central Africa than AD 1800 continued until 112 000 years ago, when virtually all of Africa, except a band along the southern border of what is now the Sahara desert and an area north of the Namib desert, received more precipitation than today (Figure 6.6). By 110 000 years ago the pattern shifted, with North and South Africa receiving more precipitation and central Africa receiving less precipitation than AD 1800.

6.4.2 The changing vegetation of Africa

The impact of this changing climate on African vegetation was complex. Proxy evidence suggests that after 115 000 years ago, Africa's rainforest decreased, but not to the extent that it did during full glacial intervals. Climate simulations show average broadleaf tree coverage and productivity was less in Africa during this interval than AD 1800. However, coverage was uneven. Broadleaf coverage observed at the end of MIS5e persisted through MIS5d, with North and South Africa experiencing more and central Africa experiencing less extensive broadleaf coverage than in AD 1800. Further, the increased broadleaf coverage and productivity in North Africa

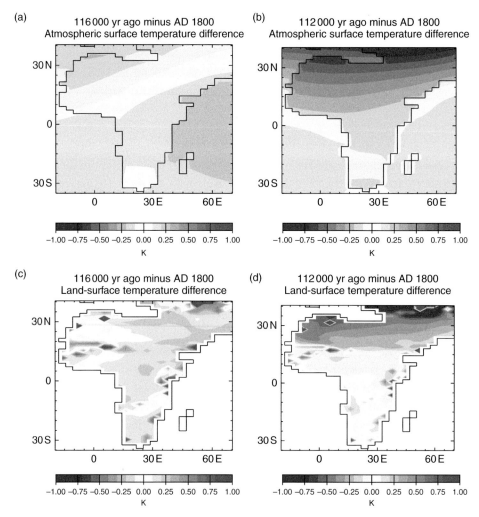

Figure 6.5   Atmospheric surface temperature anomaly for Africa (a) 116 000 years ago and (b) 112 000 years ago relative to AD 1800. Land-surface temperature anomaly for Africa (c) 116 000 years ago and (d) 112 000 years ago relative to AD 1800. Atmospheric surface temperature and land-surface temperatures derived from 122 000-year time-series climate simulation performed on the UVic Earth system climate model (UVic ESCM). (See colour version in colour plate section. Red indicates higher temperatures relative to AD 1800 and blue represents lower temperatures levels relative to AD 1800.)

stretched east into the Middle East (Figure 6.7). The band of increased broadleaf coverage and productivity observed between central and South Africa that appeared during MIS5e remained until 112 000 years ago, when it shrank to a small region along the coast of east-central Africa. By 110 000 years ago, broadleaf coverage had retreated, and this anomalous band contained less broadleaf tree coverage than in AD

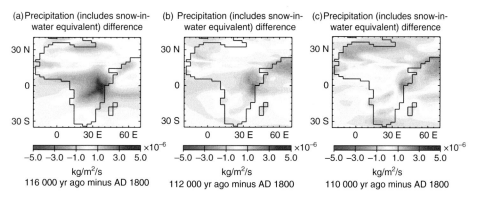

Figure 6.6 Precipitation anomaly for Africa (a) 116 000 years ago, (b) 112 000 years ago, and (c) 110 000 years ago relative to AD 1800. Precipitation anomalies derived from 122 000-year time-series climate simulation performed on the UVic Earth system climate model (UVic ESCM). (See colour version in colour plate section. Red indicates lower precipitation levels relative to AD 1800 and blue represents higher precipitation levels relative to AD 1800.)

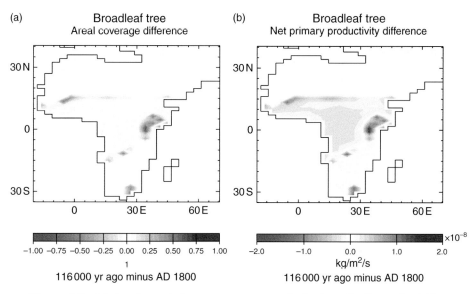

Figure 6.7 (a) Areal vegetation coverage anomaly for broadleaf trees in Africa 116 000 years ago relative to AD 1800 and (b) Net primary productivity anomaly for broadleaf trees in Africa 116 000 years ago relative to AD 1800. Areal coverage and net primary productivity anomalies derived from 122 000-year time-series climate simulation performed on the UVic Earth system climate model (UVic ESCM). (See colour version in colour plate section. Red indicates lower areal coverage and net primary productivity levels relative to AD 1800 and blue represents higher areal coverage and net primary productivity levels relative to AD 1800.)

1800. As can be seen in Figure 6.7, simulated broadleaf productivity response matched that of its coverage, with the exception of South Africa, where coverage remained higher than AD 1800, despite reduced productivity.

During MIS5d, *Podocarpus* (a montane pine tree) had its largest and most westward expansion. The last occurrence of this species in the mountains of Guinea dates to MIS substage 5a (Dupont *et al.*, 2000, p. 95, 116; Jahns *et al.*, 1998, p. 277; Willoughby, 2007, p. 87). However simulations show an expansion that was not consistent across Africa. Needleleaf trees retained higher coverage in North and South Africa and lower coverage in central Africa during MIS5d than AD 1800 (Figure 6.8). Needleleaf tree coverage in South Africa and west-central Africa expanded in MIS5d, although productivity remained lower than in AD 1800 until about 110 000 years ago. Average needleleaf productivity in central Africa remained lower than in AD 1800 throughout the entire interval. Further, the expansion of broadleaf trees along the border of central and South Africa and at the southern tip of Africa came at the expense of needleleaf tree coverage. By 110 000 years ago, needleleaf tree coverage was again expanding along this border, replacing broadleaf trees.

Simulations show average C_3 grass coverage remained consistently below that in AD 1800 throughout MIS5d despite greater distribution in parts of South and North

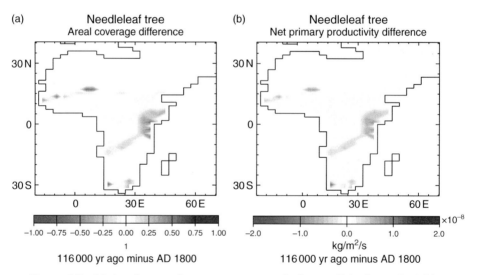

Figure 6.8   (a) Areal vegetation coverage anomaly for needleleaf trees in Africa 116 000 years ago relative to AD 1800 and (b) Net primary productivity anomaly for needleleaf trees in Africa 116 000 years ago relative to AD 1800. Areal coverage and net primary productivity anomalies derived from 122 000-year time-series climate simulation performed on the UVic Earth system climate model (UVic ESCM). (See colour version in colour plate section. Red indicates lower areal coverage and net primary productivity levels relative to AD 1800 and blue represents higher areal coverage and net primary productivity levels relative to AD 1800.)

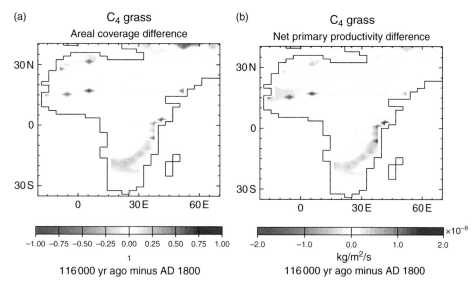

Figure 6.9 (a) Areal vegetation coverage anomaly for C_4 grass in Africa 116 000 years ago relative to AD 1800 and (b) Net primary productivity anomaly for needleleaf trees in Africa 116 000 years ago relative to AD 1800. Areal coverage and net primary productivity anomalies derived from 122 000-year time-series climate simulation performed on the UVic Earth system climate model (UVic ESCM). (See colour version in colour plate section. Red indicates lower areal coverage and net primary productivity levels relative to AD 1800 and blue represents higher areal coverage and net primary productivity levels relative to AD 1800.)

Africa. C_3 grass productivity was likewise below AD 1800 levels, except in part of central Africa, until about 112 000 years ago, when productivity increased above AD 1800 across much of North Africa. Average C_4 grass coverage remained higher than AD 1800 levels in South Africa throughout MIS5d (Figure 6.9). C_4 grass productivity was also higher than AD 1800 across much of Africa except in parts of North Africa where it began to improve around 112 000 years ago. Shrub coverage continued to be more extensive during this interval than AD 1800 except in eastern-central Africa and South Africa where shrub productivity was relatively low until 110 000 years ago and in parts of North Africa until 112 000 years ago. Barren ground coverage was generally higher during MIS5d than AD 1800 particularly in North and South Africa, whereas central Africa possessed slightly less barren ground coverage than AD 1800.

6.5 Marine isotope stage 5c

This highly variable interval between 110 000 and 100 000 years ago was warmer than the preceding MIS5d, but still 2 °C colder than today. Global eustatic sea level oscillated, rising to 20 m below present before dropping back down to about 55 m below present and then rising to 38 m below present. *Homo sapiens* were clearly continuing to migrate out of Africa and into the Levant region. This was likely stimulated by changes in vegetation and climate and by the variability of climate during this period. The stress experienced by early humans as a result of these changes likely stimulated the technological adjustments that supported a migration out of Africa (for a more detailed discussion see Chapter 7).

6.5.1 The changing climate of Africa and beyond

Our climate simulations show continued cooling in Africa with the onset of MIS5c. In north-east Africa and the Middle East, atmospheric surface-air temperatures reached more than 1.5 °C colder than in AD 1800. Land-surface temperatures rapidly dropped with the onset of MIS5c. Across Africa and the Middle East, land-surface temperatures remained more than 1 °C colder than in AD 1800. Although overall precipitation did not increase appreciably during MIS5c in Africa and the Middle East, its distribution changed. Whereas during MIS5d central Africa had been the focal area of much of Africa's precipitation, this reversed with the onset of MIS5c, with central Africa and the northern half of South Africa receiving far less precipitation than in AD 1800, and North Africa, the Middle East and the southern tip of Africa receiving far more precipitation (Figure 6.10).

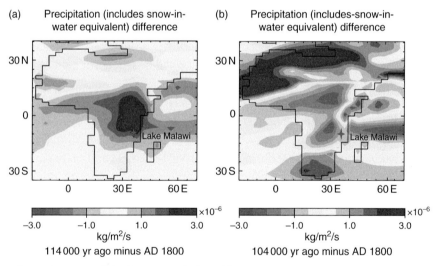

Figure 6.10 Precipitation anomaly in Africa at (a) 114 000 years ago relative to AD 1800 and (b) 104 000 years ago relative to AD 1800. Precipitation anomalies derived from 122 000-year time-series climate simulation performed on the UVic Earth system climate model (UVic ESCM). (See colour version in colour plate section. Red indicates lower precipitation levels relative to AD 1800 and blue represents higher precipitation levels relative to AD 1800.)

Lake core and seismic records from Lake Malawi indicate prolonged periods of drought and lowered lake levels between 111 000 and 108 000 years ago and between 105 000 and 95 000 years ago (Cohen *et al.*, 2007). The eastern coastal area of central Africa continued to receive more precipitation than in AD 1800 during MIS5c as it did during MIS5d.

MIS5c saw the expansion of barren ground in Africa and Eurasia. By simulating the effects of a freshwater influx mimicking that of Heinrich event 9 (H9) and the associated collapse of the Atlantic MOT, Carto *et al.* (2008) suggest all regions of Africa, with the exception of South Africa, were drier 105 000 years ago. The modelled H9 freshwater influx reduced the Atlantic MOT, completely shut down the NADW formation, and significantly reduced tropical and North Atlantic annual-mean sea-surface temperatures, surface-air temperatures, and sea-surface salinities relative to the simulations without the H9 freshwater influx. The collapse of the Atlantic MOT led to enhanced drying in the Sahel region of Africa.

6.5.2 The changing vegetation of Africa and beyond

Our 122 000-year time-series simulation and the H9 simulation by Carto *et al.* (2008) support palynological proxy data indicating that the desert in North Africa was more extensive during intervals of the last glacial cycle (Dupont and Hooghiemstra, 1989). They also support proxy records of arid episodes in the Mega Kalahari sand sea, with sand dune-building phases evident at about 115 000 to 95 000 years ago (Stokes *et al.*, 1997). Simulations also show increased barren ground coverage in North Africa and central Eurasia (Figure 6.11). By 104 000 years ago, this latter region had stretched eastward from the Mediterranean Sea, crossing three quarters of the entire continent. It bounded a narrow region in the Middle East where broadleaf tree coverage remained steady, needleleaf tree coverage slightly improved, and by 102 000 years ago barren ground coverage was being replaced by C_3 and C_4 grasses. This narrow, relatively productive band stretches across the archaeological sites of Qafzeh and Skhul Cave, just south of Nazareth, Israel, where the first *H. sapiens* found outside Africa have been dated to between 119 000 and 85 000 years ago (Valladas *et al.*, 1998). Terrestrial pollen records coincide, indicating a transition to non-forestation in France, central Italy, and eastern Macedonia and a reduction in forestation in north-west Greece around 104 000 years ago (Tzedakis *et al.*, 1997).

Our climate simulations also show that average broadleaf tree coverage in Africa contracted and landscapes became more open and fragmented. Simulations show broadleaf coverage and productivity fell until 104 000 years ago, after which they marginally improved before productivity hit an interval low at 100 000 years ago. These losses were the result of a large reduction in central Africa that overwhelmed

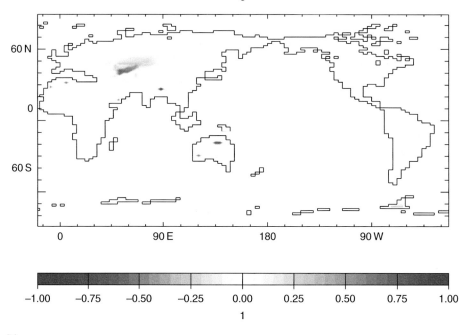

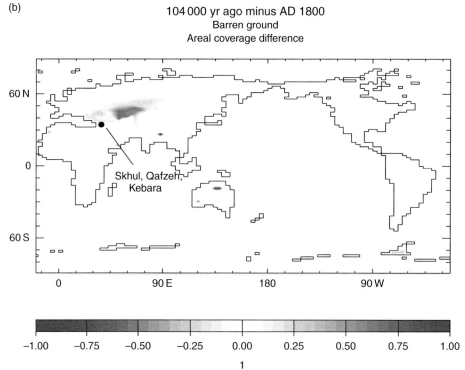

Figure 6.11 Global areal barren ground coverage anomaly at (a) 110 000 years ago relative to AD 1800 and (b) 104 000 years ago relative to AD 1800. Areal coverage of barren ground derived from 122 000-year time-series climate simulation performed on the UVic Earth system climate model (UVic ESCM). (See colour version in colour plate section. Red indicates more barren ground coverage relative to AD 1800 and blue represents less barren ground coverage relative to AD 1800.)

gains seen in the southern tip of Africa. Somewhat supportive of the interpretation by Dupont *et al.* (2000) that reclamation of rainforest lost during MIS5d occurred during MIS5c, simulations show an expansion of needleleaf tree coverage along a band crossing what is now Angola, Zambia, and Tanzania. However, needleleaf coverage generally fell steadily across much of Africa until 104 000 years ago before gradually improving by 100 000 years ago. Yet this improvement was not offset by increased productivity. In what appears to be a rather unique circumstance, needleleaf tree productivity also declined throughout this entire interval, reaching a low 100 000 years ago, and it never did replace losses in broadleaf tree productivity. Needleleaf tree coverage and productivity both remained below AD 1800 levels for the duration of this interval, as did C_3 grass coverage and productivity. However, C_3 grass coverage and productivity coincidently reached an interval low around 105 000 years ago. C_4 grass coverage steadily expanded until 105 000 years ago and remained above AD 1800 levels for the entire interval, as did C_4 grass productivity, although net primary productivity peaked at 107 000 years ago and again at 104 500 years ago. Despite these increases in coverage, C_4 grass productivity offset less than half the losses that occurred in C_3 grass productivity. Shrub coverage gradually increased from MIS5d levels and remained higher than AD 1800 during the entire interval, reaching maximum coverage 107 000 years ago. Shrub productivity in Africa dropped in half 110 000 years ago, briefly increased 107 000 years ago before dropping again and did not regain MIS5d levels until 100 000 years ago. These findings are relatively consistent with pollen evidence from Lake Malawi, in south-east Africa, which indicates sparse but diverse arboreal components and a predominance of grass between 120 000 and 75 000 years ago. Pollen evidence also shows a reduction in accumulation rates between 105 000 and 95 000 years ago and transition to drier taxa (Cohen *et al.*, 2007).

Relative to other intervals during the last glacial cycle, the repercussions of climate change during MIS5c were extremely unusual. Significant alterations in precipitation patterns combined with reduced temperatures altered the distribution of vegetation but did not generate sufficient productivity to replace that lost by species that were evident prior to the onset of MIS5c. The exception appears to be a narrow band in the Middle East.

6.6 Marine isotope stage 5b

MIS5b was a mild glacial interval spanning 10 000 years between 100 000 and 90 000 years ago. Simulations show the large area of barren ground that had developed during MIS5d in central Eurasia still persisted at the onset of MIS5b, but had drastically retreated by 98 000 years ago. Coincidently, a region with significantly less barren ground than AD 1800 had developed in the Middle East. It was during MIS5b that modern humans are thought to have continued their migration out of Africa and into the Middle East. By 92 000 years ago the large barren ground anomaly in central Asia was once again developing and by 90 000 years ago was again very pervasive (Figure 6.12).

6.6.1 The changing climate of Africa and beyond

Our climate simulations indicate that atmospheric temperatures remained between 1 and 2 °C colder than AD 1800 across Africa and the Middle East at this time. Temperatures were warmer in the south and cooler in the north-east. By 90 000 years ago, temperatures in north-east Africa and the Middle East had reached between 2 and 3 °C colder than AD 1800. Land-surface temperatures followed a similar pattern, with temperatures 2.5 to 3 °C colder than AD 1800 in north-east Africa, the Middle East, and along the northern coast of Africa 90 000 years ago. The pattern of precipitation that appeared during MIS5c did not significantly alter during MIS5b. Precipitation remained much lower than today in central Africa and higher in North Africa, the Middle East, the southern tip of Africa and along the eastern coast of central Africa.

6.6.2 The changing vegetation of Africa and beyond

Proxy evidence suggests that the African rainforest retreated during MIS5b (Dupont *et al.*, 2000). Climate simulations show broadleaf trees remain less pervasive and productive across central Africa and slightly more pervasive and productive across North Africa and the Middle East, and particularly the southern tip of South Africa (Figure 6.13). Notable increases in broadleaf tree coverage and productivity appeared along the eastern coast of central Africa between 94 000 and 90 000 years ago. Simulations indicate that needleleaf tree coverage remained below AD 1800 in central Africa particularly in the eastern regions and in the southern tip of Africa. By 92 000 years ago a band of needleleaf trees had expanded along the western half of the borders between central Africa and North and South Africa. Yet there was only a minor coincident increase in needleleaf productivity in these same regions. Needleleaf tree coverage and productivity was only very slightly above AD 1800 levels and insufficient to offset the reduction in broadleaf coverage and productivity. Simulations also indicate that C_3 grass coverage declined in South Africa and remained below AD 1800 levels along the border between North and central Africa. At the beginning of MIS5b (100 000 years ago) strong expansions of C_3 grass and an associated rise in productivity is evident along the coast of eastern-central Africa, in what is now Algeria and Tunisia in North Africa, as well as in north and eastern regions of the Middle East.

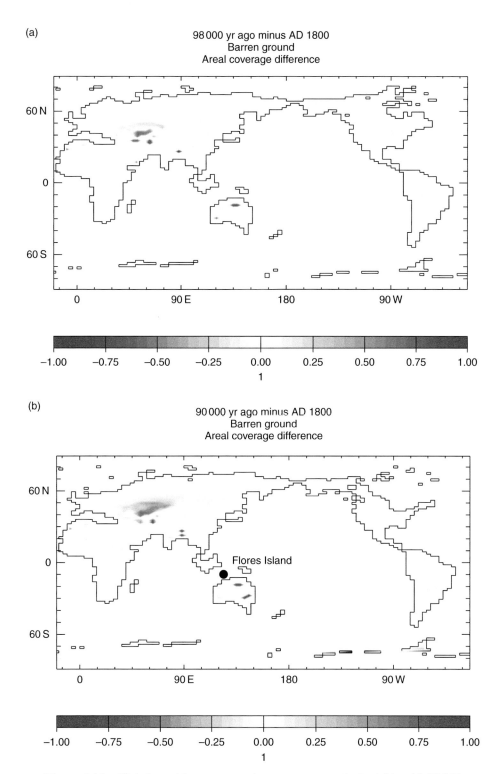

(a)
98 000 yr ago minus AD 1800
Barren ground
Areal coverage difference

-1.00   -0.75   -0.50   -0.25   0.00   0.25   0.50   0.75   1.00
1

(b)
90 000 yr ago minus AD 1800
Barren ground
Areal coverage difference

Flores Island

-1.00   -0.75   -0.50   -0.25   0.00   0.25   0.50   0.75   1.00
1

Figure 6.12   Global areal barren ground coverage anomaly in Africa (a) 98 000 years ago relative to AD 1800 and (b) 90 000 years ago relative to AD 1800. Areal coverage of barren ground derived from 122 000-year time-series climate simulation performed on the UVic Earth system climate model (UVic ESCM). (See colour version in colour plate section. Red indicates more barren ground coverage relative to AD 1800 and blue represents less barren ground coverage relative to AD 1800.)

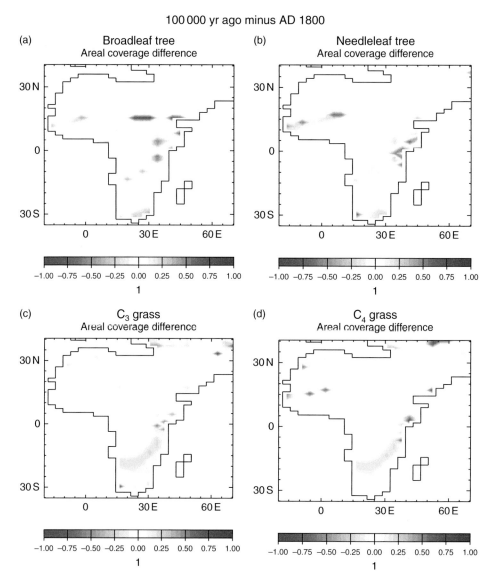

Figure 6.13   Areal vegetation coverage anomalies in Africa 100 000 years ago relative to AD 1800 for (a) Broadleaf trees, (b) Needleleaf trees, (c) C_3 grass, and (d) C_4 grass. Areal coverage anomalies derived from 122 000-year time-series climate simulation performed on the UVic Earth system climate model (UVic ESCM). (See colour version in colour plate section. Red indicates lower areal coverage and net primary productivity levels relative to AD 1800 and blue represents higher areal coverage and net primary productivity levels relative to AD 1800.)

6.7 Marine isotope stage 5a

The brief relative warming of MIS5a persisted from 90 000 to 75 000 years ago. Global atmospheric surface-air temperatures were about 2 °C colder than today. Global eustatic sea level ranged from 25 to 65 m below present levels.

6.7.1 The impact of a changing climate on global vegetation

Climate simulations indicate that the lowest global average broadleaf coverage occurred 78 000 years ago with the greatest relative reductions occurring in the northern half of South America, eastern-central South America, Central America, South East Asia, and central Africa particularly along the border between central and South Africa. Alternatively, global average broadleaf tree coverage and productivity was higher than AD 1800 in the southern tip of Africa and in central South America. Global average needleleaf tree coverage experienced substantial reductions during MIS5a north of 55° N, as well as in central Asia and eastern-central Africa, while advances were achieved in North Africa, north-western South Africa, and the southern tip of South America.

Although simulations indicate that global average C_3 grass coverage was consistently above AD 1800 levels throughout MIS5a, particularly north of 60° N, in central Asia, and in North and central Africa, C_3 grass net primary productivity remained below AD 1800 levels throughout the entire interval. This is because even where coverage expanded in central Africa, net primary productivity fell. Global average C_4 grass coverage and net primary productivity appears to have peaked around 88 000 years ago and then gradually declined until coverage fell below AD 1800 levels between 86 000 and 84 000 years ago. C_4 grass coverage and net primary productivity then rose above AD 1800 levels and remained this way through the end of MIS5a. A region on or about what is now Turkistan suffered virtual elimination of C_4 grasses relative to AD 1800 throughout this interval, but was surrounded to the north, west, and partially to the south by a region where C_4 grasses were much more pervasive than AD 1800 between 86 000 and 80 000 years ago. C_4 grasses were also much more pervasive than today in South Africa and north and south-east Australia. Despite the decline seen between 86 000 and 84 000 years ago, net primary productivity remained higher than AD 1800 levels throughout virtually all of Africa, as well as much of Australia and South America throughout MIS5a.

Global average shrub coverage remained higher than AD 1800 throughout MIS5a with the greatest extent reached at 80 000 and 78 000 years ago. In the northern hemisphere shrub coverage appears to have shifted about 10° latitude southward relative to AD 1800. Shrub coverage was higher along the central Africa border to the north and south and also in central Asia during MIS5a than in

AD 1800. The barren ground coverage that had begun to re-establish in the Middle East at 92 000 years ago continued to expand reaching its greatest extent between 86 000 and 84 000 years ago. By the end of MIS5a this region had drastically declined in extent although global average barren ground coverage still remained higher than AD 1800. This was despite reductions in barren ground coverage in much of central and southern Africa, parts of the Middle East and South East Asia, Australia and South America.

6.8 Marine isotope stage 4

The full glacial interval of MIS4 began 75 000 years ago and lasted until 60 000 years ago. Global average surface-air temperatures reached nearly 3.5 °C colder than today and global eustatic sea level fell to less than 70 m below present.

6.8.1 The impact of a changing climate on global vegetation

Climate simulations show broadleaf tree coverage and productivity remained below AD 1800 levels for the duration of MIS4 and reached their lowest levels between 66 000 and 64 000 years ago. Coverage and productivity levels were substantially less than AD 1800 levels in central Africa, South East Asia, the northern half of South America, and Central America. A few regions experienced coverage and productivity levels greater than AD 1800 including the southern tip of South Africa, south-eastern China, northernmost Central America, a small region in south-eastern North America and another in central South America. Needleleaf tree coverage and productivity remained below AD 1800 levels with reduced coverage and productivity in northern Europe, north, central and north-eastern Asia, and north-western and central North America, and central Africa far exceeding small gains made in eastern Asia, parts of Europe, east-central South America, and along the northern and southern borders of central Africa.

As MIS4 progressed, a strong reduction in C_3 grass coverage and productivity occurred along a band between about 40 and 55° N. C_3 grass coverage developed along a strip north of 55° N particularly in the Beringia region of north-east Asia, north-west North America and the exposed continental shelf between the two. C_3 grass productivity was much reduced relative to AD 1800 in most of Europe, Asia and South Africa whereas it was higher north of 60° N, as well as in North Africa, the Middle East, south-central Asia, central Australia, and along a corridor down central North America south of the ice sheets, and in Central America and western- and eastern-central South America. The region around Turkistan that suffered reduced C_4 grass coverage in MIS5a continued to expand extending across most of west and central Asia. Simulations also show that all regions around the world north of 35° N suffered reduced C_4 grass net productivity relative to AD 1800. However, these reductions were insufficient to offset the relative gains in C_4 grass coverage and productivity made across much of Africa particularly in South and west-central Africa, the Middle East, parts of South East Asia, much of Australia and northern South America. As a consequence, global average C_4 grass coverage and productivity remained below AD 1800 levels throughout MIS4.

Global average shrub coverage remained higher than AD 1800 throughout MIS4. The 10° latitude southward shift in the band of shrubs in Eurasia seen during MIS5a persisted as did the shrub development along the northern and southern border of central Africa. However, by 65 000 years ago the band of shrubs in northern Eurasia began to disappear in the centre of the continent and shrubs became very pervasive in central Asia. Although shrub productivity remained higher than AD 1800 in much of Eurasia at the beginning of MIS4, by 64 000 years ago productivity was falling in most of central Eurasia and northern and eastern Asia. At the beginning of MIS4 barren ground coverage exceeded that seen in AD 1800 and was restricted to the region that is now Turkistan. Slight increases are also apparent in North Africa and the balance of Eurasia. However, just to the south of Turkistan, in the Middle East, barren ground coverage was less than AD 1800 as it was in central and South Africa, parts of South East Asia, Australia, south-east North America and South America. However, by 66 000 years ago the barren ground in Turkistan began to expand and by 60 000 years ago it covered much of central Eurasia extending west into Eastern Europe and east into north-eastern Asia and north-west North America. Climate simulations indicate a huge expanse of the continent became far less habitable (Figure 6.14) potentially forcing early humans into the more habitable peripheral regions of Western and southern Europe, the Middle East, South East Asia and Australia.

6.9 Marine isotope stage 3

MIS3 brought cool, dry and highly unstable conditions to the world. It lasted from about 60 000 to 30 000 years ago. MIS3 was also a time of great change in human history.

6.9.1 The impact of a changing climate on global vegetation

Climate simulations show that global broadleaf tree coverage was less than AD 1800 for the duration of MIS3, making only marginal recoveries over MIS4 levels between 50 000 and 48 000 years ago. Broadleaf tree coverage and net primary productivity remained far below AD 1800 levels in central Africa, Central America and northern South America. As in MIS4, a few regions including the southern tip of

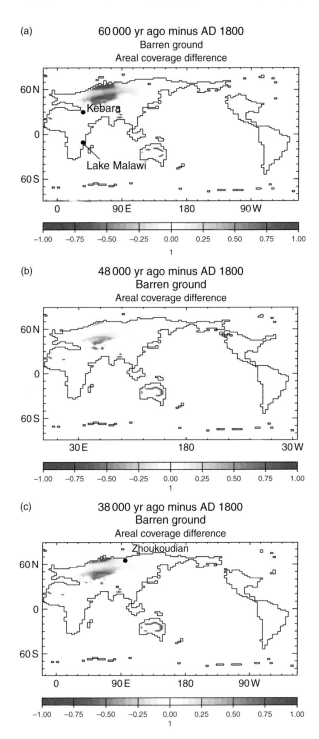

Figure 6.14 Global areal barren ground coverage anomaly at (a) 60 000 years ago relative to AD 1800, (b) 48 000 years ago relative to AD 1800, and (c) 38 000 years ago relative to AD 1800. Areal coverage of barren ground derived from 122 000-year time-series climate simulation performed on the UVic Earth system climate model (UVic ESCM). (See colour version in colour plate section. Red indicates more barren ground coverage relative to AD 1800 and blue represents less barren ground coverage relative to AD 1800.)

Africa, a couple of small regions in central South America, and a few areas in mainland South East Asia saw broadleaf tree coverage levels that were greater than AD 1800 levels (Figure 6.15). Australia experienced broadleaf tree coverage and net productivity levels that were marginally greater than AD 1800.

There was little change in global average needleleaf tree coverage or net primary productivity with the onset of MIS3. Coverage and net productivity remained below AD 1800 in northern Europe and Asia, central Africa, the southern tip of Africa, and north-west and central North America. Regions with coverage greater than AD 1800 remained in parts of eastern Asia, Europe, east-central South America, and along the southern border of central Africa. Between 52 000 years ago and 41 000 years ago needleleaf tree expansion failed to materialize in the central African border region and neither was there a coincident relative rise in net primary productivity. However, this circumstance changed 36 000 years ago when coverage expanded in this region. These findings are consistent with pollen records from the central East African mountains that show a well-developed, more widespread dry conifer forest than today at 30 000 radiocarbon years BP (35 400 years ago) (Bonnefille and Chalié, 2000). From 58 000 years ago until 30 000 years ago, simulations show the southern tip of India and east along the northern coast of the Bay of Bengal experienced greater needleleaf coverage and productivity than AD 1800.

Global C_3 grass coverage was higher than AD 1800 levels for the duration of MIS3. The strip of increased C_3 grass coverage that had developed north of 55° N during MIS4 persisted as did increased coverage in North and central Africa, the Middle East, and south-central Asia, parts of Australia particularly in the south, as well as northern Central America. By the end of MIS3, C_3 grass had become very prevalent in the Beringia region connecting north-east Asia and north-west North America, far more so than today. The band of reduced C_3 grass coverage that had developed during MIS4 between about 40 and 55° N remained although coverage began to improve somewhat in western-central Eurasia at around 50 000 years ago. By 36 000 years ago C_3 grass coverage was retreating in coastal regions of south-western and south-eastern Australia. C_3 grass net primary productivity remained below AD 1800 levels for the duration of MIS3 despite a slight increase around 50 000 years ago. Regions with C_3 grass net primary productivity levels higher than AD 1800 were limited to North Africa, the border between central and South Africa, the Middle East, parts of south-central Eurasia and a band north of 60° N and east of 75° E. This band expanded thinly westward reaching the ice sheets in northern Europe between 56 000 and 50 000 years ago, and again at 44 000 years ago, and between 36 000 and 30 000 years ago.

Simulations show global average C_4 grass coverage fell below AD 1800 levels with the onset of MIS3 as a result of reduced coverage in much of Eurasia, particularly around the Turkistan region, as well as along the border between central

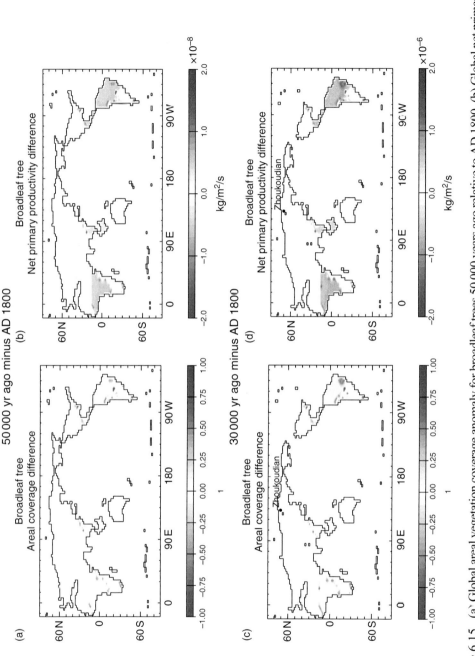

Figure 6.15 (a) Global areal vegetation coverage anomaly for broadleaf trees 50 000 years ago relative to AD 1800, (b) Global net primary productivity anomaly for broadleaf trees 50 000 years ago relative to AD 1800, (c) Global areal vegetation coverage anomaly for broadleaf trees 30 000 years ago relative to AD 1800, and (d) Global net primary productivity anomaly for broadleaf trees 30 000 years ago relative to AD 1800. Areal coverage and net primary productivity anomalies derived from 122 000-year time-series climate simulation performed on the UVic Earth system climate model (UVic ESCM). (See colour version in colour plate section. Red indicates lower areal coverage and net primary productivity levels relative to AD 1800 and blue represents higher areal coverage and net primary productivity levels relative to AD 1800.)

and South Africa. Global average C_4 grass coverage remained below that of AD 1800 until 49 000 years ago when improvements in west-central Eurasia and Australia resulted in levels above AD 1800 that remained intact until the end of MIS3. Global average C_4 grass productivity fell below AD 1800 levels between 58 000 and 56 000 years ago and again at 52 000 years ago. This was a consequence of reduced productivity in central Africa and north-eastern South America relative to MIS4. C_4 grass productivity remained below AD 1800 levels across Eurasia, in the northern tip of Africa and along the border between central and North Africa, India and North America. C_4 grass productivity was above AD 1800 in much of Australia, Africa and the Middle East.

Despite a significant reduction in shrub coverage at the onset of MIS3, simulated global average shrub coverage remained higher than AD 1800 throughout the entire interval. The most prevalent reductions occurred along the band in northern Eurasia where C_3 grass had replaced shrubs. By 50 000 years ago global shrub coverage had made a strong recovery that lasted until about 42 000 years ago. Simulated global average shrub net primary productivity continued to fall with the onset of MIS3 despite a brief recovery 50 000 years ago and again at 44 000 years ago with much of northern Eurasia experiencing less shrub productivity than AD 1800. Yet, higher productivity levels in Africa, northern Europe, and East and central Asia meant that the global average never fell below AD 1800 levels.

Although barren ground coverage was lower than AD 1800 levels in Africa, the Middle East, India and Australia, the large expansion observed in central Eurasia far exceeded these reductions allowing global average barren ground coverage to remain above AD 1800 levels throughout MIS3. It peaked at 60 000 years ago and remained close to this peak level through 58 000 years ago before declining until 48 000 years ago (see Figure 6.14 above). Global barren ground coverage increased slightly again between 46 000 and 45 000 years ago and then between about 40 000 and 38 000 years ago another large expansion occurred in west-central Eurasia potentially isolating Europe between ice sheets to the north and a barren polar desert to the east. A reduction in barren ground coverage between 50 000 and 48 000 years ago created a band of relatively more hospitable landscape stretching from eastern Asia to Europe. This band reoccurred between about 44 000 and 40 000 years ago after which an expanding barren ground landscape again closed this corridor. During intervals when barren ground was most pervasive in central Eurasia, barren ground declined in Australia and along the northern coast of the Bay of Bengal.

6.10 Marine isotope stage 2 – the last glacial maximum

The last ice age began about 30 000 years ago and maximum ice extent was reached around 9000 years later. Global sea levels dropped to about 120 m below present exposing broad continental shelves. One called Beringia spanned the shelf from north-eastern Asia to what is now Alaska and is believed to have assisted humans in their migrations to the Americas. MIS2 culminated with the Younger Dryas event, an abrupt, brief cold interval between about 12 700 and 11 650 years ago before the onset of the Holocene, a period of remarkable relative climate stability. Modern humans expanded into the Americas. By the end of the MIS2 the global human population remained around 6 million (Perreault, 2003).

6.10.1 The impact of a changing climate on global vegetation

The LGM was a time of extreme aridity in Africa when desert systems expanded to five times their present extent (Sarnthein, 1978; Sarnthein and Koopman, 1980; Street-Perrott *et al.*, 1989). Deep-sea cores taken off the coast of West Africa indicate decreasing rainfall, increasing windiness, higher evaporation rates and greater seasonality (Dupont *et al.*, 2000, p. 117; Thomas, 2000, p. 27). Cooler temperatures caused mountain taxa to move to lower elevations and grasslands to expand. Isolated pockets of equatorial rainforest west of the Congo basin in Equatorial Guinea, along the coasts of Cameroon and Gabon, east of the Congo basin, and in several small areas along the cost of Liberia and the Ivory Coast are believed to have provided isolated refugia (Dupont *et al.*, 2000, p. 117; Hamilton, 1976, 1982; Willoughby, 2007, p. 90). Arid phases again caused extreme sand dune-building phases between 26 000 and 20 000 years ago and between 16 000 and 9000 years ago (Stokes *et al.*, 1997). Central to these interpretations is that changes in aeolian activity responsible for dune-building phases are related to changes in rainfall intensity in the area, thus indicating periods of increased aridity. These dry phases were presumably punctuated by more humid intervals when rainfall was greater.

Around the world, climate simulations show global average broadleaf tree coverage levels remained below AD 1800 levels for the entire MIS2 interval. Large reductions in broadleaf tree coverage commenced with the onset of MIS2 and did not begin to steadily improve until 17 000 years ago. Global average broadleaf net primary productivity also declined with the onset of MIS2 until 26 000 years ago, when, after a brief but minor improvement, it declined again during the last glacial maximum (LGM) before gradually improving through the end of MIS2. Reductions in broadleaf tree coverage and productivity relative to AD 1800 were most apparent in central Africa, South East Asia, Central America and the northern half of South America (Figure 6.16a). Regions where broadleaf tree coverage and productivity were greater during MIS2 than AD 1800 included South Africa, central South America, south-eastern North America and what is now south-eastern China.

Simulations also show global average needleleaf tree coverage remained below AD 1800 levels for the duration of MIS2 (Figure 6.16b). With the onset of MIS2,

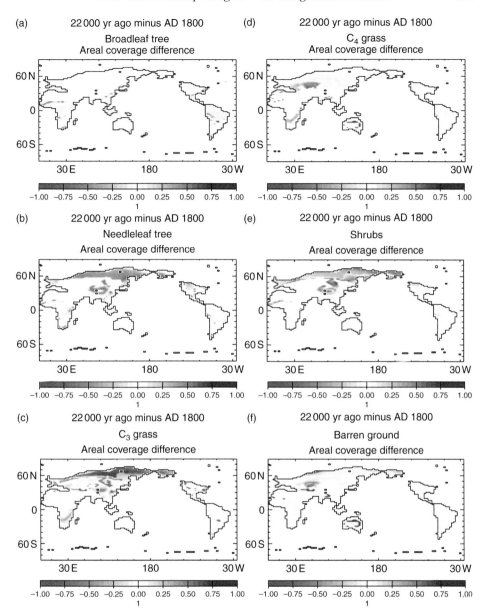

Figure 6.16   Global areal vegetation coverage anomalies 22 000 years ago relative to AD 1800 for (a) Broadleaf trees, (b) Needleleaf trees, (c) C_3 grass, (d) C_4 grass, (e) Shrubs, and (f) Barren ground. Areal coverage anomalies derived from 122 000-year time-series climate simulation performed on the UVic Earth system climate model (UVic ESCM). (See colour version in colour plate section. Red indicates lower areal coverage and net primary productivity levels relative to AD 1800 and blue represents higher areal coverage and net primary productivity levels relative to AD 1800.)

needleleaf tree coverage fell until 22 000 years ago, when coverage improved until 18 000 years ago, when it again fell to nearly the lowest level seen during MIS2. Another brief but mild recovery at 16 000 years ago interrupted further low coverage observed between 14 000 and 10 000 years ago. Simulated global average needleleaf tree productivity also remained below AD 1800 levels throughout MIS2, with brief but minor improvements at 26 000 and again at 16 000 years ago. The northern boreal forest north of 50° N was most impacted, losing up to 83% of its coverage relative to AD 1800, with parts of central Asia, South Africa, central West Africa, and southern, central, and north-western South America also experiencing large reductions. Areas where needleleaf coverage was greater than AD 1800 included parts of east-central Asia and Europe and, near the end of MIS2, also parts of south-east North America, Central America, east-central and the southern tip of South America, and east- and south-central Africa.

Global average C_3 grass coverage remained higher than AD 1800 levels as it replaced the retreating northern boreal forest and expanded into large tracts of southern Eurasia, North and central Africa, Australia, central North America and northern South America (Figure 6.16c). However, higher relative C_3 grass net primary productivity was restricted to north-eastern Asia and the Beringia region, North Africa, the Middle East, parts of central Africa, Australia and a few coastal regions in South America. Despite these expansions in C_3 grass, large retreats in central Eurasia, along the northern border of Eurasia proximal to the ice sheets, and in South Africa were prevalent until about 18 000 years ago when the climate began to warm.

Simulations show global average C_4 grass coverage remained above AD 1800 levels during MIS2 until just after the LGM when it rapidly dropped below present levels (Figure 6.16d). Large expansions in C_4 grass coverage, particularly in Australia, central and South Africa, and northern South America outpaced reductions in west-central Eurasia until 26 000 years ago when C_4 grass coverage declined somewhat until 23 000 years ago when it briefly expanded again until 21 000 years ago. It then retreated until 18 000 years ago, before rapidly falling 17 000 years ago to levels below AD 1800. By 14 000 years ago global average C_4 grass coverage had again expanded, particularly in west-central Eurasia, and remained above AD 1800 levels for the duration of MIS2. Global average C_4 grass net primary productivity, on the other hand, rose steadily, peaking at 26 000 and again at 22 000 years ago before falling below AD 1800 levels by 16 000 years ago where it remained through to the end of MIS2.

With the onset of MIS2, global average shrub coverage and productivity began to expand, particularly in north-eastern Europe, central and north-eastern Asia and along the western half of the border between North and central Africa, and did so until 25 000 years ago, when it declined until 19 000 years ago (Figure 6.16e). Global average shrub coverage briefly expanded again between 18 000 and 17 000 years ago before declining until the end of MIS2, although it never fell below AD 1800 levels.

The onset of MIS2 brought a gradual retreat in barren ground coverage, particularly in the northern Middle East, India, Australia and parts of Africa. However, by 22 000 years ago expansions in west-central Eurasia again isolated Europe between large ice sheets to the north and a broad barren landscape to the east and south (Figures 6.16f and 6.17). These 'polar desert' conditions extended into eastern China almost to the Pacific Ocean. Pollen evidence supports this simulated extensive glacial desert. Biome reconstructions of the last glacial maximum for China by Yu *et al.* (2000) highlight what they term as a 'remarkable feature' – the midlatitude 'extension of steppe and desert biomes to the modern eastern coast' (p. 659). Simulations indicate that these conditions persisted until between about 14 000 and 13 000 years ago, when this large barren expanse eroded somewhat in northern Eurasia. It then re-established its extent briefly during the Younger Dryas interval before retreating in the Holocene (Figure 6.17).

6.11 The Holocene

MIS2 culminated with the brief cool Younger Dryas interval before the warm and relatively stable Holocene epoch began around 11 650 years ago. Extensive ice sheets that had persisted across much of northern North America and parts of northern Eurasia for at least 62 000 years finally disappeared.[2] With warming came higher global average precipitation levels. Early Holocene warming was interrupted by a relatively mild cold event about 8200 years ago triggered by the release of an enormous amount of freshwater from two glacial lakes in eastern Canada. Coincident with, or just subsequent to, this event, a cataclysmic flood inundated the previously freshwater Black Sea (Euxine Lake) with salt water from the Mediterranean Sea, inundating coastal farmers along the Fertile Crescent in South East Asia (Lericolais *et al.*, 2007; Ryan and Pitman, 1999; for an alternative hypothesis see Aksu *et al.*, 2002).[3] The Holocene was also the time of a mass extinction of Pleistocene mammals in North America (Dyke, 2005). Some attribute this to human overkill (Martin, 1984), others to disease (MacPhee and Marx, 1997). Yet others attribute it to a climatic/vegetational shift. The rapid decline in body size of horses before extinction, and the survival of mammoths on St. Paul Island, Alaska, and pigmy mammoths on Rangel Island, Russia, until about 8700 and 4000 years ago, respectively, are evidence of this (for example, see Guthrie, 2003, 2004). The bodies of fossil horses indicate they shrank prior to going extinct and prior to undisputed human arrival and settlement. Shrinkage was coincident with reductions in grassland and probably reflects a shortage of food.

While some species went extinct during the Holocene, modern human populations expanded to 20 million by 5000 years ago and 100 million by the Bronze Age (Perreault, 2003). At the beginning of the Holocene there were an estimated 6 million *H. sapiens* on Earth. A relatively stable Holocene climate, combined with strategies to further reduce the impacts of changes in the environment, allowed

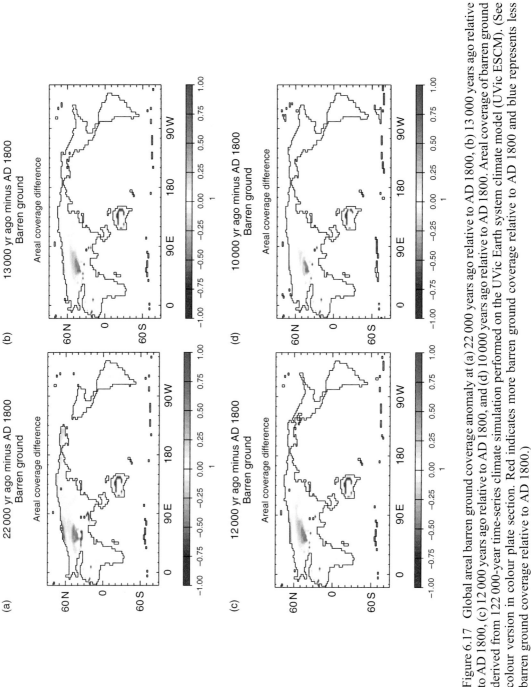

Figure 6.17 Global areal barren ground coverage anomaly at (a) 22 000 years ago relative to AD 1800, (b) 13 000 years ago relative to AD 1800, (c) 12 000 years ago relative to AD 1800, and (d) 10 000 years ago relative to AD 1800. Areal coverage of barren ground derived from 122 000-year time-series climate simulation performed on the UVic Earth system climate model (UVic ESCM). (See colour version in colour plate section. Red indicates more barren ground coverage relative to AD 1800 and blue represents less barren ground coverage relative to AD 1800.)

human populations to expand so that by AD 1900 there were 1.6 billion on the planet. Today we are 6.75 billion.

6.11.1 The impact of a changing climate on global vegetation

The warmer, wetter Holocene climate brought with it the gradual improvement in global average broadleaf tree coverage and productivity. Simulations show that although global average broadleaf tree coverage and net primary productivity fell briefly between 9000 and 8000 years ago, perhaps due to the cool event of 8200 years ago, coverage subsequently expanded, until 2500 years ago when it exceeded AD 1800 levels. However, global average broadleaf tree net primary productivity did not exceed AD 1800 levels until about 500 years later. Reductions were most apparent in central Africa, Central America, and eastern South America. Broadleaf tree net primary productivity was greater than AD 1800 in the southern tip of Africa, central South America, and parts of South East Asia until about 5000 years ago. By 5000 years ago net primary productivity was higher than AD 1800 in south-eastern North America.

During the early and mid-Holocene, greater summertime and total annual insolation resulted in warmer than present summer temperatures, and a modest northward shift in the Arctic treeline (Kaplan, 2001). Our climate simulations show that although global average needleleaf tree coverage fell between 12 000 and 10 000 years ago, it steadily improved until 2500 years ago, although never reaching AD 1800 levels, after which it again declined. With the exception of a minor reduction between 4000 and 3000 years ago, global average needleleaf tree net primary productivity gradually improved through the Holocene, exceeding AD 1800 levels by about 2000 years ago. Regional reductions were observed throughout the world excepting parts of central East Asia until 3000 years ago, the border between central and South Africa until about 7000 years ago, southern North America although conditions deteriorated after 5000 years ago, and east-central South America until about 2500 years ago. These regions experienced greater needleleaf coverage and net primary productivity than AD 1800 levels.

Although at the beginning of the Holocene, global average C_3 grass coverage was higher than AD 1800, coverage rapidly retreated particularly between 10 000 and 8000 years ago. Coverage was particularly poor in much of central and southern Eurasia, central and south-eastern North America, and central South America, as well as much of Africa except the east coast of central Africa and coastal regions of western South Africa. Climate simulations indicate C_3 grass coverage remained higher than AD 1800 levels in isolated locations in north-central Eurasia, east-central Europe, along the east coast of central Africa and coastal regions of western South Africa, eastern North America and parts of south-western North America and

Central America. Despite retreating coverage, global C_4 grass productivity improved until 9000 years ago when large reductions, particularly in central and southern Eurasia and southern North America caused net productivity to drop briefly before improving again until about 7000 years ago when it again dropped, remaining at these lowered levels until 4000 years ago when productivity improved, although remaining below AD 1800 levels until 2000 years ago.

Simulations show global average C_4 grass coverage increasing with the onset of the Holocene, particularly in south-western Eurasia and Australia, until about 8000 years ago after which declining coverage reached levels below AD 1800 by 2000 years ago. Global average C_4 grass net primary productivity declined until 9000 years ago, when improvements in south-western Eurasia, western North Africa, and Australia brought productivity levels up in excess of those seen in AD 1800. Productivity deteriorated in south-western Eurasia by about 6000 years ago and global average net productivity fell to levels below AD 1800 by 2000 years ago.

Beginning 12 000 years ago, global average shrub coverage began to improve as a result of expansions north of 55° N, occupying areas exposed from recently retreating ice. Then at 8000 years ago coverage began to retreat in those same areas, declining until 2000 years ago, but never falling below AD 1800 levels. Global average shrub net primary productivity also remained above AD 1800 levels throughout the modelled interval, but it improved between 10 000 and 9000 years ago, fell until 6000 years ago when it briefly improved before steadily declining until 2000 years ago.

Simulations show barren ground coverage expanding again 12 000 years ago as it replaced retreating ice sheets in northern Eurasia. The large expanse of barren ground in west-central Eurasia remained while barren ground coverage in Australia, isolated locations in the Middle East, north-western Africa and southern China remained far below AD 1800 levels. By 10 000 years ago global barren ground coverage was once again declining with coverage falling below AD 1800 levels 6000 years ago. However, vestiges of these anomalous coverages remained until 2500 years ago when they briefly disappeared before reappearing 2000 years ago.

6.12 Conclusion

It is clear that changing climate has had significant effects on the world's vegetation. There is no question that during the last glacial cycle, climatic volatility impacted regions where *H. sapiens* first emerged and later areas into which they dispersed. It is also apparent that the amplitude of the temperature shifts between cold glacial and warm interglacial intervals is higher in the middle and high latitudes than in the tropical and equatorial regions. Further, precipitation patterns differ globally as do vegetation shifts. Thus it is important when reviewing the impact of changing environment on early humans to pay particular attention to those regions in which humans first emerged and into which they subsequently dispersed. We have seen evidence of extensive barren landscapes or 'polar deserts' that expanded and retreated across Eurasia and Africa. We have seen tremendous latitudinal shifts in tree, grass and shrub extent. In Africa a changing climate affected vegetation, water availability and the accessibility of food, undoubtedly placing significant demands on early hominins. These pressures stimulated a response from *H. sapiens*. In the next section we take a closer look into the interaction between humans and climate and ponder how those environmental influences may have impacted behaviour and development.

Notes

1. For an in-depth discussion of the role of vegetation in glacial inception and vegetation modelling using the UVic ESCM see Meissner *et al.* (2003). For other studies involving vegetation coverage and the UVic ESCM see Matthews, Weaver, Eby, *et al.* (2003) and Matthews, Weaver, Meissner, *et al.* (2003).
2. Based on ice-sheet modelling by S. Marshall, University of Calgary, and used in climate simulations for this book.
3. For a discussion of the impact of this flood on regional farmers see Turney and Brown (2007).

Part III

The interaction between climate and humans

7

The interaction between climate and humans

7.1 Introduction

In the previous two chapters we reviewed Earth's changing climate and the impact of those changes on vegetation over the last 135 000 years. We know that over the last few million years Earth has experienced great climate variability. Massive glaciers have advanced and retreated in response to the Milankovic cycles and associated feedback cycles (Milankovic, 1998). In the early years, Milankovic cycles spanned 21 700 years, between 1 and 3 million years ago the dominant cycle shifted to 41 000 years, and over the last million years a 98 500-year cycle has prevailed. Yet during the last glacial cycle (the last 135 000 years), climate became more variable, even within the 98 500-year cycle. The amplitude of fluctuating temperature and rainfall increased (Richerson and Boyd, 2000). Climate oscillated within century and millennial cycles (GRIP, 1993). Low-latitude variations in temperature and rainfall (Broecker, 1996) and what were once centenary floods, droughts and windstorms, occurred more frequently, perhaps once a decade (Richerson and Boyd, 2000). Climate deterioration created a more patchwork global vegetation (Potts, 1996a, pp. 61–2). In contrast, the warm and relatively ice-free Holocene (the last 11 650 years) was a period of remarkable relative climate stability that allowed dominant species, particularly humans, to become more firmly established.

In this section of the text we draw attention to the interaction between climate and humans. We begin 135 000 years ago with the first modern humans and their subsequent migration out of Africa and culminate in the Holocene when human population expanded and agriculture and animal domestication proliferated. We know that the emergence of various hominin species, including those who were our early ancestors, was governed by the conditions that prevailed as a consequence of large, periodic climatic fluctuations over the last few million years. Early modern human emergence, with its unique human culture, arose later in association with rapid, extreme and repetitive shifts in climate and associated changes in habitat, fauna and flora. Early humans migrated into more hospitable regions, initially out of Africa and then around the world. They changed their behaviour and developed new cultural initiatives that would ensure a more stable living environment. These ranged from the transition from the middle Stone Age[1] to the later Stone Age[2] in Africa to the development of agriculture during the Holocene. Modern humans developed new behaviours and technologies that allowed habitation of new environments and increasingly altered their surroundings. In doing so, modern humans were able to reduce the impact of a changing and volatile environment (Potts, 1996a, p. 43).

We have already noted that proxy evidence, observational data that is gathered from ocean and lake cores, glacial and loess deposits, as well as evidence from fossil bones and stone tools and modern computer climate simulations, tell only part of the story. We do not clearly understand the entire history of climate and its impact on the environment of early modern humans during their emergence and dispersal. Huge gaps remain. Nor can we possibly perceive the many ways these changes impacted early modern human development and behaviour. This said, it is important to recognize that we must start somewhere if we are to improve our understanding about the relationship between humans and climate. If we can better grasp how and why humans developed and responded to past climate change, then perhaps we can better understand our opportunities and limitations to respond to future climate change. Thus, we begin more than 135 000 years ago and chronologically highlight key findings and interpretations that link climate change and the response of early modern humans. Of particular interest is how rapid climate change has stimulated saltatory human response.

7.2 Marine isotope stage 6 (150 000–135 000 years ago) – its impact on newly emerged modern humans

By MIS6 (150 000–135 000 years ago) modern humans had appeared in Africa. The earliest evidence of anatomically modern humans is the ~195 000-year-old *Homo sapiens* fossil remains found at Omo Kibish in Ethiopia. It is here in central Africa where *H. sapiens* had already appeared that climate and vegetation productivity deteriorated during MIS6 (refer to Figure 6.2). Climate modeling and proxy evidence indicates expanding desertification and associated dry, cool conditions. These worsening conditions potentially stimulated migration to areas outside of central Africa. Early *H. sapiens* would have likely been forced to concentrate in regions where precipitation and vegetation productivity were greater and thus more suited to their survival. According to archaeological data it is shortly after the end of MIS6, between 110 000 and 130 000, that modern humans appear in the archaeological record south of Omo, Ethiopia in Mumba, Tanzania (Bräuer and Mehlman,

1988; Clark *et al.*, 2003; McDougall *et al.*, 2005; Mehlman, 1987). *H. sapiens* also appear farther south, in the southern tip of Africa at Klasies River Mouth, by 118 000 years ago and again between 105 000 and 94 000 years ago, and at Border Cave, South Africa, between 115 000 and 90 000 years ago (McBrearty and Brooks, 2000).

The onset of MIS6 also brought global eustatic sea level to its lowest point during the entire last glacial cycle at close to 130 m below present (Lambeck *et al.*, 2002). Lowered sea level exposed the continental shelves of Africa and surrounding regions. Research indicates that these exposed shelves provided productive refugia with fresh water springs; oases from dessicated interior regions (Faure *et al.*, 2002; Hetherington *et al.*, 2008). Marean *et al.* (2007) have found evidence of the use of marine resources at Pinnacle Point on the south coast of South Africa by ~ 164 000 (±12 000) years ago. These findings indicate an early coastal adaptation along with bladelet technology, which was more typical of much later periods (Ambrose, 2002; Soriano *et al.*, 2007). During MIS6 and later glacial intervals, coastal sites may have acted as refugia, with shellfish becoming critical resources when increased aridity made interior terrestrial resources more depleted.

Thus, it is entirely possible that *H. sapiens* migrated out onto the exposed continental shelf to avoid expanding desertification. This may explain the presence of artifacts dated to 125 000 years ago discovered on the Red Sea coast of Eritrea. The Red Sea is an enclosed basin between Africa and Asia, which opens into the Indian Ocean at the Strait of Bab el Mandeb (refer to Figure 6.1). The existence of artifacts along the coast of the Red Sea implies early human occupation of coastal areas and exploitation of near-shore marine food resources in East Africa (Walter *et al.*, 2000). These finds may mean early humans expanded out of Africa and into South East Asia along the subaerially exposed continental shelf. It appears that early *H. sapiens* occupied this coastal region just prior to a rapid sea-level rise coincident with the onset of MIS5e.

7.3 The last glacial cycle and the migration of modern humans out of Africa

7.3.1 *Marine isotope stage 5e (135 000–116 000 years ago)* – Homo sapiens' *response to the Eemian interglacial*

Initially, after the cold MIS6, the changing climate associated with MIS5e (135 000–116 000 years ago) and its resultant impacts on habitat likely improved the capacity of *H. sapiens* to expand their populations and, for some, to move back into central Africa. Warm, wetter conditions caused dry forests and savanna to shrink, encouraged extensive forest development and caused vegetation to grow in areas today covered with desert. By the termination of MIS5e, a limited number of *H. sapiens* had dispersed into the Levant to the sites of Skhul and Qafzeh, Israel, where anatomically modern human remains have been dated to between 119 000 and 85 000 years ago (Valladas *et al.*, 1998, pp. 69–75; Willoughby, 2007, p. 94). African fauna also appeared in the Levant at this time (Goren-Inbar and Speth, 2004; Tchernov, 1988). Because of the biogeographic similarity between the Levant and Africa at this time, there is some doubt as to whether this early modern human expansion truly represents the first migration out of Africa, in that it may not have required a substantial adjustment to changing environmental conditions.

The MIS6 cooling and the subsequent MIS5e warming resulted in large changes in vegetation coverage and productivity, particularly in a band between central and South Africa and in a region in eastern-central Africa (refer to Figures 6.3, 6.7 and 6.8). This region in eastern-central Africa includes Kenya and parts of Ethiopia and Tanzania, where many early archaeological and hominid sites have been found (McBrearty and Brooks, 2000, Figure 7.1). Of particular note are those along the eastern edge of what is now Lake Victoria, and the archaeological sites of Mumba in Tanzania and Omo in Ethiopia, where modern human finds date to between 110 000 and 130 000 and 195 000 years ago, respectively (Bräuer and Mehlman, 1988; Clark *et al.*, 2003; McDougall *et al.*, 2005; Mehlman, 1987). These are the earliest *H. sapiens* remains yet found in Africa, slightly predating those found in Klasies River, South Africa, that date to 118 000 years ago and after (McBrearty and Brooks, 2000), and those uncovered in the Levant that date to between about 119 000 and 85 000 years ago (Valladas *et al.*, 1998). We suggest that it is no coincidence that this region of rapid and volatile environmental change was also the birth place of the first *H. sapiens*. The region experienced both rapid and highly variable environmental change that we suggest also spawned morphological, physiological and behavioural change resulting in the emergence of *H. sapiens*.

7.3.2 Marine isotope stage 5d (116 000–110 000 years ago) – the migration of modern humans out of Africa

The onset of MIS5d brought with it rapid global cooling and a drop in sea level from a high of 6 m above present during MIS5e to about 60 m below present. However, in Africa, warming observed at the end of the MIS5e remained in all but north-west Africa until suddenly, 112 000 years ago, atmospheric temperatures across all of Africa became colder. Higher precipitation levels in South and central Africa relative to AD 1800 continued until 112 000 years ago, when virtually all of Africa, except a band along the southern border of what is now the Sahara desert, received more precipitation than today (Figure 6.6). By 110 000 years ago central Africa once

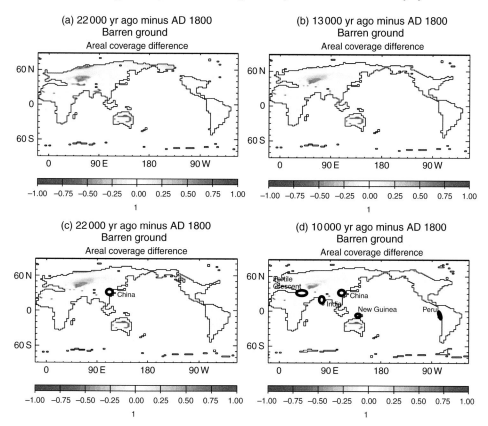

Figure 7.1 Global areal barren ground coverage anomaly at (a) 22 000 years ago relative to AD 1800, (b) 13 000 years ago relative to AD 1800, (c) 12 000 years ago relative to AD 1800, and (d) 10 000 years ago relative to AD 1800. Areas of origins of agriculture are shown. Areal coverage of barren ground derived from 122 000-year time-series climate simulation performed on the UVic Earth system climate model (UVic ESCM). (See colour version, without agriculture centres shown, in colour plate section Figure 6.17. Red indicates more barren ground coverage relative to AD 1800 and blue represents less barren ground coverage relative to AD 1800.)

again was receiving less precipitation and North and South Africa received more precipitation than AD 1800.

Broadleaf trees and C_4 grass expanded their extent and productivity between 116 000 and 110 000 years ago along a band between central and South Africa (refer to Figures 6.3, 6.4, 6.7 and Figure 6.13 above). Broadleaf tree and C_4 grass habitat also expanded in North Africa and eastward into the Middle East Levant region. Such increased fodder would have potentially promoted larger game populations within these regions. These changing habitat conditions may have stimulated what Underhill *et al.* (2001 cited in Willoughby, 2007, p. 152) indicate is an expansion of early modern humans into regions north and south of the Sahara during MIS5.

The band of increased broadleaf coverage and productivity between central and South Africa that appeared during MIS5e remained until 112 000 years ago, when it shrank to a small region along the coast of east-central Africa. By 110 000 years ago, broadleaf coverage had retreated and this anomalous band contained less broadleaf tree coverage than it did in AD 1800. This region encompasses the climatically volatile Lake Victoria region noted above.

By the end of MIS5d, 110 000 years ago, conditions, particularly in the Lake Victoria region, were deteriorating. Although initially climate change during MIS5d may have stimulated early humans to expand throughout Africa, north and south of the Sahara, when conditions deteriorated at the end of MIS5d, and habitable territory diminished, early humans would have been forced to concentrate in habitable regions that were shrinking in size. Early modern humans living in the climatically volatile Lake Victoria region, where many early archaeological and hominid sites have been found (McBrearty and Brooks, 2000, Figure 7.1), would have been required to rapidly adjust to changing living circumstances.

We stated in the chapter on human behavioural evolution that there is no evidence that the human brain has changed anatomically since the origin of *H. sapiens* around 195 000 years ago. We also stated that those brains had the potential to express the same thoughts, ideas, communication, spirituality, artistry and technical complexities as our own brains, but that it took a combination of environmental conditions, both favourable and stressful, and a social complexification to bring out that potential. We are suggesting here that the significant environmental change, which appears to have occurred in the Lake Victoria region in eastern-central Africa, combined with a concentration of populations in this area, actually stimulated saltatory change in *H. sapiens*. Population concentration combined with volatile environmental conditions likely promoted increased communication and the exchange of novel ideas that focused on dealing with the change being experienced. Such novel ideas may have included technological adjustments that would have helped *H. sapiens* to survive.

The transition from the middle Stone Age to the African later Stone Age, which must have happened prior to MIS3 when these tools were brought to Europe, is associated with drier, cooler conditions (McBrearty, 1993; Willoughby, 1993). These are just the conditions that were evident in Africa commencing 112 000 years ago. Although many archaeologists suggest that the onset of the African later Stone Age is thought to be associated with the onset of behavioural modernity, the impetus for its initiation has been difficult to determine (Willoughby, 2007, p. 95). The volatile response of the Lake Victoria region to changing climate may shed some light on this issue. The stress associated with rapidly changing conditions combined with a concentration of *H. sapiens'* populations and the potential for greater exchange and development of new ideas in this region may have created stress sufficient to stimulate the transition from African middle to later Stone Age. These and earlier adjustments may then have facilitated the capacity of early modern humans to migrate into new regions.

A more gradual progression to modern human behaviour has been suggested by McBrearty and Brooks (2000). They suggest that early transitional sites containing fossil bones that fall outside the range of modern human variation and are 'intermediate' in morphology between early *Homo* species, such as *Homo erectus* and *H. sapiens*, provide evidence of a more gradual evolution of modern human behaviour. Confirmed fossil localities that possess associations that are not problematic include Haua Fteah, Libya (dated to > 130 000 and > 90 000 years ago), Jebel Irhoud, Morocco (125 000 to 90 000 years ago and 105 000 years ago) and Ngaloba, Tanzania (> 200 000 years ago). We have not done sufficient research to be able to comment on the Tanzania site, which is dated prior to the last glacial cycle. However, the other two transitional sites are located on or close to the coast in North Africa where we would anticipate a more equable climate prevailed relative to that across the balance of North Africa during MIS5e. During MIS5e, desert limits shifted to higher latitudes, and although barren ground coverage was reduced in central Africa, it expanded in North Africa. Although needleleaf productivity was higher in North Africa than today, C_3 and C_4 grass coverage and productivity retreated to less than today. It is possible that these coastal locations provided refugia for premodern populations seeking to retreat from inhospitable territories. The concentration of early populations in these coastal locales may have stimulated the gradual development of new ideas and new technologies among a group of *Homo* species that was not yet modern but still capable of developing a number of modern human behavioural characteristics. Yet because the environment was more equable and less volatile, change happened more gradually than, for example, in the Lake Victoria region.

Early *H. sapiens* sites are also evident in South Africa at Border Cave (115 000 to 90 000 years ago), Hoedjies Punt (300 000 to 71 000 years ago), Klasies River (118 000, 105 000 to 94 000 years ago) and Sea Harvest (127 000 to 40 000 years ago) (McBrearty and Brooks, 2000). It is interesting to note that these are also coastal sites and may reflect local refugium in South Africa. Marean *et al.* (2007) have published the earliest evidence yet of a relatively advanced cultural, technological and dietary package at Pinnacle Point on the south coast of South Africa, which at 164 000 years ago is dated close to the biological emergence of *H. sapiens*. We stated earlier that there is no evidence that the human brain has changed anatomically since the origin of *H. sapiens* around 195 000 years ago. Thus, we suggest that 195 000 years ago the human brain had the potential to express complex thoughts, ideas, communication, spirituality, artistry and technical complexities, but a combination of environmental conditions, both favourable and stressful, and social complexification was necessary to bring out that potential. In this case, we propose that the concentration of populations in coastal refugium may have encouraged or stimulated the development of more advanced behaviours and technologies such as those evidenced at Pinnacle Point, South Africa. Yet the response of those populations was more gradual than in interior regions because of a less volatile environment.

7.3.3 *Marine isotope stage 5c (110 000–100 000 years ago) – expanding deserts and coastal refugia*

Continued cooling in Africa combined with reduced precipitation in central Africa resulted in a difficult MIS5c climate interval for humans living in central Africa. Prolonged periods of drought are clearly evident in the Lake Malawi region, on the border between central and South Africa, between 111 000 and 108 000 years ago and again between 105 000 and 95 000 years ago (Cohen *et al.*, 2007). However, the climate simulations show that the eastern coastal area of central Africa received more precipitation than in AD 1800 during MIS5c as it did during MIS5d. Increased precipitation was also evident in North Africa, the Middle East and the southern tip of Africa, making these regions more attractive for habitation.

As aridity intensified and desert barriers expanded, traditionally inhabited regions within Africa became reduced (Carto *et al.*, 2008). These changing conditions likely stimulated further human dispersal. However, movement into North Africa during much of MIS5c was likely limited by open and arid landscape with the desert potentially acting as a barrier to early human dispersal. But by the end of MIS5c a strong expansion of C_3 grass in Algeria and Tunisia, North Africa, indicates conditions had improved, likely facilitating emigration into these regions. Archaeological evidence indicates that by nearly 100 000 years ago, modern humans had arrived in North Africa (Willoughby, 2007, p. 197). Conditions during MIS5c also seem to have stimulated modern human movement southward from central Africa. Fossil evidence indicates that *H. sapiens* were present in the southern tip of Africa at Klasies River Mouth, South Africa by the end of MIS5c. Movement into the equable, and perhaps more easily accessible, Middle East predated the movement into North Africa and southernmost Africa. Fossil evidence indicates the presence of *H. sapiens* at Skhul and Qafzeh, Israel between 119 000 and 85 000 years ago (Valladas *et al.*, 1998; Willoughby, 2007, p. 94).

As aridity in the African interior became more pronounced during MIS5c, early modern humans may have again sought refuge in coastal regions. With sea levels falling to about 55 m below present by 103 000 years ago, continental shelves were once again more exposed. If, as Walter and others suggest, early humans occupied the Red Sea coast of Eritrea 125 000 years ago and exploited near-shore marine food resources in East Africa (Walter *et al.*, 2000), then the hyper-arid conditions of

105 000 years ago may have encouraged early *H. sapiens* to once again migrate to the coast. During periods of stress these coastal regions may have acted as 'coastal oases' when tropical regions became desiccated and subtropical lands were more arid (Faure *et al.*, 2002; Hetherington *et al.*, 2008). Thus, as interior African conditions deteriorated, modern humans may have begun their early expansion out of Africa and into South East Asia along the subaerially exposed continental shelf. Their experiences during MIS6 and MIS5e and MIS5d likely prepared them for this challenge. A more developed cultural and behavioural repertoire combined with a more equable coastal climate likely facilitated this dispersal. The coastal environment provided a more constant and diverse source of resources compared with the dwindling and likely unpredictable water and food resources available inland.

It is clear that some *H. sapiens* fled Africa during MIS5c and it is entirely possible that this was due to changing conditions in Africa. However, their capacity to inhabit regions outside of Africa was also influenced by a changing climate. While some regions of Africa, particularly central Africa were experiencing increased drought and desertification, coincidently, an expanding band of barren landscape or 'polar desert' developed stretching from southern Europe to eastern Asia. This barren ground may have restricted *H. sapiens'* movement north into Europe. It also may have stimulated the movement of some *H. erectus* and Neanderthals, who were situated in southern Eurasia, to move further southward. This may explain why an isolated Neanderthal skeleton dated to 60 000 years ago was found at Kebara 2, Israel. Migrating individuals would have been isolated from their compatriots to the north by the polar desert.

We noted above that during MIS6, *H. sapiens* migrated southward probably due to deteriorating conditions in central Africa. By 100 000 years ago the fossil evidence indicates *H. sapiens'* presence at Klasies River Mouth, South Africa. McBrearty and Brooks (2000) note the existence of an earlier appearance of upper Palaeolithic tool features evident in the South African archaeological record that was the result of 'the fitful expansion of a shared body of knowledge, and the application of novel solutions on an "as needed" basis' (p. 531). Again we suggest that Klasies River Mouth, situated on the southernmost coast of South Africa, was a refugia of sorts with a somewhat less volatile climate that confined people and encouraged the gradual development of modern behaviour.

7.3.4 Marine isotope stage 5b (100 000–90 000 years ago) – changing conditions in a mild glacial interval

Archaeologists generally agree that hominin expansion into the tropics occurred during glacial intervals or intermediate phases between glacial and interglacial when sea levels were lowered. This is supported by research that indicates climatic variance in the tropics was less variable than in the northern latitudes during glacial intervals (Tudhope *et al.*, 2001 cited in Burroughs, 2005, pp. 54–5). Archaeologists also suggest that expansion into the extra-tropical or more northern latitudes occurred during warmer interglacial intervals or during transitions from glacial to interglacial intervals (Overpeck *et al.*, 1996; Walker and Shipman, 1996). If this is true, then the Sinai Peninsula and the Levant may have acted as a 'corridor' or refugium through which early humans primarily migrated out of Africa (Bar-Yosef, 1992; Klein, 1999). Other researchers suggest that during much of the middle Pleistocene (780 000 to 126 000 years ago), ice-volume changes combined with solar insolation oscillations resulted in increased aridity in south-western Asia, central and northern Asia, and a weakening of the monsoon in South and East Asia (Dennell, 2004). They suggest that during glacial intervals, expanding desert regions may have acted as environmental barriers as habitable range contracted (Foley and Lahr, 1997; Lahr and Foley, 1998; Rolland, 1998).

Climate simulations suggest that the mild glacial conditions of MIS5b (100 000 and 90 000 years ago) likely resulted in the deterioration of early modern human habitat across much of central Africa, except for the eastern-central coastal region noted above. If simulations are correct, then improved habitat in North Africa and the Middle East, and expanding C_3 grass coverage in eastern-central Africa, North Africa and the Middle East likely encouraged movement into, and habitation of, these areas. This seems to be substantiated in North Africa where the earliest modern humans dispersed into that region close to 100 000 years ago (Willoughby, 2007, p. 197). A coincident reduction of barren ground coverage in the Middle East likely made the Levant region more habitable. It is here in the Levant where the first anatomically modern humans have been uncovered and dated between 118 000 and 85 000 years ago. However, barren ground re-established itself in the Levant 92 000 years ago.

However, with the onset of MIS5a, barren ground expanded again across much of central Eurasia limiting or blocking migration further north and potentially providing an impetus for a subsequent migration back to Africa.

7.3.5 *Marine isotope stage 5a (90 000–75 000 years ago)* – **Homo** *sp. respond to a changing global environment*

This relatively brief warm interval between 90 000 and 75 000 years ago brought with it global temperatures about 2 °C colder than today and sea levels between 25 and 65 m below present. As noted above during the relatively mild MIS5c, the barren landscape noted above in Eurasia was pervasive, likely acting as a barrier for human dispersal into the Eurasian continent. Barren ground in the Levant re-established 92 000 years ago, reaching its greatest extent between 86 000 and

84 000 years ago. These conditions may have been an impetus behind what appears to have been a subsequent retreat of modern *H. sapiens* back to Africa.

Lowered sea levels combined with expanding desertification in the Middle East likely stimulated other modern human migrations. Changing climate, sea levels and habitable environments may also have encouraged the ancestors of *H. floresiensis* to migrate to South East Asia where, although broadleaf tree coverage and productivity was less than today, C_3 and C_4 grasses, needleleaf trees and shrubs experienced coverage and productivity levels the same or better than AD 1800 levels. *H. floresiensis* may have dispersed while sea levels remained 65 m below present occupying Flores Island, Indonesia as early as 74 000 years ago, when they potentially first appear in the fossil record. If, as some scientists suggest, *H. floresiensis* was actually *H. sapiens* altered by island living, then their relations may have subsequently migrated on to Australia between 50 000 and 45 000 years ago.

7.3.6 Marine isotope stage 4 (75 000–60 000 years ago) – modern humans contend with expanding ice sheets and a polar desert

MIS4 (75 000 to 60 000 years ago) would have created very difficult conditions for early humans. Those that did not migrate back to Africa during MIS5a would have been confined to a rapidly decreasing habitat. Advancing ice sheets in northern Europe combined with what simulations show as an expanding 'polar desert' in central Eurasia would have forced Neanderthals and any other hominins in the region into a refugium in the south-westernmost regions of Europe. Potential 'refugia' also existed in the Middle East, South East Asia and Australia where simulations show higher C_3 and C_4 grass coverage and lower barren ground coverage than AD 1800. The 60 000-year-old Neanderthal skeleton uncovered at Kebara 2, Israel dates to the end of this interval. It is likely that a deteriorating climate in Europe potentially further encouraged the migration of at least some of the Neanderthal southward into the Levant.

Climate simulation findings for this interval are supported by proxy evidence interpreted by Dupont *et al.* (2000) and indicate that North African desert conditions expanded during MIS4 and rainforest cover greatly diminished. Dupont *et al.* also suggest that *Podocarpus* forests were only to be found in Angola and possibly the Congo. Deteriorating conditions outside of Africa may have encouraged a return of expatriates back into Africa as evidenced by an increase in the number of *H. sapiens* sites in Africa dating to between 90 000 and 60 000 years ago (McBrearty and Brooks, 2000). However, consistent with research suggesting conditions in interior Africa were deteriorating, two-thirds of these sites were located along the coasts of north-western Africa and South Africa. This further substantiates the hypothesis that *H. sapiens* migrated to the more equable coastal African environments during glacial intervals.

Also impacting early humans in Africa at this time was the eruption of Mount Toba in northern Sumatra, Indonesia around 71 000 years ago. This likely exacerbated global cooling and further reduced global biomass productivity (Willoughby, 2007, p. 88; see also Ambrose, 2003; Wintle, 1996). Anthropologists Stanley H. Ambrose and M. R. Rampino suggest that the Mt. Toba eruption resulted in a volcanic winter that caused a severe reduction, or 'bottleneck' in the modern human population (Ambrose, 1998b; Rampino and Ambrose, 2000). They propose that it reduced genetic diversity outside of tropical refugia, including that of equatorial Africa and may explain current differences apparent in modern humans. Others suggest that early modern humans used behavioural adjustments, including risk-management strategies to cope with a rapidly changing climate, including the purported volcanic winter caused by the Mt. Toba eruption (Gathorne-Hardy and Harcourt-Smith, 2003). In any case global modern human population is believed to have remained at about 600 000 until shortly after 40 000 years ago and the development of the upper Palaeolithic tool industry (Perreault, 2003).

Deteriorating environmental conditions may have stimulated the transition to the late Stone Age (if this transition had not already occurred during MIS5d – 116 000 to 110 000 years ago) (Willoughby, 2007, p. 88; see also Ambrose, 1998b, 2003; Rampino and Ambrose, 2000; Rampino and Self, 1992). Archaeological evidence dating to between about 70 000 and 40 000 years ago shows the development of a new upper Palaeolithic (UP) or later Stone Age (LSA) technology. It is proposed that changes in human habitat, both in terms of reduced availability and productivity of subsistence resources, and the concentration of populations in a limited number of habitable refugia stimulated the development of new technologies. Certainly the scientific evidence suggests that by the close of MIS4, early humans were forced into restricted refugia that necessitated that they interact more closely with one another and deal with habitats with reduced productivity. One refugium may have been located near Lake Malawi, where lake cores indicate dramatically wetter conditions and decreased environmental variability after 70 000 years ago (Scholz *et al.*, 2007). In any event, there is no question that the later Stone Age technology was developed and probably gave humans a means to increase the carrying capacity of their environment. It may also have facilitated a subsequent major population expansion outside of Africa (Finlayson, 2004, p. 72).

It is interesting that the number of *H. sapiens* sites in Africa fell after 60 000 years ago. Climate simulations show an extremely large decrease in global net primary productivity coincident with this reduction in number of sites.

7.3.7 *Marine isotope stage 3 (60 000–30 000 years ago) – population expansion for* Homo sapiens *and extinction for* Neanderthals

With the onset of MIS3 60 000 years ago came a dramatic decrease in global net primary productivity that was consistent with a reduction in known *H. sapiens* archaeological sites in Africa. Climate simulations show that global net primary productivity fell to a level not seen again until the last glacial maximum 21 000 years ago. Sixty thousand years ago ice sheets extended across much of the northern hemisphere and an expansive polar desert prevailed in much of central Eurasia. These conditions undoubtedly had a substantial impact on the ability of early humans to find sufficient subsistence resources. The cold and unproductive conditions of 60 000 years ago were followed by an extremely variable MIS3 (60 000 to 30 000 years ago). Climate combined with restricted habitable landscape required that early humans rapidly adjust their behaviour. They developed new technologies and migrated into eastern Asia, Australia, New Guinea, Europe and elsewhere.

By MIS3 the new later Stone Age technology had clearly been developed and an abrupt transition from middle Palaeolithic to upper Palaeolithic is evident in Europe, and considered by some to reflect a 'revolution in modern human behaviour'. More progressive stone and bone tool technologies improved the ability of humans to gather resources and, as MIS3 progressed, they combined with an overall improvement in global net primary productivity facilitating a subsequent major population expansion. By 35 000 years ago human population levels are thought to have reached about 4 million (Finlayson, 2004, p. 72; Perreault, 2003).

Extensive ice sheets in the north and an expansive barren landscape across much of central Eurasia during early MIS3 may explain why modern human presence in Australia and South East Asia was so much earlier than in Europe, where behaviourally modern humans did not appear until between 40 000 and 30 000 years ago. Further, it verifies our hypothesis that 'behavourally modern' humans did not appear with Cro-Magnon in Europe, but that the capacity to be behavourally modern goes back to the origin of anatomically modern humans.

Our climate simulations indicated that a migration out of Africa and northward into Europe was restricted by barren landscapes until ~50 000 years ago and again between 40 000 and 38 000 years ago. This is consistent with Braüer's (1984, 1989) and Clark's (1989) interpretations that the impetus for the original expansion of modern humans out of Africa and into Asia and Europe may have been primarily climate change that increased desertification. Further, it substantiates the theory that the Sinai Peninsula and the Levant potentially acted as an environmental 'corridor' or refugium and a primary migration route for hominins out of Africa when conditions ameliorated (Bar-Yosef, 1992; Klein, 1999). Alternatively, during glacial intervals the region may have behaved as an environmental barrier as desert regions expanded, habitable range contracted and sea levels fell (Foley and Lahr, 1997; Lahr and Foley, 1998; Rolland, 1998; for an alternative perspective see Fogarty and Smith, 1987). During these intervals, migration was more likely along coastal routes.

Modern humans arrived in Australia between about 45 000 and 42 000 years ago, or earlier (see Chapter 4), when simulations indicate precipitation levels were higher and global barren ground coverage was substantially lower than it was in AD 1800. Hydrological evidence at Lake Mungo, Australia, indicates the lake began filling 60 000 years ago. Between 50 000 and 40 000 years ago the lake experienced a period of intermittent change in lake levels coincident with the arrival of people (Bowler *et al.*, 2003). A major hydrological change about 42 000 years ago reflects increased aridity in the continental interior. This may have stimulated movement out onto the exposed continental shelf because between around 50 000 and 40 000 years ago modern humans arrived in lowland New Guinea and in highland New Guinea at least 5000 years later. Tasmania was likely colonized around 35 000 ka BP (40 300 years ago) when falling sea levels exposed the entire Bass Strait and resulted in intermittent connection between Tasmania and Australia (Lambeck and Chappell, 2001). A large megafauna extinction began in Australia between 50 000 and 45 000 years ago and several forest-dwelling marsupials went extinct in New Guinea after 40 000 years ago. Extinction of these species may have been the direct or indirect result of human activity. However, the extinction of the flightless bird *Genyornis newtoni*, which fed primarily or exclusively on C_3 grasses, occurred at a time when climate simulations show C_3 grass coverage retreating and being replaced by C_4 grass in coastal regions of south-western and south-eastern Australia. This supports the interpretation of Miller *et al.* (1999) that climate played a role in its extinction; particularly when the flightless emu, *Dromaius novaehollandiae*, which possessed a broader diet including C_4 grasses, survived because of its dietary adaptability to changing ecological conditions.

Late in MIS3, modern humans finally dispersed out of Africa, into the Middle East and subsequently colonized Europe. Based on non-recombining Y-chromosome analyses the M89/M213 modern human lineage dispersed out of East Africa into the Levant and expanded west, north and east about 40 000 years ago (Underhill *et al.*, 2001). This initial modern human dispersal into Europe represents the earliest upper Palaeolithic occupation, whose Y-chromosomes later became nearly extinct. It was soon after 40 000 years ago that conditions worsened in central Eurasia. The European populace became confined between persistent ice sheets to the north and an expanding polar desert to the east.

It may have been these constraining conditions that pressed *H. sapiens* in Europe into territory previously in the domain of Neanderthals. Neanderthals had survived previous glacial and interglacial intervals, but MIS3 was different. By 30 000 years ago the last known Neanderthals disappeared from Europe. Numerous climate-related hypotheses persist as to why the Neanderthals disappeared when they did. Finlayson (2004, p. 72) suggests their extinction was due to the severity of climate change around 30 000 years ago. Stringer and Davies (2001) propose that their extinction was due to competition for resources with modern humans, combined with climatic instability. Both these theories appear to be valid to some extent. Climate simulations indicate a deteriorating and highly variable climate during MIS3 that reduced the extent and productivity of human habitat. The combination of these factors would have reduced available resources. Those hominin populations most able to adjust to change would have been most successful. The fact that Neanderthals did not survive suggests to us that, like many other species that have gone extinct in the past, they lacked the capacity to rapidly adjust to changing conditions.

Certainly conditions at this time were highly variable. *H. sapiens* appeared to thrive in these conditions whereas Neanderthals did not. Genetic evidence suggests a period of non-African population growth occurred around 40 000 years ago when modern humans and Neanderthals would have still been restricted to habitable landscapes between expansive ice sheets and the broad swath of barren landscape that covered much of Eurasia. The variability of the MIS3 climate generated brief reprieves when climate improved (e.g., between 50 000 and 48 000 years ago, and 42 000 years ago). It is during these times that *H. sapiens* populations likely expanded. These relative improvements in climate were interspersed with recurring deteriorations in climate (e.g., between 40 000 and 36 000 years ago) when populations likely concentrated into more restricted areas. It was this expansion and contraction, generated by a constantly changing environment that may have stimulated the florescence of new ideas as human populations were alternately thrown together and dispersed. New ideas were likely generated when diverse peoples came together during deteriorating conditions. We suggest that new behavioural and technological innovations occurred that generated greater productivity and facilitated population expansions during subsequent climate ameliorations. These circumstances were repeated during the many climate oscillations observed during MIS3. These conditions would have necessitated that *H. sapiens* and Neanderthals develop a capacity to be flexible and adaptable. If Neanderthals were unable to physiologically, developmentally or behaviourally respond their capacity to survive would have been much diminished.

Near the end of MIS3 modern humans expanded into China. The oldest dated modern human remains in China date to between 34 000 and 29 000 from the Upper Cave at Zhoukoudian and include three complete skulls and bone fragments. With the middle Palaeolithic missing in China, the transition to late Palaeolithic occurred directly from the early Palaeolithic around 30 000 years ago. Proxy evidence indicates fauna in mainland South East Asia and southern China experienced fewer changes than those in northern China, with most fauna persisting throughout the Pleistocene until the Holocene. This persistence is thought to be due to the presence of forested tropical refugia (Schepartz *et al.*, 2000). Simulations show an area in mainland South East Asia that was less affected by expanding barren ground, had raised broadleaf tree coverage levels and where needleleaf tree coverage remained higher than in AD 1800. The northern edge of this region abuts the site of Upper Cave, Zhoukoudian, where the first modern human skeletal remains have been found in China. Although coverage levels were not always matched with increased net productivity levels throughout this region, they were near the site of Zhoukoudian (Figures 7.5a and 7.5b). It is possible that humans concentrated in this region and this concentration spawned new ideas that stimulated the transition from the early Palaeolithic to the late Palaeolithic around 30 000 years ago.

In India a refugium existed until around 59 000 years ago that facilitated *in situ* evolution of prepared core technology. However, fossil evidence indicates that no modern human existed in the area until between 31 000 and 28 500 years ago.

In Siberia, the Mousterian stone tool facies, which typically dates to MIS3, shows considerable variety. Its presence in southern Siberia could mean that it may be found elsewhere in northern Asia (Larichev *et al.*, 1987). It is thought that climate change resulted in reductions and expansions of Palaeolithic populations. It is also clear from archaeological sites dated to this interval that early peoples must have been required to adjust to arctic desert conditions.

7.3.8 Marine isotope stage 2 (30 000–11 650 years ago) – the last ice age and large-scale migrations and human behavioural change

The last ice age, MIS2 (30 000 to 11 650 years ago), was a period of large-scale human behavioural change. Modern humans migrated throughout the world and developed unique modern behavioural characteristics. A rapidly changing climate that was associated with changing habitats and lowered sea levels likely played a critical role in both the timing and direction of these shifts. Expanding deserts restricted habitation of interior regions while exposed continental shelves provided new productive habitat for a series of migrations around the world.

In South East Asia limited fossil evidence combined with a dearth of well-supported chronometric dates currently limits the ability of researchers to ascertain clear dispersal routes and the timing and impetus of modern behaviour (Lukacs, 2005). However, modern human fossil evidence has been uncovered in the Philippines and dates to between 24 500 and 24 000 years ago (Fox, 1970). Thus, it is likely that the very low sea levels evident during MIS2 played a key role in the habitation of South East Asia. Symbolic behaviour in the form of beads and 'art' first appeared in the archaelogical record on the Indian subcontinent between 30 000 and 20 000 years ago. Artifacts include engraved ostrich-eggshell pieces at the site of Patne (Misra, 2005). Their appearance is coincident with a proposed population expansion in the region (James and Petraglia, 2005; Kivisild *et al.*, 1999).

According to climate simulations, modern humans living in Asia would have been separated from others inhabiting Europe by an expansive barren landscape in west-central Eurasia. This barrier may explain why Native north-eastern Siberians show genetic similarities with Asian specific mtDNA groups, whereas the Saami aboriginals from north-western Siberia are believed to have genetic similarities with the upper Palaeolithic Europeans.

In Japan the distinct Neolithic Jomon pottery period first appeared about 13 900 years ago, near the end of MIS2. The Jomon people arose in north-east Asia and are thought to have led to the rise in the Ainu and Ryukyuan people (Tanaka *et al.*, 2004). A subsequent migration from north-east Asia, via Korea, to Japan likely brought the ancestors of the Hondo-Japanese whose genetic influence predominates in the majority of modern Japanese.

During MIS2 modern humans are thought to have migrated across the exposed Beringian 'land bridge' that stretched from north-east Asia to north-west North America and moved into the interior of North America and then down to South America via an 'ice-free corridor' east of the Canadian Rocky Mountains. If one looks at the continental configurations during the LGM, it is clear that the broadly exposed Beringian subcontinent connected north-west North America with eastern Asia (refer to Figure 5.3). During the coldest times of MIS2 modern humans living in eastern Asia were likely encouraged to move eastward onto the exposed continental shelf, which probably provided additional productive habitat, whereas previously habitable regions in northern and eastern Eurasia experienced reductions in extent and productivity (Hetherington *et al.*, 2008; Zazula *et al.*, 2003).

Early inhabitants of North America moved eastward, occupying Bluefish Caves in Yukon, Canada, by 16 100 years ago (Burke and Cinq-Mars, 1998; Cinq-Mars, 1979; Cinq-Mars and Morlan, 1999), and then attempted to move southward. However, vast ice sheets extended across the breadth of northern North America isolating the far north from the southern half of North America and Central and South America. Then as climate warmed, sea levels rose, inundating the Beringian subcontinent by about 13 000 years ago. It is clear that people must have left northern North America before 13 000 years ago in order to reach southern sites such as Santa Rosa Island, California, by 13 400 years ago and Monte Verde, Chile, by 14 700 years ago. The original theories suggested humans moved southwards along an 'ice-free corridor' that opened east of the Canadian Rocky Mountains. Yet geologic evidence indicates the 'ice-free corridor' was closed between about 24 000 and 13 500 years ago (Dyke, 1996, 2004b; Jackson *et al.*, 1997; Lemmen *et al.*, 1994; White *et al.*, 1985), implying alternative migration routes were required to get people south of the ice sheets prior to 13 500 years ago. People may have migrated along the southern edge of the exposed Beringian shelf and down the Pacific coasts of North and South America (Hetherington *et al.*, 2007, 2008). Although parts of this route were glaciated, some refugia appear to have existed along the exposed continental shelf (Hetherington *et al.*, 2003). Some researchers, including the authors of this book, also hypothesize that people, using boats, may have crossed at least portions of the open ocean to reach the Americas. For example, Montenegro *et al.* (2006) simulated prehistoric transoceanic crossings to the Americas and indicate possible ocean crossings for boats between north-east Asia and North America. These simulated crossings took 83 days when not paddling with a 3% success rate and 70 days when paddling with a 4% success rate. They also indicate possible ocean crossings between western Africa and South America.

Although the early colonizers of Australia must have utilized marine resources and thus occupied coastal regions prior to the onset of MIS2, evidence of their presence on coastal regions is scant before the LGM. Climate simulations of Australia indicate that barren ground coverage declined during MIS2, C_3 grass expanded in southern coastal Australia, and C_4 grass and shrub coverage expanded in western, eastern and north-central Australia. By the LGM, lowered sea levels connected Australia to New Guinea, Tasmania, and other islands of Melanesia forming the continent of Sahul. Large freshwater lakes developed in northern Western Australia in the Bonaparte Gulf (Yokoyama, De Dekker *et al.*, 2001; Yokoyama, Purcell *et al.*, 2001), between Australia and New Guinea in the Gulf of Carpentaria (Torgerson *et al.*, 1988) and between Tasmania and Australia in the Bass Strait (Lambeck and Chappell, 2001). Contrary to the interpretation of O'Connor and Veth (2000), that increased aridity associated with the LGM initiated a movement of people out of the interior and back to the coastal environment, it may be that improved habitat resulted in a population expansion, which, combined with lowered sea levels and an exposed continental shelf, encouraged the reoccupation of coastal environments.

In Europe, climate simulations indicate that broadleaf tree extent and productivity was less during MIS2 than in AD 1800. Needleleaf tree coverage and productivity remained below AD 1800 levels throughout Europe until about 24 000 years ago when an expansion in central Europe raised productivity to levels beyond AD 1800. Around 18 000 years ago conditions again deteriorated and remained poor until about 14 000 years ago when needleleaf tree coverage became higher than AD 1800 levels and remained so until the end of MIS2. C_3 grass extent and productivity across much of Europe remained below AD 1800 levels throughout MIS2 except for a small region in northern Iberia which persistently remained more productive than AD 1800. Although climate simulations indicate C_3 grass expanded into northern regions adjacent to the ice sheets until about 17 000 years ago, productivity remained below AD 1800 levels. C_4 grass productivity and extent remained marginally below AD 1800 levels throughout MIS2, with slight improvements after 16 000 years ago. Simulations show shrub productivity and extent continued to remain higher than AD 1800 levels with the onset of MIS2, particularly in Eastern Europe, and remained so until 10 000 years ago. These poor conditions, particularly evident between 18 000 and 14 000 years ago, may have stimulated a significant population contraction that is evident in the European archaeological record between 18 000 and 16 000 years ago. Around 16 000 years ago the expansion of C_3 grass in the Middle East, which had developed around 27 000 years ago, once again disappeared. At the same time the large expanse of barren ground that had isolated Europe and stretched across the continent receded somewhat opening a route between Europe from the Middle East. These conditions may have stimulated a movement of modern humans into Europe from the Middle East. Archaeological evidence indicates that at the close of MIS2 Neolithic farmers dispersed out of the Levant and into Europe, although the impact is relatively localized. The Y-chromosome analyses of Semino *et al.* (2004) support these Neolithic expansions into Europe from the Middle East, as well as out of the southern Balkans and from North Africa into the Iberian Peninsula. Semino *et al.* also suggest modern humans expanded from the Middle East back into East Africa, and again more recently toward North Africa and Europe.

7.4 The Holocene (11 650-AD 1800) – population expansion and the rise of agriculture and domestication

The Holocene was a time of great human technological advancement. Humans discovered how to domesticate plants and animals; agriculturalists began to settle and produce food rather than remaining mobile to hunt and collect. Independent centres of domestication in the Fertile Crescent of south-west Asia began between 13 000 and 11 500 years ago (Turney and Brown, 2007) and had clearly developed by about 10 500 years ago. Humans cultivated emmer and einkorn wheat, barley, oats and legumes and raised sheep and goats. Other centres followed: South America (10 000 years ago); northern China (9500 years ago); Africa (7000 years ago); Meso-America (5500 years ago). Agriculture appeared in Europe between 8400 and 7000 years ago; its initial appearance perhaps a consequence of flooding of south-west Asian coastal farming areas coincident with the collapse of the ice dam in Canada's Hudson Bay region about 8200 years ago (Barber *et al.*, 1999; Turney and Brown, 2007). Agriculture arrived lastly to the eastern woodlands of North America.

Climate change and the pressure from population growth are frequently cited as key stimulants for the adoption of agriculture and the domestication of animals. However, it is quite apparent that the climate during the Holocene was more equable than the earlier MIS2 interval. Yet if we review the simulated effects of climate on the landscape between MIS2 and the Holocene we recall that one of the most significant impacts was the changing extent of an enormous barren polar desert across Eurasia. Twenty-two thousand years ago it stretched from northern Eurasia where it abutted the ice sheets, south to the Mediterranean Sea and east across Asia to the exposed Beringian continental shelf. Modern humans that had left Africa during MIS2 and earlier, and migrated into Europe, would have been isolated between ice sheets to the north and this expansive barren ground to the south and east. Those who had migrated into the rest of Eurasia moved into south-west Asia, India, China and South East Asia. They would have likely concentrated in two regions in Eurasia where barren ground coverage was significantly less than today at the close of MIS2. The first is a region proximal to the Fertile Crescent where emmer and einkorn wheat, barley, oats, legumes, olives, sheep and goats were first domesticated (Figure 7.1). The other is in South Asia including the Indian subcontinent and the Indus basin, where wheat, barley, jujube, sheep and goats were domesticated around 9000 BC (Allchin and Allchin, 1997; Gupta, 2004). It is possible that modern humans concentrated in these regions during MIS2. The concentration and coalescence of hitherto independent populations likely had a twofold effect. It likely facilitated the development of new ideas and technologies, including agriculture, particularly by those living on the periphery where resources were less prolific. And it probably resulted in further population expansions as climate ameliorated. Then, despite the onset of a more equable climate, with the onset of the Holocene barren ground in these two regions expanded. This placed people living in these regions in a bit of a quandary. At the same time that barren ground coverage was expanding and thus reducing the productivity of their land, they needed to increase their resources in order to feed their expanded populations. A relatively stable climate provided some capacity to predict seasonality and temperatures, and new ideas and technologies that had begun to be implemented by those living on the less productive periphery were available. As the conditions on the periphery became those of the majority, these new technologies were then likely employed to feed an increased population trying to survive on a less productive land.

Conditions during the onset of agriculture in South and Central America were different, but we still see a pattern in that key differences between pre- and post-agricultural climate are evident. Dry cold conditions prevailed in Peru during the Younger Dryas between 12 700 and 11 650 years ago. Generally, coastal resources were more plentiful and interior terrestrial resources were depleted. However, contrary to generally accepted findings that Amazonia was more arid during the last glacial maximum, Baker *et al.* (2001) suggest that cold North Atlantic sea-surface-temperatures correspond with rising water levels in Lake Titicaca, Peru. This may mean that Lake Titicaca may have been a refugium of sorts during cold intervals. If so, populations living in the interior of Peru may have concentrated in Lake Titicaca during cold intervals. Such a population concentration would have stimulated new ideas and technologies.

With the onset of the Holocene 11 650 years ago ice began to disappear from the southern tip of South America. Between 11 650 and 8200 years ago climate rapidly warmed. The warming climate of the early Holocene prevailed until a mild cold event 8200 years ago coincided with the collapse of an ice dam holding back an immense amount of freshwater from large glacial lakes in Canada's Hudson Bay. Recently Timmermann *et al.* (2005) have shown that an influx of freshwater from melting ice sheets into the North Atlantic will trigger a collapse of the Atlantic thermohaline circulation, which impacts meridional overturning in the Pacific. A reduced meridional overturning is linked to a suppression of ENSO activity and drier, colder conditions in coastal South America.

After the brief cold event at 8200 years ago, warm conditions once again returned. Our climate simulations show that Pacific meridional overturning increased, particularly by 8000 years ago (see Figure 5.6), likely stimulating the beginning of an increase in ENSO activity. These findings are substantiated by cores taken from Laguna Pallcochoa in southern Ecuador that show an absence of ENSO events in the early Holocene and significant ENSO frequency by 7000 years ago (Rodbell *et al.*, 1999; Sandweiss *et al.*, 2001, cited in Burroughs, 2005, pp. 234–5).

ENSO variability continued to increase, peaking around 4900 years ago. It remained high until 1200 years ago, when it suddenly declined (Burroughs, 2005, pp. 234–5). When ENSO activity was reduced, upwelling of nutrient-rich waters was pervasive and coastal resources would have been good. However, interior terrestrial resources were likely depleted. During warm intervals when ENSO activity was prevalent, marine resources were depleted and coastal habitation would have been more difficult. Coastal populations that had expanded during the good times likely moved inland in search of terrestrial resources. Those who had struggled during the dry, cold years in the interior were now benefiting from mild temperatures, increased rainfall, and probably more productive conditions, but they may also have faced an influx of displaced coastal people. It is possible that the interior people developed ideas and technologies during the tough times that were subsequently employed to feed an expanded population. This may explain the development of agriculture in Peru (for a more in-depth discussion of climate and agriculture see Chapter 8).

Changing ENSO activity probably continued to play a major role in the history of development in Peru. Caral, the city of pyramids, the earliest dated major construction in Peru was occupied around 4900 years ago for about 1000 years. Its occupation coincides with a peak in ENSO activity around 4900 years ago (Burroughs, 2005, pp. 234–5). This site, located just inland from the coast of Peru, housed thousands of people who relied on a diet of both fish and agriculture.

With the onset of agriculture around the world came the rise in cultural continuity, tribal structure, and a sense of belonging to a particular tract of land. As populations expanded and concentrated in productive regions, states arose complete with hierarchical and centralized decision making, full-time religious and craft specialists, art, architecture, taxation and the military. New agricultural technologies were developed, including crop rotation, the heavy plow and irrigation. These agricultural developments and the associated sedentary lifestyle resulted in population growth, trade, competition and sometimes warfare in physically or socially confined regions. Warfare was not always present, as archaeological evidence in Caral indicates a complete lack of warfare, and instead a society built on commerce and pleasure.

Perhaps because of climatic variability these societal changes often combined with an increasing desire to control and manipulate the unpredictability of nature. Agriculture and domestication were early manifestations of this desire. By storing their harvested surplus people could 'weather' unexpected short-term changes in productivity. As further cultural developments were made, additional surpluses allowed specialists to develop and propose new ideas and practices that encouraged individuals and groups to alter their activities for further control and manipulation. With time came the Industrial Revolution. Just over 200 years ago humans developed technologies that made use of energy captured in fossil fuels like coal, oil and gas. By using increasing quantities of energy consumed by this new technology humans were able to travel greater distances more quickly, produce more food with less productive land and control our environment so we could live in hot or cold places. Thus we were able to overcome the hostile climate of previously marginal environments.

7.5 Conclusion

During the last glacial cycle, *H. sapiens* populations experienced very diverse living environments. Climate changed from being highly variable, alternating between ice ages and warm intervals, to being relatively stable with the onset of the Holocene. Rapidly changing conditions would have forced hominins to adjust. The demands of a rapidly changing climate during the early part of the last glacial cycle likely played a key role in stimulating the emergence of *H. sapiens* in Africa. As the last glacial cycle progressed, further adjustments included the migration out of deteriorating regions and into more habitable areas. The concentration of individuals in restricted habitable regions periodically placed disparate groups in social contact with one another. It is just this sort of social interaction that would have stimulated the emergence of intelligence and the development of new ideas and technologies.

During cold periods expanding ice sheets and deserts would have reduced habitable land. At the same time lowered sea levels exposed productive continental shelves onto which early humans may have migrated. Changing vegetation and desertification likely played a role in the capacity of humans to survive, migrate and develop new technologies.

Deteriorating conditions in Africa during MIS5e may have stimulated dispersal out of Africa and into the proximally located Middle East, which was biogeographically a part of Africa and where the earliest evidence of *H. sapiens* exists outside of Africa. A refugia may have existed in eastern-central Africa near Lake Victoria near the sites of Mumba, Tanzania and Omo, Ethiopia where modern human finds date to between 110 000 and 130 000 and 195 000 years ago, respectively. The dry, cool conditions of MIS5d, combined with concentrated populations may have allowed for greater exchange and development of ideas stimulating the transition from the African middle to later Stone Age. This new development may have facilitated the migration of modern humans into the Levant between 119 000 and 85 000 years ago. Falling sea levels that exposed productive continental shelves and the hyper-arid conditions in Africa's interior 105 000 years ago may have encouraged early *H. sapiens* to migrate to the coasts and from there into South East Asia. Further, an expansive polar desert that stretched from southern Europe to eastern Asia during MIS5c may have separated and isolated pre-existing populations of *H. erectus* and Neanderthals. By MIS5b improved habitat across North Africa and the Middle East likely encouraged movement into these areas. With the onset of MIS5a barren ground again expanded likely restricting movement further north and providing an impetus for a subsequent migration back to Africa. With sea levels between 65 and 25 m below present, the ancestors of *H. floresiensis* may have migrated to South East Asia where much of the vegetation coverage and productivity was the same or better than AD 1800.

The cold, dry climate of MIS4 between 75 000 and 60 000 years ago probably brought very difficult living circumstances for modern humans. Expanding ice sheets in northern Europe, an expanding 'polar desert' in Eurasia and deteriorating habitats likely encouraged movement into refugia in the Middle East, South East Asia and Australia. It is in the Middle East where the 60 000-year old Neanderthal skeleton has been found. It was also during MIS4 that Mt. Toba erupted causing what some believe was a volcanic winter responsible for a severe reduction in modern human population. Changes in human habitat combined with the concentrations of populations in restricted refugia may have stimulated the development of the later Stone Age.

The extreme variability of MIS3 saw the migrations of modern humans throughout much of the world, a rapid transition from middle Palaeolithic to upper Palaeolithic stone tool technology in Europe and a large population expansion.

The last ice age brought with it expanding ice sheets, reduced sea levels and dry, cold conditions. Expanding deserts restricted habitation of interior regions and encouraged movement onto productive exposed continental shelves. Humans migrated to the New World, developed pottery in Japan and beads and art in India, and expanded their populations. The relatively warm Holocene brought the disappearance of the great ice sheets that had persisted for more than 62 000 years. Agriculture was developed in the Fertile Crescent and elsewhere. The warming of the Holocene was interrupted by a mild cold event 8200 years ago triggered by the release of freshwater from glacial lakes in Canada. This event has been linked to the movement of agriculture into Europe. The Holocene brought in an increase in ENSO events that are linked to rain and warmth as well as reduced marine productivity in coastal Peru, which may help explain the origin of agriculture 10 000 years ago.

It is clear that the climate of the last glacial cycle was highly variable and that changing climate placed great stress on early humans. Given our current understanding about human adaptability and developmental evolution we can thoughtfully expect that these stresses probably influenced and possibly stimulated the emergence of *H. sapiens*. They also would have stimulated subsequent epigenetic, physiological and behavioural developments including the migration of humans into new habitats and the development of new behaviours including agriculture. It is the development of agriculture and its association with climate that we now address more closely.

Notes

1. According to Goodwin (1929), middle Stone Age assemblages consist of point and scraper artifacts, struck from radial, disc or Levallois cores.
2. According to Ambrose (2002), the subsequent later Stone Age artifacts include parallel-sided blade or bladelet cores, as well as retouched tools including scrapers, geometric microliths and backed tools. However, there are many middle Stone Age assemblages that possess some blade technology, most notably the flake-blades of South Africa (for more on this see Willoughby, 2007, p. 245).

8

Climate and agriculture

8.1 Introduction

Climate and agriculture have complex connections. In Chapters 6 and 7 we learned how a changing climate influences vegetation. Climate limits what can be grown. Climate change can also cause catastrophic crop failure; today, anthropogenic 'global warming' is blamed for the biggest droughts in hundreds to thousands of years and impoverished crops. These kinds of connections strongly suggest that climate change had a significant influence on the origin of agriculture and its subsequent development. The early evolution of farming had nothing to do with anthropogenic global warming, but linkages between a changing climate and the capacity to grow things in the past and present are clear.

Milankovic cycles, positive feedbacks and ENSO (El Niño–Southern Oscillation) events influence climate change and the productivity and extent of vegetation. During the post-glacial melting of ice in the late Pleistocene and early Holocene, two huge escapements of fresh water into the Atlantic abruptly slowed the northerly circulation of warm sea water and caused northern hemisphere surface temperatures to drop. The first freshwater influx, during the Younger Dryas cooling event, correlates with early attempts at farming. The second, also associated with a cooler and drier climate, affected the ranges and species of crops that could be raised. Volcanic eruptions also affect productivity and the extent of vegetation. Volcanic eruptions had local effects that early humans could step away from, provided there was somewhere to step that was not already densely populated. However, multiple eruptions and the occasional 'super volcano' could significantly alter climate patterns, creating changes that would last for several years.

Agriculture is not simply a matter of obtaining enough food. It has a major impact on human work patterns, social structure, trade, politics and warfare. We must take this into account to establish feedback connections that will provide a greater understanding of agriculture itself. This will tie in with what we have already written about climate and human behaviour. Particularly relevant is the hypothesis that we previously applied to the pre-agricultural complexification of society: catastrophe (usually climatic); communication (of hitherto isolated groups); and collaboration (if not mayhem), involving exchange of techniques.

Many of the developments that we discuss in this chapter fall within a period for which there exists an oral history, backed up by carbon isotope dating. Later, historical written records are able to provide some dates to within a year. Our model, which is on a broader scale, is here being tested by historiography, and that is one of our purposes in writing this chapter on the correlation between climate and agriculture. The assessment of similarities and contrasts are summarized at the end of the chapter.

8.2 Animal and plant domestication

8.2.1 Dogs and cats

The most parsimonious views of the origins of plant and animal domestication are inadequate to the actual history of these phenomena. However, there is merit in some of them. The first domesticated animal was probably the wolf that became 'the dog'. No doubt some wolves hung around human hunts to scavenge the remains of kills. Some bold animals may have approached temporary human camps and settlements to seek food. In the cases of the dog and cat families, the empathy between them and humans would have depended on a degree of tameness, and the playfulness of their pups and kittens. Belyaev (1979) showed that in the case of Arctic foxes that were captured for breeding for the fur trade, the selection of the tamest individuals and their interaction with humans had remarkable effects. Hormonal changes in a captive population of foxes altered reproduction patterns and produced behaviour that was similar to that found in domestic dogs, such as licking the hands of their human handlers and wagging their tails. There were also changes in coat colour and patterns.

One of the results of even the loosest relationship between humans and dogs would have been the effect on herding behaviour in sheep and goats. As anyone who has watched a border collie at work knows, sheep 'bunch up' without any stimulus except the obvious presence of the dog. It would not have taken much logical power for human hunters of sheep to realize how much easier the dogs would make the hunt, and from there to realize that the behaviour of wild herds could be controlled to make nomadic pastoralism possible. Thus far there has been no climate connection, unless this interrelationship developed in climatically harsh times, and if times got even harsher humans could always eat their dogs. Initially there had been no 'domestication' in the sheep or goats, which implies changes in the genetic makeup of the plants or animals due to deliberate selection of the most suitable types.

The domestication of dogs goes back at least 12 000 years, given their presence in human grave sites at that date. The dogs are distinguished from wolves by their smaller size, reduced muzzles and dentition. Genetic studies indicate that there was a locus of dog domestication in East Asia, and that the animals spread out from there, rather than it being a global phenomenon (Miklosi, 2008; Serpell, 1995). Cats also have a single area of origin, the Fertile Crescent (Lipinski *et al.*, 2008; Leslie Lyons is leader of the cat genetics research group). Considering the fact that the ancestors of dogs and cats were widely distributed in the northern hemisphere, the single-point-of-origin discoveries suggest two or more criteria must be met for the domestication of these animals. In the case of dogs, there must have been a particular attraction in East Asia that brought the ancestral wolves closer to humans, or perhaps there was a local epigenetic change that made the ancestral types more prone to approach humans. For example, an epigenetic mutation could have made its possessors more paedomorphic, i.e., more like the playful puppy stage carried through to adulthood. Such an epigenetic change would also be linked to anatomical developmental changes. In the case of cats, not only the presence of their ancestral wild cats was necessary, the increase of mouse and rat populations, linked to rudimentary farming, and the storage of grains, would also have been important. Epigenetic changes in wild cats that hunted rodents would also have been necessary for actual domestication. (Epigenetic effects change the course of development of organisms. They are often caused by mutations of regulator genes, as well as by environmental variations. See Appendix B on 'Developmental evolution'.) The domestic cat is much smaller than its ancestor, the wildcat, *Felis sylvestris*. Overall size changes are effected epigenetically by alterations in growth hormone production.

Dogs were beneficial in hunting, and would also be a relatively docile source of food in time of need. In fact, dog meat has always been a dietary staple of some East Asians. The survival of humans in the Arctic has been attributed, by the Inuit, as being due to dogs, and here there is a climate connection but no local point of origin. The survivability of dogs is like that of other placental mammals, dependant on the condition of warm-blooded homeostasis that makes them cold-hardy, especially if supplemented by thick hair and subcutaneous fat.

8.2.2 Population expansion and the exhaustion of resources

To continue with the parsimonious, and intuitive, line of reasoning about the origins of domestication, one argument might be that population expansion of hunter–gatherers would exhaust resources and so lead to the development of agriculture and animal husbandry. However, there is no archaeological or anthropological evidence to support this opinion. Hunter–gatherers lived isolated from each others' territories, and performed birth control by delayed weaning of the children. If the worst came to the worst, they practised infanticide. We see exhaustion-of-resource causation in action at the present time when the encroachment of farming societies into the territory of hunter–gatherers obliges the latter to switch to farming. Even worse are cases when deforestation for lumber, agriculture and access to mineral resources wipes out the traditional hunting and gathering grounds. But these did not apply in the early years of hunting and gathering.

Another factor is that, in stable conditions, life is a lot easier for small groups of hunter–gatherers than it is for large groups of farmers, who have to work long hours to achieve the same per capita benefits.[1] This also argues against the idea that the wise people of the clan might have called a moot and discussed the merits of dependable animal husbandry and agriculture and staying in one place, as opposed to exploiting fluctuating wild herds and native edible plant species. How much fluctuation would it take to make such a choice? Probably that due to major climate change, since local climate vicissitudes, such as volcanic eruptions, wildfires, tsunamis and hurricanes would simply have led to temporary migration to adjacent benign habitats, given that they were sparsely populated.

Without any doubt, some hunting and gathering groups had the sense to take on a sedentary existence and erect permanent dwellings if they enjoyed particularly good conditions, such as a combination of frequent large mammal migrations, a local supply of small game, a reliable source of water and perhaps an adjacent lake or sea with shellfish, fish and other aquatic organisms. Nevertheless, long before such conditions were available in the late Pleistocene, the Neanderthals were already sedentary cave dwellers, obliged by a harsh climate to put adequate shelter first, even if it meant roaming afar as hunters.

8.2.3 Cereals

Another parsimonious argument is that hunter–gatherers who had been marginalized by stronger and wealthier clans might have increased their use of wild cereals. In the most benign climatic period of the late Pleistocene, prior to the low temperatures and low precipitation of the Younger Dryas, the uplands of the Fertile Crescent had a variety of cereal species sufficient to sustain the daily energy needs of humans, without great effort. It would not have been a complete protein diet, and would have to have been supplemented by a regular supply of small game. Conceivably, the choice of cereals with strong rachises (or seed stems) that did not shed the ripening grains prematurely, and the likelihood of such plants establishing themselves in the vicinity of temporary settlements due to the enrichment of the soil by human excrement, might have led to the incipient stages of agriculture.

However appealing the above arguments might seem, the parsimony principle is no substitute for a reliable historical record. There is an almost universal consensus among archaeologists and anthropologists that it takes almost catastrophic conditions to make human beings change their ways. The onset of agriculture was a result of drastic changes in global climate patterns. As Childe (1951) hypothesized in his 'oasis theory', when conditions became colder and drier humans and animals retreated to the best sources of water, where hunting and gathering was still possible. There were increased local population pressures that led to agriculture. Later anthropologists argued that a general increase in population made agriculture necessary. Thus the 'oasis theory' languished until climatic evidence was found that backed up Childe.

The emergent conditions for agriculture included population concentration in refugia where water was available, such as rivers, large lakes and access to underground springs. Some refugees may have already had a rudimentary knowledge of agriculture, learned in the uplands where wild cereals were plentiful; a stock of seeds selected for their nutritious and processing qualities, and a surplus labouring population that could be directed to irrigation work as well as seeding, tending and harvesting, would lead to more sophisticated crop culture. This also presupposes a degree of hierarchical social stratification that would be able to organize the efforts.

8.2.4 Animal domestication

Now, where does animal husbandry come into this picture? Did it arise simultaneously in a coordinated way with plant cultivation? Not very likely. The early peoples who domesticated animals were nomadic pastoralists; they set up temporary settlements and collected whatever nutritious plants, insects, fungi and other scavenged foodstuffs they found along the way. But they too were affected by climate change, whether by droughts that dried up lakes, or by a cooling and drying trend that did the same. Both would have diminished the availability of fodder. Eventually the herders would have been forced towards the refugia occupied by established farmers, and their animals would have been commandeered as part of a more complex system. Farm animals are not only important as a source of milk and first-class protein, their wastes fertilize the fields, making it less necessary for farmers to seek out fertile ground when they have depleted the local area. We must nevertheless recall exceptions in communities that could depend on riverine flooding to restore soil fertility, on the Nile, for example.

The history of animal domestication and its implementation is fascinating in its own right: sheep and goat herding, domestication of pigs and chickens, and animals that could be used for bearing loads, such as oxen, donkeys, the llama and its relatives. The original domestication of the ox from the auroch was one of the most challenging tasks for early pastoralists. But once it had succeeded, oxen could be trained to draw ploughs and wagons, as well as providing milk, meat and hides. Other large Old World mammals such as horses, camels and elephants could be further used as weapons of war. The climate connections of this history are tenuous.

8.2.5 Chronology and geography of plant and animal domestication

Table 8.1 summarizes the chronology and geography of both plant and animal domestication so that we can later make whatever climatic correlations are significant.

Cattle are more mobile than sheep and goats and so were better for nomadic foraging. Fagan (2004) conjectures that the aurochs had to get used to being accompanied by humans since both need copious water, and headed to the same water holes; that humans could in fact walk among them causing less panic than lurking lions and leopards. The auroch was notorious for its aggressive behaviour, but so also was the African elephant. Yet the Carthaginian forces under Hannibal were able to tame the latter to supplement its contingent of the tamer Indian elephants. Humans have the qualities of patience and persistence when it comes to 'difficult' animal species.

One domestic organism is conspicuous by its absence from the above table, namely yeast. Since it is so ubiquitous, and causes the fermentation of carbohydrates spontaneously, it is impossible to date its first use in the making of domestic beer. Nevertheless, beer is probably the first yeast artifact that humans took advantage of, since they were using cereal mashes such as porridge and gruel long before grapes were domesticated. The production of beer, in turn, made yeast available in sufficient quantities for the baking of leavened bread. There is no distinct climate connection here, since yeasts can stand up to cold and warm weather.

8.2.6 Distribution of plants and animals from their original domestication areas

The most important point that Diamond (1999) makes regarding the development of agriculture is that latitudinal spread of the practice was much more important than its longitudinal spread. Since the Fertile Crescent was the most significant site for the domestication of plants and animals, they could disperse from there westward to Europe, and eastward to China. In Africa there was the southward barrier of the Sahara and in the Americas the barrier of impenetrable jungles and mountains. (Hence the question of whether squash was domesticated independently in Meso-America and Peruvian South America.) In this context it is not so much a matter of the ability of humans to penetrate such barriers. They had already done that as migrating hunter–gatherers. But sedentary farmers were reluctant to become

Table 8.1. *A chronology of plant and animal domestication; points of origin of domesticated species (After Diamond, 1999, with some modification from Burroughs, 2005, and Fagan, 2004.)*

Original site of domestication	Plants	Animals	Date of domestication
East Asia		dog	12 000 yrs ago
SW Asia (Fertile Crescent)	emmer and einkorn wheat, barley, oats, legumes (chick peas, lentils), olives	sheep, goats, cats	10 500 yrs ago
South America (Peru)	squash[1], manioc, peanuts, cotton		10 000 yrs ago
SW Asia		pigs[4]	9500 yrs ago
N China	rice, millet	pigs, silkworms	9500 yrs ago
SW Asia		cats	9500 yrs ago
South America	chili peppers, manioc, sweet potatoes, peanuts		9000 yrs ago
Africa, India, Fertile Crescent (independent domestications)		cattle[5]	9000–7000 yrs ago
Sahel (N Africa south of Sahara)	sorghum, African rice, peanuts, yams	guinea fowl, donkey	7000 yrs ago
South East Asia	taro, yams, breadfruit, coconuts, bananas		6000 yrs ago
East Asia (Kazakhstan)		horse[3]	5600 yrs ago
Meso-America	corn[2], beans, (squash ?)	turkey	7000–5500 yrs ago
South America	potato		5500 yrs ago
Tropical West Africa	African yams, oil palms		5000 yrs ago

Notes:

[1] According to Dillehay *et al.* (2007), squash was grown in Peru 10 000 years ago. Yet these authors point out that wild squash was not found in Peru at that time, and must have been imported by humans migrating south through wild squash habitat. The larger seeds might have been carried as a source of nutriment. The question mark against Meso-American squash is to indicate the possibility of later local domestication, rather than the transfer of the agricultural knowledge of the early Peruvians to Meso-America.

[2] Archaeological research by Pope *et al.* (2001) on the Gulf Coast of Tabasco, Mexico, indicates the earliest record of maize cultivation occurred along beach ridges and lagoons in the lowlands of the Grijalva River Delta around 7000 years ago. They suggest that Tabasco maize cultivation occurred at least 1000 years earlier than previous evidence indicates it occurred in highland Tehuacán and Oaxaca.

[3] The timing of horse domestication is difficult to estimate since the anatomy of domestic horses does not differ distinctly from that of wild horses. Were they simply hunted for meat, or bred in captivity? There is a similar problem with genetic distinctions among early

domestic horses. Since any harnesses would have been made of perishable hide and wood, archaeological evidence hinges upon the multiple skeletal remains of the same year class – young and old, but not in between. If they had been hunted, the bones of horses of all ages would have been mingled and the leg bones left behind at the site of the kill.

Ongoing research by Olsen and Capo is based on the presence of corral posts in Kazakhstan, dated at 5600 years ago (see Geological Society of America, 2006). They find that the soil within the corral has phosphate and sodium levels typical of horse manure. Many horse bone fragments, representative of the entire horse skeletons were also uncovered. Since there was no other evidence of farming, the horse was the primary source of protein, supplemented by gathered plant material.

[4] Pigs were domesticated from the Eurasian wild boar independently in North China and the Middle East. Asian pigs were introduced from China to Europe in the eighteenth and nineteenth centuries, and hybridation of the two strains led to the modern breeds of pig (Giuffra *et al.*, 2000; Lief Andersson is the research group leader).

[5] Genetics imply that there were three independent domestications of cattle from a single parental type, the auroch (*Bos primigenius*), in India, the Fertile Crescent and the Sahel (south of the Sahara) in Africa. Subsequently there was genetic mixing. The domestication of the auroch may have happened 9500 years ago, but more likely between 7500 and 8000 years ago (or maybe at different times in different locations). By 7000 years ago, domesticated aurochs (cattle) were common from the Sahara/Sahel to the Middle East.

'Johnny Appleseeds'. Even if they wanted to, they would have been unable to grow their cool-temperate crops in the changing, tropical environments that they had to pass through. Latitudinally, from Ireland to China, growing seasons were the same length, and rainfall similar enough to grow the plants originally exploited in the Fertile Crescent, so that they were universally distributed by three thousand years ago. The weather pattern of wet winters and springs followed by dry summers, suited fast-germinating crops. That they should subsequently dry out was not a problem since they could still be used as food during the winter, and stored as seed for the following season. However, detrimental climatic change could affect the entire latitudinal band of previously favourable conditions.

As Diamond points out, most of the important agricultural plant and animal species were available fairly close together geographically in the Fertile Crescent, if not all in exactly the same place. Individual plants were selected on the basis of pleasing flavour, relatively thin skins, plus in the case of grains, rachises that did not drop the seeds on the ground, and in the case of peas and their relatives, non-popping pods that retained the seeds. Human selection of such plants shifted the genetic makeup of crop populations. Another important factor was the self-pollination of these plants, so that the distinct genotypes and phenotypes would be preserved. However, the big question is, how did the whole agricultural process begin?

Ehrlich (2000) notes that the inhabitants of Abu Hureyra in the Fertile Crescent ate plant seeds while they were still hunter–gatherers between 11 000 and 8000 years ago. Out of 157 species, the most common was einkorn wheat.

He rejects the hypothesis that growing populations forced the hunter–gatherers into agriculture. Hunting and gathering was an environmentally self-regulating activity. His alternative is that climate change caused periods of scarcity with the same effect, combined with a reluctance to emigrate from home territories that had good water supplies and fishing resources. Diamond (1999) has pointed out that a farmed area could support 10 to 100 times the population that hunting and gathering would support. The availability of cereal porridges would help to wean children sooner, reduce lactation periods, and increase fertility, resulting in population expansion. Both Ehrlich and Diamond make the point that proximity to diseased farm animals would have improved human immunity over several generations, while leaving hunter–gatherer groups in other parts of the world totally susceptible to diseases carried by Eurasian farmers. Incipient agricultural practices, leading to a continued sedentary existence, made it possible for the development of agricultural tool assemblages, such as mattocks, shovels, spades, sickles and ploughs, which improved planting and irrigation, but would have been burdens to nomads.

Fagan (2004) proposes that in Abu Hureyra the harvesting of forest nuts and fruits stopped with the onset of the Younger Dryas (12 700 to 11 650 years ago). The local forests died out, to be replaced by grasslands, and the harvesting of grains became important in addition to dependence on small game. The wild cereals soon died out as well, except in well-watered areas, but einkorn wheat, rye and lentils began to be cultivated by 12 000 years ago. Harlan (1967) discovered that in the Karacadag Mountains of modern Turkey, a family group could gather enough wild einkorn wheat in three weeks to last a year. Fagan (2004) asserts that this is where the einkorn wheat phenotype was first modified by human selection, and that *the process could have taken as little as 30 years.*

A new Abu Hureyra arose about 11 500 years ago, completely dependent on cereal cultivation. Preparing grain for cooking was so time-consuming and physically demanding that female skeletons have signs of stress on knees, wrists, lower backs and toes. Males continued to hunt, fish, herd animals and tote cereal crops. Herding became more important as game (gazelle) resources diminished. By the end of the Younger Dryas, most humans had become farmers, and sheep and goat domestication was intensified. By 10 000 years ago, there is evidence from skeletons of slaughtered animals that old females and young males were culled. Animals were probably selected for milk yield and wool quality.

Fagan (2004) argues persuasively that agriculture gave rise to cultural continuity, tribal structure, a sense of belonging to a particular tract of land and spiritualism. He uses the example of the village that grew up around Jericho Springs, which produced water during the Younger Dryas. The inhabitants buried their dead and made figurines for shrines.

Perhaps, then, the greatest legacy of the great drought and the warming that followed was not food production but a completely new way of living, closely bound to the soil. People were thus exposed as never before to the harsh realities of short-term climatic changes – the floods and drought cycles that are part and parcel of the hazards of a subsistence farmer's life (Fagan, 2004, p. 100).

Thus population expansion was not universal while resources were rich. Locally, in times of need, populations expanded because of immigration into the small refugia where resources were merely adequate. Whether or not the founding populations of those areas combined some agriculture with hunting and gathering, their original possession of the territory would have allowed them to co-opt agrarian techniques, as well as the domestic animals of immigrants, and furthermore to create a class structure within which the poorest newcomers became agricultural labourers, if not slaves. There is always a possibility that the original populations might have been conquered, but since they had better hunting and killing skills than immigrating farmers or pastoralists, this was probably unlikely. Intensification of agriculture allowed the upper classes to give up working for their share of the crops, and to support some specialized non-agricultural classes such as priests and toolmakers. Once a military class emerged, conquest of the territories of hunter–gatherers was likely.

The next general question is: were farming techniques of the Fertile Crescent spread by population expansion of the farmers? There is a consensus that it was the *concept* of farming that was spread. Obviously seeds would have had to be traded, as well as some farm animals. Archaeological evidence indicates that at the close of MIS2, Neolithic farmers dispersed out of the Levant and into Europe, although the impact is relatively localized. The Y-chromosome analyses of Semino *et al.* (2004) support the idea of these Neolithic expansions into Europe from the Middle East, as well as out of the southern Balkans and from North Africa into the Iberian Peninsula. Burroughs (2005, p. 206) notes that only 20% of European males carry Y-chromosome markers typical of Levantine males, which suggests that there was no widespread emigration. However, other genetic research that looks at different Y-chromosome markers suggests a larger input, with the exception of Celts and Basques (*ibid.*, citing Chikhi *et al.*, 2002). Geoarcheological research ties the initial appearance of agriculture in Europe between 8400 and 7000 years ago to the flooding of south-west Asian coastal farming areas when the ice dam in Canada's Hudson Bay region collapsed about 8200 years ago (Barber *et al.*, 1999; Turney and Brown, 2007).

Since irrigation knowledge was important to farming communities, its application may have become part of the responsibilities of ruler/priests. The power to guarantee annual crops became part of their mystique. Time and again, although climatic challenges threatened that power and led to social failure, some surviving enclaves became more hierarchical, with improved agricultural techniques.

8.3 Climate forcing mechanisms and key events and their influence on agriculture

8.3.1 Milankovic cycles; albedo; greenhouse gases

In Chapter 5 we detailed the general forcing mechanisms that generate climate change. At this point they are worth recapitulation. At the heart of climate change are variations in the amount of solar radiation coming from the Sun relative to the amount of outgoing energy from the Earth. These variations are profoundly driven by the Milankovic cycles (Milankovic, 1998). They include the Earth's precession, which is caused by wobbling of the planet around its axis. This cycle has a periodicity of 21 700 years and creates times when Earth's northern hemisphere is closest to the Sun during the winter, and other times when it is closest to the Sun during the summer. When the northern hemisphere is closest to the Sun during the summer and furthest from the Sun during the winter seasonality is exacerbated. Alternatively periods of reduced seasonality may cause the onset of glacials as warm northern winters result in greater snowfall and cool summers facilitate the retention of snow and the accumulation of glacial ice. The next Milankovic cycle is due to Earth's eccentricity, resulting from variations in the Earth's orbit round the Sun from more to less elliptical. This cycle has a periodicity of 98 500 years. The greater the eccentricity the cooler are northern summers and the warmer are northern winters. Large eccentricity reduces seasonality, encouraging the expansion of the icefields and glaciers. The third Milankovic cycle is the Earth's obliquity, or the inclination of the planet's axis relative to its plane of orbit as it goes round the Sun. This cycle has a periodicity of 41 000 years. When the tilt of the Earth is reduced seasonality is also reduced, but reduced tilt also amplifies the equator-to-pole temperature gradient. As a result the hydrological cycle is intensified by the warming of tropical oceans; in the Pacific, a higher incidence of El Niño Southern Oscillation events would be observed relative to La Niña events (Kukla and Gavin, 2004).

The onset of ice ages is hastened by an increasing albedo effect: the bigger the snow-covered area, the greater the reflection of sunlight back into space. Changes in the levels of greenhouse gases, methane, carbon dioxide and water vapour, acting in unison, also influence climate, and have been shown to change in conjunction with the onset of ice ages and warm, melting periods. The effect of greenhouse gases in the atmosphere is to raise the overall temperature of the Earth; the greater the concentration of greenhouse gases, the more heat is trapped, and the greater the rise in the Earth's temperature and vice versa. Note that the presence of greenhouse gases is inevitable under natural conditions. They are also essential to keep temperatures adequate to support living organisms.

8.3.2 Sea level and changing marine resource productivity

Seawater levels fluctuate as an effect of changes in the amount of water locked up in glacial ice. Changing sea levels also have a feedback effect on the natural forcing mechanisms. A decrease in the area of oceans due to glaciation causes a slight temperature rise in the air over the oceans. This would, however, be offset by the cold air over the icefields spilling into general atmospheric circulation. The role of the oceans in relation to atmospheric gases is the storage of massive amounts of carbon dioxide. They also act as a temperature buffer that responds much more slowly to change than does the atmosphere.

The oceans are huge heat storage sinks, and through their circulating currents distribute some of their heat to colder waters. If the currents are slowed down, or impeded altogether, there is a significant impact on climate. Ocean currents also circulate important nutrients for marine life. Upwelling along continental shelves enhances the productivity of coastal marine regions. Thus when upwelling is reduced, for example during ENSO events along Peru's Pacific coast, marine productivity falls.

As discussed in Chapter 7, lowered sea levels expose the continental shelf, providing productive habitat particularly during dry, cold intervals. Rising sea levels inundate coastal regions and river deltas; regions where human habitation typically concentrates. Thus, changing sea level and altering the productivity of coastal marine environments can have a significant impact on the availability of subsistence resources for humans. This is relevant when we review the development of agriculture. For example, the earliest record of maize cultivation occurred along beach ridges and lagoons in the lowlands of the Grijalva River Delta on the Gulf Coast of Tabasco, Mexico around 7000 years ago, a time of prolonged and heightened ENSO activity (Pope *et al.*, 2001). This was at least 1000 years earlier than previous evidence indicates maize cultivation occurred in highland Tehuacán and Oaxaca.

8.3.3 Volcanism

Volcanic activity can have an immediate effect on climate due to the sudden emission of large quantities of sulphur-rich gas and ash into the atmosphere. This can cause global climate to cool by about 0.2 to 0.3 °C (Zielinski, 2000). It may take several years for this effect to be sufficiently reduced to return global temperatures to the previous 'normal'. Such events are particularly strong in tropical volcanic regions.

There is little doubt that in Earth's history the climatic affects of volcanic emissions, sometimes coupled with the impacts of bolides, have resulted in agricultural disasters. We reiterate that a poor growing season makes inroads into the storage of grain held both for emergencies and for future sowing. A second poor, or failed, growing season brings a culture to its knees. In the larger historical context a climatic change might only last a few years or a few centuries. Either way, farmers and their societies are devastated.

About 3100 years ago there was a major eruption of Hekla in Iceland. Fagan (2004) quotes Casper Peucer, a seventeenth-century physician, who described Hekla as the gate of hell: 'for people know from long experience that whenever great battles are fought or there is bloody carnage somewhere on the globe then there can be heard in the mountain fearful howlings, weeping and gnashing of teeth' (pp. 173–4). Volcanoes have erupted in Iceland four or five times in the last thousand years. Hekla's effect on the climate probably wiped out the majority of subsistence farmers. Cattle husbandry became more intense and deforestation and division of land into fields became extensive.

In AD 535 there was a supervolcanic eruption (or a comet impact) that spread ash into the air and had a devastating effect on farming in the northern hemisphere until AD 542. Seven years is a long time to go hungry! The site of eruption or impact is not known, but the dust and sulphuric acid evidence is clear. The Avar Mongols were on the move, this time extending further than Hungary to form their own empire.

In an essay on a late Holecene eruption of Popocatepetl in central Mexico, Plunket and Uruneula (2006) write:

Volcanic disasters often have been invoked as prime movers in the culture history of ancient civilizations. They have been used to explain large-scale migrations, the destruction of cities, famine, and demographic collapse. In this paper we explore the geological, archeological, and sociological records in order to provide insights into the complex nature of human responses to a major volcanic event of Popocatepetl in central Mexico that took place 2000 years ago. We suggest that the population implosion experienced by two emerging highland cities in the first century AD – Teotihuacan in the basin of Mexico and Cholula in the Puebla Valley – was due to both the immediate consequences of the volcanic event and the disaster-driven acceleration of social processes already underway when the catastrophe struck. We conclude that a better understanding of the relationship between human populations and volcanic hazards and disasters permits a more realistic assessment of the social and cultural significance of eruptive phenomena in the prehistoric period.

(p. 19)

They also argue that the arrival of refugees provided the labour for architectural changes related to the desire to propitiate the gods.

8.3.4 The Lake Agassiz meltwater effect and the Younger Dryas

After the last glacial maximum 21 400 years ago, there was global warming. The Fertile Crescent, which takes in Lebanon, parts of modern Turkey, Syria, Iraq and

Iran, provided refuge for hunter–gatherers from a vast interior desert that spanned much of central Asia (see Chapters 6 and 7). Another productive refugium likely existed in the Indus basin. Between 12 700 and 11 650 years ago, the Younger Dryas cooling and drying period, probably triggered by the flooding of glacial Lake Agassiz into the North Atlantic, abruptly slowed the northerly flow of warm sea water. Lowered temperatures and reduced precipitation vastly diminished the resources for most hunter–gatherers in Europe and northern Eurasia, except where there was abundant fresh water available from lakes, rivers and springs. The barren ground in central Asia expanded westward to the Mediterranean coast and eastward into eastern China. Refugia remained in the Fertile Crescent and the Indus basin. During this period, some tribes in the Fertile Crescent became primarily agrarian. And some may have been on the way to nomadic pastoralism, especially when they acquired domestic dogs, which probably happened in the wake of the Younger Dryas. By 10 000 years ago there is archaeological evidence of domestication of herd animals: only young and old animals were slaughtered. In the Indus basin, wheat, barley, jujube,[2] sheep and goats were domesticated around 11 000 years ago (Allchin and Allchin, 1997; Gupta, 2004).

8.3.5 The Laurentide meltwater effect

About 8200 years ago, Laurentide meltwater sent outflows into the North Atlantic and the Gulf of Mexico. Again the northern circulation of warm water was cut off. Farming was limited to the area of the Euxine Lake (to become the Black Sea), which was below the level of the Mediterranean, but protected from the incursion of salt water by the Bosporus Sill. The lake was much smaller than the Black Sea, its littoral shelf intersected by river deltas; altogether a suitable site for farming. According to Fagan (2004) early settlers around the Euxine Lake had combined farming with hunting and gathering, giving them a technical adaptability that allowed them to spread north and east. They did not have any choice once the Laurentide event raised the level of the Mediterranean by 1.4 metres. The Bosporus Sill was breached and the Black Sea was created by a massive influx of salt water that not only changed the fish resource, but also flooded established farmland. This story was first revealed to the general public by Ryan and Pitman (2000). Despite the controversy that the book raised, we find the recent scientific account by Turney and Brown (2007) to be very persuasive. They come down on the side of the Ryan and Pitman argument, arguing that the Laurentide event was the fundamental cause of the creation of the Black Sea.

The rise of agriculture and domestication began around 9500 years ago in China. Agricultural settlement occurred in the Huang He River valley in north-eastern

China and the Yangtze River valley further to the south between 8000 and 7500 years ago. Rice was cultivated in the latter region, and millet was domesticated in both (Higham, 1989, 1995). Although we have shown that overall climate was more equable during the Holocene than the earlier MIS2 interval, one of the most significant changes in China associated with the Holocene was the changing extent of an enormous barren polar desert. During the last ice age, expansion of this barren ground likely encouraged previously independent populations across China to move into relatively hospitable territories located south of the still extensive ice sheets and south and east of the barren ground. Two areas where people likely concentrated were the Huang He and Yangtze River valleys where relatively high precipitation levels kept the expansion of barren ground coverage at bay (Figure 8.1). A concentration of disparate populations in these valleys probably facilitated the development of new ideas, including agriculture. The expansion of agriculture likely followed during the Holocene climate that brought

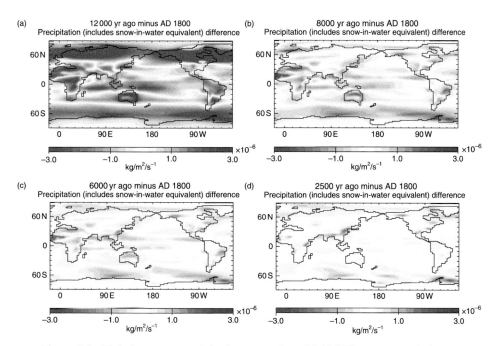

Figure 8.1 Global average precipitation anomaly at (a) 12 000 years ago relative to AD 1800, (b) 8000 years ago relative to AD 1800, (c) 6000 years ago relative to AD 1800, and (d) 2500 years ago relative to AD 1800. Average precipitation anomalies derived from 122 000-year time-series climate simulation performed on the UVic Earth system climate model (UVic ESCM). (See colour version in colour plate section. Red indicates lower precipitation levels relative to AD 1800 and blue represents higher precipitation levels relative to AD 1800.)

with it more stable conditions that allowed people to better predict seasonality and temperatures.

As the adoption of agriculture expanded, a further feature of the evolution of agriculture is noticed by Dow *et al.* (2005):

'Even if foragers have no problem conserving wild resources, agricultural societies might tend to destroy nearby habitat as a by-product of farming activities. This could lock in farming because over time the potential returns from foraging would drop. It could also give neighboring foragers more reason to adopt agriculture. This factor, coupled with the technical advantages of domesticated crops and the military advantages of large populations (Diamond, 1999), may help explain the diffusion of agriculture out of its pristine centers'

(p. 2837).

8.3.6 El Niño–Southern Oscillation (ENSO) events and droughts

El Niño was first described in 1892. It was once thought to be a local weather pattern (rain and warmth on the coast of Peru as opposed to normal, cool and dry). Now it is regarded as a manifestation of a global phenomenon called the Southern Oscillation. El Niño remains the common term, though it is formally known as ENSO (Fagan, 2004, p. 169 ff.). Normally warm water is pushed westward in the Pacific by trade winds, causing a compensatory upwelling in the eastern Pacific that increases oceanic productivity, resulting in large populations of anchovies and squid. An ENSO event is produced when the trades fail. Kelvin waves push warm water eastward.[3] The western Pacific cools, the monsoons fail and drought results in Australasia. The jet stream flows northward, bringing rainstorms to the North American west coast. The phenomenon has global repercussions. Timmermann and An (2005) have shown that an influx of freshwater from melting ice sheets into the North Atlantic will trigger a collapse of the Atlantic thermohaline circulation, which impacts meridional overturning in the Pacific. A reduced meridional overturning is linked to a suppression of ENSO activity and drier, colder conditions in coastal South America. Alternatively, during warm intervals (e.g., 8000 years ago) ENSO activity increases. The consensus is that ENSO events have had major impacts on humans for at least 5000 years.

The title of Brooks's 2006 essay, 'Cultural responses to aridity in the middle Holocene, and increased social complexity', speaks for itself. Global regions that were once well watered and vegetated became arid in the middle Holocene and are deserts today. But at the same time social complexity increased. The Holocene was punctuated by cold spells, especially the one caused by Laurentide spillwater flooding into the North Atlantic about 8200 years ago. Thereafter climatic change had more to do with 'transient climatic perturbations [i.e., ENSO events] and terrestrial feedbacks involving vegetation' (p. 32), such as the rapid desiccation of the Sahara in the wake of vegetation die-off. He also postulates that these later changes were not global, but heterogeneous. He cites Steig's (1999) conclusion that about 6000 years ago, social complexification into urban societies and states emerged where there was a shift to aridity. Lake levels in northern Africa began to drop, reaching a minimum within 2000 years. There was a similar effect in North America, where prairie vegetation replaced forests. He argues that in North Africa the drying trend reduced hunting and gathering and encouraged pastoralism.

Now, the question arises: which of these climate-forcing mechanisms can be considered immediately relevant to agriculture? The large-scale, long-term climatic factors, such as the Milankovic cycles, have durations of tens to hundreds of thousands of years. Thus, we must see many of these effects as the very gradually changing backdrop upon which changes of agricultural significance can be painted, although a caveat to the previous statement is that precessional wobbling of the Earth's axis may be more directly influential than most of the other long-term cycles. Whereas, relatively sudden and sometimes catastrophic changes have been strongly correlated with the origin of agriculture and the destruction of current farming practices as well as the emergence of new techniques.

It is the impacts the climate-forcing mechanisms have on human habitat that must be considered immediately relevant to agriculture. For example, the feedbacks associated with changing Milankovic cycles and solar insolation altered the patterns of oceanic circulation, the frequency and duration of ENSO events, and influenced temperature, precipitation and the extent and productivity of vegetation. Strong ENSOs deplete marine resources along the western coast of South America likely making coastal habitation more difficult. In Eurasia, our time-series climate simulation indicates that the rapidly warming Holocene brought with it changes to a massive polar desert. As described in Chapter 6, human populations that had concentrated in refugia in the Fertile Crescent and the Indus basin to avoid desertification elsewhere during MIS2 faced locally expanding barren ground with the onset of the Holocene. Thus, of immediate relevance is the influence that climate change had on the capacity of the land to produce food sustainably.

We have also just raised the influence of volcanic activity. It fits the latter bill perfectly. And remember that in Chapter 5 we pointed out that the effect of volcanic eruptions seems to have been greatest between 13 000 and 7000 years ago, although subsequent intermittent supervolcanoes also had important catastrophic impacts on climate and agriculture.

Here we present case histories, examining each relevant factor in chronological sequence. These are summarized in the table at the end of this chapter.

8.4 Case histories

8.4.1 Mesopotamia, Africa and Egypt

During the Laurentide meltwater cooling event 8200 years ago, northern hemisphere summertime solar radiation was at its highest level in nearly 100 000 years (refer to Figure 5.2b). Mesopotamia, or the 'Fertile Crescent', which included much of modern-day Iraq as well as parts of Syria, Turkey, Iran and Khuzestan, experienced a monsoon season. This region is also known as 'the cradle of civilization' because it was the home of a series of empires, including the Sumer, Akkadian, Babylonia and Assyrian. In Mesopotamia, around 7800 years ago, irrigation was developed to deliver water to crops via canals, with drainage ditches and levees to control flooding. The hierarchical social structure became more complex, and at this time religions were institutionalized.

Brooks (2006) attributes to Steig (1999) the conclusion that the steepest decline in northern hemisphere solar insolation since the early Holocene occurred about 6000 years ago (refer to Figure 5.2b). Rainfall in much of Mesopotamia was diminished (see Figure 8.1). It was during the following period that 'the first, complex, urban, state-level societies emerged, in regions experiencing a shift towards aridity' (Brooks, 2006, p. 32). Lake levels were relatively high in northern Africa about 6000 years ago and then began to drop to a minimum by 2000 years. Where irrigation was possible, settlements survived and local populations increased, resulting in greater bureaucratic control of the grain supply. Elsewhere, former farmers reverted to hunting, gathering and nomadic herding. At that time, a cattle-herding culture, the Badarians, had inhabited the banks of the Nile between modern Cairo and Luxor. Although part-time farmers, they were semi-nomadic, and intermingled with the herders of the eastern Sahara. Wilkinson (1999) argues that this gave rise to the pharaonic religion. Early kings and later pharaohs used bulls as symbols of power. Agriculture intensified during a drying period that began about 5800 years ago. It involved irrigation and clearing of natural vegetation. Desert nomads continued to migrate to the Nile during droughts.

By 5000 years ago, the Sumerian and Akkadian civilizations had begun to rise. 'Sumerian civilization was a mosaic of intensely competitive city states, each presiding over a highly organized hinterland, ruling over territories that butted up against those of their equally competitive neighbors' (Fagan, 2004, p. 137).

The effect of periodic droughts, apart from simply wiping out vulnerable settlements, was the growth of larger cities in the most benign environments, with more bureaucratic organization of irrigation, planting and cropping, and more religious activity to control the elements. The leaders of some civilizations took most of the cereal crops and reissued grain and oil to their citizens. Standing armies were already in place before the Sumerian and Akkadian civilizations arose. All because of the climate!

According to Fagan (2004), Tell Leilan on the Habur Plain in the south of the Sumerian sphere of influence was one of Sumer's largest cities. It was overrun by Akkadian forces about 4800 years ago, and expanded and fortified. Akkadian rule lasted only a century, at a time of relatively benign climate. Four thousand two hundred years ago a major eruption to the north of Tell Leilan imposed a volcanic winter that may have lasted several years. Then began a 278-year drought, potentially caused by the deceleration of the Atlantic northerly circulation. The Habur plain was deserted for three centuries. The same events reduced the flooding of the Nile, and the centralized Egyptian government broke down into small settlements. 'Mentuhotep I reunified Egypt in 2046 B.C. He and his successors, having learned their lesson, invested heavily in agriculture and centralized storage, and redefined themselves less as gods than as shepherds of the people. They had learned that doctrines of royal infallibility could be a political liability and a literal death sentence' (p. 144). After 3900 years ago Tell Leilan was reinhabited, becoming the centre of an Amorite state.

The social implications of drought are further illustrated by changes in the behaviour of pastoral groups in the Sahel and on the Nile. As Wilkinson (2003) sagely remarks, 'Egypt may be the gift of the Nile; but ancient Egyptian civilization was the gift of the deserts' (cited in Fagan, 2004, p. 147). The Sahara once received much more precipitation than today (see Figure 8.1) and was once (12 900 to 7800 years ago) more vegetated, with temporary pools of standing water and permanent lakes, and megafauna including crocodiles and hippopotami. This was due to the northerly disposition of the intertropical convergence zone (ITCZ), which provided monsoon rains. But for most of the subsequent time, when cattle were herded, the northern Sahara was largely desert. In good years the herds were driven north to the temporary forage. Some small herding settlements persisted around oases supplied by aquifers. In times of severe drought, the herds headed towards the Nile. (Hence Wilkinson's aphorism.)

As Fagan (2004) says, 'The Nile was a huge oasis in 4,500 B. C., with plenty of fertile soil, ample pasturage, and many hectares of ponds, swamps, and marshes where fish teemed and edible food abounded. Compared with their contemporaries in southern Mesopotamia, who spent months every year laboring on the simple irrigation canals on which their survival depended, the Egyptians had it easy' (p. 160).

The 300-year drought in Mesopotamia, beginning 4200 years ago, had devastating effects. The pharaonic state of Egypt foundered, in part because of the senility and death of the Pharaoh Pepi. After a century of chaos, King Mentuhotep I of Thebes reunited the state and the Middle Kingdom began, despite continued diminution of the Nile floods. The pharaohs gave up on divine infallibility and concentrated on improved irrigation. However, Fagan notes that the drought did not invoke new agricultural techniques. Propitiating the gods remained the dominant response. Short generational memory and low life expectancy meant that past droughts were forgotten and no contingency plans were developed (*ibid.*, p. 177).

In his 2006 essay on cattle cults in relation to climatic change, Di Lernia notes:

Abrupt climatic changes in marginal areas, such as the central Sahara in the early and middle Holocene were among the major environmental constraints on prehistoric human groups. Social responses to these events were different, with different paths and outcomes. The spread of a 'cattle cult' – animals buried in 'megalithic' stone structures – in the Sahara at the end of the 7th millenium BP (ca. 6400–6000 yr BP) [7300 to 6800 years ago] is seen here as a collective ritual that emerged, within Saharan pastoral societies, to face uncertain climate and socially relate to 'superhuman' entities. The type of rite – slaughtering of precious domestic livestock – reveals a shared identity in coping with catastrophic episodes – i.e., abrupt droughts. The spread of this 'cult' over large parts of present day Sahara is interpreted as the result of rapid movements of nomadic groups in search of pasture and water. Dramatic climatic deterioration at 5000 yr BP [5700 years ago] is one of the causes of a further major social shift in the rituals archaeologically detected by stone structures; these monuments become human burials, underlining a shift from social to individual identity, as mirrored in the funerary traditions of later pastoral groups.

(p. 50)

In her essay on Holocene climatic changes and human societies, Jousse (2006) focuses on northern Africa, asking, for example, how humans reacted to the southern migration of the Sahel and desiccation of lakes? One fishing community in the Hassi el Abiod region stayed where it was and switched to hunting and herding cattle.

In north-east Africa, the human responses to environmental change are described in terms of four phases of occupation of the Western Desert of Egypt and Libya:

… a first period of *recolonization* of the Sahara with the first cattle herders; a period of *development* with the introduction of sheep and goats and the development of nomad pastoralism; and then *regionalization* and *marginalization* when people start to leave the desert area as it becomes too arid, diversifying or specializing their subsistence activities. The dynamics of western Neolithic populations recall the four stages of Kuper, but the diet patterns do not.

(Kuper 2004 cited by Jousse, 2006, p. 70)

In the early West African Neolithic, populations already exploited less diversified relict fish populations … and hunted the few local ungulates. When human populations migrated southward, they encountered waters and landscapes with more diversified fish and mammalian fauna and opportunistically exploited these resources. This model of opportunistic behaviour can be used to characterise the main West African Neolithic populations (Van Neer 2002). The faunal diversity also explains why no preferential hunting focusing on a single taxon is known in these regions, as is the case in the Levant during the Neolithic with gazelles.

(Jousse, 2006, pp. 70–1)

Jousse concludes:

In West Africa, the first desiccation pulse of the mid-Holocene period forced human populations to migrate and expand into newly explored regions where they found more varied food resources. Climate and environment, which determine the basis of what can be selected and eaten, changed progressively or in punctuated stages until 2000 yr BP [~2000 years ago]. Humans exploited these resources in an opportunistic, flexible manner. No cultural revolution concerning subsistence strategies can be identified during the Neolithic. Only when resources become scarce, as climatic crises strongly disturb ecosystems during the transition to historical times, do people begin to specialise. In other words, it is when a catastrophic event occurs, where catastrophic denotes rapid and extreme changes, that perceptible impacts on human dietary customs become apparent. Aridity therefore appears as a forcing factor leading to cultural changes and social development.

(p. 71)

She notes that this is the reverse of ideas that associate cultural *collapse* with either aridity or increased rainfall.

When comparing East and West Africa, the dietary patterns differ significantly even if the climatic evolution is similar, indicating also that the zoological and cultural heritage from more ancient times provides a context within which further cultural evolution and adaptation to climatic stress occur, cautioning us against simple environmental determinism (Jousse, 2006, p. 71).

The next major series of droughts began 3200 years ago, and brought down the Hittite empire which had expanded into the northern Levant and the Mediterranean Mycaenian kingdom. Only the Egyptian kingdom remained completely intact, despite incursions from the Libyans. Piratical refugees, known as the Sea People, began to ravage coastal towns as far as Sicily and Tunisia. They probably became the Phoenicians. Fagan (2004) attributes the droughts to a shifting intertropical convergence zone (ITCZ) that made Greece, Turkey and the northern Levant arid for several years with devastating effects on civilization (see also Brooks, 2006, p. 32).

The Garamantes were a desert 'civilization' that began around the Wadi al-Hayat, Libya, about 3000 years ago. Groundwater was more important to keep the land watered than rainfall runoff. There is archaeological evidence of numerous shallow wells. The Garamantians probably started with pastoralism and then developed agriculture, using wells to tap groundwater that could be used for irrigation. 'Numerous settlements and fortifications, stone architecture, quarrying activity, evidence of agriculture, abundant communal cemeteries, and of course the dense foggara networks [irrigation channels fed by horizontal wells into escarpments] all point to a complex, organised society' (Brooks, 2006, p. 35). The system was already fragmenting when the first Arabs arrived, probably due to a fall in ground-water levels. This was coincidental with the decline of the Roman Empire and the collapse of trading, perhaps two millennia ago. Brooks concludes this is an example of population agglomeration in areas where water was available, during an overall period of desiccation.

This is compared to the rise of Egyptian civilization on the Nile, with immigration from the Sahara and the Nubian deserts between 5000 and 4800 years ago. As with the Garamantian civilization, 'an emerging elite controls the trade in raw materials and a class of skilled workers appears which achieves elevated social status through its association with the "Royal" authority of the early Pharaohs' (Brooks, 2006, p. 36). The increase in population density drives innovation in agricultural technology as modes of food production are adapted to a new socio-ecological environment, and also provides an impetus for the eventual expansion of the Naqada culture of Upper Egypt that leads to the unification of the country around 5200 years BP (~6000 years ago). The Naqada expansion and the consequent formation of the pharaonic state are linked to environmental change (*ibid.*, citing Midant-Reynes, 1992). Agriculture was already being practised in the south of Egypt around oases, and those populations expanded northward. Immigration provided a labour force for irrigation projects, buildings and pyramids, as well as military might.

In Mesopotamia, widely regarded as the first place for complex, urban states, the first evidence for social stratification occurred about 7000 years ago. Brooks (2006) further remarks: 'Algaze (2001) has postulated a combination of changes in river courses and increasing aridity as drivers of increasing social complexity, stimulating social instability, regional competition and conflict, and population agglomeration. These processes appear to have been operating in Mesopotamia since the 8th millenium BP, and to have accelerated towards the end of the 6th millenium BP, encouraging what Hole (1994) refers to as the "urban revolution"' (p. 38).

According to Hole (1994), 'These innovations take place in the context of a demographic crash which began about 5500 BC. Settlement in some regions disappears, and in others declines rapidly. No region shows an increase in sites and no region was newly opened to settlement. The pattern of gradual expansion across the arable landscape of the near east by small villages, begun by 7700 BC, had ended. Now an increasing proportion of people resided in the larger towns and much of the landscape was devoid of settlement' (quoted in Brooks, 2006, p. 37). Brooks adds that there was considerable trade between the Uruk group and other places, including gem stones such as lapis lazuli and carnelian, plus stamp seals that indicate the development of administrative systems to control trade. The first true states appeared within Uruk culture in the 6th millenium BP.

'The Uruk period is associated with the mass production of pottery, the style of which differed significantly from the Ubaid, the further development of accounting systems introduced before the Ubaid, the introduction of writing, and explicit representations of violence and authority (Matthews, 2003)' (Brooks, 2006, p. 37). Also, the drying trend made rich, former swamp land available for agriculture, but there was competition for it. These multiple effects of climate change demonstrate the difficulty of simply relating the evolution of farming to isolated cases.

8.4.2 The Indus-Sarasvati region

The Indus or Harappan civilization flowered in the 5th millennium BP. Climate dehydration began in the 6th millennium BP. Lake Lunkaransar was at its highest between 7000 and 5700 years ago and the lake was dry by 5500 years ago. The inference is that the civilization flowered as a result of the drying trend. Human habitation in small villages and pastoral camps of nomadic herders began in the early Holocene. The 'early Harappan' stage of regional population concentration and emigration began from 5200 to 4600 years ago, marked by innovative technology and changes in burial practices. The regionalization is presumed to have concentrated on the most benign areas (in terms of water supply). This was followed by rapid urbanization over a century to become the 'mature Harappan'. This period is distinguished by public architecture, town planning, bureaucracy, a system of weights and measures, social and political stratification and the disappearance of small, haphazard villages, some of which were replaced by new towns.

According to Possehl (1990), 'the Harappan civilization created a distinctive set of signs and symbols that can easily be differentiated both from what came before it and from the material culture of the contemporary peoples in adjacent regions' (quoted in Brooks, 2006, p. 40). Possehl infers that the changes had an ideological component, and that trading with Mesopotamia may have had significant sociological impact. Brooks (2006) remarks, 'Whether organised trade, formal religious ideology, and social stratification and differentiation are the *driving factors* behind the emergence of complex civilization, *or more likely products of it*, is a matter of some debate … Once the transition to such a society is underway, we can be confident that these factors will interact, and it is the nature of this interaction, mediated by the broader environmental context, that gives a complex society its unique identity' (p. 40) [emphasis added]. Brooks favours increasing aridity as the most significant cause. It forced pastoral groups to settle along the rivers, where populations grew larger, with resulting social change that he compares with developments along the Nile.

8.4.3 *Meso and South America*

Based on cores of Lake Chichancanab on the Yucatan Peninsula, Hodell *et al.* (1995) find a series of three droughts that affected Maya culture. 'The first was from 475 to 250 BC, when Maya civilization was first forming. The next lasted from 125 BC to AD 180, which is contemporary with a widespread abandonment of larger Maya settlements. But the most severe drought of all, from AD 750 to 1025, coincides with the great Maya collapse of the southern lowlands' (cited in Fagan, 2004, p. 236). There is evidence from geological cores that these droughts were typical of ENSO shifts. If one sets the history of Maya civilization against the background of recurring drought, there are some remarkable coincidences. The first of three dry cycles occurred while Maya agriculture was still providing sufficient flexibility to accommodate dryer years. The second cycle descended on the Maya just as the first efflorescence of cities and civilization appeared in the lowlands. Cities such as El Mirador were situated in low-lying areas, where water could be trapped and stored. At first the system worked, but soon the city grew too large, the vulnerability threshold was crossed. El Mirador's lords lost their spiritual credibility in the face of environmental disaster, and the people dispersed – there was still enough space for them to do so. When the drought ended, growth resumed and Maya civilization entered an astoundingly rapid expansion. By the time the greatest drought of all settled over the lowlands, essentially all the arable land was under cultivation and Maya agriculture was very close to the critical threshold where even a slight drop in productivity was critical. The final Maya drought lowered the water table, produced inadequate rains and ravaged an agricultural economy that already had trouble satisfying the accelerating demands of the nobility. The Lake core evidence is backed up by a marine core from the Carioco basin in the southern Caribbean that correlates the Mayan droughts with ENSO events. Between 890 and 910 AD the final cities fell and there is evidence of survival cannibalism (Fagan, 2004, p. 237).

A similar fate befell Tiwanaka on the Altiplano of what is now part of Peru, Chile and Bolivia. It was farmed from *c.*400 BC. By AD 650 a sumptuous city had arisen, with an advanced system of irrigation and water storage. There was also an agricultural complex of raised fields that resisted the infiltration of saline water as well as providing protection from frost for winter vegetables such as potatoes. A system of outlying settlements was responsible for continued maintenance of the land as well as agriculture. There were three cycles of drought between AD 540 and 1450. Over four centuries during the third cycle the raised fields disappeared completely.

In the 3rd millenium before the present in Peru, in the river valleys of the Norte Chico region, there were 20 residential centres with monumental architecture. These towns were preceded by small hunting–gathering centres with some horticulture. Brooks (2006) notes that the timing approximates to the foundation of complex societies in the changes in the Afro–Asiatic desert belt, and was probably due to a prior drying period that reduced rainfall in the Altiplano and falling water levels in Lake Titicaca and its tributaries. Brooks's sources attribute this to changing ENSO patterns that reduced the effect of coastal warm waters, followed by increases in El Niño events. Yet, as we noted in Chapter 7, dry, cold conditions that are associated with reduced ENSO activity may correspond to rising water levels in Lake Titicaca, Peru (Baker *et al.*, 2001), making Lake Titicaca an interior refugium of sorts during cold, dry intervals. If so, during dry, cold intervals, populations living in the interior of Peru may have concentrated in the Lake Titicaca region. Such a population concentration would have stimulated new ideas and technologies. Then with warm intervals when ENSO activity increased and coastal resources were depleted, coastal communities would have become more dependent on terrestrial resources. This likely stimulated the trade between interior areas and coastal habitations. This substantiates Brooks's (2006) suggestion that isolated towns specialized in different resources, but were 'symbiotic'.

His conclusions are worthy of full quotation:

It is now well established that the earliest complex, highly organised, state-level societies emerged at a time of increasing aridity throughout the global monsoon belt. This trend towards desiccation commenced around 8 kyr BP [8000 years ago] as a result of declining solar insolation associated with changes in the Earth's orbital parameters, but accelerated around 6 kyr BP [~6000 years ago] after a widespread centennial-scale arid episode that may have been the result of transient cooling in the North Atlantic. The following millenium was a time of profound cultural change that saw the development of the world's first states in Mesopotamia and Egypt, and laid the foundations for similar developments in South Asia, northern China, and northern South America. The available data suggest further abrupt changes in climate in the late 6th millenium BP, when regional records indicate environmental and cultural discontinuities. The period around 5.2 kyr BP [5200 years ago] seems particularly significant. At this time a unified Egyptian state emerged, and the Uruk culture of Mesopotamia collapsed and gave way to the transitional Jemdet Nasr period, characterised by fragmentation and regionalism. In South Asia the beginnings of the early Harappan phase have been placed at 5.2 kyr BP [5200 years ago] (Possehl, 2002) and the beginnings of urbanisation are evident at the site of Harappa (McIntosh, 2002). In Northern China there is evidence for an abrupt drop in temperature and accelerated aridity coinciding with the Yangshao-Longsan transition in the final centuries of the late 6th millenium BP (Liu, 1996). In South America this period was characterised by profound changes in the ENSO cycle and an increase in coastal upwelling (Reitz and Sandweiss, 2001; Andrus *et al.*, 2004) prior to the emergence of large urban centres exhibiting monumental architecture. Changes in the behaviour of monsoon systems appear to be one aspect of a wider reorganisation of the global climate.

... there is abundant evidence of regional environmental desiccation and, while the trajectories of social change in Mesopotamia and Egypt are very different, in both cases increasing social complexity is associated with increases in local population densities, the congregation of populations along rivers, and evidence of competition or conflict at the end of the 6th millenium BP. In Egypt this occurs within a context of political unification, whereas in Mesopotamia the context is one of fragmentation. In both cases environmental deterioration coincides with trends towards greater urbanisation and the development of ideologies of political power.

(Brooks, 2006, pp. 42–3, reprinted with permission
from Elsevier/INQUA, copyright 2006.)

A recurring theme in this discussion has been population agglomeration in environmental refugia, necessitating the development of new social institutions and relations, and techno-logical and institutional adaptations related to water extraction, food production and dis-tribution. As suggested by Fagan (1999), these developments are not necessarily dependent on the prior emergence of an organizing elite ... It might be argued that the emergence of elite groups is a by-product of, rather than a prerequisite for, such adaptations, and [their emergence] results from the exploitation by certain groups of both geographic advantage and emerging social relations and institutions that provide opportunities for exerting control over other, relatively disadvantaged groups ... providing a pool of 'human capital' available for exploitation through organized labour or military activity by emerging elites. Ideological systems developed to legitimise the power of elites and support the emerging social hierarchy (Yoffee, 2005), while innovations in administration, production and distribution would have to be required in order to maintain the emerging social system. In the model presented here, increases in social complexity leading to the emergence of urbanization and state-level societies are not driven by surpluses in food production. Instead, the capacity of agricultural systems to generate surpluses (from which high-density, specialized, urban societies can be supported) is harnessed out of necessity. While surpluses may have played a role in the first steps towards complexity, for example in the Ubaid and earlier periods in Mesopotamia and in the development of village agriculture in pre-Harappan times in South Asia, the rapid increases in complexity in the late 6th and early 5th millennia BP are interpreted as *precipitated by hardship rather than abundance* [emphasis added]. This interpretation reflects the main conclusion of Marshall and Hildebrand (2002) regarding the domestication of plants and animals in Africa, which they argue was driven by a desire for greater predictability in the food supply rather than a deliberate attempt to increase yields and thus volumes of available foodstuffs. They point out that 'The end results of agriculture – visible today as larger yields, higher carrying capacity, denser stands of crops, larger seeds and seed heads, or greater animal productivity – were not necessarily realized during the earliest phases of the domestication process' (Marshall and Hildebrand, 2002, p 101). As with agriculture, so with urbanization and the development of the institutions of state – the consequences of such developments would not have been anticipated by those populations that unwittingly precipitated them.

... the diversity of the societies discussed above, and the different trajectories leading to them, should caution us against reductionism. While environmental desiccation provided the context for, and was arguably the principal driving force behind, their emergence, a multitude of other factors influenced their development. Each complex society discussed here emerged from a different cultural context. For example, while the Garamantes ultimately emerged from the cattle herding cultures of the Sahara, and mobile cattle-based societies appear to have formed a component of proto-Dynastic Egyptian society; there was no significant tradition of agriculture in the prehistoric Sahara. This may be contrasted with the long-established and geographically widespread tradition of village agriculture in Mesopotamia, and livelihoods based on marine resources on the pre-urban Peruvian coast. The different physical environmental contexts would have provided each precursor society with different constraints and opportunities as they became more complex. Ideological systems would have developed with reference to the physical environment, but would also have been influenced by personalities and by contacts with other cultures. Trade would also have played a role in shaping livelihoods, in the development of economic and social structures, and in shaping world views. Each of the mature complex societies discussed here was distinct from the others, exhibiting its own unique 'personality'; the unified Egyptian state in which the bulk of the population remained rural may be contrasted with the much more urbanised society of Mesopotamia, characterised during much of its existence by competing and cooperating city states. The high levels of social stratification in Egyptian and Mesopotamian societies, dominated by powerful ruling elites and individual leaders, suggest very different social structures to the apparently far less hierarchical Indus culture (McIntosh, 2002). In terms of responses to 'abrupt' climate change, it should be noted that the climatic changes apparently associated with the emergence of complex societies are qualitatively if not quantitatively similar to those blamed for societal collapse.

(Brooks, 2006, pp. 44–5, paraphrasing Fagan, 1999, and citing
Marshall and Hildebrand, 2002, p. 101)[4]

Rather than being the result of deliberative processes, past adaptations have as a rule emerged from upheavals triggered by environmental change. Past adaptation has occurred out of necessity, after damages have already been incurred, and itself is not without cost. We may be justified in viewing civilization as a form of adaptation to climate change, but the negative aspects of the transitions with which its development is associated, for example increased inequality and violent conflict, suggest that it should perhaps be viewed as a 'sub-optimal' adaptation by the standards of present-day aspirations.

(Brooks, 2006, p. 45).

8.4.4 China

In northern China, state-level societies began in the late 5th to the early 4th millenium before the present. The period, which came after a diminution of the monsoons, was a time of 'profound climatic and environmental change … from a warm humid environment to cool, arid conditions across the region with the exact nature and timing of this transition varying with location' (Brooks, 2006, p. 40, citing Wu and Liu, 2004). The concentration of populations along major river courses as a result of the drying trend was responsible for social stratification. The original sources of this research, Wu and Liu (2004), suggest that changes in the intensity of the East Asian monsoon-related rain belts in eastern China resulted in a significantly altered hydro-logical regime that generated drought in the north and flooding in the south of China, which was mainly responsible for the collapse of Neolithic cultures around the Central Plain (p. 153). Zhang *et al.* (2000) identify an extreme dry spell, which they link to the end of fluvial-lacustrine activity along the Hongshui River in northern China around 3000 years ago. Around the same time the climate of the Mongolian Steppe became colder. Mongolians migrated to China and were repulsed. Those who moved to the Hungarian plain settled. Horse culture spread into Europe within a few generations.

8.4.5 Europe

Fagan (2004) suggests the warm climate of Europe, that had been persistent since about 8000 years ago, began to cool about 5500 years ago when village-based agriculture had been established (p. 180).[5] Broad beans became more widely used. (This is the Old World pulse, *Vicia faba*, often confused with the New World lima bean, *Phaseolus lunatus*, which was confined to the Americas and not introduced to Europe until the sixteenth century.) Millet became the major grain for making porridge, bread and beer.

Around 3100 years ago there was a major eruption of Hekla in Iceland; 12 km^3 of volcanic ash covered about 80% of the country. Its effect on the climate probably wiped out the majority of subsistence farmers in Iceland. Cattle husbandry became more intense. Deforestation and division of land into fields was extensive. About 3000 years ago the climate of the Mongolian Steppe became colder. Around 2800 years ago a drop in temperature was coincident with the ecotone between Mediterranean and temperate climates shifting to North Africa. Farmers retreated from high elevations. In the Low Countries there was an increase in the availability of water-borne iron oxide precipitates, in sufficient quantities for making iron tools, including ploughshares as well as swords. This became associated with the increase in the size and organization of agricultural settlements, and a culture of raiding neighbours and fortifying the villages. Fortification required lumber, and deforestation increased (Fagan, 2004, pp. 192–9 ff.).

By 2300 years ago the Mediterranean ecotone had drifted as far north as central France. Existing agriculture in Italy was more suited to the new conditions and the Roman sphere of influence expanded. Celtic societies did not have the central organization needed to resist the Romans, who built roads to facilitate moving and supplying the army, and also had shipping (*ibid.*, pp. 203–5).

The Roman advantage began to diminish as the Mediterranean ecotone drifted back to North Africa, and the breadbasket of Gaul failed. The Germanic tribes had learned to organize themselves sufficiently to invade Italy. When the Roman Empire collapsed, villagers and townspeople who had been heavily taxed were free to invest their own time and money in things that were of importance to them (Levine, 2001, p. 149). Subsistence farming became universal in Europe. Previously migrating groups came together, resulting in urbanization and centralized control of agricultural surpluses.

In Europe, by AD 900 the Mediterranean ecotone had shifted north again, bringing in four centuries of the Mediaeval Warm Period, with weather warmer and more predictable than at present. However, during this interval there were still many famines in Charlemagne's reign (768–814), and more in the early eleventh century. Levine (2001) suggests the recovery from these was due to better farming techniques and better crops, but he does not make a direct climatic link. He does note that after AD 1300 there was a marked decrease in temperatures over a few generations, reducing the area of cultivable land. For example, the farms of the Scottish Lammermuir Hills were probably abandoned because of oat-crop failures.

In Roman times the Gallo-Roman, Germanic and Celtic tribes utilized thin, well-drained soils that could be easily worked. This had persisted into the Carolingean period beginning about 1200 years ago.[6] 'The introduction of the iron-flanged, moldboard plow was the key to the spatial transfer of agricultural production. This was accompanied by the Carolingian shift from the thin, sandy loess soils to the heavy clay lands of the river valleys' (Levine, 2001, p. 153). Experienced farmhands began to migrate outside their past boundaries to use their new agricultural techniques. Levine suggests that by AD 1200, 200 000 Germanic emigrants had moved into the Slav lands, escaping the kind of feudal system that reigned in France and England. They brought with them the new plough and heavy felling axes. One of the rewards was the discovery of new metal deposits. The replacement of oxen with more versatile horses was also important from the turn of the millennium. In turn, the new techniques and sources of metal made the blacksmith culturally important.

The climate connection for these changes is tenuous. Depletion of soil, migration of parts of an expanded agrarian population, especially those with improved tools and the ability to use work horses occurred at a time when the Mediaeval Warm Period had been ushered along in with a northward shift of the Mediterranean ecotone. Perhaps the causal chain began with improvement in the weather, followed by population expansion, resulting in soil depletion, the development of heavy axe blades and plough shares and a quest for lands that were agriculturally less exploited.

Levine (2001) is also of the opinion that when the Mediaeval Warm Period cooled, land that was cultivable shrunk, and a combination of dense population, deforestation for building and tool-making and soil depletion, contributed as much to the fourteenth-century 'deserted villages' as the Black Death (bubonic plague = *Yersinia pestis* infection) (p. 161). He does not establish a climate connection for this, but notes that deforestation can have a negative local effect on rain levels, and a global effect as well if the clearing is on a large scale. He suggests that growing hay, clover and stall feeding during the winter helped to reduce the negative effects. Later, the cultivation of turnips was to help both farmers and their animals through severe winters.

Fagan (2004) argues that the golden age of gothic cathedral construction was stimulated by confidence in God's approval. The Mediaeval Warm Period ended in AD 1315 with continuous rain throughout spring and summer in Europe. By the time the rains subsided in AD 1321, 1.5 million had died of starvation and related diseases (p. 248). This was a minor toll compared to the deaths to come of the Black Death, smallpox, rampant syphilis and fourteenth-century wars. In contrast, new husbandry techniques were developed, such as growing hay, and stall feeding cattle and horses.

Fagan (2004) remarks that the name the 'Little Ice Age' is a misnomer, although there were occasional severe winters. Despite the lowering of temperatures, an agricultural revolution began in continental Europe in the fifteenth century and then spread to England. Aspects of this were land enclosure, the use of clover fodder and turnips as a winter feed for cattle and people. By the beginning of the Industrial Revolution most European countries were self-sufficient in grain. 'Only France remained agriculturally backward in an era when deteriorating climate made bad harvests more frequent. As it has done throughout the Holocene, increasing hunger led to social dissolution and a loss of legitimacy for society's rulers. In this case, civil chaos joined with a philosophical enlightenment to produce the French Revolution, which in turn influenced the American ideal of democracy and the rise of the United States as an economic and industrial power' (pp. 248–9).

The cooling associated with the 'Little Ice Age' continued into the nineteenth century, exacerbated by the supervolcanic eruption of Mt. Tambora in AD 1815, which brought the 'year without a summer' in AD 1816. Warming trends occurred in the 1820s and 1830s, with cold blips in 1829 and 1837–38. After 1850 the warming trend continued until the present time, which has been attributed in part to the Industrial Revolution and the human contribution to greenhouse effect and global warming. As an additional political effect of climate, crop failures in Scotland prior to 1707 contributed to the decision to unify Scotland and England.

To summarize Fagan's (2004) general conclusions, he distinguishes three major effects of climate change on culture.

1. People move to more suitable environments, a process possible when populations are low.
2. Some migration provided cheap labour, improved social organization and central control of irrigation and agriculture.
3. Collapse of civilization occurred because of starvation, rejection of rulers' claims to be able to call on the gods for improvement, and the lack of anywhere else to go, therefore war.

We can see that climatic catastrophe has the negative effects of disrupting societies through starvation and war. In contrast, some areas benefited from improved crop-growing conditions. There are also positive effects of diasporas that spread the technology of the disrupted societies, changed local social organization, and made it possible for the new societies to become more complex and organized.

8.5 Conclusions

As we wrote at the outset of this chapter, the development of agriculture is certainly relevant to climatic change, but is difficult to extricate from other social consequences, such as the development of hierarchical societies that have embraced religion and militarism. Such consequences might be locally and temporarily undone by sudden catastrophes, such as volcanic eruptions, tsunamis and plagues. However, in the long term, it has been climatic change that has had the most influence on civilization based on agriculture.

Our assessment of the evolution of agriculture seems to confirm our 3c's (catastrophe–communication–collaboration) hypothesis. In evolutionary terms it was saltatory, or emergent, predictable only by hindsight. The need for rulers to continue to control their people depended upon using methods that would feed them. In the wake of catastrophe there came population expansions because of immigration into refugia. Such mingling allowed the different agricultural techniques of the immigrants to be applied under the new conditions. A surplus of labour allowed the development of cultural specializations in art, artisanship, religion, militarism, philosophy and education. And these different roles combined symbiotically to achieve new sociopolitical hierarchies.

Not all agricultural advances were saltatory emergences. The need of complex societies to keep a record of their produce for the sake of lawful use of such produce, including trade practices, required a system of numeration and alphabetization that would evolve into mathematics and a written language. Practical experience and the application of logic would bring new approaches to tillage, such as the use of heavy ploughs to exploit rich clay soils in the face of the depletion of thin, easily workable, sandy soils. There were of course minor revolutions such as those brought about by the switch from oxen to horses as the primary draught animals, the notion of crop rotation, and investment in winter feeds for the benefit of enclosed domestic animals; and humans too in the case of potatoes and turnips. Most of these advances were the products of individual farmers who had identified the problems and taken a scientific approach to their solution.

Table 8.2 summarizes a number of climatic events, their global distribution and their effects on agriculture. It is clear that climate had an influence on the

Table 8.2. *Summary of climatic events, their global distribution and effects on agriculture (Based largely on Brooks 2006, Diamond 1999, Fagan 2004, and Levine 2001.)*

Date	Type of climatic event	Effect on agriculture and civilization
12 700– 10 000 yrs ago	Lake Agassiz meltwater spill – Younger Dryas; cooling and drying.	Domestication of dog – herding of sheep and goats in Fertile Crescent. Incipient cereal culture. Populations concentrated in refugia. Crop culture in the Americas and China. [See Table 8.1.]
8200 yrs ago	Laurentide meltwater spill; cooling and drying in northern hemisphere.	Advancement of farming around Euxine Lake (Black Sea). Emigration of farmers. Complexification of Chinese settlements around major rivers.
7800 yrs ago	Expansion of monsoon regions.	Spread of farming with more sophisticated tools.
7000 yrs ago	Consistent higher frequency of ENSO events – drying of N African lakes.	Increased social stratification; better irrigation techniques in Fertile Crescent.
6000 yrs ago	Decrease in solar insolation begins.	Rise of cattle cults. Badarian cattle herders settle on west bank of Nile. Rise of Sumer: specialization of military organization.
5500–4900 yrs ago	Peak in ENSO events.	Increased dependency on irrigation. Occupation of Caral, Peru.[7] Rise of Chinese and Harappan states.
4200 yrs ago	300-year drought in Mesopotamia.	Changes in major crops in Europe. [See Table 8.1]
3100 yrs ago	Eruption of Hekla; cooling and drying.	Extermination of Icelandic farming; intensification of deforestation for farming and timber constructions.
3000 yrs ago	Extreme drying in China.	Migration of Mongols to Hungarian Plain – horse culture expands in Europe.
2300 yrs ago	Migration of Mediterranean ecotone to central Gaul.	Extension of benign climate; increase in Roman influence in Europe.
~2485–2260 yrs ago	Droughts related to ENSO shifts.	Early Maya settlements – local growth of agriculture.
2000 yrs ago	Eruption of Popocatépetl.	Crash of Meso-American cities. [Possibly linked with top-heavy social changes in pre-eruptive period, and ENSO events.]
~2135–1830 yrs ago	Droughts related to ENSO shifts.	Survival of cities with irrigation.
~1260–985 yrs ago	Final drought that lowered the Mayan water table.	Collapse of Maya cities. Arable areas desertified.

Table 8.2. (*cont.*)

Date	Type of climatic event	Effect on agriculture and civilization
~1500 yrs ago	Supervolcano eruption (or comet impact.); 7-year cold period.	South-eastward migration of Mongols.
~1100–700 yrs ago	ENSO events; 'Mediaeval Warm Period'; southern migration of Mediterranean ecotone.	Decrease in soil quality; use of heavier agricultural tools; eastward migration of European farmers. Increased deforestation.
700 yrs ago	Cooling associated with the 'Little Ice Age' began.	Crop yields diminished by cooling and drying; widespread starvation.
195 yrs ago	Eruption of Mount Tambora – 'Year without a summer'.	Widespread crop failures.
160 yrs ago	Gradual warming period until present time. (Global warming from industrial revolution?)	Agricultural revolutions – more selective breeding of plants and animals; improved survival of over-wintering cattle

development of agriculture, and yet a great many unknowns and inconsistencies remain. Researchers possess contradictory evidence and theories as to the extent of climate change and its impact on human development. For example, it is believed that hunter–gatherers lived in a rich forest environment in the Fertile Crescent after the last glacial maximum and prior to the Younger Dryas cooling event. However, our climate simulations do not show this although they do indicate far less barren ground in the Fertile Crescent than today. Another example is evident when Fagan speaks about the warm climate of Europe beginning to cool about 5500 years ago when village-based agriculture had been established. Our climate model simulations show little atmospheric surface-air temperature change in Europe between 9000 and 7000 years ago, with warming of between 0.5 and 1.0 °C beginning between 5000 and 6000 years ago. However, interestingly, simulations do indicate coastal region land-surface temperatures remained warmer than 1800 AD and interior land-surface temperatures cooler than 1800 AD until about 6000 years ago when all of Europe became about the same as 1800 AD.

Thus, there is much we have learned about the complex connections between climate and agriculture, but there is much we do not know. It is clear however, that agriculture allowed humans to directly manipulate the unpredictability of nature.

Humans expanded their populations and developed centralized states with full-time religious and craft specialists. These societal changes combined with a human desire or need to control their environment eventually led to the Industrial Revolution which began just over 200 years ago. The use of energy became critical and we developed fossil fuels like coal, oil and gas. This allowed us to overcome hostile climates and marginal environments. Today the planet's more than 6.75 billion people have become very adept at utilizing and controlling Earth's resources. Yet, this manipulation has its own consequences: although we are beginning to understand the relationship between humans and climate in the past, what of the present, and what of the future?

Notes

1. For a good discussion on the work effort and caloric requirements of subsistence foraging for a living see Lee (1993), particularly pp. 56–60.
2. A tree whose fruits provide both nutritious and medicinal benefits.
3. Kelvin waves can be either coastal or equatorial and are strongly influenced by a change in winds, for example a shift in trade winds associated with the onset of an El Niño event.
4. For a thorough discussion of societal collapse read Diamond (2005).
5. Our climate model simulations show little atmospheric surface-air temperature change in Europe between 9000 and 7000 years ago, with warming of between 0.5 and 1.0 °C beginning between 5000 and 6000 years ago.
6. The Carolingian period, frequently referred to as the Carolingian Renaissance (~AD 780–900) began during the reign of King Charlemagne and is associated with a resurgence of art, literature and education and their reform in greater Europe.
7. See Chapter 7 for a discussion on the occupation of Caral, Peru.

9

Climate and our future

9.1 What then of the effects of climate change?

Climate change is not new. The Earth has experienced significant climate variability over the last 650 000 years and more particularly within the time of modern human existence. However, relatively speaking, global climate over the last 10 000 years has remained relatively stable. Modern humans took advantage of this stability. We developed agriculture and domesticated animals. We further enhanced the stable environment by storing food in good times for use during lean times. Irrigation enabled modern humans to grow crops in regions otherwise prohibitive. As a result, growing seasons could be lengthened and water stored for use in times of drought. We migrated to more distant and less hospitable regions, building shelters and subsequently moved into more extreme environments by further stabilizing and controlling our environment. The modern human population expanded as did our dominance over the land and other species. Yet this dominance required an increasing dependency on energy and technology. Today we have become so heavily dependent on energy and technology that we expect them to protect us from all manner of natural and climatic vagaries.

Our growing obsession with, and economic dependency on, fossil fuels, combined with our penchant for consumerism, has resulted in humans becoming a climate-change mechanism. Our behaviour is now influencing global climate, both today and in the future. As a result, the relatively stable climate that persisted over much of the last 10 000 years and facilitated the proliferation and dominance of the modern human species appears to be over. Atmospheric greenhouse gas levels are rising and so, coincidently most scientists agree, are global average surface-air temperatures. The world is experiencing an increase in the number and intensity of extreme weather events, sea levels are rising, glaciers and sea-ice are melting and biodiversity is reduced. These changes are in addition to depletion of the ozone layer, acid rain, rising toxic pollutant levels, soil erosion, deforestation, increased desertification, reduction in potable fresh water and an increased likelihood of global epidemics. Concern is growing over what this means for the welfare of people around the world. For small island states, climate change has become an issue of national survival.

History shows us that catastrophic environmental change can result in the extinction of previously dominant species, and on numerous occasions has done so. The space vacated by dominant species is subsequently filled by species that previously lived on the periphery – species nimble enough to adjust to a changed circumstance. The evolution of vertebrates is a case in point. Today, *Homo sapiens* is the dominant species, and scientists are forecasting rapid climate change within decades with catastrophic consequences.

In this chapter we look at the climate of today and tomorrow. What is the extent of climate change to which we must adjust? Based on our understanding of the human relationship with climate throughout our history, we seek to understand whether all or some portion of our species is capable of adjusting. Will humans go extinct or can we survive?

9.1.1 The climate of today

Ice core records from Greenland and Antarctica indicate that greenhouse gas levels in the atmosphere increase and decrease naturally as the Earth progresses through ice ages and warm interglacial periods. Over the last 650 000 years atmospheric CO_2 levels have varied between 180 parts per million (ppm) during ice ages and 290 ppm during warm interglacial periods. Today, atmospheric CO_2 levels exceed 390 ppm, far above those naturally evident in the more than 650 000 years of ice-core records. This recent anomalous rise in atmospheric CO_2 has been directly attributed to human activities – primarily our use of fossil fuels as well as agriculture, deforestation and other activities that change the makeup of the land. The atmospheric levels of other greenhouse gases, particularly methane, are also escalating rapidly.

Although firewood remains a major source of energy in developing countries, over the past half a century it has increasingly been displaced by coal, oil, natural gas, gasoline, diesel, nuclear and hydro energy – the energy sources of developed countries (Kimmins, 2008). The production and use of some of these non-renewable energy resources particularly coal, oil and gasoline has resulted in a significant increase in emissions of CO_2 into the atmosphere. As global CO_2 emissions rise and the coincident atmospheric CO_2 levels increase, the average global surface-air temperature also rises. In 2005, the average surface-air temperature on Earth was about 1.0 °C higher than it was in 1850, during the early years of the Industrial Revolution. However, in some regions of Asia, Africa, southern Europe and North America, the average temperature increased by 1.0 °C to 2.0 °C between 1970 and 2004.

Evidence of the impacts of these changes in climate is being seen today.

- Glaciers are melting around the world.
- Permafrost is becoming increasingly unstable.
- Global average sea levels rose by an average of 3.3 millimetres per year between 1993 and 2006.[1]
- There have been heavy rainfall events – for example, in July 2005 there was a rainfall of 94.4 centimetres in some parts of Mumbai, India's financial centre, stranding 150 000 people in train stations, closing airports and breaking communications linkages. One million people lost their homes. Before that, the heaviest single rainfall was 83.82 centimetres on 12 July 1910, at Cherrapunji, India, one of the rainiest places on Earth.
- In the northern hemisphere, the amount of snow cover that persists through March and April has been reduced.
- Many people in large cities are dying during increasingly frequent and intense heat waves – for example, 26 000 died during the European heatwave of 2003, which caused US$ 13.5 billion in direct costs.
- Droughts and forest and wild fires have intensified – during the 2006–7 summer season Australia experienced the worst drought in 1000 years and the summer of 2008–9 brought horrendous wildfires.
- Tropical cyclones have increased in intensity over the past three decades. The 2008 cyclone Nargis in Myanmar resulted in an estimated 100 000 deaths.
- River and coastal flooding in Bangladesh has resulted in a million lives lost and the loss of worldly possessions for even more.
- The geographical distribution of species has changed as plants and animals move pole-ward and/or to higher elevations in search of a cooler climate.
- Some species have become extinct, including the golden toad of Costa Rica and the Baiji dolphin of China, the 'goddess of the Yangtze', a species that had existed on Earth for 20 million years.

Having harvested what we believed were endless resources from our oceans and forests, we are now experiencing the consequences. When healthy, the Earth's vast natural resources act as carbon 'sinks' – they sequester carbon from the atmosphere. However, numerous expanding 'dead' zones now exist in our oceans that are virtually devoid of living things. Scientists have noted that 'dead zones' are increasing in size and number.[2] Located predominantly along the continental shelves of populated regions, these zones have low levels of oxygen. During the last 50 years there has been an exponential increase in the number of these oxygen-starved zones. And net global deforestation now accounts for more than 18% of the worldwide greenhouse gas emissions (Betts, 2008).

Changing annual weather patterns are causing agricultural and fishing difficulties. The biggest causal effects result from El Niño Southern Oscillations (ENSOs). Increases in ocean temperatures through global warming are likely to cause more

ENSO events. Furthermore, if global warming reduces glacial volume in the north-ern hemisphere, as it is predicted to do, a necessary source of meltwater for agriculture will be diminished and will potentially result in flooding at the wrong time in the wrong places. This is particularly true in the Tibetan plateau, where if temperatures reach 5 °C over pre-industrial levels, glaciers will melt and all major south-eastern Asian rivers previously fed by the glaciers will deplete, leaving little to no water available for agriculture, the products from which currently feed 2 billion people.

Our current dependence on monoculture crops that grow in ever-expanding fields has opened the possibility of diseases, pests and climate change effects that would cause crop failure. In the past, farmers who relied on multiple crops raised in small fields could always rely on alternatives if one failed. However, having become dependent on a limited number of alternative crops or options, our risks escalate. For example, the people of Ireland who subsisted on potatoes in the mid-nineteenth century died by the millions of starvation and consequent diseases when the potato blight was introduced. The same is true today with intensive raising and international transportation of animals. One only has to think of millions of 'factory-farmed' chickens and cattle being slaughtered and burned because of infections.

In British Columbia, Canada's western province, where forested coastal moun-tain ranges dip into the Pacific Ocean to the west and spread into an interior desert to the east, a mountain pine beetle epidemic has devastated the landscape. Since the year 2000 the pine beetle larvae have destroyed an area representing one fifth of the province's forested landscape (Betts, 2008). This beetle infestation has reduced the forests' ability to uptake carbon from the atmosphere and will increase future emissions as killed trees decay. The extent and severity of these and other beetle infestations are directly linked to climate change and the milder winters the region is experiencing (Kurz *et al.*, 2008). Warmer temperatures have reduced larvae mortality and reproduction time, while at the same time expanding accessible range (e.g., allowing beetles to invade forests at higher elevations and move north into the higher latitudes of the boreal forest). Yet large-scale climate model simula-tions do not currently account for these types of outbreaks, despite the fact that pest infestations can severely impact net primary productivity. Thus, although these types of incidences are directly related to climate change, their severity and sudden impact are only just beginning to be understood (*ibid.*).

Other climate-related issues are expected in Canada. The mountain pine beetle is not the only insect that will cause problems. Ticks, which are the Lyme-disease vector, are expected to expand northward 1000 kilometres in Canada and increase in abundance two- to four-fold by the 2080s (Parry *et al.*, 2007, p. 47). The combina-tion of warmer summer temperatures, pest infestations and disease will extend the

Canadian fire season by 10 to 30% and increase the area burned by 74 to 118% by 2100 (*ibid.*, p. 62). The livelihood of the Inuit in Nunavut in northern Canada will continue to be impacted by melting permafrost (*ibid.*, p. 65), reduced seasonal Arctic sea ice and the effect of climate change on northern organisms. Melting Arctic sea ice will also open the Arctic for transport and resource extraction.

Perhaps most critically, the growing global population is demanding more of Earth's resources than can be replenished. Even without climate change, many of our freshwater aquifers are in the red zone, depleting faster than they are replenished. Over the last century the population has grown by a factor of three, but our water use has grown by a factor of six.[3] We are using oil at a rate greater than new resources are being discovered. The increase in the number of people on the planet, combined with increasing pressure on resources, is pushing us closer to a critical threshold, beyond which the Earth will no longer have the capacity to sustain us in the manner to which we are accustomed. These factors all result in our increasing vulnerability in the face of impending future rapid climate change. This situation limits our options for change. For example, the over-exploitation of aquifers means that the traditional way that people dealt with droughts is no longer available. Underground freshwater resources have been depleted.

What will it mean if we cross the threshold, and how will we know when we are close? What does the future hold for humankind? What will climate be like in our and our children's future?

9.1.2 What tomorrow brings

Using a variety of scenarios based on various assumed CO_2 emission rates, scientists predicted in a report that was published by the Intergovernmental Panel on Climate Change (IPCC) in 2007 that global average temperatures would increase by about 0.6 °C per decade beginning in 2020.[4] They also predicted that by 2029 global average temperatures would reach about 2.0 °C above pre-industrial levels. Warming is expected to be most significant over land and in the northern latitudes. Temperatures will rise more slowly over the Southern Ocean and parts of the North Atlantic Ocean.

Changing precipitation and evaporation patterns will accompany rising temperatures because precipitation, evaporation and temperature are interdependent. For example, warmer winters are associated with greater high-latitude precipitation in the form of snow because the warmer atmosphere holds more moisture. Warmer summers mean shorter intervals of snow-covered landscapes and sea-ice coverage; this makes the surface of the Earth duller, reducing solar albedo, which further intensifies the warming global climate. As a result of climate warming, precipitation will likely increase in the high latitudes and decrease in subtropical regions. Further, rain will come in more extreme events; frequent and/or intense storms will replace gentle showers.

The clouds that are associated with precipitation blanket the Earth, generating an effect similar to that of greenhouse gases. However, the warming effect is offset by their reflectivity so that, on average, clouds tend to have a cooling effect on climate (IPCC, 2007b, p. 95). Cloud cover remains the least understood feedback mechanism in climate change. Our climate models do a relatively poor job of representing cloud cover and simulating the response of cloud cover to climate change. Thus, current research is focused on better understanding how clouds change in relationship to climate warming (*ibid.*, pp. 99, 118).

Nearly one-sixth of the world's ecosystems are expected to be transformed by climate change, with anywhere from 5% to 66% of them shrinking. Many animal and plant species will not be able to survive.

If humans continue to dump CO_2 into the atmosphere, it will not take long at current rates of emissions before levels of 450 ppm will be reached and temperatures warm beyond 2 °C above today. With these changes, the North Atlantic meridional overturning is expected to be reduced by about one-third. The cooler temperatures this brings to Europe will be offset by global warming.

In the 2007 IPCC report, scientists predicted that by 2020, between 75 and 200 million people in Africa will be vulnerable to water stress as climate change causes countries to draw down their naturally renewable water resources. (When a country uses more than 20% of its water reserves, it is said to be suffering from water stress. As a result, its development is limited. Countries that withdraw 40% or more of their water reserves are under high stress.)[5] If their reserves are not naturally replenished through rainfall or snowmelt within about two and a half years, their water resources may be permanently depleted. By 2020, yields from rain-fed agriculture could be reduced by up to 50%. Throughout Asia, particularly in large river basins, there will be less fresh water, affecting more than 1 billion people. Ironically, these same people, as well as those in Africa and those living on small islands, will be vulnerable to extreme flooding around river deltas and along coastlines. This has already been made evident in Bangladesh. By 2050, high-latitude regions can expect an increase in water availability of between 10 and 40%; however, more of that water will come as heavy precipitation events that will result in flooding. In dryer regions and the midlatitudes, as well as in the dry tropics, the water available for drinking, agriculture, health, sanitation and energy uses is expected to fall by 10% to 30%.

Should global average surface temperatures increase by 4.6 °C the Greenland and West Antarctic ice sheets are likely to melt, raising the average sea level by between 7 and 13 m or more.[6] Rising sea levels of this magnitude will flood many of the world's coastal cities, regions that during the twentieth century have experienced considerable growth in population and capital infrastructure investment. A seven-metre rise in sea level would require the relocation or protection of between 10% and 17% of the world's population over the next several centuries. Although it may take hundreds or thousands of years for us to experience the full impact of that melting, it is our actions over the next few decades that will determine whether these ice sheets melt (Lowe *et al.*, 2006).

The level of warming that the world experiences as we move farther into the twenty-first century will be a direct consequence of the amount of greenhouse gases that we dump into the atmosphere today, tomorrow and over the next few decades. Even if we stopped emitting CO_2 today, past emissions commit the world to an estimated warming of at least another 0.6 °C by the end of this century. If we reduce our CO_2 emissions but still emit more than can be removed by natural systems, the level of CO_2 in the atmosphere will continue to rise, but it will do so more slowly.

If CO_2 concentrations reach 550 ppm, temperatures are expected to reach about 3 °C higher than pre-industrial levels. Scientists agree that when escalating temperatures reach 2.0 to 3.0 °C above the average global temperatures of AD 1850, or when atmospheric levels of CO_2 surpass 450 ppm, numerous damaging effects will ensue (see Table 9.1). Few of the world's ecosystems will be able to adjust to an increase in temperature of 3.0 °C, but even the most 'green' scenarios put forward by scientists predicting future climate change show temperatures rising beyond 3 °C in the northern latitudes by the end of the twenty-first century. Worst-case scenarios show global temperatures rising beyond 7.5 °C.

9.1.2.1 Recent updates

When scientists put together their findings for the 2007 Summary Report by the International Panel on Climate Change (IPCC), they were predicting with near certainty a 2 °C increase in global average surface-air temperature within just decades based on a number of alternative scenarios. However, by March 2009 it was clear that we were already on the path of the worst-case scenario put forward by the IPCC. Emissions are now growing and climate is now changing faster than the scientists anticipated and the climate models predicted. Sea and land ice are now melting faster and sea level is rising 50% more rapidly than projected. Global average surface-air temperature has already increased by about 0.75 °C since the beginning of the twentieth century. Based on new evidence, scientists now feel that their prediction of a 2 °C rise in global average surface-air temperatures above pre-industrial levels was too low. Instead they feel that a 3 °C rise is far more likely. If we want a reasonable chance of limiting the rise in global average surface temperatures to 2 °C then developed countries must reduce their greenhouse gas emissions by 80% by 2050 and developing countries must decrease theirs by 15 to 35%.

Table 9.1. *Global effects of rising temperatures and atmospheric CO$_2$
concentration levels*

Temperature increase	Atmospheric CO$_2$ concentration levels	Effects
1.0 to 3.0 °C	380 to 550 ppm	Agricultural yields will begin to fall in Africa In Peru, there will be shortages of water for drinking, agriculture and energy Steppic regions around the world will experience greater drought, generating water stress and crop failure Some regions in the northern hemisphere will experience larger crop yields One-tenth of global ecosystems will lose between 2 and 47 percent of their extent The Great Karoo grass flats in South Africa will shrink The Queensland rainforest in Australia will potentially lose 50 percent of its extent The Dryanda forest of Australia will experience species extinctions The Kalahari dunefields will become activated Oceans will become more acidified resulting in dieback of corals There will be a higher incidence of flooding in low-lying areas around the world
1.5 to 2.5 °C	400 to 500 ppm	One-fifth to one-third of plant and animal species will face extinction The Greenland ice sheet may begin to melt, causing the beginning of an expected long-term 7 m rise in sea level Clean fresh water will be increasingly unavailable due to droughts and flooding Water availability is expected to fall by between 10 and 30 percent in dry regions and midlatitudes By 2020, between 75 and 200 million people in Africa will experience water stress, and rain-fed agriculture could decline by as much as 50 percent In Asia, more than 1 billion people will face reductions in fresh water and be vulnerable to flooding
2.0 °C	450 ppm	Between 1.0 and 2.8 billion people will have trouble getting drinking water 97 percent of the coral reefs in the world will die 16 percent of the world's ecosystems will be transformed, with between 5 and 66 percent losing some of their extent Rising sea levels combined with cyclones will force millions of people in coastal regions to relocate

Table 9.1. (*cont.*)

Temperature increase	Atmospheric CO_2 concentration levels	Effects
beyond 2.0 °C	> 450 ppm	The world's cereal crop yield will fall by 30 to 180 million tons; up to 220 million people will risk hunger
		There will be large-scale displacement of people in Africa as a result of poverty, starvation and thirst; millions more people will be at risk of malaria
		The total loss of summer ice in the Arctic will mean the destruction of Inuit hunting culture, with severe impacts on walrus populations and a drop of 30 percent in the polar bear population by 2050, a predicted 60 percent decline in the lemming population, and a decline in the stability of the Arctic tundra (less than half of it will remain stable), endangering high Arctic shorebirds and geese
		In the Americas, vector-borne diseases are anticipated to expand poleward; more cases of malaria will be seen in North America as the mosquitoes that carry the disease extend their range
		Inter-regional tensions will likely mount in Russia when crop productivity falls as temperatures continue to rise
		In Asia, 1.4 to 4.2 billion people will face water shortages, and vector-borne diseases will start appearing farther north and south
		China is expected to lose half of its boreal forest
		Half of the Sundabans wetland in Bangladesh will also disappear
		In Australia, the threat of extinction will expand; half of the Kakadu wetland will be lost
2.0 to 3.0 °C	450 to 550 ppm	The overturning of the North Atlantic Ocean will be reduced by about one-third[*]
		The Amazon rainforest will collapse
		Hundreds of millions of people will suffer from increasing water shortages
		Between 20 and 30 percent of plant and animal species will be at risk of extinction
		In Africa, 80 percent of the Karoo grass flats will potentially be lost, endangering 2800 plants; five of South Africa's parks will lose more than 40 percent of their animals; fisheries will be lost in Malawi; three-quarters of South Africa's crops will end in failure; and all of the dunefields in the Kalahari will be active, threatening the ecosystems and agriculture of the sub-Sahara

Table 9.1. (*cont.*)

Temperature increase	Atmospheric CO_2 concentration levels	Effects
above 3.0 °C	> 550 ppm	Maple trees, the source of Canada's world-renowned maple syrup, will be threatened
		In Australia, the Kakadu wetlands and alpine zone will be lost
		The Tibetan plateau in Asia will undergo desertification and a shift in the permafrost
		The Chinese boreal forest could disappear
		The gross domestic product of 65 countries will fall by 16 percent
		As malaria continues to expand, so too will dengue fever, with potentially more than half of the world's population exposed to the disease, compared to one-third in 1990
4.0 °C	700 ppm	Entire regions will be unable to produce agricultural crops, resulting in more people at risk of hunger
		Australia will no longer be able to produce agricultural crops and will no longer have an alpine zone
		In Africa, 70 to 80 percent of the human population will face hunger
		In Russia, a 5 to 12 percent drop in agricultural production is predicted to occur over 14 to 41 percent of the country's agricultural regions
		In Europe, 38 percent of the alpine species are projected to lose 90 percent of their alpine range
		Malarial regions are predicted to expand by 25 percent
		20 percent of the perennial zones will disappear, along with more than 40 percent of the taiga-producing regions
		The extent of tundra regions will potentially be reduced by 60 percent
		Timber production will increase by 17 percent
		North Atlantic thermohaline circulation will slow significantly
2.0 to 4.5 °C	450 to 775 ppm	A 20- to 400-million-ton reduction in global cereal production is anticipated to result in up to 400 million additional people at risk of hunger; 70 to 80 percent of those people will be in Africa
		An additional 1.2 to 3.0 billion people will suffer from water stress
		There is a 50 percent chance that the west Antarctic ice sheet will begin to melt; about one-third of global coastal wetlands will be lost
		In the Americas, half of the world's migratory bird habitat will potentially be lost

Table 9.1. (*cont.*)

Temperature increase	Atmospheric CO_2 concentration levels	Effects
		Alpine species in Europe and Russia are projected to be near extinction, along with 60 percent of Mediterranean species
		In China, rice yields may fall by 10 to 20 percent
		Australia is predicted to lose half of its eucalyptus

* A. J. Weaver, M. Eby, M. Kienast, O. A. Saenko. (2007). Response of the Atlantic meridional overturning circulation to increasing atmospheric CO_2: sensitivity to mean climatic state. *Geophysical Research Letters,* **34**, no. 5 (2007): L05708.

Note: There are many uncertainties inherent in climate-change projections, including the level of future emissions, the climate response to those emissions, future population levels, economic development, technological development and changes in behaviour. These need to be added to climate modelling uncertainties. Scientists have assigned a series of confidence levels to their predictions from very high (90 percent chance of being correct) to very low (less than 10 percent chance of being correct). Researchers are often wary of making dire predictions because they do not like to be responsible for initiating actions that turn out to be unnecessary. However, most individuals and policy makers should be more worried about hedging their bets and taking precautionary steps in case serious climate change does occur. The events listed above may happen at higher or lower temperatures or CO_2 concentration levels, and some are predicted with greater or lesser confidence than others. All are based on published research done by scientists from around the world (see below for sources). Although there is some question as to exactly what will happen when, there is no doubt that if we fail to alter our attitude and reduce our CO_2 emissions, these things will begin to happen. For a popular reference into descriptions of what each degree of warming would mean for humanity see Mark Lynas's book *Six Degrees: Life on a Hotter Planet.*

Source: Much of the scientific data presented in this table has been obtained from three main sources:

Intergovernmental Panel on Climate Change (IPCC). (2007). *Climate Change* 2007: *The Physical Science Basis.Contribution of Working Group I to the Fourth Assessment Report of the Intergovernmental Panel on Climate Change*, ed. S. Solomon, D. Qin, M. Manning, Z. Chen, M. Marquis, K. B. Averyt, M. Tignor, and H. L. Miller. Cambridge: Cambridge University Press, available online at www.ipcc-wg1.ucar.edu/wg1/wg1-report.html.

IPCC. (2007). *Climate Change* 2007: *Impacts, Adaptation and Vulnerability. Contribution of Working Group II to the Fourth Assessment Report of the Intergovernmental Panel on Climate Change*, eds. M. L. Parry, O. F. Canziani, J. P. Palutikof, P. J. van der Linden, and C. E. Hanson. Cambridge: Cambridge University Press, available online at www.gtp89.dial.pipex.com/chpt.htm.

Warren, R. (2006). Impacts of global climate change at different annual mean global temperature increases. In *Avoiding Dangerous Climate Change*, eds. H. J. Schellnhuber, W. Cramer, N. Nakicenovic, T. Wigley, and G. Yohe. Cambridge: Cambridge University Press, pp. 93–131.

Even if we are able to accomplish these reductions we can expect major climatic changes to occur. We know from the past, that even small changes in global average surface-air temperature, in the order of 1 to 2 °C can have a dramatic impact on human habitat. For instance, it was only a 5 to 6 °C change in global average atmospheric surface-air temperature that altered the landscape of North America from evergreen and broadleaf forests and grassy plains to an expanse of massive ice more than 2 kilometres thick that spanned from the Pacific coast to the Atlantic. Today, a change of 3 °C will mean, in many regions of the world, the difference between a climate that is suitable for agriculture and one that is not.

9.1.2.2 Tipping elements

Recent human activity has altered the Earth system. It is pushing some parts into new modes of operation with large-scale impacts on humans and ecological systems (Lenton *et al.*, 2008). Earth's climate and ecological systems possess a number of 'tipping elements', which under certain circumstances can be switched into a different state by relatively small perturbations. Some of these may reach their critical point within this century. A group of leading experts combined their knowledge with contributions from 52 members of the international scientific community to compile a list of tipping elements most relevant to global warming, future environmental and economic uncertainty, and policy development. Tipping elements high on this list, along with their transition time, include retreat of the Arctic summer sea-ice resulting in an ecosystem change (~10 years); melting of the Greenland and West Antarctic ice sheets generating a 7 to 12 m rise in sea level (>300 years); reduction in the Atlantic thermohaline circulation causing cooling North Atlantic regions (~100 years); more intense El Niño–Southern Oscillations (ENSOs) leading to drought in South East Asia and elsewhere (~100 years); reduced rainfall during the Indian summer monsoon resulting in drought and decreased carrying capacity of the region (~1 year); reduction in the Amazon rainforest and the boreal forest causing biodiversity loss (~50 years) (*ibid.*).

Another 'tipping element' is Earth's capacity to provide the necessary resources to support humans. Modern 'agribusiness' relies heavily on petroleum products for farm machinery (which adds to greenhouse emissions) and for the production of fertilisers. Petroleum is a non-sustainable resource. There are many geological experts who feel we have already reached 'peak oil' production.[7] Add to this limitation the likelihood that future climate change will significantly alter those areas currently able to produce agricultural products. Large tracts of currently productive agricultural land are expected to diminish in size and productive capacity. The subtropics are expected to be reduced in their capacity for agricultural production because of lack of water and failure of the monsoons. Although some regions are expected to be suitable for farming, precipitation patterns are not confidently modelled and thus there is limited confidence about the availability of water in lakes, rivers and acquifers. And let us not draw a veil over the problems arising from the lack of clean drinking water and the availability of irrigation water in the face of rising populations. Another problem related to higher temperatures in higher latitudes is an expanding population of disease vectors such as mosquitoes, which would affect not only domestic animals, but also wild animals and humans.

One of the serious limitations of our current economic market system is that it depends on an economy that requires expanding populations and markets. However, the Earth's climate and ecological systems are already being heavily taxed and may in fact be beyond their natural carrying capacity and unable to continue to sustain a modern human population of 6.75 billion people in the manner to which we are accustomed. But what do we mean by 'beyond their natural carrying capacity?'. It might be argued that everybody alive at present could be sustained by currently available produce. All it would take would be a global distribution system, and the reduction or removal of agricultural tariffs. This, however, would require major political cooperation. Farming expansions in Africa and South America have resulted in deforestation of huge tracts of equatorial jungles, negatively impacting climate. But even if sanity and sympathy prevail, it comes back to the fact that present levels and methods of agricultural production depend on non-renewable petroleum products, fertilisers, pesticides and a diminishing supply of water for irrigation.

There is no question that our behaviour is severely impacting the Earth. We appear to have lost a critical connection to nature that would have allowed us to recognize imminent danger. Alternatively, we have failed to heed the warning or chosen to ignore it. Our ability to control and dominate the natural world in order to stabilize our external environment and grow our economies has resulted in dangerously accelerating human-induced climate change and related consequences. What does that mean for the future of human societies and the human species? A review of our past relationship with climate provides some insights into our capacity to evolve and adjust.

9.2 Modern humans' capacity to evolve and adjust

9.2.1 Past and future human evolution

In the past, human evolution occurred in small, isolated populations, in which novel variants would be most likely to persist, rather than being absorbed by the multitudes of the mundane. Going back over the history of hominin evolution as far as *Homo habilis*, there were epigenetic changes in anatomy such as posture and brain size, accompanied by behavoural novelties. These trends were continued by *H. erectus*, which ultimately formed a large, widely dispersed population, though it would be more accurate to call it a mosaic of smaller, semi-isolated populations, where genetic drift was more important than the build-up of naturally selected traits. It is difficult to say whether these changes were continuous or saltatory, but the fossil evidence indicates the latter condition. It certainly appears that the changes in posture and brain size found in hominins were saltatory emergent phenomena. It is not clear whether or not there were hormonal, interneuronal novelties, or relative changes in the cerebrum. We have taken the parsimonious view that the anatomical and physiological qualities found in the first anatomically modern humans were all that was necessary to evolve socially, to finally produce *H. sapiens*, in the form that is usually called 'behaviourally modern human'.

The emergence of the latter was probably due to a series of interactions between groups that had been isolated and developed their own unique technical skills. The communication and cooperation of the newly coalesced population was in many instances the result of environmental changes that stimulated migration into refugia.

In the Appendix we illustrate how environment can play a role in altering genes rapidly – within a generation or so. Epigenetic evolution might have occurred: by mutations (including tandem repetitions) of regulatory genes, shifts in the internal environment of the mother and behavioural changes. Stress such as starvation, heat shock and cold shock can also change the course of development, though known examples prove to be detrimental.

9.2.1.1 *Human developmental evolution*

During the last 135 000 years of human existence climate variability, water availability and changing vegetation placed significant demands on early humans. It is just these types of environmental influences that can typically impact development, both behavioural and epigenetic.

Epigenetic evolution can be effected by adding new features to an organism late in their life cycle after most development is complete. An environmental influence that affects an organism's development can be genetically assimilated so that it becomes heritable. Changes in temperature and day length, the presence of members of the same species, the presence of members of other species and the availability of food, can all influence epigenesis or development.[8]

In the past, there were genomic mutations and epigenetic shifts that might trigger improvements in thermoregulation, such as anatomical changes from a gracile form to a stocky stature. Accompanying these were alterations in the distribution and numbers of sweat glands and the production of variant enzymes and hormones, through gene duplication and mutation. This mechanism arose in the early history of the vertebrates.

Physical changes once incorporated, whether as a consequence of environment or not, may further dictate a change in behaviour and environment. For example, if human legs became longer, humans would likely incorporate running skills into their hunting behaviour, and begin to hunt new kinds of prey. This may have allowed them to migrate into and successfully occupy new regions.

How much might these ideas apply to human evolution in the future?

All of the causal biological factors of emergent evolution might still apply in the future. However we live in a global village. Biological changes in development, functional anatomy, physiology and behaviour might occur locally, but no community is sufficiently isolated to be immune to the homogenization of novelties. Genetic mixing on a global scale is possible, albeit inhibited by religious and ethnic taboos. New symbioses, particularly between human hosts and bacteria, are not out of the question. Endogenous retroviruses, which have worked their way into our genomes without detriment, could still bring about significant novelties. Local emergences in cultural evolution, which in our opinion has been more important to human history than biological evolution, can now spread through transcontinental travel, and instantaneously through the internet. New ideas, aesthetic and technical, can be subjected to close scrutiny and distributed universally, although conservative biases can hinder such progress. So let us take a closer look at human physiological and behavioural adaptability.

9.2.2 Physiological and behavioural adaptability

A newly altered organism, perhaps changed by an altered climate, might be a hopeful monster, noticeably different from its parents and maladaptive under the prevailing conditions (Goldschmidt, 1940).[9] Yet it might have the ability to exploit and thrive in a new environment. This would depend on its physiological and behavioural adaptability. Adaptability, in contrast to the inflexibility of genetic adaptation, allows an individual organism to modify or adjust its physiology or behaviour to changing environmental conditions. In essence adaptations are gene determined, and adaptabilities are organism determined.

However, for humans, and indeed all mammals, exploratory behaviour and the coincident exposure to new environmental conditions, and all the evolutionary consequences associated, depends on a reliable, stable physiological state. You cannot live in an extreme climate unless you are physiologically capable of keeping your internal environment within safe limits. Thus, maintaining a stable state – homeostasis – has been critical for human survival. Homeostasis is the foundation of human behavioural adaptability. It allows us to change what we do according to circumstances. We can explore new habitats and/or retreat from them. Further, we can do so without imposing harmful internal changes.

In a broad sense, human intelligence could be included as part of our homeostasis. Humans possess the ultimate independence from the vicissitudes of the external environment because we can complement physiological self-maintenance with intelligent behaviour; for example, the use of fire, and development of artifacts such as clothing, housing and air conditioning. Humans are able to think about the potential dangers and rewards of our behaviour without committing ourselves.

Moreover, we have the added advantage of teaching the inexperienced so that they do not have to repeat our mistakes and ordeals.[10]

Yet it is not just humans that possess cognitive abilities and physiological adaptability. Placental animals in general possess both cognition and physiological adaptability and both have played a major role in the progressive improvement of exploratory behaviour in general. The progressive evolution of the primates involved expansions of various parts of the brain, rewiring some of the central nervous system and improving the capacity of brain cells to communicate with each other. These internal evolutionary changes were linked to the environment through behaviour. Thus, it is important that we carefully consider the degree to which features that are often taken to be characteristic of modern humans may be found in earlier humans. A case in point is brain size. As outlined in Chapter 2 and Appendix B, *H. floresiensis*, a putative relative of *H. erectus*, was discovered in fossil form on the Indonesian island of Flores. Despite its dwarf size and a brain the size of a lemon, it managed to somehow drift or sail in makeshift 'boats' to Flores Island and survive alongside the Komodo dragon and in close proximity to *H. sapiens*, for between about 18 000 and 38 000 years.

Compared to genetically fixed and specialized forms of placental mammals, the more adaptable human form performs relatively well. Specialized placental mammals, based on a foundation of physiological adaptability, developed innovative behaviour and then chose the conditions of life to which they would subsequently adapt, in the sense of acquiring genetically fixed traits. Alternatively, our own ancestors opted to be generalists. Human generalists combined their innovative behaviour with intelligence and education to undertake a wide variety of activities that other placentals could only outdo in their specific area of specialization. Thus, humans were able to explore and improve their conditions of life. For example, humans lit fires and put on the skins of animals to keep warm as climate cooled. Cold-adapted placental animals developed thick insulating fur coats. When rapid environmental changes were imposed on early hominins *in situ*, for example when climate rapidly changed and altered the vegetation and reduced the number of food species available to early *Homo* species in Africa, they epigenetically developed, and become more physiologically and behaviourally adaptable. These changes then permitted early humans to explore new environments; they moved out of Africa at the beginning of the last glacial cycle. The exploration of new environments likely induced novel responses and as a result additional new environments were occupied. As a consequence modern humans subsequently dispersed around the world.[11]

It is unlikely that humans will undergo future genetic adaptational specializations unless we are catastrophically subjected in small, isolated populations to extended periods of confinement in stressful conditions. Any changes will appear instead as epigenetic, physiological and behavioural developments. Thus, for individual organisms, behavioural, physiological and developmental adaptability is highly significant in the capacity to change. These factors operate above the gene level. Actions of the individual organism also act in a feedback loop directing the impact of environmental change, and the linkages between behaviour, physiology and epigenetics. These feedbacks throughout human history have generated a more complex and intelligent human brain.

Neocortical expansion of the forebrain is common to all placentals. But only in primates does the cerebrum dominate the rest of the brain in size and fold in such a way to extend the space available for neurons. This increase in neurons potentiated higher brain functions such as logic, language, memory and intelligence. It was aided by the multiplication of dendrites, thread-like roots that connect nerve cells and integrate specialized regions of the brain and an increase in the variety of chemical messengers in the brain, with a corresponding increase in the types of receptor molecules so that logic, speech and grammar were integrated into language. Normal development of the brain requires a reliable homeostasis and source of nutrients. Brain development, including dendritic complexification depends on the experiences of the individual. Developmental stimulation from the outside environment, particularly from individuals of the same species, maximizes the ability to make dendritic connections, and thus the ability to make intelligent connections between various pieces of information. In this way, humans have interacted with, and responded to, a series of climate and environmental changes through history.

Our capacity to stabilize our external environment progressively developed through human history. As the *H. sapiens* population expanded and inhabited most of the Earth's land surfaces we learned how to adjust and manipulate the varied habitats we encountered, whether they were coastal marine environments, grassy plains, or mountains and river valleys. We found in the forests wood for heating and cooking. Later we used wood for housing, industrial and transportation needs. We discovered that the river valleys and deltas provided exceptionally fertile conditions for growing crops and grazing livestock. The rivers and seas provided plentiful marine resources. And even the cold Arctic environment, with seasons that oscillate between 24 hours of darkness and 24 hours of light, provided seal, whale and other Arctic life that, when understood and properly harvested, could sustain a healthy human population. Yet, one negative consequence of our increasingly stabilized environment may be a reduction in our ability to recognize environmental change, behaviourally adjust to it and to make the necessary intelligent connections between changes in our environment and our behaviour.

9.2.3 The clash between humankind and nature

Our capacity, need and penchant to control our environment have resulted in a significant problem. As so aptly put by Dr. Anwar A. Abdullah, 'the real problem is the clash between nations and nature… We have imposed our will on nature through technology. Yet nature is silent, and humans speak. We are both the question and the answer'.[12]

In fact, the merging breakdown between modern economic development and surrounding ecosystems could be reached back to the deep past of our civilizations. Since the early building of Mesopotamian city-walls, and of the French revolution, mankind has suffered from the illusion of nations; each against all other and all in all against nature. Here, nature is considered as stock of economic goods while the rising societies; of what has become known as nations of mighty urban-civilizations, have inflected their technological must upon silent flora and fauna. Paradoxical themes like these do reflect indeed the instability of cultural orientation, and of rather a permanent expression of mastery: of 'man-and-nature' is little else but of 'master-and-slave'. And thus, to acquire a place on Earth, each nation has to justify its means of mastery over the rest; men and nature, whilst to feel right even when doing wrong. Here, the silent song of nature has never been subjected to a heartily detection. Nor its quality is praised beyond the claim of dry equations. So the current dilemma seems almost as a deep clash between nation and nature. And still, nature has its own stratagems as to absorb our blows and fool us whilst turning them upon mankind an utter wrath.

(Personal communication from Dr. Anwar A. Abdullah, 6 April 2009)

Dr. Úrsula Oswald Spring sees the clash more generally between humankind and nature. Both Spring and Abdullah feel we fail to hear nature's voice. We have lost the capacity to listen and to understand nature's language. Nature responds as nature does. If we do not wish to experience the wrath and ferocity of nature's response, we need to relearn nature's language and remember how to listen.

Physicist Dr. Ursula Franklin states:

If you look at the world as it is now, nobody in their right mind intended it to be like that and it is not solely lack of foresight, of which there is a lot, but also genuine lack of understanding of how things work. And that is not stupidity, but it is the limitation of what the human mind can grasp … It wasn't that someone single handedly delivered the mess in a plain brown envelope. It evolved as nature's response to step-by-step stupid decisions and responses not read. It is as if nature constantly sent emails and nobody opened them of those that could.

Humans are only a very small part of nature. In certain ways humans function best the better they understand the workings of nature and the more they respect it. They have the right, as every creature has, to modify nature so that they survive. That is why some birds migrate and some birds stay … And that is a perfectly rightful cycle in nature. Now it doesn't modify the climate, it modifies one's behaviour in order to live in the existing climate, and that is a very great difference, understanding nature so that one can modify one's own behaviour and survive.

(Personal communication with Dr. Ursula Franklin, 16 April 2008)

During the relatively stable climate of the Holocene, humans modified their behaviour to adjust to climate variability as it occurred. Yet as time progressed, we increasingly used technology to protect us from change and adjusted ourselves less. We are now immensely capable of developing technological instruments to intervene in nature in ways we previously thought incomprehensible. However, we do not always understand the repercussions of new technology and, as Ursula Franklin notes, constant innovation can be a real drawback for humanity (*ibid.*). The constant emphasis on technological innovation never allows people to experience the new change long enough to know how the latest innovation is going to work out. It creates continuous societal imbalance that does not allow action based on long-term experience. Further, it distances us from nature and removes the requirement to adjust our behaviour. Perhaps we need to depend less exclusively on technology and recognize the importance of environmental change in stimulating human behavioural change.

Further, Franklin notes that it is important to recognize that there exists an appropriate structure, shape and size for a particular social ecology and to recognize that social ecology functions within a world filled with many other things with their own natural ecologies. Franklin suggests that the crisis we are experiencing is that of human conceit augmented by the inventory of instruments that is so totally out of step with both our intellectual and emotional capacity and our assessment of our role and place in the world. 'What stares us in the face is that utterly hyperbolic arrogance that damages, that kills. That kills the innocent bystanders whether they are people, or nature, or rivers, or birds' (*ibid.*). Yet, few people, if anyone, appear to be taking seriously into account that 'nature in the biggest sense of the word has a voice and has a veto. And nature speaks and nature has spoken…Nature is an ecology in which humans are a small part' (*ibid.*). What we need to do is 'check the instruments and throw out the ones that are harmful. Do not develop any where one cannot, in some way, either allow the time to look at their consequences, or go slowly enough not to do more harm than can possibly be done and sustained' (*ibid.*).

9.3 The climate connection: human vulnerability to rapid climate change and adaptability

It is clear that there is much we do not know. There are large gaps in the archaeological record, the climate proxy record and in our understanding of what those records mean. And even though our climate simulation and economic models are the most advanced in human history, they are not the world. There is 'a real and important difference to understand between even the best and most careful of all our models…and what actually happens in the world. There are great gaps and they are fundamental basic gaps and it is pretty important that we understand how little of the totality is open for our understanding' (*ibid.*).

Yet it is also clear that something significant is amiss. Climate model simulations predict that the relatively stable climatic environment we have enjoyed for the last 10 000 years or so is becoming highly unstable once again. The devastation of forests caused by pests, together with fires, and that of melting Arctic sea ice and perma frost, along with droughts, floods, and hurricanes provides additional insights into the rapid impact that a changing climate can have on human communities. In response to these, along with tipping elements and as yet unforeseen changes, we will likely see large impacts on humanity.

9.3.1 Recognizing past efforts

Recognizing that the world is facing impending climate change is not new. It is clear that scientists were aware that the world was facing serious climate change issues nearly 25 years ago. In the World Meteorological Organization's report on the 1986 International Conference on the Assessment of the Role of Carbon Dioxide and Other Greenhouse Gases on Climate Variations, scientists state, 'As a result of the increasing concentrations of greenhouse gases, it is now believed that in the first half of the next century, a rise of global mean temperature could occur which is greater than any in man's history' (quoted in Roots, 1989a, p. 224).

Scientists were also aware that plant and animal specialization helps species to cope with the distinctive environmental conditions imposed by their respective climates, that rates of change were critical, and that, in reference to polar and sub-polar regions,

… when the climate changes, these specializations may put the organism or the population at a disadvantage … [polar and sub-polar] ecosystems tend to be sensitive to physical disturbance, by natural or human causes … Marine ecological conditions in high latitudes are on the whole more stable than those on land; but arctic biological populations nevertheless seem to be subject to environmental variations more sudden and violent than of those in lower latitudes … The rate of change, rather than the absolute amount or extent of change, would appear to be a particularly critical factor in the ability of high-latitude ecosystems to adjust to new conditions.

(Roots, 1989b, pp. 14–15)

In 1989, scientists were also trying to draw our attention to the interconnectivity of the Earth's climate and ecosystems.

Because of the critical role that the arctic regions, together with the Antarctic, play in determining the global heat balance and the flow of energy from the tropics to the polar regions, not only will changes in the physical, chemical and biological conditions almost anywhere on the globe have an effect on the environment of the Arctic, but changes in the Arctic will in many cases have repercussions in non-arctic lands and seas.

(ibid., pp. 16–17)

Scientists postulated about what predicted climate changes would have on regional climatic conditions. Though not as clearly defined as today, those effects were substantial and included strong warming in continental and coastal areas, drier mid-continent summers, milder winters in the North American Arctic and Greenland, more stormy weather, retreat of northern hemisphere sea ice, a long-term reduction in glaciers and ice sheets and a warming of average surface-air temperatures by 1.5 to 4.5 °C by 2030 due to the accumulation of greenhouse gases; rising temperatures would be accompanied by rising sea levels of between 20 and 200 centimetres. Scientists noted that atmospheric CO_2 levels had reached 350 ppm by 1980 and methane levels were increasing rapidly. Scientists also predicted significant shifts in habitat. This was exemplified by the northward extension and then decline in the west Greenland cod fishery, apparently the result of about a 1.5 °C change in the upper mixed layer of Baffin Bay water (Roots, 1989b, p. 28).

It is also clear that scientists have understood and made clear to policy makers since at least 1993 that the capacity of the Earth to support human activities is being seriously damaged by human activities (Roots, 1993). By 1993 it was evident that expanding human populations and their use of energy and materials was disrupting local and planetary environmental processes and ecosystems, resulting in human conflict. 'Deteriorating environmental conditions in many parts of the world are creating a new class of distressed human – the environmental refugees – whose rights, and the rights of those who must support them or make room for them, raise many ethical, political, and economic questions' (*ibid.*, pp. 529–30). These issues were making it difficult to fulfill many of the basic human rights recognized by the United Nations, including the right to food, clean water, a healthy environment and the ability to earn a living.[13] The rights of future generations and the rights of non-human species in light of 'accelerating global environmental change' had also been recognized. In 1993 it was clear that humanity's existence was dependent on Earth's finite resources and limited by the constraints of the natural environment. Further, this knowledge had been evident for thousands of years. For, as Roots states,

From the earliest days of cultivation of crops or domestication of animals, there was knowledge of a practical limit to biological productivity, and thus a limit to how many people could be supported. As knowledge became more systematically organized, the obvious and sensible notion of a limit to the amount that could be produced or supported within any semi-closed system became known as *carrying capacity* (Ehrlich and Ehrlich, 1972). A good current example is the river Nile. No river on earth is better known. For 5000 years, clever people have been studying the Nile, using its vagaries and seasonal eccentricities to support, through most of human history, a vigorous society whose successive technologies were put to use as civilization advanced. The carrying capacity of the river system to support people was developed to its then practical limit in the Stone Age, increased again in the Bronze Age, once again in the Iron Age up to nearly modern times. But in the Electric Age, the Aswan High Dam was built to expand the carrying capacity to a still higher level, to provide power for the industrialization of Egypt, to control flooding, to irrigate new 'reclaimed' desert lands, and, it was hoped, to support a population that had more than doubled in a century. Technically, the dam has worked, just as it was planned. But, in Egypt's most fertile area, the Nile delta, the cessation of annual flooding has led to [an] alarming increase of the parasitic disease schistosomiasis [or] bilharzia; farther upstream the salt content of irrigated desert soils has run wild; a centuries-old sardine fishery in the eastern Mediterranean has disappeared and the number of fish species in the river reduced from more than 130 to about 10. Egypt, long an exporter of food to the surrounding region, now is dependent on imported fertilizer and has become a net food importer. There is little doubt that the vulnerability of the entire socio-economic complex in the lower Nile valley to a change in climate or to mismanagement is now greater than before the dam was built. The production of food has increased, for a few years, but at what cost? Has the net carrying capacity increased? Has the industrial development, designed to increase regional economic capacity, maintained the vigour of the nation and improved the 'rights' of all its citizens? One could also ask, what might have been the situation in Egypt today had the Aswan High Dam not been built? Such questions are at the heart of the issues of carrying capacity and the rights of people or nations to use or manipulate natural resources. There are similar examples, on every continent, where the manipulation of carrying capacity has led to new or larger problems.

(Roots, 1993, pp. 532–3)

In 1993, when the population of the planet was little more than 5 billion people, scientists knew that the Earth's physical, chemical and biological processes were not changed and could not be changed by humans (*ibid.*, p. 536). Nature acts as nature does. Yet they also felt that demanding more of Earth's resources to satisfy the growing human population would cause damage to Earth that would inflict a large future price on humanity and would lead to greater subsequent human tragedy. It was also clear that the ability of the world's forests to continue to balance and regulate the environment was 'very seriously declining' (*ibid.*, pp. 538–9). Since then, the population of the Earth has grown to more than 6.75 billion and we recognize that humans are a driving mechanism in generating present and future climate change. As Roots (1993) stated more than 15 years ago, 'the only areas where humans can do anything to restore the balance are to change the numbers of people, to use less energy, or to refrain from using environmentally-damaging technologies' (p. 542). Yet the human population has grown, we use more energy not less and we continue to use environmentally damaging technologies at an escalating rate.

Why, when we know that maintaining our current behaviour will result in future devastation, do we fail to change our behaviour? Diamond (2005) says that 'creeping normalcy' or 'landscape amnesia' are a major reason why people fail to recognize a problem until it is too late (p. 426). Creeping normalcy refers to changes that happen in small increments over time and therefore go unnoticed. Yet if the cumulative effect of the same change occurred as one major change, it would solicit a major response. Diamond further states that even when societies do perceive a problem, they frequently fail to solve it. This is because of 'rational behaviour' – behaviour that is in an individual's best interest. He suggests that people are highly motivated to reap 'big, certain, and immediate profits, while the losses are spread over large numbers of individuals' (p. 427).

Roots (1993) suggests that changing our behaviour has been perceived as seriously infringing 'on the rights of individuals to pursue their own immediate short-term or culturally-determined interests and desires' (p. 542). In the past, we were able to keep our economic engines running by gathering resources farther and farther away, as was evident in the European colonial enterprises of the nineteenth century. Yet, by at least 1993, when Roots wrote his environmental paper for the background documents for the UN Conference on Human Populations in Vienna, it was clear that our economic systems were expanding across the planet and impoverishing the place and people whence new resources came. Our economic systems 'call into question the basis of welfare and the responsibility of one group of humans toward their fellow humans' (*ibid.*, p. 544; referencing Sachs, 1976). Roots (1993) outlined six issues that must be addressed in the twenty-first century (pp. 545–7).

- 'The right to use some natural resource or amenity, exercised by one individual or a group, reduces the availability to someone else who, in theory, has equal rights. In the twenty-first century, *people must acknowledge that they are no longer consumers, but exchangers.*'
- Human population growth places two human rights in conflict: 'the individual and collective rights to natural resources and a healthy environment versus the right of humans as individuals to reproduce'. Failure to voluntarily and collectively address this situation will likely result in 'severe human tragedy'.
- What are the individual rights of a growing number of *environmental refugees*, 'wherever they arrive, compared to the rights of people into whose region they come and whose resources they will want a part of?'
- What rights do people have to maintain their cultures, their distinct societies, and their right to be different when these rights have been so impaired by the infringement of others? How are those rights guaranteed locally, nationally and internationally?
- 'How do we identify and defend the rights of generations yet to come?'
- What are *'the rights of humans compared to the rights of non-human species?'* (This question highlights the importance of the human–nature conflict that was discussed above and Ursula Franklin's comment that 'humans function best the better they understand the workings of nature and the more they respect it'.)

There have been numerous attempts to incorporate climate-change issues into government and industrial policies, including the Kyoto Protocol. In 1998, 37 industrialized countries, including the United Kingdom and the European Community, signed this protocol and agreed to binding targets to reduce their overall emissions of greenhouse gases 'by at least 5% below 1990 in the commitment period 2008 to 2012'.[14] Yet, it is already clear that these efforts have met, and continue to meet, with little success. The world now faces an even more precarious and rapidly changing climate than was presented at the end of the last century. Clearly it is not only difficult to gain agreement across countries, regions, industries, societies and cultures on climate-change policy, but it is even more difficult to implement and effect change. There are numerous explanations for this. It is difficult to get governments to focus on long-range planning policies when their political life is based on short-term objectives and initiatives. Climate change policies can exacerbate or emphasize local or regional differences, particularly those between developed and developing countries. Those who are generating most of the emissions are feeling few of the consequences, at least thus far. This frequently leaves many people in developed countries, including influential business people and policy makers, in a state of denial that impending rapid climate change actually exists. Others, who recognize that it does, feel there is little they can do that would make a difference, cannot generate sufficient initiative or alternatively, need more support to begin to do so.

At the Smithsonian Tropical Research Institute (STRI) in Panama City, Panama, between 25 and 29 February 2008, a major international symposium, 'Climate Change and Biodiversity in the Americas', was held 'to provide a forum for leading scientists to present the results of research and monitoring activities of climate change and biological diversity throughout the Americas'. Its aim was 'to establish a co-operative science, research, and monitoring network of activities that interlink biodiversity conservation and sustainability, policy responses, and adaptation to climate change throughout the Americas'.[15] One hundred and fifty people attended from every North, Central and South American country and the Caribbean as well as senior representatives from the Convention on Biological Diversity, the World Meteorological Organization, UNESCO (MAB), the InterAmerican Institute (IAI) and the Caribbean Community Climate Change Center. Thomas Lovejoy, senior scientist from the Smithsonian, opened the scientific sessions with a 'chilling summary of what was happening to ecosystems around the world'. Canadian scientist Fred Roots was there, and in a personal communication he noted one impression that was hard to shake:

Almost everywhere where there has been careful monitoring or analysis for more than a few years, the biodiversity is declining, whether it be in wet tropical forest, dry forest, coastal mist forest, or chapparal or pampas. A common finding from many papers was that land-use changes were having a more destructive effect on forests than for instance was pollution or intensive land use. The most destructive practice of all, on a large scale, is the conversion of tropical forest and cotton or coffee plantations to palm oil plantations, in the so called move to reduce the demand on fossil petroleum. Two thirds of the rainfall in Brazil and Argentina is water substance from forest transpiration. With half of the forests of a century ago gone, and half of the remainder expected to go in the next fifteen years, the outlook for long-term agriculture or water supply is grim. There were many calls for protected areas at all latitudes to be designed with the trends of climate change and biodiversity in mind. Protected areas and reserves need to be large – a series of small ones will do nothing, for plants or animals'.

Although the final technical event of the symposium focused on 'building a network on climate change and biodiversity in the Americas', in Roots's words, the immensity of the problem is 'sobering' and 'the outlook is very disturbing'.

We can expect that the world will look back on 2009 as the critical year in the global effort to address climate change. The Secretary General of the United Nations termed 2009 the 'year of climate change'. In his address at the opening plenary of the World Bank Group's Energy Week, with the theme 'Energy, Development and Climate Change', in Washington on 31 March 2009, Yvo de Boer, executive secretary of the United Nations Framework Convention on Climate Change, stated: 'Copenhagen 2009 will be the moment in history in which humanity has the opportunity to rise to the challenge and deal decisively with climate change.'[16] On 23 and 24 March 2009, experts met in Bonn, Germany, to establish 'reference emission levels for deforestation and forest degradation, sustainable forest management practices and the enhancement of forest carbon stocks'.[17] The meetings continue. There is no question that awareness about this issue is growing worldwide. Yet, the much anticipated United Nations Climate Change Conference, held in Copenhagen between December 7th and 18th 2009, was anticlimactic, resulting in very limited progress. We cannot help but wonder whether our world leaders will be able to cooperate and instigate the kind of behavioural changes that are necessary to take us down the road to recovery.

It is abundantly clear that changing human behaviour is not simply a matter of asking people to do things differently. In our past, real changes in behaviour occurred when humans experienced significant environmental stress. Future environmental stress is clearly predicted in our future, so behavioural change may potentially be on the way. The problem is that we must change now before climate changes sufficiently enough for us to feel those stresses. The following section outlines what some of our future stresses are likely to be and what are some of our options for change. It is followed by a section forewarning us about the added difficulties to be faced by complex societies.

9.3.2 Future efforts

As climate change progresses, humans will attempt to migrate to congenial regions as other regions become less habitable. However, unlike in the past, when humans could migrate to new regions relatively unobstructed, in the future with 6.75 billion people on the planet and this number growing to 9 billion, rapid migrations will generate huge sociopolitical stresses. The world is not as it was 10 000, 1000, or even 150 years ago. The global human population has skyrocketed and political borders and economic limitations block the easy movement of people from non-productive to productive habitats. Yet, climate change will make some people's current homes uninhabitable.

If we are going to minimize the impacts of climate change on displaced people and environmental refugees then developed countries need to support climate change action plans in the developing world. Further, the developed world cannot expect the developing world to adjust to climate change by limiting their incomes.[18] Assistance will be required. Around the world people will need a better understanding of responsible behaviours that will assist them and the rest of the world adjust to rapid change. We will need to work together closely in a global community. We will benefit from sharing innovative ideas, behaviours and technologies that are beneficial for human societies and the planet.

If the authors of this book are correct in our understanding of the relationship between rapidly changing environmental circumstances and greater human interaction and sociopolitical stress, then we will see revolutionary change. Whether that revolutionary change is positive or negative for the future of humanity and other species depends on whether we have sufficient foresight to realize we are all in this together. If we unite under a common objective and pool our combined intellectual resources there is far more likelihood that we will develop a beneficial solution for all humanity and the other species on Earth. Alternatively, we can divide and seek to conquer. But given our honed capacity to destroy ourselves with nuclear, chemical and biological weapons, our wisdom will have to be great to overcome the chance of us becoming the next extinct dominant species.

When it comes to behavioural change, as suggested by Dr. Ursula Franklin, not all innovation and technology will necessarily be good. Nor is change for change's sake the answer. Inaction in the case of climate change is inexcusable. But that action must be accompanied by thoughtful and inclusive discussion and behavioural adjustment where possible. We have already begun to develop new green technologies that range from wind farms and geothermal energy to low-cost solar panels. If implemented thoughtfully and equitably and combined with direct support for those who need it we will go a long way in resolving the climate crisis. However, we must also change our behaviour. We must curb human population growth and address poverty. If, when faced with the intense pressures of climate change and economic difficulties we allow ourselves to choose conflict over cooperation we risk adopting aggressive behaviours that will escalate our already precarious situation. We risk developing a siege mentality that will lead to the adoption of innovative aggressive surveillance, tracking, communication and warfare technologies. For instance, we risk choosing surveillance and militarization over human rights. We risk choosing cell-phone tracking systems and remote-control and robotic weapons for borders. The technology already exists to track DNA and modify the weather.[19] If we implement these new aggressive technologies, particularly if we do so without being fully aware of their consequences, we risk everything. We must be very cautious in our development and implementation of technologies that control or dominate nature or individuals or sectors of human society. Many of our choices in the recent past have dominated and controlled nature. Many of these choices have generated the mess in which we currently find ourselves. Repeating our mistakes and continuing past behaviours will not work as we experience rapid future change.

This applies equally to our economic system as to our climate system. Large bailouts made in 2008 and 2009 to the financial and auto industries are extremely risky. If these funds are used to maintain the status quo there is little hope that the necessary adjustments will be made to prepare for and adjust to future change. If we support industries that are reluctant or unable to change in the face of impending major environmental and economic change, we risk losing on two fronts. If those companies fail to use those funds to create change in their organizations or are unable to make changes because they are too fixed in their ways and instead use those funds to continue past behaviours and reward individuals for past behaviour, the funds and a critical opportunity are lost. Rewarding past behaviour simply encourages that behaviour in the future. When circumstances are rapidly changing, encouraging behaviour that resulted in a crisis will simply stimulate a renewed crisis. So not only will we have lost valuable funds we will also have risked losing a critical opportunity to use those funds to stimulate new innovation, research, development and behavioural changes that would have helped us to adjust to future change. When the next crisis hits, where will we find the funds, the time and the incentive to help us to adjust?

Evolution in the traditional sense is not likely to help us out, but social evolution will. Although we can anticipate that climate change may result in stress-induced genomic changes, there are few, if any, examples that have a positive (or 'adaptive') effect. Thus the genetic impact of changing temperatures is likely to be nil in the future. Alternatively, social evolution of humans was the result of our 3 C's syndrome: catastrophe, communication and cooperation. We do and will continue to suffer catastrophes, and hitherto isolated populations might communicate and share their ways of dealing with emergencies. In the modern world, the actions of aid workers frequently far outweigh local responses, where they are permitted. Note the contrast between Myanmar and China. In the wake of a destructive cyclone, Myanmar would not allow the entry of aid workers, and behaved as if totally indifferent to the fate of its people. China, on the other hand, though not completely open about the full impact of the Sechuan earthquakes, admitted aid workers and western journalists. When the response to disaster is positive among the victims and their national government, the system works well. If local governments participate in assisting people in crisis this frequently leads to the implementation of positive 'adaptation' strategies.[20]

But this too is diminished by international politics and the reluctance to provide sufficient aid. In the longer term, the removal of agricultural tariffs that had been applied to the farmers of developing nations would go some way to strike a balance between self-aid and foreign aid. Úrsula Oswald Spring suggests that if developed countries provided relief for the billions of dollars of debt developing nations are required to pay back to developed countries, they would be far more able to deal with local issues of poverty, development and climate change. The external pressure on developing countries to pay back their outstanding debt needs to be reduced in order for them to generate the internal capacity to respond. Further, Spring states that 'poor countries lose 180 billion dollars each year due to the unjust trade system, subsidies and existing non-trade barriers. Debt services and trade barriers are reducing the money urgently required for development, disaster risk management and infrastructure to deal with climate change and risks' (personal communication, March 2009).

For developed countries to protest that subsidizing weaker economies is detrimental to our market system is an illusory argument. Given the degree to which current global governments bailed out indebted and failing financial institutions, it is difficult to use a 'laissez-faire' argument that by giving a free hand to the markets things will naturally sort themselves out. Our national and international financial and political and economic institutions will have to change. It is clear that social hierarchies have not yet evolved to the point of political and economic homeostasis. The global monetary system is not perfect, and ongoing and future imperatives suggest that further change is on the way.

One effect of global warming is a rise in sea levels. At particular risk are small, low-lying islands and coastal environments. Even a very small rise combined with related ENSO events can have a major impact, with cyclones that involve loss of life and the destruction of boats, buildings and infrastructure. These are the kinds of catastrophe that entered into human social evolution in the past. They might force previously isolated surviving populations together in refugia, e.g., the tsunami in Indonesia and points west, where they could exchange techniques such as fishing and building practices. But communication of these variations is already in place, with the internet accessible to teachers, fishers and artisans. Again, overcrowding of potential refugia, nationalistic prejudices and environmental change that diminishes the area of suitable occupation, militate against local solutions to these problems. An effective international coalition of the UN, the International Money Fund, the best intentions of the G20 nations, combined with a global agreement to reduce developed countries' greenhouse gas emissions by 80% over 1990 levels and developing countries' emissions by 15% to 35% by 2050 may be adequate to deal with global warming. To deal with future human migration we will need a global agreement that facilitates the movement of people out of uninhabitable regions and into countries where they can be included as productive, contributing members of society. Cooperation between international governmental bodies could offer a powerful advance in dealing with such problems. This is the kind of development that constitutes social evolution.

This is precisely the kind of agreement that was reached on 2 April 2009 at the London Summit. During the worst economic crisis the world had experienced since the Second World War, world leaders met in London, UK, and unanimously agreed to work together to restore growth, jobs, reject protectionism, deliver on development aid pledges and work together to build a green and sustainable global recovery.[21] It is clear, however, that to enact these decisions will require a new way of doing business.

The United Nations Environment Programme's Global Environment Outlook 4 recently released a study showing the implications of four different economic scenarios based on different global and regional policy approaches and societal choices (United Nations Environment Programme, 2007). In the 'Markets First' scenario, the private sector, with active government support, pursues maximum economic growth and relies on technological fixes for environmental challenges. In the 'Policy First' scenario, the government, with civil and private support, initiates strong environmental and human well-being policies, while still emphasizing economic development. Competition between the private sector and the government in the 'Security First' scenario focuses efforts on maintaining the well-being of the wealthy and powerful in society with limited regard for environmental sustainability. Government, civil society and the private sector work together in the 'Sustainability First' scenario to increase human well-being, equity and environmental protection. In this scenario, it is understood that results will take time.

Economists and politicians from the developed world may find the results surprising. Under 'Markets First', 13% of species go extinct by 2050, compared with 8% under 'Sustainability First'. 'Markets First' generates 560 ppm atmospheric CO_2 by 2050, compared to 475 ppm for 'Sustainability First'. But, interestingly, investment in social and environmental sustainability does not hinder economic development. GDP per capita in nearly all less-developed regions is higher in the 'Sustainability First' and 'Policy First' scenarios than in 'Markets First' and 'Security First'. There can be a greater investment in health, education and the environment under 'Policy First' and 'Sustainability First' 'without sacrificing economic development in most regions'. Inclusion of input from all levels of society in the 'Sustainability First' scenario leads to greater buy-in; it reduces degradation, strengthens local rights and builds capacity and legitimacy.

This example makes it clear that our existing policies and institutional arrangements are inadequate to deal with current environmental problems (United Nations Environment Programme, 2007, p. 459). Innovative, flexible and diversified policy options are required. Potential options include green taxes, using environmental accounting, and shifting the environment to centre stage in decision making in all economic and political sectors. Determined action must begin immediately, as the cost of waiting far exceeds that of immediate action. As individuals, communities, governments, policy makers and business leaders, we must seek new ideas from diverse peoples and organizations and recognize that much greater success will result if instead of competing against one another, we work together towards global sustainability.

In their recent book, Homer-Dixon and Garrison (2009) suggest that the 2008–9 global economic crisis 'will pale beside the longterm consequences to humankind of energy scarcity and climate change' (p. 204). They argue that peak oil and climate change converge into a single problem – a carbon problem. Peak oil production has been reached and the cost of finding and producing more oil is rising. This situation is encouraging countries to utilize more carbon-intensive fuels like coal, and is exacerbating the climate-change problem. On the other hand, green initiatives aimed at mitigating climate change remove the focus on finding new oil and will make the economic impact of limited oil supplies even greater. What the world needs is innovation in particular, we suggest, behavioural innovation.

An enormous opportunity exists. If we, the authors of this book, are correct that much of our past development arose in response to rapidly changing environmental conditions and as a consequence of concentrating populations and the resultant stimulation of revolutionary ideas, then there exists hope for our future. This is because our future will hold just these set of circumstances. In the past, stable conditions encouraged conventional behaviours that brought the greatest benefit to the dominant social group.[22] However, when conditions rapidly change, conventional behaviours are no longer as beneficial. Maintaining conventional behaviours often puts those individuals or groups in a straitjacket restricting their opportunity for change and increasing their risk of extinction. Alternatively, when stressful environmental conditions appear, humans seek options and congregate in the most congenial conditions where environmental effects are minimized and where human interaction is amplified. As a result, human behaviour typically changes, which then alters the conditions that might generate future change. Thus, rapidly changing circumstances (e.g., climate change, economic change, disease) are more likely to stimulate the emergence of bright ideas and revolutionary change than are stable conditions. The development of fire, agriculture and domestication were likely responses that were meant to maintain long-term environmental stability. These developments were followed by specialization, which enhanced our technological and educational capabilities, and the development of fossil fuels, which brought with it central heating and air conditioning, further stabilizing our environment. Today humans can descend into the marine abyss deeper than whales in our submersibles and ascend beyond the atmosphere in spacecraft – without much stress on our internal homeostasis. We are capable of so much, yet we continue to alter the external environment for short-term benefit, often in ways that threaten our long-term survival.

9.3.3 Forewarning: the vulnerability of complex societies

The idea that complex societies are more vulnerable to environmental and climatic disruption was raised by a panel of scientists and humanists who participated in a workshop on Civilization and Rapid Climate Change organized by the University of Calgary's Institute for the Humanities in August 1987 (Dotto, 1988). The relatively stable Holocene environment allowed modern human societies to specialize, expand their trade and communication networks, develop interdependencies and become increasingly complex. This group recognized that in complex societies, decisions are centralized, resources are pooled and tasks are specialized. There is a high degree of interconnectedness and dependency between elements of society.

Today, with increased globalization, local economies have become more dependent on imports of food, energy and technology and have increasingly lost their capacity for self-sufficiency. Further, local problems increasingly stem from distant causes. Think here of the bankruptcy of Wall Street investment bank Lehman Brothers in September 2008 and the subsequent domino effect that generated a global economic crisis. There is also a tendency for those dominant parts of complex societies to resist changing behaviour the more standardized they become. This is because their way of doing business worked in a previously predictable, stable environment. Consequently there is a reluctance to change behaviour that worked well in the past.

However, it is precisely these characteristics that make modern complex societies vulnerable to collapse during times of rapid environmental change.[23] For example, we depend on large uniform agricultural crops (monoculture) and the use of large energy-consumptive technologies such as fertilizers and pesticides to feed our growing population. But as a result we are making ourselves more vulnerable to even small changes in climate. A climate change that destroys one crop in a 12-crop monoculture system has a far greater impact on its consumers and producers than that same change has on a diverse agricultural system with one thousand varieties.

It is also true that societies that are more self-reliant and possess greater cultural, behavioural and intellectual diversity have a better chance for survival under conditions of mass disruption than those which are highly dependent and lack diversity. This is because when life support networks are cut off individuals and communities have a greater chance of fending for themselves. This was highlighted by Dr. Úrsula Oswald Spring at the Climate/Security conference in Copenhagen in 2009 when she noted that in the aftermath of Hurricane Katrina in 2005 in New Orleans, 250 000 illegal Latinos were 'not there' so could not depend on the support system to help them. As a result they self-organized and survived by helping each other. A key factor in their survival was cooperation – with each other; flexibility – to adjust their behaviour; and resilience – the ability to persevere under difficult and perilous conditions. These realities highlight the importance of maintaining and teaching the skills of human survival, e.g., local crop production, tool and shelter manufacture and basic first aid skills. The ability to innovate, think outside the box, adjust to change at a moment's notice and choose cooperation rather than conflict when circumstances become trying are skills that are imperative if we are to successfully adjust to future climate change.

Nature and humankind have an adaptive capacity for change, but this is limited by the degree and intensity of change. How close an individual, society or natural system is relative to their tipping point or their capacity to adjust to change within a limited time is also critical. What influences that tolerance for change?

Our understanding of saltatory evolution supports the idea that dominant specialized species and societies have greater difficulty adjusting to rapid change than generalist nimble species that must live by their wits. These species have lived more on the periphery relative to specialized dominant groups and species must constantly respond to change in order to survive. Modern humans – *Homo sapiens* are no different and we are currently the dominant species. Within the overarching *H. sapiens* family there are individuals and groups that are more flexible and open to novelty and diversity than others. This is generally because they have had to be just to survive. Our capacity to adjust to future climate change will be better in those groups and societies that are open and willing to accept difference and change. Further, impending climate change can and will stimulate adjustments in both development and behaviour, but the degree of acceptable change depends on how sensitive the organism, group or society is to change and how well the change is recognized and understood. If change is not recognized or understood, or if the change required is too great for the individual, organism or society to manage, or alternatively if they are not willing to adjust, then decline and even extinction prevails. As such, humans are not invulnerable to extinction.

Thus, understanding our past evolutionary relationship with climate allows us to better prepare for our future. It is our hope that this book helps its readers understand these relationships and the importance of its messages. Economic and environmental stress will result in behavioural change. Thus, change is on the way. Those who are open to change and diversity and start sooner will fare better than those who are not. Further, those who choose cooperation over confrontation will have a greater chance of adjusting and avoiding the negative consequences of war, famine and poverty. The risk of confrontation is extreme – the escalation of nuclear warfare could spell the end of *Homo sapiens* and many other species as well. The risk of cooperation is far less extreme – diversity will have to be embraced, those with more will likely have to consume less and share more. We will have to work together. Yet there is much to be gained in an evolutionary sense by participating in this era of human evolution. Perhaps the greatest of which are future generations.

Imperative in humanity's successful adjustment to impending rapid climate change will be a capacity for natural and social scientists, writers, artists and musicians to communicate complex information and ideas to the public. This will nourish the public's thirst for information about the frightening prospect of future climate change and provide a foundation on which we can build future individual and group behavioural change and from which can stem new government policy, political action hope and change. A drastic shift is needed in our behaviour to generate an immediate and global reduction in greenhouse gas emissions. If we seek solutions to a common goal we are far more likely to stimulate an advancement in human intelligence that will not only reduce our fossil fuel consumptive behaviour and alter the impact of present and future climate change, but will also reduce the extent and impacts of future climate and environmental change on *H. sapiens* and all other species on the planet on which we depend for our survival.

Notes

1. See Brahic (2007).
2. Information from comments made by Dr. Lisa Levin, Scripps Institution of Oceanography, University of California San Diego, during her presentation 'Why Are Dead Zones in the Sea Proliferating?' (presented at the forum 'Are We Killing the World's Oceans?' University of Victoria, 21 and 22 February 2007).
3. Comments about water-use rates and aquifer depletion were made by roundtable participant Úrsula Oswald Spring, Mexico City at the Climate/Security conference on 9 March 2009 at the University of Copenhagen. Úrsula Oswald Spring is a professor and researcher at the Regional Centre of Multidisciplinary Research at the National University of Mexico (CRIM-UNAM).
4. Much of the scientific data presented in this section has been obtained from three main sources: Intergovernmental Panel on Climate Change (IPCC) (2007a and 2007b) and Warren (2006).
5. Nigel Arnell and Chunzhen Liu, 'Hydrology and Water Resources,' Chapter 4 in IPCC (2001, p. 213), www.grida.no/climate/ipcc_tar/wg2/180.htm.
6. Although the most recent IPCC 2007 report estimates sea levels will rise at least seven metres over the next several centuries with the melting of the Greenland ice sheet and another six metres with the melting of the West Antarctic ice sheet, James Hansen, head of the climate science programme at the National Aeronautics and Space Administration's Goddard Institute for Space Studies in New York, suggests those estimates are conservative. He believes sea level will rise far more quickly, possibly by as much as twenty-five metres by 2100. See Hansen 2007.
7. See for example Hughes (2009).
8. For a complete discussion of epigenesis and human developmental evolution see Appendix B – Developmental evolution.
9. For a complete discussion of hopeful monsters see Appendix B – Developmental evolution.
10. For a complete discussion of human adaptability and homeostasis see Appendix C – Human adaptability: the physiological foundation.
11. For a further discussion on human adaptability see Appendix C – Human adaptability: the physiological foundation.
12. Statements made by Dr. Anwar A. Abdullah, Senior Advisor to the Prime Minister on Sustainable Development, Kurdistan Regional Government, at the Climate/Security conference on 9 March 2009 at the University of Copenhagen.

13. These rights are set out in the Universal Declaration of Human Rights (1948) and in the International Bill of Human Rights (1978), which includes the Universal Declaration of Human Rights, the International Covenant on Economic, Social and Cultural Rights, and the Optional Protocol to the International Covenant on Civil and Political Rights. Texts of all documents are available at www2.ohchr.org/english/law/.

14. The text of the 1998 Kyoto Protocol to the United Nations Framework Convention on Climate Change is available at www.unfccc.int/resource/docs/convkp/kpeng.pdf.

15. Comments about this symposium are based on personal communications and notes made by Fred Roots. Also see Fenech *et al.* (2008), a book based on this symposium.

16. See Yvo de Boer's statement to the opening plenary of World Bank Group Energy Week, 'Energy, Development and Climate Change', 31 March–2 April 2009 at unfccc.int/files/press/news_room/statements/application/pdf/090331_speech_energy_wb.pdf.

17. See Outcomes of Expert meeting on reference emission levels at: http://unfccc.int/methods_and_science/lulucf/items/4799.php.

18. Based on comments made by Sir Nicholas Stern on 12 March 2009 at the Climate Change: Global Risks, Challenges and Decisions Congress in Copenhagen, 10–12 March 2009.

19. Based on a presentation by Steve Wright, Leeds Metropolitan, UK, at the Climate/Security Conference on 9 March 2009 at the University of Copenhagen.

20. Based on comments by roundtable participant Úrsula Oswald Spring, Mexico City. Úrsula Oswald Spring is a professor and researcher at the Regional Centre of Multidisciplinary Research at the National University of Mexico (CRIM-UNAM).

21. See 'Global plan for recovery and reform: the Communiqué from the London Summit' at londonsummit-stage.londonsummit.gov.uk/en/summit-aims/summit-communique/.

22. For a more extensive discussion see Appendix C – Human adaptability: the physiological foundation.

23. For an in-depth look at what caused the collapse of early civilizations see Diamond (2005).

Appendices: The biological background to the story of evolution

Homo sapiens is the main evolutionary player in the text of this book. To biological evolutionists, however, evolutionary change in humans is minor. Adaptational features such as gracile bodies, alterations in sweat patterns and dark skin are connected to living and working in hot, sunny conditions. A tendency to lay down thick adipose tissues, and a stocky stature, are characteristic of humans living in Arctic conditions and the high plateaux of the Andes and Himalayas. These features, as well as those associated with humans living in less stressful conditions, are also correlated with human physiological, developmental, behavioural and social variations, as we have demonstrated in the main text.

Although all of the 'serious' evolution of living organisms (usually seen as major anatomical changes) occurred before the emergence of *Homo sapiens*, the appearance of the new species is itself of great interest. Moreover, to return to adaptational changes in our species, it is important to note that they reinforce physiological aspects of homeostasis, the dynamic state of equilibrium that evolved over 500 000 years, providing major improvements in adaptability. Therefore it is vital to understand the evolution of adaptability, as well as to know what humans have made of it, the latter being strongly featured in the body of this work. Anthropological and psychological treatments of human evolution usually neglect the important aspects of evolution in general, or simply refer in passing to natural selection as the one true cause of evolution. Our Appendices therefore provide the big evolutionary picture.

Appendix A: Evolutionary theory

A.1 Aspects of evolutionary theory

A.1.1 Lamarckism

Before the end of the seventeenth century, little thought had been given to the possibility of evolution. Then, in eighteenth-century France, several biologist–philosophers began to revive Greek ideas about the origin of life, speculating about the phenomenon of adaptation, and the possibility that species might be mutable. Georges-Luis Leclerc de Buffon was the first to grasp the logic that if species were mutable there was no reason why they might not all have evolved from a simple, common ancestor.[1] This influenced two important characters in the drama of evolutionary theory: Charles Darwin's grandfather Erasmus Darwin, and Jean-Baptiste Lamarck. Erasmus Darwin's 1794 book *Zoonomia* had a large chapter on evolution, and anticipated the ideas of the late nineteenth-century neo-Lamarckists. Lamarck found different inspiration in Buffonian evolution, publishing his thoughts in the 1809 book *Philosophie zoologique*, and in his subsequent books on invertebrate zoology.

Lamarck's most general precept was that evolution is a gradual, continuous gradation of organisms from the simple to the complex. This was the result of a progressive, inherent drive in all organisms. As organisms became more complex, they underwent adaptation to their way of life through the inheritance of acquired characteristics. Interestingly, Lamarck thought that adaptation could be in conflict with progressive complexification and slow the latter down.

The 'four laws' of Lamarckism are[2]:

1. There is a general trend for an increase in size in evolutionary lineages. (This is often true, but not invariable. This law prompted little discussion or criticism.)
2. Organisms respond to 'needs' by evolving novelties of behaviour, physiology and anatomy. (This is equivalent to the neo-Darwinist idea of selection pressure producing

evolutionary change. The difference is that Lamarck thought the organism was striving to change; Darwinists think that biological change is random and non-universal.)

3. Use and disuse. Organs that are used more get bigger; organs that are used less get smaller. (Darwin adopted this for his own theory, and it is not a debatable issue among modern evolutionists.)

4. The inheritance of acquired characteristics. (Darwin finally adopted this too, but his descendants rejected it.)

At this point, it might very well be asked: if there is to be a historical introduction to evolutionary theory, why not start with Darwin, and ignore earlier misconceptions; even modern evolutionary theory is Darwinian, is it not? The reasons why not are as follows.

1. Lamarck put progressive evolutionary complexification at the centre of his theory; Darwin acknowledged that it was problematical, but did not know how to deal with it, and neo-Darwinists treat it as an accident of adaptation and natural selection. So they have no constructive explanation of how the human condition evolved.

2. Lamarck made the organism an agent of evolution. Although Darwin understood that the organism was important, natural selection was paramount. One of the most fundamental aspects of all evolution has been the action and interaction of individual organisms. Yet, neo-Darwinists ignore the organism, defining evolution as the redistribution of alleles in population-wide gene pools. The organism is no more than a temporary agglomeration of genes. In contrast, some modern biologists and philosophers take the view that the whole organism in its environment must be taken into account, as well as its interacting hierarchical components, from molecules to mind. For example, a change of climate might make you change your mind about how best to live, and turn you from a wandering nomad to a cave-dweller. Also, environmental stimuli such as light and sound affect the nervous system, which affects the hormonal system, which can affect the DNA–RNA complexes that contribute to protein synthesis.

3. The inheritance of acquired characteristics was a universal idea in Lamarck's time and it should not be taken to exclusively characterize Lamarckism. Even now, a substantial number of educated members of the population of North America, including university professors, regard it as true.[3] One way in which the idea can be taken literally is in regard to human culture. What is discovered by an individual is passed from one person to the next and learned by a generation. It is then passed down through the generations by education. In that sense, intellectual acquisitions become heritable.

4. For better or for worse, the idea of the inheritance of acquired characteristics, largely abandoned for over a century, is now finding favour with molecular biologists. They observe that patterns of methylation and histone-binding that repress DNA can be imposed and changed by environmental factors, and although there is no change in the genes themselves, these patterns are heritable. What is changed is gene expression during development; i.e., the *epigenetic* process. These will be discussed in Appendix B: Developmental evolution.

In view of the aforementioned four points, it might be argued that Lamarck had a better intuitive grasp of the nature of evolution than Darwin, in addition to a chronological priority of fifty years over Darwin's *The Origin of Species* (1859).

A.1.2 Darwin's theory

Darwin's advances over Lamarck were twofold. The first is in the sound structure of his historical theory. It is important to realize that any evolutionary theory consists of two independent theories; the historical and the causal. The former is the portion that deals with the evidence for the reality of evolution as a historical phenomenon. Here, Darwin presented facts pertaining to changes wrought by farmers in their crops and herds by selective breeding. He wrote that the fossil record indicated evolutionary change: simple, primitive organisms were in the oldest rocks, and the higher vertebrates, such as the mammals, were in the youngest deposits. Darwin also interpreted biogeography as evidence that related groups evolved in a particular locale, e.g., the great apes in Africa, and most marsupials, such as kangaroos, in Australia. Embryological and homological similarities were also important evidence. For example, the underlying anatomy of the hand of a human and the forelimb of a whale is similar; and in early embryonic development they are almost identical. But because the subsequent developmental pathways diverge, the finished products look dissimilar. Darwin saw that these 'homologies' demonstrated an evolutionary relationship. He then drew a parallel between the results of 'artificial' selective breeding, and the process he called 'natural selection', which was to be the heart of his mechanistic or causal theory.

The second perceived advance over Lamarck was that Darwin did not make the mistake of giving every organism an inherent trend to evolve progressively to more complex states. For Darwin the process of variation was random, and nature selected only the variations that were fit, in other words, best adapted to their environments. This is the basis of his causal theory.

A.1.3 Causal theory

Lamarck's causal theory is usually treated as the inheritance of acquired characteristics, and left at that. However, if we make a synopsis of all of the Lamarckian causes of evolution we must add an inherent drive to increase in complexity and size; and the individual organism's ability to respond to new contingencies by anatomical, physiological and behavioural change. As a consequence, proportionate size changes occurred, i.e., as a result of use and disuse. For example, the length of the legs of wading birds increased, compared to their ancestors that foraged on the ground.

Darwin's fundamental causal mechanism was natural selection or survival of the fittest. However, since he had no idea of what caused the random variations that were the material to be selected, he finally adopted the Lamarckian principles of use and disuse, the inheritance of acquired characteristics and the neo-Lamarckist notion that the environment could impose change on the organism. So, he was careful to write, in the final edition of *Origin of Species*, that natural selection was the most important, but not the exclusive mechanistic process of evolution. Since he had no theory of how variations that were selected originated, a major component of a mechanistic theory – namely a generative hypothesis – was missing. Waiting respectfully until Darwin had died, the self-proclaimed 'neo-Darwinists' asserted that natural selection alone was the all-sufficient mechanism of evolution, and that evolutionary theory needed cleansing from Darwinian dithering. They also claimed that a generative hypothesis was unnecessary. It did not matter how variations originated, as long as natural selection was able to pick the good ones.

Because of the logical redundancy of natural selection defined as 'survival of the fittest', modern neo-Darwinism redefined it as 'differential survival and reproduction'. Affecting not to notice that this seemed more an effect than a cause, neo-Darwinists then redefined evolution as 'changes in the distribution of alleles in populations'.[4] More recently the changes in the alleles themselves, in the form of DNA mutations, were tacitly added to the definition. However, whether by the original or modified definition, tacit assumptions and all, what most people think of as evolution, i.e., a process of complexification from simple bacteria to complex humans – i.e., progressive evolution – is absent from neo-Darwinism. It is present in Lamarckism. But Lamarck made such a fundamental error in stating its nature that he poisoned the branch of progressive evolution to an extent where neo-Darwinism, even now, feels safe in ignoring it.

A.1.4 Neo-Lamarckism

Despite Lamarck's errors, neo-Lamarckism flourished for the later part of the nineteenth century, and well into the twentieth. This was for a number of reasons. The neo-Lamarckist idea that the environment imposes change directly on the organism is correct. What is at issue is whether or not that change is evolutionary. Secondly, neither natural selection nor the inheritance of acquired characteristics had been empirically tested in the laboratory or the field. Not only was the matter wide open to dispute, Darwin himself had espoused some of the neo-Lamarckist principles.

According to its dean, A. S. Packard, neo-Lamarckism consisted of Lamarck's four laws, plus the direct effect of the environment on the organism, plus natural selection. Like Erasmus Darwin and Geoffroy before them, the neo-Lamarckists felt that the environment might have negative influences, and they adduced natural selection as an arbiter of failure and success. Neo-Lamarckism dominated evolutionary theory in the United States in the early twentieth century until gene theory was adequately developed. In Europe its demise may be marked by the suicide of Paul Kammerer, who was an accomplished neo-Lamarckist at the Vienna Institute of Experimental Biology. He had been accused of faking his experimental results, though somebody other than Kammerer was probably the culprit.[5] Nevertheless, the belief that change imposed by the environment in organisms might become heritable persisted to the present.

A.1.5 Neo-Darwinism and the modern synthesis

Neo-Darwinism began with August Weismann's 1893 declaration that natural selection was 'all-sufficient'. That has been its Ariadne's thread ever since. However, neo-Darwinism is now identified with several theoretical population geneticists: Ronald Fisher, and J. B. S. Haldane in the United Kingdom, Sewall Wright in the United States, and, in the Soviet Union, Sergei Chetverikov, whose ideas were popularized in the United States by his student Theodosius Dobzhansky. There were two movements during the 1940s towards a grander synthesis of evolutionary theory. Julian Huxley's, as expressed in his 1942 book *Evolution: The Modern Synthesis,* had chronological priority as well as being more comprehensive than rival books in the United States. The American version attempted to amalgamate ecology, genetics and population biology with palaeontology. It did not embrace physiological and developmental evolution, unlike Huxley's modern synthesis.[6] For half a century, the dominant American synthesis was stripped down to population biology as it putatively related to evolution. Also, it was responsible for the dubious redefinitions of key evolutionary terms mentioned above. Only in the last two decades has the dominant American 'evolutionary synthesis' reopened its doors to palaeontology. More recently, and with considerable reluctance, it has issued an invitation to developmental evolutionists – the 'evo-devos'. Physiology remains out in the cold. Behavioral biology or ethology is also excluded. However 'evolutionary psychology', which has infiltrated both ethology and anthropology with a programme of strict selectionist interpretations of all aspects of human behaviour, has written its own membership card for the modern synthesis. It is important to realize that we are here considering matters that are more political than scientific. Unfortunately, because the combatants use scientific terminology, onlookers are often deceived.

A generative theory is not provided by the modern synthesis. Yet, evolutionary biology still requires one. It should explain the causes of change, and understand that differential survival and reproduction are inherent in such changes, and that selection theory is therefore irrelevant to evolution. Instead it is relevant to the dynamic equilibria that have dominated geological time, since life first evolved three and a half billion years ago. Biological innovations may evolve at any time, given appropriate generative conditions. But if robust dynamic equilibria already exist they present an obstacle to evolution by resisting change. Nevertheless, since dynamic equilibria have dominated the history of life, their consideration must be part of a larger general theory, or synthesis of biology.

Within the modern synthesis there are some that would agree with this point of view, to some extent. For example, the idea of 'punctuated equilibrium' proposes that there are episodes of fairly rapid evolutionary change (punctuations) followed by long phases of stasis (equilibria). The progenitors of punctuated equilibrium, Niles Eldredge and Stephen Jay Gould (1972), came around to the view that while the punctuations might have happened quite quickly, on a timescale of tens of thousands of years, the mechanisms were fundamentally the same as those proposed by neo-Darwinism; natural selection acting on random variations throughout the periods of change and stasis. We take a stronger position: evolutionary change is rapid on a biological timescale. For example, a genetic mutation takes only micro-seconds, and may change a species' characteristic in a few generations. This is relevant to human evolution if we consider that environmental change can be so rapid that the climatic conditions of life can change drastically over a human lifetime. Evolutionary theory should concentrate on how innovations come into being, and leave selection theory to deal with demographics and dynamic equilibria in ecosystems.

A.1.6 Saltatory or gradual evolution

Both Lamarck and Darwin believed that evolution occurred gradually on a geolo-gical timescale. Slow, gradual evolution was theoretically necessary to validate Darwin's causal mechanism of natural selection. As an extreme illustration, suppose a hairy baby mammal were to hatch from the egg of a scaly lizard. There would be no role for natural selection as the cause of this event. Instead, Darwin and the neo-Darwinists believed that the process of the evolution of mammals from reptile ancestors consisted of multiple small steps, each of which had been approved by natural selection, retained as a new species characteristic, and subsequently built upon by further small, but fit variations. However, a perennial criticism of Darwinian gradualism has been that the fossil record shows little evidence of continuity of intermediate forms between the highly diverse organisms we find living at the present time.

Because of apparent discontinuities in the fossil record, many palaeontologists, such as Darwin's strong supporter T. H. Huxley, believed that nature does make leaps, and these leaps are usually referred to as 'saltations'. The original 'mutation theory' of Hugo de Vries[7] proposed that a new species might come into existence in a single step. Affiliated with such saltatory theories is geneticist Richard Goldschmidt's concept of the 'hopeful monster'.[8] Monstrosity implied a radical difference of the offspring from the parental type. The hope lay in the new type being able to find an appropriate mate for it to breed true. It also hoped that it might outcompete the parental population through novel advantages, or alternatively find a new environment appropriate to its qualities. There is no valid logical reason to discard the hopeful monster concept in its entirety. Saltatory embryological 'errors' often produce 'monsters' that survive, and breed true. Giantism and dwarfism are common amongst humans, and not so disadvantageous as to prevent reproduction. Pygmyism is an even more viable condition in humans, and pygmies regard the rest of us as hopeful monsters. Processes and examples of saltatory evolution involving life-cycle changes will be discussed further in Appendix B (Developmental evolution).

The neo-Darwinist argument against the emergence of hopeful monsters through saltatory developmental changes is logical enough: there would be too many uncoordinated, dysfunctional consequences, and so the monsters would be hope-less. But there are too many examples of successful saltations for this to be generally valid. They are especially obvious in species studied by plant and animal breeders. As to the advantage of developmental saltations, it has been historically recognized that they can help a lineage escape the rigidity of the highly adapted or over-specialized mature form, which is locked into dynamic ecological stasis.

Another kind of major saltation is the all-or-nothing emergence of endosymbio-sis, a condition in which a microorganism enters another cell – either another microorganism, or the cell of a many-celled organism. The mutual advantages that are brought to the symbiosis create a whole greater than the sum of its parts. Such endosymbioses probably happened frequently among primitive microorgan-isms, culminating in the creation of the eukaryotic cell. That is the kind of cell of which we are composed, and typically has a nucleus with chromosomes, and various organelles such as mitochondria, derived from other microorganisms that once lived independently.[9] Once eukaryotic cells began to associate with each other in multicellular organisms, there was an increase in 'evolvability'; the potential to give rise to new and more complex interactions.

In conclusion, the evidence for saltatory change is so strong that to ignore it for the sake of bolstering an inadequate mechanistic theory of natural selection is intellectually bankrupt. A more holistic treatment would recognize that both gradual and saltatory changes occurred during evolutionary history, and that saltatory emergences were responsible for the more significant punctuations and subsequent diversifications.

If we are to give the matter of saltatory change a broader context, we must also consider sudden environmental changes. Some of the big ones in evolutionary history include the oxygenation of the seas, lakes and atmosphere, the climatic effects of plant evolution, the consequences of bolide impacts on the Earth's surface, of geological catastrophes, of rapid temperature oscillations, and finally, the influence of a certain species of hominins.

The emergence of photosynthesis has had an important effect, since it resulted finally in the universal distribution of oxygen, which in the early eras of prokaryote evolution was toxic to such organisms. Photosynthesis did not emerge in a single step: several endosymbiotic saltations brought photosynthesizing microorganisms into a state of biochemical complementarity until the condition of the familiar eukaryotic plant cell was reached. The build-up of oxygen was gradual, but finally reached a stage where it was almost universally distributed in water and air. Only organisms capable of detoxifying oxygen, and especially those that could convert the detoxification pathway to an energy-producing function, could succeed in the oxygen zone. The symbiotic eukaryotic cell became particularly adept at this.

At that point an upper-atmosphere ozone layer emerged. It reduced the intensity of ultraviolet radiation reaching the Earth's surface. There was also a build-up of atmospheric carbon dioxide, both from volcanic sources, and from plant and animal respiration. Its greenhouse effect warmed the Earth's surface, and increased evaporation of surface waters, and subsequent precipitation. To assess the climatic importance of oxygen, ozone and carbon dioxide, we can examine what happened in their absence. The 'snowball Earth' theory proposes a series of intense glaciations that froze all surface water during the late pre-Cambrian. The effect on life and its evolution at the surface of the oceans, except perhaps in a narrow equatorial zone, is obvious. Periodically, large volcanic eruptions, spontaneous, or perhaps triggered by bolide impacts, emitted enough heat and carbon dioxide to create a temporary greenhouse effect and melt the ice. Life had emerged prior to these events, but did not undergo major diversification until after temperatures stabilized at relatively high temperature. The invasion of land by plants had another climatic impact; namely the increase in surface temperature by a darkened albedo.

There can be no more sudden climatic change than that induced by the impact of bolides – meteorites, comets and asteroids. The high-energy radiation, heat and pressure, kill everything at the point of impact. In the splash-zone of vaporized and molten rock and solid debris there is also total fatality. Surface waters of lakes and seas are acidified. Smoke and dust clouds then cause an overall temperature decrease. Bolide impacts are also believed to trigger large-scale volcanic disruptions that may in the longer term have the greater climatic influence. Five geological epochs were terminated by major climate catastrophes. In the geological stratification record they are indicated by major biological changes. Large numbers of previously common species became extinct and new or previously rare ones became more widely distributed.

Some biologists suggest that novel types might have emerged as a result of heat stress produced by bolide impacts. However, in the most important examples, the type that diversified after the catastrophe had already existed. For example, flowering plants and mammals had evolved by the Cretaceous, but after an initial setback, underwent large-scale diversification after they were liberated by the sudden climate change from the dynamic stasis dominated by cycads, ferns and reptiles. And the placental mammals existed long before an Eocene catastrophe that gave them the freedom to diversify.

The question then arises, what quality allowed those that survived and diversified to do so? The simple answers are adaptability or luck.

A.1.7 Adaptation and adaptability

In the arena of human action, we commonly use the word 'adaptation' to mean behavioural change. We 'adapt to the cold' by putting on more clothes and lighting fires. However, the word *adaptation* can apply to two distinct phenomena. The first is the Darwinian natural selection of genetic change appropriate to particular conditions. For example, 'adaptation to the cold' in such a case might refer to hereditary changes in subcutaneous adipose tissue, hair structure and general anatomy. The second is adaptability: the physiological and behavioural self-modification made by individuals in the face of changing conditions – donning clothes and lighting fires when the temperature drops. However, common usage does not often distinguish between adaptation and adaptability. This lack of discrimination carries over into the literature of biology and anthropology, so that the two distinct categories tend to remain confused. The result is that adaptability is falsely assumed to be a subset of Darwinian adaptation. Consequently, in biological writing, the evolution and importance of adaptability are underplayed. In some anthropological writing, the importance of physiological, behavioural and cognitive adaptability is recognized. However, it is often assumed to have evolved by an accumulation of Darwinian adaptations. There are no sound grounds for that assumption. Yet, its application by evolutionary psychologists to anthropology further assumes that contemporary humans are adapted in the Darwinian sense to the particular environmental conditions of the Pleistocene.[10]

Darwin himself excluded adaptability from the category of adaptation, referring to it as an 'innate flexibility of constitution' that had *not* been the product of natural selection.[11] Yet the progressive improvement of that innate flexibility is what has made organisms in the vertebrate lineage more *evolvable*, in the sense of potentiating the more visible diversifications of functional anatomy. For example, the emergence of a placenta rounded off a series of advances in physiological adaptability that allowed the placental mammals to diversify into tigers, horses, bears, bats, whales and humans. The life-long ability to maintain a dynamic internal homeostasis under changing conditions, coupled with exploratory behaviour, permitted placental mammals to penetrate cold, hot, wet and dry environments. At this juncture, we make the more general theoretical assertion that a generative theory of evolution would in large part be a theory of adaptability. We will also return to it in Appendix C (Human adaptability).

Our fundamental questions are: to what extent did climatic influences produce Darwinian adaptation in hominids? And, to what extent was cognitive and social adaptability the emergent result of climatic conditions? We have responded to these questions in Chapter 2 Section 2 (The emergence of anatomically 'modern' humans).

A.2 Emergence theory

Throughout this book we have been using expressions such as 'the emergence of the eukaryotic cell, flowering plants, placental mammals, modern humans' etc. In this appendix we have proposed that a generative theory of evolutionary change is necessary and that in large part it would be a theory of adaptability: adaptability of physiology and behaviour increase with greater size and complexity of structure, cell differentiation and self-organization. An information theorist would boil it down to 'increase in information'. This adaptability is not a correlate of any particular habitat or way of life, but has a general blanket utility. It has tended to become more sophisticated in some lines of evolution, especially in the vertebrates, which is the only one that developed the adaptabilities of complex homeostasis and cognition. It has failed to make progress in some lineages, and it has sometimes regressed. Yet, when it has progressed, it is improved adaptability that has made it possible for organisms to invade new environments and to diversify in the relative absence of competition and predation. The emergence of modern humans is the best exemplar of the principle. Our adaptabilities allow us to change our environments: to descend into the marine abyss, live in deserts, climb the highest mountains, fly, launch ourselves into outer space … And they might yet result in the regression of the human condition.

To the theoretician, the big question is how did adaptability evolve? The problem of progressive evolution of adaptability is central to Julian Huxley's *Evolution: The Modern Synthesis* (1942), but the reluctance of his co-evolutionists to deal with that central issue sidelined his best efforts. Progressive evolution is also the subject of *Biological Emergences: Evolution by Natural Experiment* (Reid, 2007). It puts the evolution of adaptability into the larger biological context of an 'emergence theory'.

In any sense, emergence suggests the sudden appearance of something new. In the case of biological emergences, for example, a mutation occurs in a few microseconds; its effect in terms of protein synthesis takes milliseconds; physiological consequences take seconds; developmental changes result in hours, days or months. There are no emergences on a geological timescale, nor even on the thousand-year scale of punctuated equilibria. However, it might take a very long time for the appropriate generative conditions to be built up. For example, when the first green cells emerged and began to photosynthesize, the build-up of oxygen to significant levels in the sea was on a geological timescale. Further emergences, dependent on free oxygen as a generative condition, had a long wait.

There are two kinds of emergence; the most striking is the saltatory kind, sometimes referred to as 'strong emergence' by philosophers. Symbiosis is a saltatory emergence. When the independent prokaryotes that represented the protomitochondria, for example, entered their host cells, it was an all-or-nothing event, with immediate beneficial consequences arising from the biochemical complementarity of the previously independent cell types. The other kind may work more slowly to a threshhold and then emerge: 'critical-point emergence' sometimes referred to as 'weak emergence'. As an example, take the evolution of flight. A gradual amplification of proto-wing size and muscular strength might see a potential bird lineage through a period of gliding and soaring. But there will be a critical point at which lift exceeds drag and true flight emerges.

What kind of emergences might we be dealing with in human evolution? A mutation that resulted in a change in the efficiency of mandibular muscles might have had a multiplicity of phenotypic effects. (Multifunctionality is a common feature of major emergences.) A change in the migration of neural crest cells could certainly have effected a single-generation change in skull anatomy. Whether by mutation or DNA binding-pattern changes, alterations in gene expression, again all-or-nothing events were responsible for changes in everything from skin colour, hair texture, skeletal morphology, fat deposition, lung size, haemoglobin titre, the size and plasticity of the birth canal, the onset of cranial bone fusion, shifts in the proportions of regions of the brain, new potentials for dendritic connections of brain cells, expansions of consciousness and cognition – the list might be endless.

The evolutionist wants to know the particular mechanisms of these processes and also wants to know how they were triggered. What, in other words, were the generative conditions of evolutionary emergences? For the evolution of symbioses and societies it is self-evident that the potential associates have to be in the same place at the same time. For physiological evolution, stress conditions at the fringes of environments provide for the generation of emergences. This presumes that the organisms are able to survive the stress in the first place, i.e., that they are robust, or physiologically and behaviourally adaptable. These are the kinds of organisms that are also able to penetrate new environments and take advantage of their resources. In new environments there may also be a different range of factors that affect organismal development, physiology and behaviour. Therefore, in seeking an explanation for a particular aspect of evolution we have to cast a wide net because of the interactions of many causal elements.

For the particular purpose of the present work, we want to know what adaptations in the strict sense, i.e., genetically fixed traits, might have resulted from climatic change, what aspects of human adaptability might have exposed humans to such change, and what developmental, physiological, behavioural, cognitive and social interactions were involved. With answers to these questions we will be in a better position to assess what the future holds. The following section deals with the development of the embryo and maturation of the juvenile. Many of the most obvious aspects of evolution are anatomical and these are largely brought about by changes in development. With the emergence of molecular biology, developmental studies are amongst the most rapidly advancing areas of evolutionary research.

A.3 Contrasts between the selectionist and emergentist views of evolution

Julian Huxley, one of the founders of the modern (selectionist) synthesis was dubious about the universality of natural selection as a universal cause of evolution. Nevertheless, he concluded that human beings could not evolve further, since they had removed the influence of natural selection. Medical treatment and social support of the 'genetically inferior', guarantees 'survival of the unfittest'. This is still part of the selectionist canon, and some extremists, such as social Darwinists and eugenicists, have urged reproductive laws that would allow only the mentally and physically fit to have children, i.e., a state-approved 'natural' selection.

The emergentist view is that evolution in the form of (usually) heritable change can occur anywhere at any time. Moreover, the successful survival and reproduction of novel organisms is best assured in the *absence* of the agents of natural selection, such as competition and predation. And *that* is what the more enlightened human societies guarantee. Therefore, continued human evolution and the success of its creations are theoretically possible.

Moreover, despite current religious and ethical objections to interference in human biology, it is technically possible to modify ova and embryos in such a way as to alter their phenotypes, and, in a sense, produce 'designer' humans and clones – which would also have to be protected from natural selection, in the sense that they might lack robustness. This may be the wave of the future, instead of random evolution correlated with the exposure of natural populations to unusually stressful environmental conditions. However, this is more of a philosophical exercise than a means of forecasting the future of our species.

The notion of designing new types of humans incorporates a thorough knowledge of gene action – not simply the molecular structures provided by the human genome project, but also the effects of changes in the molecular regulatory mechanisms of development. Beyond these molecular processes, there are also physiological, behavioural and environmental affects on development. All have to be known before directing human evolution becomes a reliable reality. We also have to learn for ourselves as much as is known of such factors if we are to make biological judgements that pertain to the moral issues. Furthermore, for the purposes of understanding the future, whether through random or human-directed changes, we also have to understand the past. In the next section of the appendix we will undertake to provide as much general information on developmental evolution as is relevant to these requirements.

Notes

1. Buffon as translated by Samuel Butler (1879).
2. We have paraphrased various translations. The 'four laws' are drawn from several of Lamarck's publications.
3. This is based upon informal surveys conducted by R. G. B. Reid.
4. The definition is usually attributed to Theodosius Dobzhansky.
5. See Koestler (1971).
6. See Betty Smocovitis (1996) for a comparison of the various versions of the synthesis, and Schwartz (1999) and Reid (2007) for a further analysis.
7. de Vries, H. (1901–3) and (1909–10).
8. Goldschmidt (1933). A proper reference is not available since the term appeared in his lecture at the Chicago World Fair. See Goldschmidt (1940) and Goldschmidt's (1960) autobiography.
9. See *Acquiring Genomes* by Margulis and Sagan (2002), and numerous earlier works by Margulis as the sole author.
10. See Pinker (1997) and Cosmides and Tooby (1996).
11. Darwin C. (1859).

Appendix B: Developmental evolution

B.1 Introduction

At the close of Appendix A we emphasized that during the course of evolution, changes in organismal development from a fertilized egg to an adult result in altered anatomical features. Some changes during the development of an individual organism result from changes in behaviour, for example exercise alters the size and strength of muscles and can also affect skeletal structure. Although it is not widely believed that such changes are passed on to later generations, we cannot rule out the possibility altogether. Also, it is almost axiomatic that, as Aristotle said, 'Nature is true to type.' Cats beget cats, dogs beget dogs and humans beget humans. To explain this, some modern biologists use the metaphor of a computer programme, or algorithm, which is responsible for such consistency in development. In any case, during the course of evolution, consistency of development was periodically interrupted, and developmental pathways were altered. Otherwise, the most complex organisms would still be microscopic single cells, or small aggregates of several identical cells. Therefore, to understand how humans emerged from ape-like ancestors we must be as familiar as possible with normal developmental processes, and all of the factors that cause them to change, from mutations of DNA to environmental influences.

B.2 Epigenesis and epigenetics

Even prior to Darwin, it was clear to some evolutionists that the diversity of living forms had to have resulted largely from modifications that occurred during embryonic and juvenile development, a process known as 'epigenesis'.[1] It was also obvious to them that the earlier a change occurred in epigenesis the greater would be its impact on the anatomy of the mature organism. However, Darwin played down the importance of developmental evolution because it implied 'saltations', i.e., large-scale alterations that happened suddenly, as opposed to slow changes through the constant and gradual application of natural selection. Because Darwin relied on the mechanism of natural selection to explain evolution, anything that involved 'leaps of nature' as he called them, would contradict his causal hypothesis. The early proponents of developmental evolution were largely dissenters from the causal supremacy of natural selection. Despite them, the modern synthesis of evolution, founded by neo-Darwinists who mainly studied population biology, took the Darwinian view that all evolutionary change occurred gradually in very small steps, each one vetted by natural selection. Nevertheless, with the rise of molecular biology, developmental evolutionary studies have become more acceptable to neo-Darwinists, although the implication that molecular evolution is both sudden and episodic is something they would rather not countenance.

Developmental evolution implies changes in embryonic pathways that result in organisms that are different from their parents. 'Epigenetics' is a broader category, which includes anything that involves the regular process of development, as well as evolutionary events that change the path of development. The word is not derived from 'genetics', which is an early twentieth-century word for the study of heredity. It comes from the word 'epigenesis', an early Greek term for the generation of a complex, coordinated organism from simple, microscopic beginnings. Although molecular biologists consider epigenetics to apply only to DNA-determined, and hence heritable, developmental phenomena, there are many important, non-heritable, environmental influences on epigenesis. In addition there are some environmental influences, such as those that change methylation patterns, which are heritable *without* involving DNA directly. Therefore we prefer not to restrict epigenetics to heritable, DNA-determined mechanisms.

Methylation of DNA represses gene expression. This is important during normal development for differentiating cells and organs. For example, our gut cells produce digestive enzymes and mechanisms that absorb the products of digestion. Our liver cells synthesize detoxifying enzymes and bile, but no digestive enzymes. Our nerve cells are specialized to transmit electrical signals, and to make neurotransmitter molecules, but they cannot digest nor detoxify food. Each type of cell is specialized for a narrow range of functions, in part because most of their DNA is repressed by methylation during embryonic development. Other examples of non-DNA epigenetic change are cited below (Section 4) in the segment entitled 'Neoteny and foetalization in humans'.

B.3 Epigenetic modes

Epigenetics has a large vocabulary of specialized terms for developmental changes, some of which are essential for understanding the topic as it applies to human evolution. The definitions given here are applicable across the zoological board, as well as to human developmental evolution. Here we draw largely on Stephen Gould's treatment of the subject in *Ontogeny and Phylogeny* (1977), which is straightforward and familiar to most biologists.

Epigenetic evolution can be effected by adding new features late in the life cycle, when most development is almost complete. This is called 'peramorphosis', and it is a phenomenon that does not interfere with early development. An example of an added-on feature in the human lineage is the emergence of bipedalism in early childhood. This is an epigenetic process that affects the way the skeleton and muscles and ligaments develop through childhood. It is partly hereditary, and partly the result of behaviour. And it usually involves a non-hereditary element of learning.

The extension or exaggeration of particular adult characteristics is called 'hyper-morphosis'. The increase in size of the forebrain in monkeys, apes and humans is an example. In the case of humans, the expansion of the forebrain's neocortex, and specialization of its components are associated with the emergence of higher intelligence, memory, mind and language.

'Deviation' is the most radical kind of alteration during the early development of an organism. It is the kind of phenomenon that geneticist Richard Goldschmidt believed could result in a 'hopeful monster', noticeably different from its parents (Goldschmidt, 1940).[2] The problem with this concept, as many conventional thinkers surmised, was that an early epigenetic change would disrupt development so much that the organism would never survive. However, there are some known ways of buffering deviations and preventing harmful results. Also, some legitimate examples of deviation are known in sea-urchin development, and they might have occurred early in the vertebrate lineage. The emergence of lungs from gills is a possible example. But no radical deviations are known in the evolution of the primates.

Harmful disruption of normal development can be avoided if the later stages of anatomical development are simply dumped from the life cycle. 'Paedomorphosis' is persistence of a juvenile form into adult life. 'Neoteny' is the variant of paedo-morphosis where the animal grows to sexual maturity and adult size at the normal rate, but retains the juvenile form instead of undergoing gradual or metamorphic change to the old adult form. One of the best known and non-controversial examples of this is the axolotl, the large, sexually mature tadpole of the tiger salamander. The failure of the axolotl to metamorphose to the adult form is due to changes in the synthesis of hormones such as thyroxine. The young larvae can be made to metamorphose if they are treated with thryoxine. Since we are interested here in how the environment as well as internal factors can affect developmental evolution, it is also worth considering what happens in other salamanders. In some, a decrease in temperature causes neoteny. In others the availability of iodine in the diet for the synthesis of thyroxine is important. In yet others, metamorphosis cannot be induced. The neotenous condition has been genetically assimilated; a process explained later in the chapter.

The other kind of simplification in form is known as 'progenesis'. This is usually a heritable phenomenon, in which the simple early juvenile form becomes sexually mature precociously, remains very small, and fails to complete growth and development into the fully adult anatomy. However, this condition does not apply to human evolution. It is common in crustaceans such as sandhoppers, and shrimps, spiders, insects and molluscs, such as sea snails and clams. Since the time taken to reach maturity is reduced, several generations can fit into the time formerly taken for one. Nevertheless, success in this epigenetic arena depends on conditions in the larger environment.

B.4 Neoteny and foetalization in humans

Some simple cases of paedomorphosis are products of a single biochemical or physicochemical event. It might be light or temperature change, or the presence or absence of an epigenetically stimulating or inhibiting molecule, or the amount of yolk deposited by a female animal in her eggs, as Søren Løvtrup (1974) and Ryuichi Matsuda (1987) have demonstrated. Some may result from multiple factors of this kind, as seems to be the case in human evolution. L. Bolk's essay, 'The origin of racial characteristics in man' (1929), proposed that *Homo sapiens* evolved by 'foetalization', because adult humans look a lot more like the foetuses of the great apes than their mature forms. This process would be classified as neotenic: ancestral characteristics fail to develop, the juvenile stage is extended and sexual maturity delayed.

Neoteny in humans results in the following: a flat face, simplified, late-erupting teeth, the loss of heavy skull ossification and exaggerated bony crests and ridges, late closure of skull sutures and the persistence of the epicanthic eye fold. The latter is only prominent in some oriental peoples; in others the eye fold is embryonic. Neoteny has also shifted the foramen magnum, the hole in the skull through which the spinal cord passes, to a ventral position that suits an upright posture. The relative hairlessness of humans and anatomical changes in hands, feet and secondary sexual organs, i.e., penis, vulva and mammary glands are also neotenic. Along with these features come a prolonged period of infantile dependency, growth and lifespan. The most important anatomical changes in our evolution gave us an upright stance, longer legs, increased manipulative skills and continued growth of the brain long after birth. The last feature is not neotenic, but a hypermorphic extension of an old adult condition, as are our elongated limbs, some details of skull anatomy, and most importantly, nose and throat modifications that make speech possible.

Simultaneously, the enlargement of the female pelvis allowed the birth of big-headed babies. Since most of these changes involve molecular biology, and the actions of cells that organize normal development, we need to understand epigenetics at those fundamental levels, in order to understand embryology as well as developmental evolution.

For the sake of understanding evolutionary history it is desirable to correlate developmental changes with their historical life cycles, and associated behavioural and environmental factors. Now, what of the earlier phases of hominin evolution? Palaeontologists warn against selective sampling of the fossil record to meet a preconceived agenda – despite the fact that they do it themselves. Nevertheless, even if it contains a lot of wishful thinking, the chronologically arranged fossil remains of hominins roughly suggest a phyletic series involving several species of apes, and leading through *Australopithecus* to *Homo*. No one would now try to argue that humans emerged through a single foetalizing saltation of an ape-like ancestor. On the other hand, even if some of the 'missing links' that tie us to our early ancestors have been retrieved, the multiple changes are anything but continuous and gradual. Within the same embryo, some developmental processes can be accelerated while others are retarded, without mutual interference. Many of the anterior features of skull, jaw, tooth and neck formation can be altered by mutations of genes that regulate epigenesis. And the expression of those regulatory genes can be harmoniously modulated by the redistribution of organizing cells in space and time during development.

B.5 The role of neural crest and nerve cells

A particularly important embryonic organizing feature of vertebrates is the system of neural crest cells. These migrate from the neural crest of the early embryo to distant points where they become a variety of organ components. Some only migrate as far as the developing brain, where they become glial cells that provide a packing material for the nerve cells as well as influencing their functions. The neural crest cells not only migrate to a target location, they can also affect the development of tissues through which they pass. The evolutionary emergence of neural crest cells in the early vertebrates greatly increased their potential for developmental evolvability, as well as adding a dimension of complexity that is not easy to reduce to simple causes. Therefore, although human head anatomy is more like the foetal gorilla's than the adult's, it is an oversimplification to say that it is due to foetalization, and more accurate to say that it is the product of many epigenetic causes including neoteny and some novel epigenetic additions.

Changes in neural crest cellular effects, along with epigentic regulator gene mutations, are also responsible for the highly plastic jaw structures that change periodically in domestic dogs. These appear in a saltatory manner in 'sports', the archaic term used by domestic animal breeders for offspring that differ markedly from the parental type. Sports are the origin of new strains, and their novel traits are subsequently exaggerated by selective inbreeding. There is no a priori reason to suppose humans emerged differently. In other words, the hominin lineage was capable of producing modified forms in a single generation. In contrast, conventional wisdom has novelty appearing in very small variations that require natural selection for their accumulation.

Although neural crest cells are unique to the vertebrates, developing nerve cells have similar epigenetic effects in all multicellular animals. Although they do not migrate, they send out axons and dendrites to meet target organs. For example, a motor axon must be sent to a developing muscle, so that the brain can instruct it to contract. And in order for the brain to be aware of its surroundings it must send axons to the developing eyes and ears. It is known that the growth of the axons and dendrites can be attracted or repulsed by determinants in the local cellular environment. This is common to all animals. Yet, it has special implications for human biology. In the evolution of a brain that has intelligence and memory, an increase in the complexity of the local cellular environment is necessary. This is accomplished by an increase in the number of hormone-like neurotransmitters, and an increase in the number of specialized molecular receptor sites that are sensitive to the neurotransmitters. In effect, the brain is capable of making more complex connections between its cells during the lifetime of the individual human. This permits a great cerebral plasticity, since learning, and memorization processes, depend upon the continuous formation of new neural connections. Such plasticity would not be possible if the structure and function of the brain were entirely gene determined; an assumption that is made by evolutionary psychologists.

B.6 Bipedalism

The evolution of bipedalism, a characteristic emergent feature of the genus *Homo*, reinforces the contrast between the conventional selectionist view of evolution and an emergentist interpretation. Bramble and Lieberman (2004) argue that one of the important consequences of bipedalism was the development of an ability for sustained, long-distance running. They develop their case in the context of selection theory, suggesting that the selective advantage of bipedalism was the improved ability to compete with other predators for prey. Certain co-adaptations were necessary, such as improving the processes for cooling the body from the heat of pursuit. The putative origins of long-distance running go back two million years in the *Homo* line, and the authors infer that it was acquired under the selection pressure of competition for nourishment. Recent finds of 4.4 million year-old *Ardipithecus ramidus* from the Afar Rift region of northeastern Ethiopia raise doubts about some of these interpretations. *Ar. ramidus* appears to have lived before *Australopithecus* in a mixed woodland and open environment and engaged in a primitive form of bipedality (see White *et al.*, 2009). Interestingly however, Bramble and Lieberman also support the hypothesis that a number of unique characteristics developed contemporaneously over a short period of time, which fits an emergentist interpretation rather than a selectionist one.

Now, anyone who is interested in bipedalism and its evolution ought to become familiar with Slijper's little goat.[3] Some kind of epigenetic effect altered the timing of development of the forelegs of this animal, so that it was born with little more than a set of hooves coming out of its shoulders. The little goat taught itself to walk on its hind legs. What is especially cogent is that the post-partum development of the goat's functional anatomy was dramatically altered, with changes in postural muscles and the bony skeleton that were appropriate to its novel gait. Here is an animal that became bipedal in a saltatory manner and its late development accommodated the change harmoniously. This further emphasizes that there is no good a priori reason to suppose that human bipedal evolution took thousands of generations to develop miniscule changes under the scrutiny of natural selection. And there is plenty of evidence from human development that shows how rapidly individuals who are anatomically challenged can accommodate to such conditions. The changes might ultimately become genetically fixed, but the originating event is rapid, and the first adjustments are made in individuals' developing functional morphology, rather than in their genes.

B.7 Genetic assimilation

How does something non-heritable, such as an environmental influence that affects an organism's development, become heritable? An answer to this question lies in the principle of 'genetic assimilation'.[4] As well as environmental and behavioural effects on development, the genome is in a constant flux of 'natural experimentation' involving random and non-random point mutations, codon duplications, gene transpositions from one chromosome to another, gene duplications and chromosome duplications. These may interact in a way that alters gene expression during development.

A non-hereditary factor that causes change in the individual may join forces with such a genetic experiment in a way that improves the overall quality of life. The end result is a diminished role for an external stimulus, such as temperature change, that originally was required for normal development. The stimulus has become 'internalized'. An example of this is seen in the effects of temperature on salamanders. Some species still require a temperature increase to metamorphose, but in others the stimulus has become internal and is governed by a hereditary biochemical mechanism. Such animals are no longer at the mercy of the weather for regular development. Mammals are homeothermic, i.e., they have a constant body temperature that is maintained by metabolic heat. Therefore a temperature change would not affect them epigenetically. Yet, it might make them change their behaviour, with effects on anatomy, followed by genetic assimilation, and hereditary consequences. Also, problem solving affects brain maturation, and challenges to intelligence might indirectly affect brain evolution.

B.8 The genome as a generator of evolutionary potential

Our view of the genome has changed from a static system, subject to a gradual accumulation of minor changes in DNA structure, to one in which an alchemist's cauldron is constantly bubbling and frothing. Or think of the genome as a laboratory run by inventors who are constantly experimenting and producing innovations that are offered up for human welfare. Some of them might have an immediate use for activities that are already occurring. Some of them might suggest a novel way to conduct old practices, or ways to do things that were never done before. Some might suggest no such thing, but might sit on a shelf until a use was found for them, as was the case for diode valves before their use in primitive radiotransceivers was conceived. This is now a historical example, since we have moved from diodes through transistors to computer chips in the evolution of signal-amplification technology. Our view of the genome as a system that can produce change almost as fast as it is needed, instead of waiting for hundreds of thousands of years for point mutations that are more likely to hurt than to help, is only part of the whole story. The other part is what individual organisms do in their environments. If the organisms have the physiological adaptability to support innovative behaviours, they will explore different environments that will have an effect on development. The larger context is feedback loops between physiology, behaviour, the environment and the genome. These implications will be explored further in the following chapter.

   The primary requirement for epigenetic (or developmental) evolution is that the novel organism must remain a harmonious, if changing whole, in a dynamic, changing relationship with its environment. Genetic assimilation brings a heritable consistency, after the fact. The genes are not where evolution starts, but where it finishes. Keeping this in mind, we can take a more detailed look at how epigenetic change has contributed to human evolution.

B.9 Humanness

The crucial quality of humanness is the wholeness that arises from anatomy and physiology, especially brain structure and function. Mind may have emerged as an epiphenomenon of allometric brain enlargement and complexification. By this we mean that the neocortex and its components probably expanded as a result of an increased effect of growth hormone in the forebrain. Here we are referring to proportional changes in the regions of the brain, especially the frontal lobes, not an overall increase in brain capacity. That a variety of brain functions were improved as a result was not caused by natural selection, but was simply the consequence of neo-cortical expansion and the potential for more sophisticated neural wiring. Moreover, it required generative and ancillary conditions such as the prolongation of later stages of foetal development, complementary acceleration of cranial bone growth and female anatomical accommodations. Bigger is not necessarily better, and one of the new qualities of the primate lineage was progressive packaging of the brain tissues into organized layers, and cortical folds, to achieve functional efficacy without having to keep on enlarging the brain.

Susumu Ohno (1970) suggested that the emergence of the complex wiring of the brain was due to duplication and mutation of genes responsible for the recognition sites on neurons. During development, the growing dendrites of adjacent cells home in to these targets, to complete and expand the neural circuitry. Dendrites are the microscopic threads that allow a single neuron, i.e., nerve cell, to communicate with many other nerve cells: the greater the number of differentiated sites the greater the number of circuits that can be established and the greater the organized complexity of the brain.

However close Ohno may have come to identifying a key emergent feature, i.e., increase in complexity of signalling molecules and their receptors in the brain, other generative conditions had to be in place, including the metabolic physiology for the support of intelligence and the developmental plasticity of the neocortex, cranium and ancillary structures. The emergence of humanness also conferred multiple features such as the potential to perceive and analyze problems, artistic abilities, empathy with other members of the species and the development of a complex structured language. In combination with the curiosity that is a characteristic of the primates, logic is valuable to the individual and its immediate kin, since inferences can be illustrated by example, without a structured language – 'monkey see, monkey do'.

The problem of language is trickier, since there is no question that the primates in general, as well as other mammals or birds that lack an expanded neocortex, communicate vocally. Some birds can add the songs of other species to their repertoire, usually without significant meaning; but starlings can do a bald eagle imitation that might help survival of the flock. Parrots can acquire large, meaningful vocabularies, but their ability to construct new phrases from their words is limited. Coming closer to humans, vervet monkeys have numerous sounds indicating different types of predators as well as foods. But they cannot combine those sounds to invent new meanings. Chimpanzees can sometimes invent novel word combinations. However, only humans have the ability to, without rehearsal, combine sounds in sequences that convey complex meanings that other humans can understand. Once humans had the emergent logical or grammatical system to accommodate complex language, fine-tuning of vocalization anatomy could then have provided for a large and phonetically subtle vocabulary.

Logic and the potential for language are not the only emergent properties of humanness. Others include the talent of image-making, both symbolic and realistic. As H. J. Jerison (1973) argues, the evolution of the brain in primitive nocturnal mammals involved the ability to construct a mental model of the visible, olfactory and auditory environments. These images could then be used to make short-term predictions about the location of prey and predators. The human brain emerged with an ability to amass multitudinous long-term memories as well. Teaching and learning were pre-human mammalian and bird qualities, but the emergent human brain and mind facilitated the development of a tradition of human experience that could be passed on for thousands of years, long before the invention of writing. Its ineffable emergent qualities also allowed it to interact with logical functions to produce art, myth, metaphysics and religion. Individuals with a particular endowment of such properties had no talent for confronting 'nature red in tooth and claw'. Instead, their societies had to shelter them from natural selection, as artists, shamans, priests and scientists.

One ancient debate is long dead. We are not descended from the apes, in the sense that one of the extant ape species was our ancestor; although DNA evidence indicates a common ancestor for the human phyletic line and chimpanzees. The question of whether *Homo sapiens* arose multiregionally in several semi-isolated, large populations of *H. erectus* or *H. antecessor*, or just once in a small population, is still being debated. Although Darwin had his limitations, we agree with him and his followers that different regions place different demands on their inhabitants, and that this would lead to different kinds of evolutionary adaptations. Hence the isolated regions would produce different species rather than only one new type. Therefore we prefer the alternative, known as the 'out-of-Africa' hypothesis.

The recent discovery on Flores Island of the fossil remains of a putative dwarf species most closely related to *H. erectus* has a number of evolutionary implications, including epigenetic ones. The species, named *H. floresiensis*, supposedly died out 18 000 years ago because of a major volcanic eruption, although there has been some speculation that the species might still be alive as the Indonesian Orang Pendek![5] In any case, *H. floresiensis* must have overlapped with *H. sapiens*. Its presence on an island that was cut off from the mainland for the entire duration of its stay, suggests that *H. erectus* had the ability to build rafts or boats, as well as to use tools. The dwarfing of *H. floresiensis* might have occurred after its arrival on Flores Island. But its continued use of tools suggests that although its brain was much diminished in volume, it retained much of the functional efficacy of the normal sized brain of *H. erectus*. Moreover, rafting or boat building must reflect the intellectual capacity of *H. erectus* as well as *H. sapiens*.

Conventional wisdom takes the self-contradictory view that natural selection has produced both dwarves or giants as having a high fitness for island living. Well-known examples are the pygmy elephants and hippopotami found as fossils on islands of the Mediterranean, and the Wrangel Island pygmy mastodons that died out only 4000 years ago. Islands certainly seem to have more than their share of dwarves and giants. The dwarf humans of Flores Island were accompanied by primitive pygmy elephants of the genus *Stegodon*. Some giantism also occurs among island fauna, e.g., the Komodo dragon in the case of Flores Island. Giantism can be explained in terms of a high production of growth hormone and its peptide receptors due to gene duplication. Dwarfing can be a saltatory epigenetic event due to a curtailment of growth-hormone production, such as happens in pygmies belonging to the human species.

However, dwarfing and giantism can occur at random anywhere at any time. They are not *caused* by island environments. If a dwarf or a giant appears in a stable population in equilibrium with its community – especially food and predator species – it might be at a distinct disadvantage compared to the established or 'normal' types. However, new immigrants to an isolated island environment may find adequate resources and perhaps no predators. Therefore the competition that is believed to be the major agent of natural selection is *absent* until populations expand sufficiently. And in the meantime, it does not matter whether dwarves or giants emerge, and if they do, the molecular processes involved occur quite suddenly.

B.10 Epigenetic algorithms

Earlier in this chapter we raised the issue of 'epigenetic programmes', or 'algorithms'. Both terms imply that development is driven by a set of instructions similar to those found in a computer. This metaphor has a severe limitation in as far as computer programmes require a human programmer. In epigenetics we have to assume that if such algorithms are present, they must have evolved. They must have increased in complexity to produce the vertebrates from their simple ancestors that lacked backbones, and to produce humans from their primate ancestors. It is widely but falsely assumed that 'genetic programmes' reside in the genome. The structural gene only contains a code for the sequence of amino acids that makes up a protein. So, does the algorithm exist in the modifier genes that tell the DNA when the protein synthesis must be initiated? They might be regarded as an algorithic switch that can be turned on and off. In fact those modifier genes are often very complex algorithmic switches, since they can respond to many different signals and control several structural genes. But the signals do not come from the structural genes or the modifier genes. They come from the cytoplasm of the nucleus, which gets them from the cytoplasm of the cell, which gets its chemical messengers from the blood, which gets them from the glands that secrete hormones, which in some cases are secreted by the brain. In turn, the brain is influenced by external environmental conditions. Changes in temperature and day length, the presence of members of the same species, the presence of members of other species, and the availability of food, can all be affective. *If there is an epigenetic algorithm it is the whole organism in its environment.*

B.11 Environmental causes of epigenetic change

If epigenesis, or development, is normally modulated, and occasionally changed by elements of the internal and external environment, we must develop a better understanding of these processses, in addition to learning about the developmental roles of molecules. By educating ourselves about this we can better evaluate the evolutionary influence of climate as well as related environmental impacts. We have already considered examples of how the metamorphosis of salamanders is influenced by their food and changes in temperature. Salamanders overcrowded by members of their own species sometimes develop into aggressive carnivorous forms. In mammals the presence in the intestinal environment of a symbiotic bacterium, *Bacteroides thetaiotamicron*, influences early juvenile gut development, triggers mucus and enzyme production and activates the immune system (Hooper *et al.*, 2001). While the genes of the bacterium are certainly involved in the processes of stimulating changes in the host, there is no known genetic transfer from the symbiont to the host. Thus these important epigenetic influences originate outside the affected individuals.

B.12 Evolutionary changes through changes in methylation patterns

In their landmark work, *Epigenetic Inheritance and Evolution* (1995) Eva Jablonka and Marion Lamb describe many mechanisms that influence development and its evolution that do not affect the DNA directly, but alter the patterns of repression of genes through methylation as described above. The way in which the DNA is bound by 'histones' (i.e., the chromosome proteins) also affects gene expression. Some of these DNA-repressions are heritable, even though the DNA itself is not altered. A more recent rekindling of interest in these phenomena brings Leslie Pray (2004) to discuss a significant example involving the impact of malnutrition on humans. During the 'Dutch hunger winter' in 1944–45, 30 000 people died of starvation. Of those that survived many developed diabetes, heart disease and cancers, and bore babies with abnormally low birth weights. The latter is of particular interest for evolutionary studies, since the trend continues after several generations, despite the fact that the nutrition of pregnant and nursing mothers had returned to and remained normal in each subsequent generation. This is all due to the heritability of alterations of methylation patterns during the time of famine (Stein and Lumey, 2000). The unfortunate 'starvation experiment' has been successfully recreated with mice and rats as the subjects.[6] At the heart of the process is a molecule called a heat-shock protein. Under such stress as high temperatures and abnormal metabolic rates, heat-shock proteins can stimulate the expression of normally repressed genes by changing methylation patterns. The effect causes fruitflies to develop such abnormalities as bristles in the eyes, and even if the stress is of brief duration, the consequent changes in methylation patterns persist for many generations, without continuous or repeated heat stress. Because of the evolutionary effects of stress proteins, we must be open to the possibility that catastrophic changes in climate caused by plate tectonics, volcanic activity and bolide impacts might have had direct evolutionary consequences.

Although the best known examples of persistent methylation pattern change are detrimental, there is no reason to exclude the possibility that they also can cause radical epigenetic shifts that could produce novel organisms. The same can be said for a self-amplifying process that increases the number of copies of particular codons in DNA. As a result, the proteins for which they code have extra copies of specific amino acids in their primary structure. The phenomenon was first observed in pathological human conditions such as myotonic dystrophy, Huntington's disease and fragile-X syndrome. Prior to the discovery of the underlying molecular mechanism, these were characterized as 'anticipation' diseases. Because of the self-amplifying nature of the codon repetition the diseases affect subsequent generations earlier and with worse symptoms.[7] Now it is known that a similar self-amplifying mechanism occurs in regulatory genes in dog breeds, affecting changes in the relative size of the snout and legs, and presumably acting through neural crest cells as intermediates (Fondon & Garner, 2004). Clearly, anticipation is not harmful in such cases, though the new traits might only be of interest to dog breeders. However, similar effects are found in a range of wild animals including wolves, foxes, coyotes, otters, walruses, rabbits, bats and humans. The sizes of their snouts and the length of their limbs are affected. In a bat, for example, codon repetition in regulatory genes might have been responsible for longer 'fingers', i.e., the digits that support the wings, and therefore make their flight more effective. Furthermore, Fondon and Garner consider their finds to fit the category that Goldschmidt called the 'hopeful monster'.

B.13 Self-amplifying genomic changes as evolutionary processes

The existence of self-amplifying systems that become more and more exaggerated from generation to generation is relevant to the epigenetic phenomenon of 'allometry'. We have already discussed this in the context of proportional growth shifts in the evolution of the human brain. The classic illustration is giraffe evolution from the okapi form, which is deer-like. The necks of both have the same number of vertebrae, but in the giraffe the bones are much longer, and accompanied by larger, stronger neck muscles. The heavy neck of the giraffe is supported by a stouter pectoral girdle, and longer, stronger legs. Therefore there has been an allometric expansion of the anterior of the giraffe, with some adjustments to the rear end as well. The only means of achieving such allometry, without extending the gestation period of the ancestral type, is to increase the relative rate of growth and reproduction of the cells involved. This can be achieved by differential distribution of growth hormone, or, more likely, by localized increased receptivity of the cells to the effective hormones. An overall increase in growth hormone would simply produce a larger okapi. If the DNA of giraffes and okapis were compared there would be insufficient difference to explain the different appearances of the two animals. In other words, the differences are not genetic but epigenetic. Starting out with very similar DNA to the okapi, the giraffe modifies its gene expression and its local cellular functions differentially. We must be able to explain the differences between human brain structure, and that of our close relatives, in the same way. Somehow the epigenetic hormone and receptor distribution is locally changed to make the forebrain grow faster than the rest of the brain. Its components, such as the ones associated with speech, grow differentially as well.

What other processes might cause allometric shifts in body structure, especially the brain and associated bones and muscles? In the vertebrates, some allometric growth shifts are caused by changes in the behaviour of neural crest cells. At the gene level, craniofacial changes are affected by repetitive sequence mutations in regulatory genes that influence development, probably correlated with neural crest cell function, as outlined earlier in this chapter.

Allometric growth is part of all diversifying changes in body form. The conventional explanation is that the partial exaggeration of a limb, lengthening of the legs in humans, for example, is due to the point mutation of a structural gene or its modifier gene. If such a change brings advantage to the organism it is naturally selected. Therefore the form of natural selection known as 'directional selection' supposedly drives allometry. If the conditions of life remain stable, a further exaggeration through a second mutation would have even more selective value. However, there is no known mechanism whereby a second random mutation of a gene would cause a further exaggeration in the same direction as before, as opposed to a change to a different direction or simply a detrimental effect. Let us assume instead that allometry is caused by self-amplifying molecular processes such as codon or gene duplication. This assumption is based on genetic evidence, unlike the assumption made by neo-Darwinism. If we are correct, the allometric shift will continue in the affected lineage, whether or not it has selective advantage. This makes self-amplifying allometry into 'orthogenesis'. This means allometric evolution due to some kind of inherent drive – in this case, well-established molecular mechanisms. We have established that some of these self-amplifying changes have environmental causes such as stress. However, some are probably generated internally without external stimuli. Both kinds have to be taken into account to understand developmental evolution.

Orthogenesis was once a popular explanation for lines of evolution that followed a straight line, according to the fossil record. However, the idea was abandoned in favour of directional selection; more for ideological reasons than obedience to any real evidence. Orthogenesis offers a way around one of the problems with which Darwin and his followers wrestled. In the earliest, imperceptible stages of a change such as the exaggeration of the length of a limb, or the size of the skull crests in a dinosaur – or a human – how is there enough useful advantage for natural selection to recognize, and so drive the process forward? But in the case of a self-amplifying drive, natural selection does not matter. Eventually the inherent drive might bring the allometric exaggeration to the point where it would have a distinct advantage in relation to the animal's behaviour and environment, and so it might spread beyond the lineage in which it started to the whole population and become a species' characteristic. But the process would continue regardless of advantage.

Associated with orthogenesis was the corollary that the inherent drive might take the affected lineage past the point of extinction, something that natural selection would not countenance. We know that in self-amplifying anticipation diseases (for example, Huntington disease); extinction of the lineage does occur. Eventually a family line suffering from Huntington's disease can no longer reproduce. This obviously has not happened in most known examples of allometry, since they are alive and well around us in the form of giraffes, for example. Molecular drives do have off switches. The allometric shift in the length of human leg bones did not proceed to an unstable giantism.

In his discussion of human evolution, L. Bolk (1929) surmised that orthogenesis might be involved as well as foetalization. This was one of the reasons why his whole hypothesis was rejected by orthodox Darwinists. Now that we know of the real existence of orthogenetic mechanisms we might think of applying the idea to aspects of human epigenetic evolution, such as the elongation of limbs and brain allometry. One thing we would have to take into account is as follows. We have in this chapter been thinking in terms of behavioural and environmental change being prior to epigenetic and genetic change. An orthogenetic drive might, however, when it got to the point of significant anatomical change, dictate a change in behaviour and environment. If human legs get longer, humans are going to incorporate running skills into their hunting behaviour, and maybe go after new kinds of prey. It is always necessary to have a holistic view, and see how the biological factors such as anatomical changes and behaviour and environment interact, instead of making exclusive assumptions about evolutionary causation.

In the next section, Appendix C: Human adaptability, we will be discussing physiological evolution. Some authors have taken the stance that the physiological functions of mature organisms are the end result of epigenetic change, and are therefore unimportant in their own right. However, especially in placental mammals, including humans, normal epigenesis is dependent on the physiology of the mother, who provides a uterine environment that is consistent with her own mature, physiological homeostasis. It is necessary to see epigenesis and physiology as 'system interactions'. Although we separate the two for the sake of simplicity, we have to be able finally to put them back together again, which is the purpose of this book, connecting climate change with human evolution.

Notes

1. Of particular note is the French anatomist and developmental evolutionist Étienne Geoffroy de Saint-Hilaire, usually referred to as Geoffroy.
2. The 'hopeful monster' was first conceived for a lecture given at the Chicago World Fair in 1933.
3. Slipjer's goat (Slijper, 1942a, 1942b) is described in detail by Mary Jane West-Eberhard (2003).
4. Mary Jane West-Eberhardt (2003) elaborates on all the nuances of genetic assimilation.
5. See Brown *et al.* (2004) and Morwood *et al.* (2004) on *Homo floresiensis*. The Orang Pendek (little man of the forest) is a half-human, half-ape creature that is part of the folklore of the Indonesian tribespeople.
6. See Burdge *et al.* (2007) for a recent report and review.
7. A complete historical and biomedical account may be found in Friedman (2008).

Appendix C: Human adaptability: the physiological foundation

La fixité du milieu intérieur, c'est la condition de la vie libre.

(Claude Bernard, 1873)[1]

C.1 Introduction

In our discussion of developmental evolution we concluded that the physiology of the mature organism merits separate attention from the physiology of the developing organism. Particularly important for the evolution of humans, and indeed for all mammals, is the fact that exploratory behaviour, and hence exposure to new environmental conditions, and all the evolutionary consequences of such activities, depends upon a reliable, stable physiological state. You cannot live in an extreme climate unless you are physiologically capable of keeping your internal environment within safe limits.

In this section we intend to establish how most aspects of human physiology are fundamentally similar to that of the placental mammals, and how that basic physiology enhanced the evolvability of the placental mammals, including humans. We also go further back in time to show how environmental changes have had very significant links with physiological evolution. All of this evolution had already happened by the time humans emerged, but it is best to know the foundations if the upper storeys are to be understood. This should not however divert our attention from our emergent physiological and psychological characteristics, such as mind, memory, language and arts. Nor should that uniqueness divert our attention from our ability to abuse our talents.

Before proceeding we wish to re-emphasize the distinction that we made in 'Evolutionary theory' (Appendix A), regarding adaptation and adaptability. Adaptation is defined as a genetically fixed trait, such as the structure of a head. Because

335

the anatomy is genetically fixed, very little modification is possible, except through gene mutation to a differently fixed form. Once that happens, the organism cannot return to the original condition, so the evolutionary result is inflexibility. All genetically-fixed changes are not necessarily adaptations in the sense that they are suited to the prevailing conditions of life. The newly altered organism might be a hopeful monster, maladaptive under the prevailing conditions. Yet it might have the ability to exploit and thrive in a new environment. This would depend on a degree of physiological and behavioural adaptability. Adaptability, in contrast to inflexible adaptation, allows an individual organism to modify or adjust its physiology or behaviour to changing environmental conditions. We might say that adaptations are gene determined, and adaptabilities are organism determined.

Regrettably, the word 'adapt' in everyday language is used both for evolutionary adaptation and for the use of the individual's adaptabilities. This supports the unwarranted assumption that the two distinct qualities are evolutionarily identical, and leads to confusion, even amongst professional physiologists, psychologists and especially anthropologists. We will continue to strictly apply the distinction between adaptation and adaptability in this book. Most physiological adaptabilities adjust automatically to change, and they arise from a general state of dynamic internal equilibrium known as homeostasis. Homeostatic modifications are also modulated by both instinctive and intelligent behaviour.

C.2 Homeostasis

In the late nineteenth century, French physiologist Claude Bernard stated that the stability of the internal environment frees an organism from changes in the external environment. The chapter epigraph is Bernard's original aphorism. He had already established that blood sugar, ions and body temperature are maintained within narrow limits. Subsequently the regulatory roles of the hormonal and nervous systems were elucidated by other physiologists. In his book, *The Wisdom of the Body* (1932), Walter Cannon coined the word 'homeostasis' for the self-maintaining physiological stability to which Bernard had referred to half a century earlier as 'la fixité du milieu intérieur'. Both Bernard and Cannon were aware that homeostasis is not permanently fixed, but is instead dynamic. Physiological parameters vary with conditions, but homeostatic mechanisms bring them back within the narrow range of viability. For example, if the ambient temperature falls suddenly, body temperature also falls. In homeothermic animals, i.e., those that can maintain a constant body temperature, such as mammals, this is registered by a biological thermometer in the hypothalamus of the brain, and a hypothalamic thermostat cuts in. The hormones norepinephrine and epinephrine are secreted, and fat metabolism is increased, thereby raising the internal temperature. Skin capillary circulation is reduced, cutting heat loss. And in furry mammals, piloerector muscles in the skin cause the hair to stand upright, increasing the insulating quality of the fur.

The ultimate independence from the vicissitudes of the external environment is found in human beings, because we can complement physiological self-maintenance with intelligent behaviour, the use of fire and development of artifacts such as clothing, housing and air conditioning. We can descend into the marine abyss deeper than whales in our submersibles, and ascend beyond the atmosphere in spacecraft – without much stress on our internal homeostasis. And we can also alter the external environment for short-term benefit, but often in ways that threaten our long-term survival.

Homeostasis is the foundation of our behavioural adaptability. It allows us to change what we do according to circumstances, explore new habitats and retreat from them, without imposing harmful internal changes. In a broad sense, human intelligence is a unique part of our homeostasis. We can think about the potential dangers and rewards of our behaviour without committing ourselves. Moreover, we have the added advantage of teaching the inexperienced so that they do not have to repeat our mistakes and ordeals. This is not to say that what we teach is error-free, or that how we conduct our everyday life, as well as our explorations of new conditions, is harmless to us and our environment. To examine the physiological foundations of these homeostatic abilities, it is necessary to analyze their components and to understand how they emerged from simpler systems. Apart from special human qualities that arise from a sophisticated brain, our homeostatic qualities are common to all placental mammals. These are the group whose females protect their embryos and foetuses in their uteri and nourish them through a large placenta that allows the useful components of the mother's blood to pass into the embryo's circulation, until a late stage of development.

C.3 The homeostasis of placental mammals

The regulatory mechanisms of homeostasis range from simple chemical interactions to multi-stage enzymatic, hormonal and neural pathways. These often interact cooperatively. Most are feedback mechanisms that limit their own action when sufficient output is achieved. They are found locally within organ systems. They may tap into endocrine organs which provide hormones that regulate body functions, and so maintain a stable internal environment. The most sophisticated regulatory mechanisms are found in the brain, including the evolutionarily ancient medulla, which coordinates heartbeat and respiration with other activities, the cerebellum, which ensures the coordination of balance, walking, talking and intermittently chewing gum (this statement was made by Lyndon Baines Johnson, referring to Gerald Ford, during the United States presidential campaign in 1974). These brain regions go back to the origins of vertebrates in the late Cambrian, over four hundred million years ago. More recent physiological acquisitions that regulate homeostasis are found in the hypothalamus of the forebrain. It sends releasing hormones that act through the adjacent pituitary gland to secrete hormones that stimulate the adrenal gland. Epinephrine from that gland increases metabolic rate and heightens muscle tone, readying the mammal for rapid locomotion, either as an escape response from predators, or for a speedy attack mode. The pituitary gland secretes hormones that regulate salt and water balance and portions of reproductive cycles. The hypothalamus also responds to changes in blood acidity arising from an excess of respiratory carbon dioxide, by increasing the breathing rate.

Detection of low oxygen levels is a more primitive function, carried out by the carotid glands, in the carotid arteries, which carry blood to the brain. It is less sensitive in terrestrial animals where carbon dioxide levels give a better indication of potential respiratory problems. As already mentioned, an essential hypothalamic function in warm-blooded animals is the detection of body core temperature, and reaction to excessive changes through the neural thermostat. Warm-bloodedness, or homeothermy, or the maintenance of a relatively constant body temperature is not unique to mammals, being also found in birds. What birds and mammals have in common is thermogenesis (the ability to generate sufficient body heat to survive in cold conditions), hypothalamic thermostats and effective insulating systems in the form of feathers or hair. Some of the other physiological components of homeothermy are found in the reptile ancestors of birds and mammals, and even feather-like and hair-like scales have been found in fossil reptiles. It is necessary to go that far back in time to realize that physiological features that we associate with the mammals are not unique, and did not emerge full blown with them.

In the mammals, the signature physiological function is nurture of the young by the secretion of mothers' milk. Strictly speaking, this is not a homeostatic mechanism, but is a necessary complement to homeostasis in the young as a supply of fat and sugar energy to fuel it, as well as providing amino acids that provide for energy, growth and repair. The qualities of milk are not unique to placentals, having emerged prior to the divergence of the placentals from the lineage that had already given rise to the egg-laying monotremes, such as the duck-billed platypus and the spiny anteater. Although not a direct component of homeostasis, milk is certainly a major aspect of human adaptability, since it frees mothers from constant attention to the provision of other foods for a period after birth. The environmental control of lactation leads to changes in birth rate that may be beneficial to hunter–gatherers in straitened circumstances, who cannot be burdened with carrying newborns and juveniles during their wanderings. On the other hand, the mother–child bond that forms during lactation is also an important part of nuclear family life, not forgetting the secondary bond that must persist between father and mother if the family unit is to succeed.

An extreme case of such physiological bonding has been discovered by Kathy Wynne-Edwards, in her study of the Djungarian hamster. And it points us to the role of hormones in human evolution. The Djungarian hamster of lower Siberia lives in cold, dry conditions, and has an effectively insulating, fine, fur coat, as well as furry feet. When the female has borne her pups and commenced lactation, the fur coat becomes a liability. Contact with the young, and the loss of water and nutrients in milk, cause over-heating and dehydration. Periodic escape from the offspring is necessary for her survival. But how, then, would the pups themselves survive? Very unusually, for hamsters, and most other mammals, the father is there to babysit. He keeps the young warm while the mother goes off to forage for food and cool off. He stays at home because his mate has an extended period of post-partum secretion of the sex hormone progesterone, sufficient to keep him interested and present.[2]

This study has a message for humans trying to survive glacial conditions: the closer their family units, whether by hormonal attraction or by conscious choice, the better they can stay warm and also cooperatively deal with emergencies. Wynne-Edwards has now undertaken a study of human male bonding during the early phase of childrearing. It is already widely known that the higher primates mate throughout the year instead of during a short rutting season. But in modern humans the degree of family closeness is greater, and the hormonal propensities of both fathers and mothers must both be evaluated and taken into account, in addition to other interpersonal factors.

C.4 How placental physiology relates to *la vie libre*

The foetus in the womb shares the mature homeostasis of its mother. Her ability to adjust her body temperature, and levels of blood respiratory gases, salts, water, nutrients, hormones, as well as altering the path of circulation if necessary, makes it unnecessary for the foetus to develop such a sophisticated level of self-maintenance while in the uterus. Nevertheless, it has its own circulation, and foetal haemoglobin that loads and unloads oxygen at the tensions found within the placenta, and the foetus itself. These are much lower than in the body of the mother, and adult haemoglobin could not work as a molecule for transporting oxygen in the foetus. Prior to birth, adult haemoglobin, appropriate to air-breathing, is synthesized. After birth, close body contact helps to complement the inefficient thermogenesis of the baby. But there is an organ unique to the placentals that helps to sustain the temperature of the young. 'Brown fat' is so-called because of the presence of large numbers of mitochondria, coloured brown by cytochromes; iron-containing, energy-exchange molecules. In these mitochondria, a decoupling protein causes the generation of heat. In contrast, mitochondria in other tissues convert the energy of sugars, fats and amino acids to adenosine triphosphate, which is the most common immediate source of chemical energy for fueling metabolic reactions. Brown fat depots, found in the vicinity of the shoulders, are suitably placed to heat blood flowing to the brain. And the tissues themselves are well supplied by minor blood vessels, so that the heat dissipates, instead of building up to lethal levels in the brown fat.

Failure of a mother's lactation during the post-natal period could be disastrous, therefore, one benefit of an extended family is the possibility that another lactating female might be available as a wet nurse. Another neonate fail-safe is physiological: a healthy mother has babies that have normal fat depots, convertible to metabolic energy rather than thermogenesis, i.e., self-generation of heat, as in the case of brown fat. Newborns' tissues contain more water than those of older children. This improves survival from dehydration in the face of lactation failure, or temporary absence of the mother. In the aftermath of the 1985 Mexico City earthquake, the recovery of eight live babies from a destroyed maternity ward, up to eight days after the quake, is testament to the adaptability of humans at the most vulnerable stage of their life.[3] However, for some disasters there are no such physiological respites. Under famine conditions, babies born to starving mothers lack the reserves to survive for long, and may not even have completed normal healthy development. Mental retardation is a typical consequence. (See previous section on the Dutch winter famine.)

C.5 How *la vie libre* relates to diversifying evolution in placental mammals

Despite, or perhaps because of the tendency of male mammals to wander off on their own much of the time, the placentals have done more than any other mammal to diverge anatomically as they exploited a large variety of habitats and experimented with a wide range of behaviours. Females do their own exploring in the company of their offspring, increasing the chance of reproduction. In any case, both genders had to be present for geographic extension of the population through reproduction and migration. Flexible exploratory behaviour progressed in the mammals as the neo-cortex expanded in size. Behaviour may be the most important causal factor in divergent evolution, but it does not fossilize. Therefore it is often overlooked in favour of anatomical diversity, which is striking both in living animals and in the fossil record. Early evolutionary theory was dominated by the wonderment at how the placental forelimb, with very similar basic anatomy in every case, could have diverged into such varied organs as the hands of primates, the flippers of seals and whales, the hooves of the large herbivores, the clawed paws of carnivores, bats' wings and the spades of moles. The explanation that is offered by anatomically fixated evolutionists is that if, somehow, an animal exposes itself to different environmental conditions, any random, but appropriate anatomical changes are selected and then accumulate. This misconception is compounded by the belief among evolutionary psychologists and some biologists that all behaviours are gene determined, and adaptationally selected. In contrast, the progressive evolution of the vertebrates, and especially the mammals, is characterized by *increasing freedom* from gene determination and the ability to choose amongst a variety of behaviours.

Since exploratory behaviour has an evolutionary significance in getting animals into novel situations, and responding in appropriate ways, we must also remember that without a reliable homeostasis and other physiological adaptabilities there could be no survival at the extremes, or at the edges of a species' normal habitat. And extremes there are, ranging from cold mountain tops, polar conditions, desiccating deserts, ocean deeps, isolated islands and the aerial environment as well, if we remember the placental bats. Even mammals remaining within their ancestral habitat can experience extremes when climate changes.

Obvious examples of such sudden changes are the violent fluctuations in temperature following large-scale volcanic eruptions, impacts by bolides from outer space and combinations of the two. Sixty-five million years ago, the K-T impact at what is now Chicxulub Mexico, did not immediately wipe out the dinosaurs, but the subsequent drop in global temperatures due to volcanic outpourings and dust clouds, probably finished the job. Birds and mammals that could generate their own body heat, insulate themselves with feathers and fur, and also behave appropriately by hiding in nests and burrows, were much better able to survive the colder temperatures. G. G. Simpson, one of the architects of the modern synthesis of evolution, wondered at the *rapid* divergence of whales, bats, hooved browsers, clawed predators and all the rest of the placental zoo. Claude Bernard would have known where to look: to the emergence of the physiological and behavioural adaptability of the placental mammal. But he was not an evolutionist, and even physiologists brought up later in the era of the modern synthesis of neo-Darwinism have made scant contribution to evolutionary theory. Physiologists are good at taking things apart to see how they must work as a whole organism, but they tend to specialize in particular aspects of organ function, rather than theorizing about how the physiological adaptability in general evolved.

We know that the progressive improvement of exploratory behaviour with a cognitive content and the physiological adaptability to back it up has been of great evolutionary significance to the placental mammals in general. Therefore we have to think carefully about the degree to which features that are often taken to be characteristic of modern humans may be found in earlier humans. As mentioned in the previous section on 'Developmental evolution' (Appendix B), a putative relative of *Homo erectus* was discovered in fossil form on the Indonesian island of Flores. The most recent fossil remains of *H. floresiensis* date from before 38 000 years ago until at least 18 000 years ago. It is a dwarf species with a brain the size of a lemon. Anthropologists engaged in the study of the new fossil species note that:

> ... the survival of *H. floresiensis* into the Late Pleistocene shows that the genus *Homo* is morphologically more varied and flexible in its adaptive [sic] responses than is generally recognized. It is possible that the evolutionary history of *H. floresiensis* is unique, but we consider it more likely that, following the dispersal of *Homo* out of Africa, there arose much greater variation in the morphological attributes of this genus than has hitherto been documented. ... In this instance, body size is not a direct expression of phylogeny ... Explanations of the island rule have primarily focused on resource availability, reduced levels of interspecific competition within relatively impoverished faunal communities and absence of predators ... Dwarfing in LB1 may have been the end product of selection for small body size in a low calorific environment, either after isolation on Flores, or another insular environment in southeastern Asia. ... Anatomical and physiological changes associated with insular dwarfing can be extensive, with dramatic modification of sensory systems and brain size ...[4]

Despite its small brain, was *H. floresiensis* smart enough to sail to Flores Island, and either compete successfully with, or avoid competition with, *H. sapiens*? Perhaps it was its ancestral species *H. erectus* that made the passage, and *H. floresiensis* emerged on Flores, and was the only one of the two that survived for a long period. Fossil proof of the presence of *H. erectus* has yet to be found. There was no land bridge for either species to traverse to reach their destination. Fortunately for both there were no significant predators other than the Komodo dragon, *Varanus komodoensis*. In fact, the stone tool-making *H. floresiensis* seems to have hunted that species. That a dwarfed relative of *Homo erectus* had such 'adaptive responses', for which read 'physiological and behavioural adaptability', suggests that *H. erectus* too might have been more adaptable than previously thought, and that dwarfing was less detrimental than Brown *et al.* infer. They do however recognize that 'neurological organization is at least as important as EQ [encephalization quotient – a measure of relative brain size]' (*ibid.*, p. 1060).

C.6 The history of physiological evolution and environment[5]

One of the primary questions that we have posed ourselves is 'What impact has environment had on human evolution?' Up to this point we can assess our progress thus: we know that physiological and behavioural adaptability permits exploration of new environments, which induce novel responses. Finally the new environments are occupied. The reverse of the coin is that the same adaptabilities allow survival if rapid environmental changes are imposed *in situ*, for example sudden climatic changes and reduction in the availability of food species. A mammal can survive temperature fluctuations better than a dinosaur. An omnivore such as a human can survive the loss of particular animal prey by eating insects or snails or worms or fruit or tubers; whereas food specialists such as the giant panda or koala would die without the leaves of bamboos or particular gum trees. Thus it is useful to examine a broader physiological and palaeontological context to reach a fuller understanding of how environment might have directly influenced physiological and behavioural evolution.

C.6.1 Dilution physiogenesis

We begin in the sea, when the first fishes were emerging in the late Cambrian. At the time, almost 500 million years ago, the salinity of the sea was the same as it is now, and the variety of food resources similar in variety to the range of modern marine food material: seaweeds, invertebrates and other fish. The body fluids of the primitive fish, like those of their invertebrate ancestors, were almost identical to the sea in their water, salt and pH (hydrogen ion concentration, or acidity). Eventually, during the late Ordovician, Silurian and Devonian, $\sim$ 444–360 million years ago, marine fish began to explore the brackish waters of river estuaries and lagoons.

Thick scales helped them survive being diluted by the less salty water. However, no amount of impermeable epidermis or scales can protect fishes from the direct effect of the environment, since their gill and gut surfaces must remain permeable to absorb oxygen, and the products of digestion. There may have been some reduction of freshwater osmosis by the action of an ancestral hormone that can reduce membrane permeability. This hormone seems to have been ancestral to the growth hormones and prolactins of the more recent vertebrates. The name 'growth hormone' speaks for itself. The prolactins have different functions in different groups. Basically they are associated with water transfer across cell membranes and osmoregulation in fish, amphibians and reptiles, but in female birds they cause broodiness, and in female mammals they contribute to milk production. Returning to the adventures of primitive freshwater fish, although hormonal accommodation might have extended survival in brackish conditions, the fish that made the transition initially were simply able to *tolerate* the environmentally imposed change. This had widespread consequences, since most enzymes are affected by the electrolyte concentration of the body fluid, and there would have been a period of internal, adaptational adjustment lasting much longer than the transition period, and dependent upon the rates of effective point mutations in the genes coding for the relevant enzymes. However, the structure and function of an enzyme can be changed in an instant by a single point mutation.

Environmentally imposed physical and chemical change in organisms is called 'physiogenesis', an expression coined by neo-Lamarckist Edward Cope (1887). Body fluid dilution was one of the first physiogenic changes to affect vertebrates during their exploration of new environments. Interestingly, the post-physiogenic internal adaptations produced a system that could not tolerate re-salinization. Those fish that then wanted to return to the sea had to first reinforce their hormonal accommodatory mechanisms in such a way that the seawater could not physiogenically re-impose high salinity. There are a variety of complex, but well-understood processes that allow animals with a dilute body fluid to survive in the sea, but these are beyond our present mandate. However, another general physiological evolutionary principle arises from this. It is widely known that if a goldfish is put in seawater it dies. If a marine fish like a herring is put in fresh water it dies. Yet the ancestors of those fish could probably move back and forth between the sea and rivers, and survive. The ancestral adaptability has, in other words, undergone 'regression'. This is a common fate of adaptabilities that are not periodically put into action, and the principle is sometimes known to biologists as the 'use-it-or-lose-it' principle. To take an example closer to home, the abilities of humans and the great apes to synthesize vitamin C (ascorbic acid), and their ability to degrade uric acid have regressed. Uric acid is the breakdown product of nitrogenous bases derived from DNA and RNA. It is usually excreted by the kidneys, but in great apes and humans its build-up can cause gout.

C.6.2 Oxygen physiogenesis

The next major environmental effect on fish came from atmospheric oxygen. The present atmosphere contains 21% oxygen; although estimates of atmospheric oxygen are as low at 5% for the time when fish began the transition to the terrestrial environment, that is considerably higher than the 0.9% maximum that can be dissolved in cold, clean, fresh water. Due to decomposition of vegetation, the oxygen content of the rivers and ponds inhabited by transitional fish was probably less than 0.1%. Exposure of permeable surfaces, such as the oral cavity and pharynx to air would have immediately saturated fishes' blood with oxygen. In addition to that sudden availability of oxygen to fish who broke the water–air interface, oxygen has a known epigenetic effect, i.e., it changes the developmental processes. High oxygen levels cause greater vascularization, i.e., more growth of epidermal blood capillaries, thus improving the role of exposed tissues as crude respiratory organs. Eventually the posterior gill pouches with their vascularized epidermis became a pair of lungs.

C.6.3 Gravity physiogenesis

A third environmental state imposed on fish during their transition from water to land was gravity physiogenesis. The transitional lobe-finned fish that became the ancestors of the amphibians and all other terrestrial tetrapods heaved themselves onto land by means of their pectoral forelimbs and pelvic hindlimbs. This deprived them of the support of water, and caused gravitational mechanical stress. This is a physiogenic influence that acts both on the developing organism and on the mature adult. If we consider the human condition, it is obvious that a newborn child cannot support itself. Yet, early maturation and exposure to mechanical stress strengthen the neck and back muscles so that the baby can sit up, then the arms, legs and pectoral and pelvic girdles are altered, so that the child can crawl. A transitional upright posture changes the skeletal structure and muscular insertions in the legs, so that walking is possible. These are not gene-determined developments. They depend largely on brain development, skeletal development, exploratory behaviour, and, in humans, and to a lesser extent in other placentals, parental encouragement. This point is well illustrated by Slijper's little goat, which was born with truncated forelegs. As mentioned in the previous section, the animal taught itself to become bipedal, triggering many of the anatomical changes that we associate with humans.

C.6.4 The emergence of thermoregulation

The final major physiogenic change experienced by the reptiles, protobirds and mammals was probably the imposition of a relatively high body temperature by an extended, stable warm climate. Under such conditions biological thermogenesis is unnecessary. As with enzymes that are adapted to particular salinity levels, many muscle kinases, i.e., the enzymes that help to provide muscular energy, and neural enzymes, are adapted to high temperatures. Our immediate pre-mammalian reptilian ancestors, like many extant snakes and lizards, had rudimentary abilities to sense their own body temperatures, and to thermoregulate by behaviour modification. When they were cold, they basked in the sun, and when they were too warm they sought shade. But they lacked the ability of endotherms such as birds and mammals to actually generate heat on demand. That adaptability is necessary if the metabolism of a sophisticated nervous system is to be reliable enough to potentiate the emergence of intelligence. If intelligence is to be reliable it has to be constantly available, and brain cells malfunction or die below viable body temperatures. Intelligence is also affected by the environment, though in a more complex way than simple physiogenesis. The emergence of intelligence not only depends upon homeostasis, and adequate nutrition to sustain physiological functions and growth, it also depends on a range of environmental stimuli that provoke curiosity, trial-and-error experimentation and memory. The mind continues to develop throughout the life cycle by the extension of dendritic connections in the brain. These are qualities developed under unique circumstances in individual animals, and are, as far as we know, not heritable. What is heritable is the ability to form and develop a brain with the propensity to be positively affected by experience of the external environment. Genes contain the codes for protein synthesis; protein synthesis is regulated according to local, internal physical, biochemical and cellular conditions. But there are no genes for particular brain structures and functions. We will return to this aspect of evolution in a subsequent section of this chapter.

C.7 Environment, diet and development

Going beyond the series of vertebrate ecological encounters that finally established the internal milieu, there is the matter of environmental sources of food and how they can be exploited. The tetrapods that first emerged from freshwater lakes or brackish lagoons were in all likelihood carnivores, but there was a rich fauna of land crustaceans and insects to be exploited, in addition to luxuriant vegetation. Whatever resource was chosen, the diners already had a complement of protein, fat and carbohydrate-digesting enzymes. No further novelty was necessary, although the proportions of the different classes of enzymes could be altered to suit the diet – something that is possible in an individual animal. The most difficult to digest potential food material was cellulose, a major component of terrestrial plants. Emergent vertebrates lacked cellulases, enzymes needed to break down the primary energy-containing carbohydrate of land plants. To exploit plant tissues completely, symbiotic bacteria, fungi and protozoans were needed. These were probably picked up at first with partially composted humus, which is more easily digested than fresh vegetation, and the free-living microorganisms then established the symbiotic relationship. The alimentary tract of tetrapods is much of a muchness, a series of compartments: buccal, oesophageal, gastric, intestinal, colonic and rectal. And they all have the same digestive enzymes in common. Symbiotic cellulose-utilizers are usually accommodated in *novel* gut compartments, the multiple 'stomachs' of cattle being the best-known example. Although we do not regard the primates as typical exemplars of cellulose exploitation, colobus monkeys do depend on symbiotic microorganisms, gorillas have cellulose-digesting bacteria in their enlarged hindguts, and the appendix of the human gut is taken to be the vestige of a compartment of the hindgut that may have housed symbionts as well as acting as a temporary storage compartment for plant food.

A further anatomical corollary of a vegetarian diet is a jaw and tooth structure suitable for continual mastication of tough stems and seeds. In contrast, a more elongated snout, with pronounced canine teeth, is to be found in primates such as baboons, whose eating habits are more carnivorous than most. The primitive primates were probably dietary generalists, feeding on insects, digestible leaves, fruits, nuts, and possibly tree-frogs, small reptiles and birds. Jaw and tooth development can be modified during the juvenile and early adult portion of the life cycle according to the choice of diet. It has been argued that in modern humans, over the last thousand years, there has been a reduction in the size of the jaw, due to a shift in diet from badly ground grains and tough game meat to the more easily masticable soft breads, and cooked domestic meats. How much of this change is gene determined, epigenetically determined or sexually selected has not been established. However, a claim has been staked for a point mutation in a gene that codes for muscle protein. The mutation weakens the jaw muscles, which supposedly results in a shift in feeding behaviour and a diminution of the mass of the sagittal crest of the skull and cheekbones (Stedman *et al.*, 2004). Alternatively, a paedomorphic change could have had similar effects, i.e., the condition where the mature organism does not develop ancestral anatomy but is still similar to the juvenile form by the time it matures sexually. Brown *et al.* (2004) note that *H. floresiensis* has 'facial and dental proportions, post-cranial anatomy consistent with human-like obligate bipedalism and a masticatory apparatus most similar in relative size and function to modern humans'.

Although the above discussion of the impact of environment has taken us back far earlier than the short period of hominin evolution with which we are mainly interested in this book, it illustrates some of the major steps in the evolution of physiological adaptability, and the impact of environment on that evolution. It also raises some principles pertinent to human evolution. Clearly, body fluid salinity, oxygen levels and temperature that contribute to the homeostasis of the placental mammals were environmentally set prior to their origination, but together they constituted the generative conditions for that emergence. Recollect from our discussion of emergence in Appendix A: 'Evolutionary theory', emergence may be catalyzed by a single change, but it only works in the context of other features that also have to be in place. These are the generative conditions.

C.8 The homeostasis paradox

Homeostatic organisms are able to continue to function when the environment changes, and to alter their actions when the environment stays the same. Homeostasis is an internal physiological state that resists environmentally imposed change. How then, if evolution is defined loosely as biological change, does the change-resistant homeostatic condition evolve? And evolve it does, in terms of its salinity, organic solutes, circulatory complexity and temperature regime.

We have just been considering how this might be brought about. Whatever levels of homeostasis were possessed by our early vertebrate ancestors, they were disequilibrated by the environment, and subsequent re-equilibration involved greater complexity of self-organization. This is not a general rule, since the emergence of novel physiological organs, such as the placenta, is not related to external environmental change. They were *endogenously* produced, i.e., by internal processes. The anatomical diversification of the placentals did involve changes in behaviour and environment. However, the latter changes would not have been possible unless the placentals already possessed a dynamic, homeostatic condition whose physical and chemical parameters have not subsequently changed. Beyond that foundation, progressive evolution of the primates involved expansions of certain parts of the brain, rewiring some of the central nervous system, and improving the range of chemical messengers that allow brain cells to communicate with each other. These internal evolutionary changes were linked to the environment through behaviour. Also, there are some clear examples of climatic change that affected hominin evolution in its early stages, and shortly we will consider just how important such changes were and are for modern humans.

C.9 The primate lineage: neurophysiology; neocortical expansion; foetalization of hominids

This section is meant to focus on physiology and environment. Nevertheless, we must return to the epigenetic causal arena, introduced in Appendix B, to understand how the major structures that were capable of physiological/behavioural functions such as intelligence came into existence. Neocortical expansion of the forebrain is common to all placentals. Biologists call such differential size changes 'allometry'. They have been integral to the evolution of body forms of all multicellular plants and animals. In the previous section we looked at the famous example of the evolution of the giraffe form. But here we are more concerned with allometric shifts in brain structure and function.

Only in the primates does the cerebrum dominate the rest of the brain in size, and fold in a way that extends the space available for neurons. This kind of progressive packaging makes the most of spatial distribution, and also includes lamination of the neocortex into layers of neurons. Sometime in the hominid lineage a novelty emerged that permitted specialization in the two halves of the brain. This was the 'corpus callosum', a transverse connecting trunk of nervous tissue that maintained the functional integrity of the forebrain despite the growing independence of its left and right hemispheres. Within the neocortex, it was not only the large increase in the number of neurons that potentiated higher brain functions such as logic, language, memory and intelligence. This was also achieved by the number of dendritic connections between the nerve cells. The dendrites are a multiplicity of thread-like 'roots' possessed by each neuron. The more the cell has of these, the more cells it can communicate with. How the links are established depends on the experience of the individual organism. Such connections integrated the specialized regions of the brain, so that although logic, speech and grammar may occupy separate loci, they are all integrated in the final output of language.

The normal development of the brain depends on a reliable homeostasis and source of nutrients. There is some heritable variation of cerebral abilities among individuals. For example, mathematical and musical prowess seems to run in families. However, the development of brain functions, including dendritic complexification, depends also upon the experiences of the individual, and especially on education. On top of that there is a positive feedback effect. Such brain enrichment not only adds to memory and the ability to access it and use it intelligently, it increases the potential for absorbing more experience, leading to further enrichment. With such developmental stimulation from the outside environment – usually from other individuals of the same species – dendritic connections, and hence the ability to make intelligent connections between independent scraps of information, can be maximized. They can be further complexified by the differentiation of neurotransmitters. These are the molecules that pass along neuron axons and stimulate a signal in the adjacent nerve cells. Many are small peptides; amino acid polymers. Gene duplication, combined with point mutations, can give rise to families of these peptides whose members have different functions.

Evolution of the functional anatomy of the brain must be seen in a larger anatomical and behavioural context. A large brain signifies a large head, whose size is limited by the girth and flexibility of the mother's pelvis. If a large brain is to develop further, it requires an extensible cranium. Therefore the infant with an unprotected brain is physically vulnerable, and needs a nurturing and protective family unit. Thus the physical emergence of a human child at birth demanded the simultaneous evolutionary emergence of multiple anatomical, physiological and behavioural qualities, combined with appropriate generative factors, including physiological and behavioural adaptabilities that we have already considered. Nor should we lose sight of the epigenetic adaptabilities discussed in the previous chapter. Moreover, epigenetic changes do not only act in utero, they continue after birth until the animal is fully developed. In fact, once we re-focus upon human evolution, we will find that the most obvious responses to environment, in particular to tree dwelling, walking on open ground, and living at high altitudes and temperature extremes are a combination of epigenetic and behavioural changes, and physiological adjustments.

C.10 Comparison of the adaptability and adaptations of humans and other placentals: generalization vs. specialization

First, we should remember that adaptability is defined as the individual's ability to modify itself, physiologically, behaviourally and, to some extent anatomically – i.e., organism determined. Adaptation is defined as a genetically fixed characteristic that is not subject to self-adjustment – gene determined. The human form is adaptable after birth to the conditions of development and maturation. Even in mature humans, anatomical adjustments can be made to fit the requirements of such extreme behaviours as gymnastics, rock climbing, running, diving, swimming, ballet dancing, and piano playing; body building being one of the most obvious examples.

Amongst the diversity of genetically fixed and specialized forms of other placental types the adaptable human form performs comparatively well. Some humans can outrun horses, for example, and present some competition to swimming and diving mammals. As anatomical generalists we are in a position to undertake a wide variety of activities that other placentals can only outdo in their own areas of specialization. Those behavioural specialists have body forms that are usually said to be 'adapted to their way of life', which is almost a truism, were it not for exceptions like Slijper's little bipedal goat whose anatomical development was *adaptable* to a novel way of life. Even flying insectivorous specialists can show a remarkable behavioural adaptability. The short-tailed New Zealand bat roosts and flies like other bats, but descends to the ground to forage among leaf litter for insects and worms, using its folded wings as legs, instead of preying upon flying insects like most other bats.

The adaptational consequences of diversification (sometimes known as 'adaptive radiation'), found in specialist placentals, are all based on a foundation of physiological adaptability. It was because they were initially physiologically adaptable that they could become behaviorably adaptable, and then choose the conditions of life to which they would subsequently adapt – again in the sense of acquiring genetically fixed traits. Among the specialists, innovative behaviour that would take them *away* from the conditions to which they are adapted is usually of little value. Quite the opposite is true for the human generalist whose innovative behaviour can be combined with intelligence and education to explore and improve the conditions of life. Cold-adapted placentals have thick, insulating fur coats. Generalist humans put on the skins of such animals as clothing in cold climates, and take them off again when it gets warm. This is not to say that humans lack special adaptations that can be correlated to geographical location or climate. Dark skin colour protects the epidermis from ultraviolet radiation; thick layers of sub-cutaneous fatty tissues are extra insulation in cold climates. The length of limbs and form of the trunk are correlated with temperature extremes. Short limbs and stocky bodies tend to retain body heat. Thin bodies and long limbs extend the surface area for cooling by radiation and the evaporation of sweat. High lung capacity and blood haemoglobin titre are considered to be adaptations to altitude. However, in modern humans, these adaptations have not become so specialized as to result in speciation due to different ways of life. Some neo-Darwinists argue that small populations that have unique adaptations to the local environment will become new species if isolated from the larger parental population. Clearly, isolation and adaptation have been insufficient to lead to mutual sterility among different human groups.

These examples do, however, provide us with some inkling of what is biologically possible for the human species, as a result of exploratory behaviour that takes us to the climatic extremes. And it is unlikely that further adaptational specializations will occur in the future, regardless of environmental change – unless humans are somehow catastrophically subjected to extended periods of confinement in stressful conditions. *Adaptability*, both physiological and behavioural, is a far more important quality for humans – past, present and future. On the whole, physiological adaptabilities have remained the same since the placentals first emerged. However, the special organism-determined quality of humans is intelligent individual behaviour, combined with familial and social interactions, both in their stable, consistent states and also in the form of revolutionary changes. Under stable conditions, conventional behaviours may bring the greatest benefit to the social group, and may result in the genetic assimilation of minor anatomical changes. Revolutionary changes may be due to the emergence of bright ideas, or of sociopolitical stress such as war, disease or climate change. If, as a result, we begin to behave differently, we alter the conditions that might generate further change or the re-establishment of equilibria.

Further human speciation through special gene-determined adaptations is presently unlikely, due to the 'global village' aspect of the modern human condition. Instead of seeking or remaining in stressful isolation, we now know there are options, and tend to congregate in the most congenial conditions where environmental effects are minimized. This can lead to other kinds of stress, but with little of the evolutionary potential that comes of being restricted to hot, dry climates, rainforest conditions or glaciations.

C.11 Adaptability or variability?

At first glance, adaptability might literally be regarded as a kind of variability. However, as used by neo-Darwinists, variability is a population characteristic. Conventional theory treats individual organisms as ephemeral aggregates of genes which do not evolve and are therefore evolutionarily insignificant. For such a theory the gene pool is a greater reality than the organism. The concept of the gene pool, meaning the totality of genes in a population, is a product of the American version of the modern synthesis, championed by Ernst Mayr and others.[6] However, the gene pool is a mental construct, a metaphorical and mathematical convenience, and, no matter how persuasive, metaphors are not mechanistic causes. The contents of gene pools reside only in the individual organisms that make up breeding populations, and it is a serious error to ignore how individual organisms act and interact. What has been set out as a highly significant evolutionary factor in the earlier portion of this appendix is the adaptability of individual organisms: behavioural, physiological and developmental. This quality operates above the gene level, and although it involves gene expression, it cannot be detected in the putative gene pool. Therefore we must avoid confusing adaptability with neo-Darwinist variability.

Some authors find it difficult to discriminate. Richard Potts, in 'Evolution and climate variability' (1996), which addresses hominid evolution, suggests that human qualities emerged in response to natural conditions, and that evolutionary response required flexibility and the capacity 'to adjust and diversify our behavior, physiology and overall way of life',[7] enabling us ultimately to modify our surroundings. We could not agree more. He also argues that 'variability selection' is important during changing climatic conditions. He bases his case for variability selection on the variability of gene pools, yet, in addition to novel gene combinations, he refers to 'complex behaviours' as things that can be selected (presumably novel gene-determined and hence heritable behaviours arising from the novel gene combinations). He further begs the question by inferring that bipedality, relative brain size, technological innovation, manipulation of symbols, dental morphology and anatomical adaptations to climate are types of 'flexibility' that arise from gene-pool variability. What is wrong with this proposition?

First, we have demonstrated earlier that almost all of the kinds of flexibility, or as we call it, *adaptability*, are expressions of the phenotype, not the genotype. This case was made strongly by David Rollo in *Phenotypes: Their Epigenetics, Ecology and Evolution* (1994). The genetic code is involved in the operation of adaptability, since it contains the information for the structure of proteins that are necessary for anatomy, as well as enzymes and peptide hormones. However adaptability is not caused by genes, but arises from the interaction of the internal hierarchical systems of the organism in its environment. And those higher level hierarchies have a downwardly causal effect on gene expression. This kind of holism or interactionism was proposed by Susan Oyama in 1985. Unfortunately, the hold of neo-Darwinism is such that most anthropologists, psychologists and even philosophers accept its premises without question, and discussions of evolutionary physiology are rare in scientific literature.

Furthermore, individual adaptability is not proportional to gene-pool variability. Whatever contribution gene-pool variability might make to evolution, the flexibility/

adaptability of the individual hominin far outweighs it. And the physiological foundation for such flexibility already existed in the placental mammalian lineage that emerged as the primates. Further additions to those adaptabilities depended on the actions of individual organisms, the direct impact of environmental change, and the feedback links among behaviour, physiology and epigenetics. Out of those interactions came a more complex and more intelligent brain. Not only that, exploratory groups of any kind of animal testing the nature of new environments are very small in comparison to the parent population, and carry a small sample of the gene pool. Therefore, if those exploratory groups settle under new conditions, their gene pools are bound to be *less* varied than those of the parent population.

Note on 'stress': up to this point we have been using this term loosely. 'Stressful conditions' are readily comprehensible as conditions under which we, and other organisms, would prefer not to live. However, to a physiologist, stress relates to energy expenditure. Under stressful conditions an animal has to adjust frequently or continuously in its physiological and behavioural responses to the environment. This requires greater energy use than would be necessary in more restful conditions. For example, if we go out into a winter's day without adequate insulation we shiver constantly, using a lot more stored energy than usual. Stress has the additional connotation of distraction from normal activities. For example, constant shivering distracts us from work that requires skilled manipulation of small objects. In the case of psychological stress, the physiological principles also apply. There may be a constant drain on some kinds of hormones, such as epinephrine. The distraction element is even more significant in psychologically stressful conditions.

C.12 Summary of environmental impacts on humans – from molecules to mayhem

In this appendix on adaptability, and in the previous one on epigenetics, we have introduced evolutionary mechanisms, processes and principles that have a bearing on one of our major goals, namely understanding human evolution. In some instances, for example, regarding how our blood salt content was initially determined by environmental conditions, we have had to go back hundreds of millions of years. Furthermore, since the basic adaptability of humans is common to all placental mammals, we have returned to their past to understand our present. Without such undertakings, we would leave the false impression that knowledge of the human condition is sufficient to address our future.

At this point it will be helpful to summarize all the known ways in which the environment can have an impact on humans that might be heritable, and relate these to gene-determined adaptations and organism-determined adaptability.

1. Point mutations: mutations of the nitrogenous bases of DNA that alter the code for protein amino acid sequences. A classical example from human genetics is the single point mutation in the DNA coding for haemoglobin. It produces sickle-cell anaemia, which though debilitating in its most extreme expression, provides some protection from malaria.

   Point mutations are spontaneous, but increase under environmental physicochemical extremes such as hard radiation including ultraviolet light, high temperatures, the presence of mutagenic chemicals and possibly electromagnetic influences.

   Point mutations may be more likely in some regions of the DNA than others (Caporale, 2003). For example, hypermutation in antibody genes has been universally valuable for the immune system, ever since it emerged in early vertebrates.

   Point mutations typically come under the *adaptation* category, since they are genetically fixed. However, point mutations in a series of duplicated and reduplicated genes can provide *adaptability* as well i.e., a system that can be modified by the individual organism to suit current conditions.

2. Heat-shock proteins, also known as stress proteins, normally function in protein molecular folding. Under environmental stress, these functions are diverted, and the mutability of the DNA increases. Methylation patterns, i.e., patterns of gene repression, are also changed.

3. Toxic chemical stress has been correlated with duplication of genes involved in the synthesis of enzymes that break down the toxins. In other words, more copies of the genes mean more copies of the enzymes.

4. Humans exposed for generations to prolonged sun exposure have had alterations in skin pigmentation, involving the epigenetic intensification of melanin production in skin cells, which protects from cellular damage, including skin cancer. Tanning is an example of adaptability in an individual. However, when the skin tone is heritable, it is considered an *adaptation*.

5. The evaporation of sweat cools the body surface during prolonged exercise. However, individual humans acclimatize to high temperatures by reducing perspiration rates, thereby saving water. Thus, sweating is an *adaptable* physiological function.

6. In hot climates, long limbs extend the skin area, and greater evaporative cooling of sweat is possible. Limb-length, being heritable, is considered an *adaptation*.

7. Hairlessness is correlated with sweating during strenuous exercise in hot conditions. The amount of body hair in humans is heritable, hence *adaptational*. (In animals such as dogs whose hair can be partially shed for the summer, the hair loss is an example of *adaptability*.)

8. Humans exposed for generations to high altitudes have developed high haemoglobin counts and enlarged lungs. The amount of red blood cells can be increased when lowland mammals take to dwelling on high mountains, hence is *adaptable*. An increase in lung capacity is however, epigenetically fixed.

9. Humans exposed for generations to low temperatures have increased subcutaneous adipose layers, shorter than average appendages, and stocky trunks, adaptations to heat loss. These are typically *adaptations* to the cold environment, although the amount of fat is modifiable.

10. The social environment of extended families includes hormonal responses cued by a variety of environmental stimuli, and probably involves pheromones, i.e., airborne

hormones that can affect both the person who secretes them and a person who smells them.

11. 'Togetherness' and its social consequences are increased by limitation of habitation, such as the small number of suitable caves in cold environments.

In Part I of the text we focussed on the recent evolutionary history of the genus *Homo*. While there are some imponderable factors in its interpretation, we do at least know from solid evidence that our species and its immediate ancestors did evolve. Also, we are now equipped with the basic information regarding human development and physiology that will help us to make more sense of the process. The key points are as follows.

In venturing into the unknown, humans have depended upon the physiological homeostasis characteristic of all placental mammals. That homeostasis allowed most of the placental mammals to diverge and specialize, behaviourally and anatomically. However, our own ancestors opted to be generalists. As a consequence they created the generative conditions for the emergence of the bipedalism, skillful manipulative abilities and the higher brain functions characteristic of *Homo sapiens*. To be realized, these emergences required a causal matrix involving environmental, behavioural (individual and social), physiological and molecular interactions. The result is that our species can enter and survive in the greatest environmental extremes, and potentially create such extremes by our own folly.

We have taken pains to distinguish between genetically fixed *adaptation*, and organism-determined *adaptability*, which implies the individual's ability to self-modify or adjust to environmental changes. In everyday language that distinction is blurred or non-existent, despite its importance for evolutionary theory. The problem is particularly evident when people talk about human 'adaptation'. For example, 'Humans have adapted to hot climates by having black skins' and 'We adapt to the cold by putting on warmer clothes'. Behavioural adaptability is particularly prone to being equated with genetic adaptation. Throughout this book we have stressed the distinction between adaptation and adaptability as we have defined them. These are important considerations when we look toward the future of the human species and our capacity to adjust to future climate change.

Notes

1. The origin of Bernard's aphorism is supposed to have been a lecture given in 1873.
2. See references for Wynne-Edwards (1998, 1999), and her co-worker H. J. McMillan and Wynne-Edwards (1998, 1999).
3. This information is gleaned from contemporary Mexican newspaper reports. We were unable to find scientific reports that presented statistical analyses.
4. See Brown *et al.* (2004, pp. 1060–1).
5. Details of this history are elaborated by Reid (2007).
6. See Mayr and Provine (1980).
7. Potts (1996b, p. 12).

References

Abramova, Z. A. (1985). Must'erskiy grot v Khakasii (The Mousterian shelter in Khakasia). *Kratkiye Soobschcheniya Instituta Arkheologii AN SSSR*, **181**, 92–8.

Adovasio, J. M., Pedler, D. R., Donahue J., and Stuckenrath, R. (1998). Two decades of debate on Meadowcroft Rockshelter. *North American Archaeologist*, **19**, 317–41.

Aguirre, E., and Carbonell, E. (2001). Early human expansions into Eurasia: the Atapuerca evidence. *Quaternary International*, **75**, 11–18.

Aksu, A. E., Hisott, R. N., Mudie, P. J., Rochon, A., Kaminski, M. A., Abrajano, T., and Yasar, D. (2002). Persistent Holocene outflow from the Black Sea to the eastern Mediterranean contradicts Noah's flood hypothesis. *GSA Today*, **12**, 4–10, doi: 10.1130/1052–5173(2002)012<0004.PHOFTB>2.0.CO;2.

Algaze, G. (2001). The prehistory of imperialism: the case of Uruk Period Mesopotamia. In *Uruk Mesopotamia and Its Neighbours: Cross-Cultural Interactions in the Era of State Formation*, ed. S. Rothman. Santa Fe, NM: School of American Research Press, pp. 27–84.

Allchin, B., and Allchin, R. (1997). *Origins of a Civilization: The Prehistoric and Early Archaeology of South Asia*. New Delhi: Penguin Books.

Ambrose, S. H. (1998a). Chronology of the Later Stone Age and food production in East Africa. *Journal of Archaeological Science*, **25**, 377–92.

(1998b). Late Pleistocene human population bottlenecks, volcanic winter, and differentiation in modern humans. *Journal of Human Evolution*, **34**, 623–51.

(2001). Paleolithic technology and human evolution. *Science*, **291**, 1748–53.

(2002). Small things remembered: origins of early microlithic industries in sub-Saharan Africa. In *Thinking Small: Global Perspectives on Microlithization*, ed. R. G. Elston and S. L. Kuhn. *Archaeological Papers of the American Anthropological Association*, **12**, 9–29.

(2003). Did the super-eruption of Toba cause a human population bottleneck? *Journal of Human Evolution*, **45**, 231–7.

An, Z. S., Gao, W., Zhu, Y., Kan, X., Wang, J., Sun, J., and Wei, M. (1990). Magnetostratigraphic dates of Lantian *Homo erectus*. *Acta Anthropológica Sinica*, **9**, 1–7.

An, Z. S., and Ho., C. K. (1989). New magnetostratigraphic dates of Lantian *Homo erectus*. *Quaternary Research*, **32**, 213–21.

An, Z. S., Kutzbach, J. E., Prell, W. L., and Porter, S. C. (2001). Evolution of Asian monsoons and phased uplift of the Himalaya-Tibetan plateau since Late Miocene times. *Nature*, **411**, 62–6.

357

Anderson, D. (1990). *Long Rongrien*. Philadelphia: University of Pennsylvania Museum.

Andrus, C. F. T., Crowe, D. E., Reitz, D. H., and Romanek, C. S. (2004). Otolith $\delta^{18}O$ record of mid-Holocene sea surface temperatures in Peru. *Science*, **22**, 1508–11.

Antón, S. C. (2002). Evolutionary significance of cranial variation in Asian *Homo erectus*. *American Journal of Physical Anthropology*, **118**, 301–23.

Antón, S. C., and Indriati, E. (2002). Earliest Pleistocene *Homo* in Asia: comparisons of Dmanisi and Sangiran. *American Journal of Physical Anthropology Supplement*, **34**, 38.

Antón, S. C., Leonard, W. R., and Robertson, M. L. (2002). An ecomorphological model of the initial hominid dispersal from Africa. *Journal of Human Evolution*, **43**, 773–85.

Antón, S. C., and Swisher, C. C. (2004). Early dispersals of *Homo* from Africa. *Annual Review of Anthropology*, **33**, 271–96.

Ariai, A., and Thibault, C. (1975). Nouvelles precisions à propos de l'outillage paléolithique ancien sur galets du Khorassan (Iran). *Paleorient*, **3**, 101–8.

Arnold, J. E. (1995). Transportation innovation and social complexity among maritime hunter-gatherer societies. *American Anthropologist*, **97**, 733–47.

Arrhenius, S. (1896). On the influence of carbonic acid in the air upon the temperature of the ground. *Philosphical Magazine and Journal of Science* (5th series), **41**, 237–75.

Arz, H. W., Pätzold, J., and Wefer, G. (1998). Correlated millennial-scale changes in surface hydrography and terrigenous sediment yield inferred from last-glacial marine deposits off northeastern Brazil. *Quaternary Research*, **50**, 157–66.

Asfaw, B., Beyene, Y., Suwa, G., Walter, R. C., White, T. D., WoldeGabriel, G., and Yermane, T. (1992). The earliest Acheulean from Konso-Gardula. *Nature*, **360**, 732–5.

Asfaw, B., Gilbert, W. H., Beyene, Y., Hart, W. K., Renne, P. R., WoldeGabriel, G., Vrba, E. S., and White, T. D. (2002). Remains of *Homo erectus* from Bouri, Middle Awash, Ethiopia. *Nature*, **416**, 317–20.

Avis, C., Montenegro, A., and Weaver, A. J. (2007). The discovery of Western Oceania: a new perspective. *Journal of Island and Coastal Archaeology*, **2**, 197–209.

Ayala, F. J. (1995). The myth of Eve: molecular biology and human origins. *Science*, **270**, 1929–36.

Ba Maw (1995). The first discovery of the early man's fossilized maxillary bone fragment in Myanmar. *East Asian Tertiary Quaternary Newletter*, **16**, 72–9.

Ba Maw, Than Tun Aung, Pe Nyein, and Tin Nyein (1998). Artifacts of Anyathian cultures found in a single site. *Myanmar Historical Research Journal*, **2**, 7–15.

Baba, H., Aziz, F., Kaifu, Y., Suwa, G., Kono, R. T., and Jacob, T. (2003). *Homo erectus* calvarium from the Pleistocene of Java. *Science*, **299**, 1384–8.

Baker, P. A., Seltzer, G. O., Fritz, S. C., Dunbar, R. B., Grove, M. J., Tapia, P. M., Cross, S. L., Rowe, H. D., and Broda, J. P. (2001). The history of South American tropical precipitation for the past 25,000 years. *Science*, **291**, 640–3.

Baksi, A. K., and Hoffman, K. A. (2000). On the age and morphology of the Reunion Event. *Geophysical Research Letters*, **27**, 2997–3000.

Balter, M. (2002). Becoming human: what made humans modern? *Science*, **295**, 1219–25.

Barber, D. C., Dyke, A., Hillaire-Marcel, C., Jennings, A. E., Andrews, J. T., Kerwin, M. W., Bilodeau, G., McNeely, R., Southon, J., Morehead, M. D., and Gagnon, J.-M. (1999). Forcing of the cold event of 8,200 years ago by catastrophic drainage of Laurentide lakes. *Nature*, **400**, 344–8.

Barker, G. (2005). The archaeology of foraging and farming at Niah Cave, Sarawak. *Asian Perspectives*, **44**, 90–106.

Bartstra, G.-J. (1982). *Homo erectus erectus*: the search for his artifeacts. *Current Anthropology*, **23**, 318–20.

Bar-Yosef, O. (1992). The role of western Asia in modern human origins. *Philosophical Transactions of the Royal Society of London, Series B, Biological Sciences*, **337**, 193–200.

(1994). The lower Paleolithic of the Near East. *Journal of World Prehistory*, **8**, 211–66.

Bar-Yosef, O., and Belfer-Cohen, A. (2001). From Africa to Eurasia – early dispersals. *Quaternary International*, **75**, 19–28.

Bar-Yosef, O., and Kuhn, S. (1999). The big deal about blades: laminar technologies and human evolution. *American Anthropology*, **101**, 322–38.

Baskaran, M., Marathe, A. R., Rajaguru, S. N., and Somayajulu, B. L. K. (1986). Geochronology of Palaeolithic cultures in the Hiran Valley, Saurashtra, India. *Journal of Archaeological Science*, **13**, 505–14.

Batista, O., Kolman, C. J., and Bermingham, E. (1995). Mitochondrial DNA diversity in the Kuna Amerinds of Panama. *Human Molecular Genetics*, **4**, 921–9.

Behl, R. J., and Kennett, J. P. (1996). Brief interstadial events in the Santa Barbara basin, NE Pacific, during the past 60 kyr. *Nature*, **379**, 243–6.

Behrensmeyer, A. K., and Bobe, R. (1999). Change and continuity in Plio-Pleistocene faunas of the Turkana Basin, Kenya and Ethiopia. In *The Environmental Background to Hominid Evolution in Africa*, INQUA XV International Congress, Book of Abstracts, eds. J. Lee-Thorp and H. Clift. Rondebosch, South Africa: University of Cape Town, p. 20.

Bellwood, P. S. (1989). The colonization of the Pacific: some current hypotheses. In *The Colonization of the Pacific: A Genetic Trail*, eds. A. V. S. Hill and S. W. Serjeantson. Oxford: Oxford University Press, pp. 1–59.

(1992). Southeast Asia before history. In *The Cambridge History of Southeast Asia*, ed. N. Tarling. Cambridge: Cambridge University Press, pp. 55–136.

(1996). Early agriculture and the dispersal of the southern Mongoloids. In. *Prehistoric Mongoloid Dispersals*, eds. T. Akazawa and E. J. E. Szathmary. Oxford: Oxford University Press, pp. 287–302.

Belmaker, M., Tchernov, E., Condemi, S., and Bar-Yosef, O. (2002). New evidence for hominid presence in the Lower Pleistocene of the Southern Levant. *Journal of Human Evolution*, **43**, 43–56.

Belyaev, D. K. (1979). Destabilizing selection as a factor in domestication. *Journal of Heredity*, **70**, 301–8.

Bergen, A. W., Wang, C. Y., Tsai, J., Jefferson, K., Dey, C., Smith, K. D., Park, S. C., Tsai, S. J., and Goldman, D. (1999). An Asian-Native American paternal lineage identified by RPS4Y resequencing and microsatellite haplotyping. *Annals of Human Genetics*, **63**, 63–80.

Berger, W. H., and Jansen, E. (1994). Mid-Pleistocene climate shift: the Nansen connection. In *The Polar Oceans and Their Role in Shaping the Global Environment, Geophysical Monograph Series 85*, eds. O. M. Johannessen, R. D. Muench, and J. E. Overland. Washington, DC: American Geophysical Union, pp. 295–311.

Bermúdez de Castro, J. M., Martinón-Torres, M., Carbonell, E., Sarmiento, S., Rosas, A., Van der Made, J., and Lozano, M. (2004). The Atapuerca sites and their contribution to the knowledge of human evolution in Europe. *Evolutionary Anthropology*, **13**, 25–41.

Betts, J. (2008). Western Silviculture Contractors' Association: time to act decisively on MPB. *Canadian Silviculture*, May, 17.

Bhattacharjee, Y. (2004). In the courts: posthumous victory. *Science*, **305**, 605.

Bianchi, N. O., Catanesi, C. I., Bailliet, G., Martinez-Marignac, V. L., Bravi, C. M., Vidal-Rioja, L. B., Herrera, R. J., and Lopez-Camelo, J. S. (1998). Characterization of ancestral and derived Y-chromosome haplotypes of New World native populations. *American Journal of Human Genetics*, **63**, 1862–71.

Bickerton, D. (1995). *Language and Human Behavior*. Seattle: Washington University Press.

(1998). Catastrophic evolution: the case for a single step from protolanguage to full human language. In *Approaches to the Evolution of Language*, eds. J. R. Hurford, M. Studdert-Kennedy, and C. Knight. Cambridge: Cambridge University Press, pp. 341–58.

Binford, L. R. (1985). Human ancestors: changing views of their behavior. *Journal of Anthropological Archaeology*, **4**, 292–327.

(1987). An interview with Lewis Binford (edited by A. C. Renfrew). *Current Anthropology*, 28, 683–94.

(1989). Isolating the transition to cultural adaptations: an organizational approach. In *The Emergence of Modern Humans: Biocultural Adaptations in the Later Pleistocene*, ed. E. Trinkhaus. Cambridge: Cambridge University Press, pp. 18–41.

(1992). Hard evidence. *Discover*, February, 44–51.

Bird, M. I. (1995). Fire, prehistoric humanity and the environment. *Interdisciplinary Science Reviews*, **20**, 141–54.

Bird, M. I., Hope, G., and Taylor, D. (2004). Populating PEP II: the dispersal of humans and agriculture through Austral-Asia and Oceania. *Quaternary International*, **118–119**, 145–63.

Bischoff, J. L., Shamp, D. D., Aramburu, A., Arsuaga, J. L., Carbonell, E., and Bermudez de Castro, J. M. (2003). The Sima de los Huesos hominids date to beyond U/Th equilibrium (> 350 kyr) and perhaps to 400–500 kyr: new radiometric dates. *Journal of Archeological Science*, **30**, 275–80.

Bischoff, J. L., Soler, N., Maroto, J., Julia, R. (1989). Abrupt Mousterian/Aurigancian boundary at c. 40 ka bp: accelerator 14C dates from L'Arbreda Cave (Catalunya, Spain). *Journal of Archaeological Science*, **16**, 563–76.

Bitterman, M. E. (2000). Cognitive evolution: a psychological perspective. In *The Evolution of Cognition*, C. Heyes and L. Huber. London: MIT Press, pp. 61–79.

Black, D. (1931). On an adolescent skull of *Sinanthropus pekinensis* in comparison with an adult skull of the same species and with other hominid skulls, recent and fossil. *Palaeontologia Sinica Series D*, **7**, 1–145.

Bobrowsky, P., and Rutter, N. W. (1992). The Quaternary geologic history of the Canadian Rocky Mountains. *Géographic Physique et Quaternaire*, **46**, 5–50.

Boesch, C., and Boesch, H. (1984). Possible causes of sex differences in the use of natural hammers by wild chimpanzees. *Journal of Human Evolution*, **13**, 415–40.

Boesch-Acherman, H., and Boesch, C. (1994). Hominisation in the rainforest: the chimpanzee's piece to the puzzle. *Evolutionary Anthropology*, **3**, 9–16.

Bolk, L. (1929). The origin of racial characteristics in man. *American Journal of Physical Anthropology* 13, 1–28.

Bolnick, D. A. (2004). Using Y-chromosome and mtDNA variation to reconstruct eastern North American population history. *American Journal of Physical Anthropology Suppl.*, **123**, 65 (Abstr.).

Bolnick, D. A., and Smith, D. G. (2003). Unexpected patterns of mitochondrial DNA variation among Native Americans from the southeastern United States. *American Journal of Physical Anthropology*, **122**, 336–54.

Bonatto, S. L., Redd, A. J., Salzano, F. M., and Stoneking, M. (1996). Lack of ancient Polynesian-Amerindian contact. *American Journal of Human Genetics*, **59**, 253–8.

Bond, G., Showers, W., Chezebiet, M., Lotti, R., Almasi, P., deMenocal, P., Priore, P., Cullen, H., Hajdas, I., and Bonani, G. (1997). A pervasive millennial-scale cycle in North Atlantic Holocene and glacial climates. *Science*, **278**, 1257–66.

Bonnefille, R., and Chalié, F. (2000). Pollen-inferred precipitation time-series from equatorial mountains, Africa, the last 40 kyr BP. *Global and Planetary Change*, **26**, 25–50.

Bortolini, M-C., Salzano, F. M., Thomas, M. G., Stuart, S., Nasanen, S. P., Bau, C. H., Hutz, M. H., Layrisse, Z., Petzl-Erler, M. L., Tsuneto, L. T., Hill, K., Hurtado, A. M., Castro-de-Guerra, D., Torres, M. M., Groot, H., Michalski, R., Nymadawa, P., Bedoya, G., Bradman, N., Labuda, D., and Ruiz-Linares, A. (2003). Y-chromosome evidence for differing ancient demographic histories in the Americas. *American Journal of Human Genetics*, **73**, 524–39.

Bosinski, G. (1995). The earliest occupation of Europe: Western Central Europe. In *The Earliest Occupation of Europe: Proceedings of the European Science Foundation Workshop at Tautavel (France), 1993*, eds. W. Roebroeks and T. van Kolfschoten. Leiden: University Press, pp. 103–28.

Bowdler, S. (1990). Peopling Australasia: the 'Coastal Colonization' hypothesis re-examined. In *The Emergence of Modern Humans – An Archaeological Perspective*, ed. P. Mellars. Ithaca, NY: Cornell University Press, pp. 327–43.

Bowen, B. W., Abreau-Grobois, F. A., Balazs, G. H., Kamezaki, N., Limpus, C. J., and Ferl, R. J. (1995). Trans-Pacific migrations of the loggerhead turtle (*Caretta caretta*) demonstrated with mitochondrial DNA markers. *Proceedings of the National Academy of Sciences of the United States of America*, **92**, 3731–4.

Bowler, J. M., Johnston, H., Olley, J. M., Prescott, J. R., Roberts, R. G., Shawcross, W., and Spooner, N. A. (2003). New ages for human occupation and climatic change at Lake Mungo, Australia. *Nature*, **421**, 837–40.

Brahic, C. (2007). Sea Level Rise Outpacing Key Predictions, *New Scientist Environment* (February 1, 2007), www.environment.newscientist.com/article/dn11083.

Brain, C. K., and Sillen, A. (1988). Evidence from the Swartkans cave for the earliest use of fire. *Nature*, **336**, 364–6.

Bramble, D. M., and Lieberman, D. E. (2004). Endurance running and the evolution of *Homo*. *Nature* **432**, 345–52.

Bräuer, G. (1980). Die morphologischen Affinitäten des jungpleistozänen Stirnbeines aus dem Elbmündungsgebiet bei Hahnöfersand. Zeitschrift für *Morphologie und Anthropologie*, **71**, 1–42.

  (1984). A craniological approach to the origin of anatomically modern *Homo sapiens* in Africa and implications for the appearance of modern Europeans. In *The Origins of Modern Humans: A World Survey of the Fossil Evidence*, eds. F. H. Smith and F. Spencers. New York: Alan R. Liss, pp. 327–410.

  (1989). The evolution of modern humans: a comparison of the African and non-African evidence. In *The Human Revolution: Behavioural and Biological Perspectives on the Origins of Modern Humans*, eds. P. Mellars and C. Stringer. Princeton, NJ: Princeton University Press, pp. 123–54.

Bräuer, G., and Mehlman, M. J. (1988). Hominid molars from a Middle Stone Age level at the Mumba Rock Shelter, Tanzania. *American Journal of Physical Anthropology*, **75**, 69–76.

Brinkhuis, H., Schouten, S., Collinson, M. E., Sluijs, A., Damsté, J. S. S., Dickens, G. R., Huber, M., Cronin, T. M., Onodera, J., Takahashi, K., Bujak, J. P., Stein, R., van der Burgh, J., Eldrett, J. S., Harding, I. C., Lotter, A. F., Sangiorgi, F., van Konijnenburg-van Cittert, H., de Leeuw, J. W., Matthiessen, J., Backman, J., Moran, K., and the

Expedition 302 Scientists. (2006). Episodic fresh surface waters in the Eocene Arctic Ocean. *Nature*, **441**, 606–9.

Britten, R. J. (2002). Divergence between samples of chimpanzee and human DNA sequences is 5%, counting indels. *Proceedings of the National Academy of Sciences of the United States of America*, **99**, 13633–5.

Broadhurst, C. L., Wang, Y., Crawford, M. A., Cunane, S. C., Parkington, J. E., and Schmid, W. E. (2002). Brain-specific lipids from marine, lacustrine, or terrestrial food resources: potential impact on early African *Homo sapiens*. *Comparative Biochemistry and Physiology Part B: Biochemistry and Molecular Biology*, **131**, 653–73.

Broca, P. (1861). Sur le volume et la forme du cerveau suivant les individus et suivant les races. *Bulletin Société d'Anthropologie Paris*, **3** (part 2), 13 pp.

   (1873). Sur les crânes de la caverne de l'Homme-Mort (Lozère). *Revue d'Anthroplogie*, **2**, 1–53.

Broecker, W. S. (1995). Chaotic climate. *Scientific American*, **273**, 62–8.

   (1996). Glacial climate in the tropics. *Science*, **272**, 1902–3.

Broecker, W. S., and Hemming, S. (2001). Climate swings come into focus. *Science*, **294**, 2308–9.

Brooks, N. (2006). Cultural responses to aridity in the Middle Holocene and increased social complexity. *Quaternary International*, **151**, 29–49.

Brown, M. D., Hosseini, S. H., Torroni, A., Bandelt, H.-J., Allen, J. C., Schurr, T. G., Scozzari, R., Cruciani, F., and Wallace, D. C. (1998). mtDNA haplogroup X: an ancient link between Europe/Western Asia and North America? *American Journal of Human Genetics*, **63**, 1852–61.

Brown P. (2001). Chinese Middle Pleistocene hominids and modern human origins in East Asia. In *Human Roots – Africa and Asia in the Middle Pleistocene*, eds. L. Barham and V. Robson Brown. Bristol: Western Academic, and Specialist Publishers, pp. 135–45.

Brown, P., Sutikna, T., Morwood, M. J., Soejono, R. P., Jatmiko, Saptomo, E. W., and Due, R. A. (2004). A new small-bodied hominin from the Late Pleistocene of Flores, Indonesia. *Nature*, **431**, 1055–61.

Brunet, M., Guy, F., Pilbeam, D., Mackaye, H. T., Likius, A., Ahounta, D., Beauvilain, A., Blondel, C., Bocherens, H., Boisserie, J-R., De Bonis, L., Coppens, Y., Dejax, J., Denys, C., Duringer, P., Eisenmann, V., Fanone, G., Fronty, P., Geraads, D., Lehmann, T., Lihoreau, F., Louchart, A., Mahamat, A., Merceron, G., Mouchelin, G., Otero, O., Campomanes, P. P., Ponce De Leon, M., Rage, J-C., Sapanet, M., Schuster, M., Sudre, J., Tassy, P., Valentin, X., Vignaud, P., Viriot, L., Zazzo, A., and Zollikofer, C. (2002). A new hominid from the Upper Miocene of Chad, Central Africa. *Nature*, **418**, 145–50.

Bui Vinh. (1998). The stone age archaeology in Viêt Nam: achievements and general model. In *Southeast Asian Archaeology 1994*, vol. 1, ed. P.-Y. Manguin. Hull, UK: Centre for Southeast Asian Studies, University of Hull, pp. 5–12.

Burdge, G. C., Slater,-Jefferies, J., Torrens, C., Phillips, E. S., Hanson, M. A., and Lillycrop, K. A. (2007). Dietery protein restriction of pregnant rats in the F_0 generation induces altered methylation of hepatic gene promoters in the adult male offspring in the F_1 and F_2 generations. *2007 British Journal of Nutrition*, **97**, 415–39.

Burke, A., and Cinq-Mars, J. (1998). Paleoethological reconstruction and taphonomy of *Equus lambei* from the Bluefish Caves, Yukon Territory, Canada. *Arctic*, **51**, 105–15.

Burroughs, W. J. (2005). *Climate Change in Prehistory: The End of the Reign of Chaos*. Cambridge: Cambridge University Press.

Butler, S. (1879). *Evolution Old and New*. London: Hardwicke and Bogue.

Byrne, R. (1995). *The Thinking Ape: The Evolutionary Origins of Intelligence*. Oxford: Oxford University Press.

Cacho, I., Grimalt, J. O., Pelejero, C., Canals, M., Sierro, F. J., Flores, J. A., and Shackleton, N. (1999). Dansgaard-Oeschger and Heinrich event imprints in Alboran Sea paleotemperatures. *Paleoceanography*, **14**, 698–705.

Calvin, W. (2005). *The Creative Explosion*. Draft chapter for Academia Sinica book. See also WilliamCalvin.com/2005/CreativeExplosion.htm.

Cameron, D., Patnaik, R., and Sahni, A. (2004). The phylogenetic significance of the Middle Pleistocene Narmada hominin cranium from Central India. *International Journal of Osteoarchaeology*, **14**, 419–47.

Cann, R. L. (2001). Genetic clues to dispersal in human populations: retracing the past from the present. *Science*, **291**, 1742–8.

Cann, R. L., Stoneking, M., and Wilson, A. C. (1987). Mitochondrial DNA and human evolution. *Nature*, **325**, 31–6.

Cannon, W. (1932). *The Wisdom of the Body*. New York: Norton.

Caporale, L. H. (2003). *Darwin in the Genome: Molecular Strategies in Biological Evolution*. New York: McGraw-Hill.

Carbonell, E., Mosquera, M., Rodríguez, X. P., and Sala, R. (1999). Out of Africa: the dispersal of the earliest technical systems reconsidered. *Journal of Anthropological Archaeology*, **18**, 119–36.

Carto, S. L., Weaver, A. J., Hetherington, R., Lam, Y., and Wiebe, E. C. (2008). Out of Africa and into an ice age: on the role of global climate change in the Late Pleistocene migration of early modern humans out of Africa. *Journal of Human Evolution*, **56**, 139–51.

Cavalli-Sforza, L. L. (1991). Genes, peoples and languages. *Scientific American*, **265**, 72–8.

Cavalli-Sforza, L. L., Piazza, A., Menozzie, P., and Mountain, J. (1988). Reconstruction of human evolution: bringing together genetic, archeological and linguistic data. *Proceedings of the National Academy of Sciences of the United States of America*, **85**, 8002–6.

Chakravarti, A. (2001). Single nucleotide polymorphisms…to a future of genetic medicine. *Nature*, **409**, 822–3.

Chamberlain, J. G. (1996). The possible role of long-chain, omega-3 fatty acids in human brain phylogeny. *Perspectives in Biology and Medicine*, **39**, 436–45.

Chapman, M. R., and Shackleton, N. J. (1998). Millennial-scale fluctuations in North Atlantic heat flux during the last 150,000 years. *Earth and Planetary Science Letters*, **159**, 57–70.

Chatters, J. (2000). The recovery and first analysis of an early Holocene human skeleton from Kennewick, Washington. *American Anthropologist*, **65**, 291–316.

Chen, Q., Wang, Y., and Liu, Z. (1998). U-series dating of stalagmite samples from Hulu cave in Nanjing, Jiangsu Province, China (in Chinese with English abstract). *Acta Anthropológica Sinica*, **17**, 171–6.

Chen, T. M., Yang, Q., Hu, Y.-Q, Bao, W.,-B., and Li, T.-Y. (1997). ESR dating of tooth enamel from Yunxian Homo erectus site, China. *Quaternary Science Reviews*, **16**, 455–8.

Chen, T. M., Yang, Q., and Wu, E. (1994). Antiquity of *Homo sapiens* in China. *Nature*, **368**, 55–6.

Chen, T. M., Yuan, S., and Gao, S. (1984). The study of uranium series dating of fossil bones and an absolute age sequence for the main Paleolithic sites of north China. *Acta Anthropológica Sinica*, **3**, 259–68.

Chen, T. M., and Zhang, Y. Y. (1991). Paleolithic chronology and possible co-existence of *H. erectus* and *H. sapiens* in China. *World Archaeology*, **23**, 147–54.

Chikhi, L., Nichols, R. A., Barbujani, G., and Beaumont, M. A. (2002). Y genetic data support the Neolithic demic diffusion model. *Proceedings of the National Academy of Sciences of the United States of America*, **99**, 11008–13.

Childe, V. G. (1951). *Man Makes Himself*. New York: New American Library.

Chinnappa, C. C. (1997). Cytogeographic studies on the vascular plants of Brooks Peninsula. In *Brooks Peninsula: An Ice Age Refugium on Vancouver Island*, Occasional Paper No. 5, ed. R. J. Hebda and J. C. Haggarty. Victoria: Royal BC Museum, pp. 6.1–6.7.

Chown, S. L., and Gaston, K. J. (2000). Areas, cradles, and museums: the latitudinal gradient in species richness. *Trends in Ecology and Evolution*, **15**, 311–5.

Cinq-Mars, J. (1979). Bluefish Cave 1: A Late Pleistocene East Beringian cave deposit in the northern Yukon. *Canadian Journal of Archaeology*, **3**, 1–32.

Cinq-Mars, J., and Morlan, R. E. (1999). Bluefish Caves and Old Crow Basin: a new rapport. In *Ice Age Peoples of North America*, eds. R. Bonnichsen and K. L. Turnmire. Corvallis: Center for the Study of the First Americans, Oregon State University Press, pp. 200–12.

Ciochon, R., Long, V. T., Larick, R., González, L., Grün, R., de Vos, J., Yonge, C., Taylor, L., Yoshida, H., and Reagan, M. (1996). Dated co-occurrence of *Homo erectus* and *Gigantopithecus* from Tham Khuyen Cave, Vietnam. *Proceedings of the National Academy of Sciences of the United States of America*, **93**, 3016–20.

Clague, J. J. (1989). Cordilleran ice sheet. In *Quaternary Geology of Canada and Greenland.*, ed. R. J. Fulton. Ottawa: Geological Survey of Canada, pp. 40–2.

Clark, G. A. (1999). Highly visible, curiously intangible. *Science, 283*, 2029–32.

Clark, J. D. (1989). The origin and spread of modern humans: a broad perspective on the African evidence. In *The Human Revolution: Behavioural and Biological Perspectives on the Origins of Modern Humans*, eds. P. Mellars and C. Stringer. Princeton, NJ: Princeton University Press, pp. 565–88.

(1994). The Acheulian industrial complex in Africa and elsewhere. In *Integrative Paths to the Present*, eds. R. S. Corruccini and R. L. Ciochon. Englewood Cliffs, NJ: Prentice-Hall, pp. 451–69.

Clark, J. D., Beyene, Y., WoldeGabriel, G., Hart, W. K., Renne, P. R., Gilbert, H. G., Defleur, A., Suwa, G., Katoh, S., Ludwig, K. R., Boisserie, J. R., Asfaw, B., and White, T. D. (2003). Stratigraphic, chronological, and behavioural contexts of Pleistocene *Homo sapiens* from Herto, Middle Awash, Ethiopia. *Nature*, **423**, 747–52.

Clark, J. D., De Heinzelin, J., Schick, K., Hart, W., White, T., WoldeGabriel, G., Walter, R. C., Suwa, G., Asfaw, B., Vrba, E. S., and Haile Selassie, Y. H. (1994). African *Homo erectus*: old radiometric ages and young Oldowan assemblages in the Middle Awash, Ethiopia. *Science*, **264**, 1907–10.

Clark, P. U., Alley, R. B., Keigwin, L., Licciardi, J. M., Johnsen, S. J., and Wang, H. (1996). Origin of the first global meltwater pulse following the last glacial maximum. *Paleoceanography*, **11**, 563–78.

Clark, P. U., Mitrovika, J. X., Milne, G. A., and Tamisiea, M. E. (2002). Sea-level fingerprinting as a direct test for the source of global meltwater pulse 1A. *Science*, **295**, 2438–41.

Cohen, A. S., Stone, J. R., Beuning, K. R. M., Park, L. E., Reinthal, P. N., Dettman, D., Scholz, C. A., Johnson, T. C., King, J. W., Talbot, M. R., Brown, E. T., and Ivory, S. J. (2007). Ecological consequences of early Late Pleistocene megadroughts in tropical Africa. *Proceedings of the National Academy of Sciences of the United States of America*, **104**, 16422–7.

Collard, M., Kemery, M., and Banks, S. (2005). Causes of toolkit variation among hunter-gatherers. *Canadian Journal of Archaeology*, **29**, 1–19.

Collard, M., and Wood, B. (2000). How reliable are human phylogenetic hypotheses? *Proceedings of the National Academy of Sciences of the United States of America*, **97**, 5003–6.

Conard, N. J., and Bolus, M. (2003). Radiocarbon dating the appearance of modern humans and timing of cultural innovations in Europe: new results and new challenges. *Journal of Human Evolution*, **44**, 331–71.

Conard, N. J., Grootes, P. M., and Smith, F. H. (2004). Unexpectedly recent dates for human remains from Vogelherd. *Nature*, **430**, 198–201.

Conrad, N. J. (2009). A female figurine from the basal Aurignacian of Hohle Fels Cave in southwestern Germany. *Nature*, **459**, 248–52.

Cope, E. D. (1887). *The Origin of the Fittest*. Chicago: Open Court.

Cordaux, R., and Stoneking, M. (2003). South Asia, the Andamanese, and the genetic evidence for an 'early' human dispersal out of Africa. *American Journal of Human Genetics*, **72**, 1586–90.

Cortijo, E., Labeyrie, L., Elliot, M., Balbon, E., and Tisnerat, N. (2000). Rapid climate variability of the north Atlantic Ocean and global climate: a focus of the IMAGES program. *Quaternary Science Reviews*, **19**, 227–41.

Cortijo, E., Labeyrie, L., Vidal, L., Vautravers, M., Chapman, M., Duplessy, J. C., Elliot, M., Arnold, M., Turon, J. L., and Auffret, G. (1997). Changes in sea surface hydrology associated with Heinrich Event 4 in the north Atlantic Ocean between 40° N and 60° N. *Earth and Planetary Science Letters*, **146**, 29–45.

Cosmides, L., and Tooby, J. (1996). Think again. In *Human Nature*, ed. L. Betzig. Oxford: Oxford University Press.

Cotter, J. L. (1937). The occurrence of flints and extinct animals in pluvial deposits near Clovis, New Mexico. Pt. 4, Report on the excavations at the gravel pit in 1936. *Proceedings of the Philadelphia Academy of Natural Sciences*, **89**, 1–16.

Cranbrook, Earl of (2000). Northern Borneo environments of the past 40,000 years: archaeozoological evidence. *Sarawak Museum Journal*, **55** (n.s. 76), 61–109.

Crawford, M. A., Bloom, M., Broadhurst, C. L., Schmidt, W. F., Cunnane, S. C., Galli, C., Gehbremeskel, K., Linseisen, F., Lloyd-Smith, J., and Parkington, J. (1999). Evidence for the unique function of docosahexaenoic acid during the evolution of the modern hominid brain. *Lipids*, **34**, 39–47.

Crowley, T. J. (1992). North Atlantic deep water cools the southern hemisphere. *Paleoceanography*, **7**, 489–97.

Cruciani, F., Santolamazza, P., Shen, P., Macaulay, V., Moral, P., Olckers, A., Modiano, D., Holmes, S., Destro-Bisol, G., Coia, V., Wallace, D. C., Oefner, P. J., Torroni, A., Cavalli-Sforza, L. L., Scozzari, R., and Underhill, P. A. (2002). A back migration from Asia to Sub-Saharan Africa is supported by high-resolution analysis of human Y-chromosome haplotypes. *American Journal of Human Genetics*, **70**, 1197–214.

Cunningham, S. A., Kanzow, T., Rayner, D., Baringer, M. O., Johns, W. E., Marotzke, J., Longworth, H. R., Grant, E. M., Hirschi, J. J.-M., Beal, L. M., Meinen, C. S., and Bryden, H. L. (2007). Temporal variability of the Atlantic meridional overturning circulation at 26.5° N. *Science*, **317**, 935–8.

Currat, M., and Excoffier, L. (2004). Modern humans did not admix with Neanderthals during their range expansion into Europe. *PLoS Biology*, **2**, 2264–74.

Currie, P. (2004). Muscling in on hominid evolution. *Nature*, **428**, 373–4.

Cutler, K. B., Edwards, R. L., Taylor, F. W., Cheng, H., Adkins, J., Gallup, C. D., Cutler, P. M., Burr, G. S., and Bloom, A. L. (2003). Rapid sea-level fall and deep-ocean temperature change since the last interglacial period. *Earth and Planetary Science Letters*, **206**, 253–71.

Dalton, R. (2005). Caveman DNA hints at map of migration. *Nature*, **436**, 162.

Dansgaard, W., Johnsen, S. J., Clausen, H. B., Dahl-Jensen, D., Gundestrup, N. S., Hammer, C. U., Hvidberg, C. S., Steffensen, J. P., Sveinbjrnsdottir, J. E., Jouzel, J., and Bond, G. (1993). Evidence for general instability of past climate from a 250 kyr ice core. *Nature*, **364**, 218–9.

Dansgaard, W., Johnsen, S. J., Clausen, H. B., Dahl-Jensen, D., Gundestrup, N. S., Hammer, C. U., and Oeschger, H. (1984). North Atlantic climatic oscillations revealed by deep Greenland ice. In *Climate Processes and Climate Sensitivity, Geophysical Monograph 29*, ed. J. E. Hansen and T. Takahashi. Washington, DC: American Geophysical Union, pp. 288–98.

Dansgaard, W., White, J. W. C., and Johnsen, S. J. (1989). The abrupt termination of the Younger Dryas climate event. *Nature*, **339**, 532–4.

Darwin, C. (1871). *The Descent of Man, and Selection in Relation to Sex*, 2 vols. London: Murray.

(1872). *On the Origin of Species by Means of Natural Selection*, 6th edn. London: John Murray.

Darwin, E. (1794). *Zoonomia*. London: Johnson.

Davis, N. Y. (2001). *The Zuni Enigma*. New York: Norton.

Dawkins, R. (1999). *The Extended Phenotype: The Long Reach of the Gene*. Oxford: Oxford University Press.

Deacon, T. W. (1990). Problems of ontogeny and phylogeny in brain-size evolution. *International Journal of Primatology*, **11**, 237–282.

(1992). Brain-language coevolution. In *The Evolution of Human Languages*, vol. 11, eds. J. A. Hawkins and M. Gel-Man. Reading, MA: Addison-Wesley, pp. 49–83.

Dean, M. C., Stringer, C. B., and Bromage, T. G. (1986). Age at death of the Neanderthal child from Devil's Tower, Gibraltar, and the implications for studies of general growth and development in Neanderthals. *American Journal of Physical Anthropology*, **70**, 301–9.

de Castro, J. M. B., Martinón-Torres, M., Carbonell, E., Sarmiento, S., Rosas, A., van der Made, J., and Lozano, M. (2004). The Atapuerca sites and their contribution to the knowledge of human evolution in Europe. *Evolutionary Anthropology*, **13**, 25–41.

de Garnie, I., and Harrison, G. A., eds. (1988). *Coping with Uncertainty in Food Supply*. Oxford: Clarendon.

de Lumley, H. (1969). A Paleolithic camp at Nice. *Scientific American*, **220**, 42–50.

deMenocal, P. B. (1995). Plio-Pleistocen African climate. *Science*, **270**, 53–9.

(2004). African climate change and faunal evolution during the Pliocene-Pleistocene. *Earth and Planetary Science Letters*, **220**, 3–24.

Dennell, R. W. (1997). The world's oldest spears. *Nature*, **385**, 767–8.

(2003). Dispersal and colonization, long and short chronologies: how continuous is the Early Pleistocene record for hominids outside East Africa? *Journal of Human Evolution*, **45**, 421–40.

(2004). Hominid dispersals and Asian biogeography during the Lower and Early Middle Pleistocene, *c.* 2.0–0.5 Mya. *Asian Perspectives*, **43**, 205–26.

Dennell, R. W., Rendell, H., and Hailwood, E. (1988a). Early tool-making in Asia: two-million year-old artefacts in Pakistan. *Antiquity*, **62**, 98–106.

(1988b). Late Pliocene artefacts from Northern Pakistan. *Current Anthropology*, **29**, 495–8.

Dennell, R. W., and Roebroeks, W. (1996). The earliest colonization of Europe: the short chronology revisited. *Antiquity*, **70**, 535–42.

(2005). An Asian perspective on early human dispersal from Africa. *Nature*, **438**, 1099–104.

Deotare, B. C., Kajale, M. D., Rajaguru, S. N., Kusumgar, S., Jull, A. J. T., Donahue, J. D. (2004). Palaeoenvironmental history of Bap-Malar and Kanod playas of western Rajasthan, Thar desert. In Special Section on Quaternary History and Palaeoenvironmental Record of the Thar Desert in India, ed. A. K. Singhvi, *Proceedings of Indian Academy of Science (Earth and Planetary Sciences)*, **113**, 403–25.

Department of Informatics, University of Zurich. (2003). Comparing Neanderthals and modern humans. In *Computer-assisted Paleoanthropology (CAP)*, http://www.ifi. unizh.ch/staff/zolli/CAP/comparingNeand.htm.

Deraniyagala, S. U. (1992). *The Prehistory of Sri Lanka: An Ecological Perspective*. Colombo: Department of the Archaeological Survey, Government of Sri Lanka.

Derbeneva, O. A., Starikovskaya, E. B., Wallace, D. C., and Sukernik, R. I. (2002). Traces of Early Eurasians in the Mansi of Northwest Siberia revealed by mitochondrial DNA analysis. *American Journal of Human Genetics*, **70**, 1009–14.

Derenko, M. V., Gryzbowski, T., Malyarchuk, B. A., Czarny, J., Miscicka-Sliwka, D., and Zakharov, I. A. (2001). The presence of mitochondrial haplogroup X in Altaians from South Siberia. *American Journal of Human Genetics*, **69**, 237–41.

Derenko, M. V., Lunkina, A. V., Malyarchuk, B. A., Zakharov, I. A., Tsedev, Ts., Park, K. S., Cho, Y. M., Lee, H. K., Chu, and Ch., H. (2004). Restriction polymorphism of mitochondrial DNA in Koreans and Mongolians. *Human Genetics*, **40**, 1562–70.

Derenko, M. V., Malyarchuk, B. A., Dambueva, I. K., Shaikhaev, G. O., Dorzhu, C. M., Nimaev, D. D., and Zakharov, I. A. (2000). Mitochondrial DNA variation in two south Siberian aboriginal populations: implications for the genetic history of North Asia. *Human Biology*, **72**, 945–73.

Derev'anko, A. P. (1986). *Sibir': Kul'turniye istoki i gorizonti* (Siberia: Cultural sources and horizons). Novosibirsk1: Sibirskiye ogni.

d'Errico, F., and Sánchez Goñi, M. F. (2003). Neandertal extinction and the millennial scale climatic variability of OIS 3. *Quaternary Science Reviews*, **22**, 769–88.

de Vries, H. (1901–3). *Die Mutationstheorie*. Leipzig: Von Veit.
(1909–10). *Mutation Theory*. Chicago: Open Court.

Diamond, J. (1999). *Guns, Germs, and Steel: The Fates of Human Societies*. New York: Norton.
(2005). *Collapse: How Societies Choose to Fail or Succeed*. New York: Viking.

Dikov, N. N. (1979). *Drevnie Kul'tury Severo-Vostochnoi Azii: Aziia na Styke s Amerikoi v Drevnosti* (in Russian). Moscow: Nauka.

Di Lernia, S. (2006). Building monuments, creating identity: cattle cult as a social response to rapid environmental changes in the Holocene Sahara. *Quaternary International*, **151**, 50–62.

Dillehay, T. D. (1989). *Monte Verde: A Late Pleistocene Settlement in Chile*. Washington, DC: Smithsonian Institution Press.
(1997). Monte Verde: A Late Pleistocene Settlement in Chile. vol. 2, *The Archeological Context*. Washington, DC: Smithsonian Institution Press.
(1999). The Late Pleistocene cultures of South America. *Evolutionary Anthropology*, **7**, 206–16.

Dillehay, T. D., Ramirez, C., Pino, M., Collins, M. B., Rossen, J., Pino-Navarro, J. D. (2008). Monte Verde: Seaweed, food, medicine, and the peopling of South America. *Science*, **320**, 784–6.

Dillehay, T. D., Rossen, J., Andres, T. C., Williams, D. E., (2007). Preceramic adoption of peanut, squash, and cotton in Northern Peru. *Science*, **316**, 1890–3.

Dixon, E. J. (1999). *Bones, Boats and Bison: Archeology and the First Colonization of Western North America*. Albuquerque: University of New Mexico Press.

   (2001). Human colonization of the Americas: timing, technology and process. *Quaternary Science Reviews*, **20**, 277–99.

Dolan DNA Learning Center, Cold Spring Harbor Laboratory. (2003). *Our Neandertal reconstruction: a world first.* http://www.dnalc.org/neandertal.html.

Dotto, L. (1988). *Thinking the Unthinkable: Civilization and Rapid Climate Change*. (Based on the Civilization and Rapid Climate Change Conference, University of Calgary, 22–24 August 1987). Waterloo, ON: Wilfrid Laurier University Press, for the Calgary Institute for the Humanities.

Dow, G. K., Olewiler, N., Reed, C. G. (2005). *The transition to agriculture: climate reversals, population density, and technical change*. Simon Fraser University Economic Discussion Paper.

Driver, J. C. (1998). Human adaptation at the Pleistocene/Holocene boundary in Western Canada, 11,000 to 9000 BP. *Quaternary International*, **49/50**, 141–50.

Duarte, C., Mauricio, J., Pettitt, P. B., Souto, P., Trinkaus, E., van der Plicht, H., and Zilhao, J. (1999). The early Upper Paleolithic human skeleton from the Abrigo do Lagar Velho (Portugal) and modern human emergence in Iberia. *Proceedings of the National Academy of Sciences of the United States of America*, **96**, 7604–9.

Dunbar, R. (1996) *Grooming, Gossip, and the Evolution of Language*. London: Faber and Faber.

Dupont, L. M., and Hooghiemstra, H. (1989). The Saharan-Sahelian boundary during the Brunhes chron. *Acta Botanica Neerlandica*, **38**, 405–15.

Dupont, L. M., Jahns, S., Marret, F., and Shi Ning (2000). Vegetation changes in equatorial West Africa: time-slices for the last 150 ka. *Palaeogeography, Palaeoclimatology, Palaeoecology*, **155**, 95–122.

Dyke, A. S. (1996). *Preliminary paleogeographic maps of glaciated North America*. Geological Survey of Canada, Open File 3296.

   (2004a). Vegetation history, glaciated North America. Text to accompany Dyke, A. S., Giroux, D., and Robertson, L. *Paleovegetation maps of northern North America, 18000 to 1000 BP*. Geological Survey of Canada, Open File 4682.

   (2004b). An outline of North America deglaciation with emphasis on central and northern Canada. In *Quaternary Glaciations – Extent and Chronology*. Part 2, *North America: Developments in Quaternary Science*, vol. 2b, eds. J. Ehlers and P. L. Gibbard. Amsterdam: Elsevier, pp. 373–424.

   (2005). Late Quaternary vegetation history of northern North America based on pollen, macrofossil, and faunal remains. *Géograhie Physique et Quaternaire*, Vic Prest Special Volume, **59**, 211–62.

Dyke, A. S., Giroux, D., and Robertson, L. (2004). *Paleovegetation maps of northern North America, 18000 to 1000 BP*. Geological Survey of Canada, Open File 4682.

Dzaparidze, V., Bosinski, G., Bugianisvili, T., Gabunia, L., Justus, A., Klopotovskaja, N., Kvavadze, E., Lordkipanidze, D., Majsuradze, G., Mgeladze, N., Nioradze, M., Pavlenisvili, E., Schmincke, H.-U., Sologasvili, D., Tusabramisvili, D., Tvalcrelidze, M., and Vekua, A. (1989). Der altpaläolithische Fundplatz Dmanisi in Georgian (Kaukasus). *Jahrbuch des Römisch-Germanischen Zentral-museums Mainz*, **36**, 67–116.

Easton, N. A. (1992). Mal de mer above terra incognita, or, 'what ails the coastal migration theory?' *Arctic Anthropology*, **29**, 28–42.

Echo-Hawk, R. C. (2000). Ancient history in the New World: Integrating oral traditions and the archaeological record. *American Antiquity,* **65,** 267–90.

Edgar, B. (2005). The Polynesian connection. *Archaeology,* **58,** 42–5.

Ehlers, J., and Gibbard, P. L. (2004a). *Quaternary Glaciations – Extent and Chronology, Part I: Europe.* Amsterdam: Elsevier.

(2004b). *Quaternary Glaciations – Extent and Chronology, Part II: North America.* Amsterdam: Elsevier.

(2004c). *Quaternary Glaciations – Extent and Chronology, Part III: South America, Asia, Africa, Australia, Antarctica.* Amsterdam: Elsevier.

Ehrlich, P. R. (2000). *Human Natures: Genes, Cultures, and the Human Prospect.* Washington, DC: Island Press.

Ehrlich, P. R., and Ehrlich, A. (1972). *Population, Resources, Environment: Issues in Human Ecology,* 2nd edn. San Francisco: W. H. Freeman.

Eldredge, N., and Gould, S. J. (1972). Punctuated equilibria. An alternative to phyletic gradualism. In *Models in Paleobiology,* ed. T. J. M. Schopf. San Francisco: Freeman, Cooper, pp. 82–115.

Ellis, C., Goodyear, A. C., Morse, D. F., and Tankersley, K. B. (1998). Archaeology of the Pleistocene-Holocene transition in eastern North America. *Quaternary International,* **49/50,** 151–66.

Endicott, P., López, J. C., Gonzalez, S., and Cooper, A. (2004). *Human mtDNA and the origins of the first Americans: what can archaeological specimens add?* Paper presented at the Second Symposium El Hombre Temprano en América, Mexico City.

Endicott, P., Thomas, M., Gilbert, P., Stringer, C., Lalueza-Fox, C., Willer Siev, G., Hansen, A., and Cooper, A. (2003). The genetic origins of the Andaman Islanders. *American Journal of Human Genetics,* **72,** 178–84.

Erlandson, J. M. (1994). *Early Hunter-Gatherers of the California Coast.* New York: Plenum Press.

Esat, T. M., McCulloch, M. T., Chappell, J., Pillans, B., and Omura, A. (1999). Rapid fluctuations in sea level recorded at Huon peninsula during the penultimate deglaciation. *Science,* **283,** 197–201.

Estrada, E., and Meggers, B. J. (1961). A complex of traits of probable transpacific origin on the coast of Ecuador. *American Anthropologist,* **63,** 913–39.

Etler, D. A. (1996). The fossil evidence for human evolution in Asia. *Annual Review of Anthropology,* **25,** 275–301.

Ewen, T., Weaver, A. J., and Schmittner, A. (2003). Modelling carbon cycle feedbacks during abrupt climate change. *Quaternary Science Reviews,* **23,** 431–48.

Fagan, B. (1999). *World Prehistory: A Brief Introduction,* 4th edn. New York: Longman.

(2004). *The Long Summer: How Climate Changed Civilization.* New York: Basic Books.

Fairbanks, R. G. (1989). A 17,000-year glacio-eustatic sea level record: influence of glacial melting rates on the Younger Dryas event and deep-ocean circulation. *Nature,* **342,** 637–42.

Fairbanks, R. G., Mortlock, R. A., Chiu, T-C., Cao, L., Kaplan, A., Guilderson, T. P., Fairbanks, T. W., and Bloom, A. L. (2005). Marine radiocarbon calibration curve spanning 0 to 50,000 years B.P. based on paired $_{230}$Th/$_{234}$U/$_{238}$U and $_{14}$C dates on pristine corals. *Quaternary Science Reviews,* **24,** 1781–96.

Falguères, C., Bahain, J-J., Yokoyama, Y., Arsuaga, J. L., de Castro, J. M. B., Carbonell, E., Bischoff, J. L., and Dolo, J-M. (1999). Earliest humans in Europe: the age of TD6 Gran Dolina, Atapuerca, Spain. *Journal of Human Evolution,* **37,** 343–52.

Faure, H., Walter, R. C., and Grant, D. R. (2002). The coastal oasis: ice age springs on emerged continental shelves. *Global and Planetary Change,* **33,** 47–56.

Fenech, A., MacIver, D., and Dallmeier, F., eds. (2008). *Climate Change and Biodiversity in the Americas*. Toronto: Environment Canada.

Fenton, M. M., Moran, S. R., Teller, J. T., and Clayton, L. (1983). Quaternary stratigraphy and history in the southern part of the Lake Agassiz Basin. In *Glacial Lake Agassiz*, Geological Association of Canada Special Paper, eds. J. T. Teller and L. Clayton, **26**, pp. 49–74.

Ferring, C. R. (1994). The role of geoarchaeology in Paleoindian research. In *Method and Theory for Investigating the Peopling of the Americas*, eds. R. Bonnichsen and D. G. Steele. Corvallis: Center for the Study of the First Americans, Oregon State University, pp. 57–72.

Fiedel, S. J. (2000). The peopling of the New World: present evidence, new theories, and future directions. *Journal of Archaeological Research*, **8**, 39–103.

Field, J. S., and Lahr, M. M. (2006). Assessment of the southern dispersal: GIS-based analyses of potential routes at oxygen isotope stage 4. *Journal of World Prehistory*, **19**, 1–45, doi: 10.1007/s10963–005–9000–6.

Finlayson, C. (2004). *Neanderthals, Modern Humans*. Cambridge: Cambridge University Press.

Fitzhugh, W. W. (1988). Comparative art of the North Pacific Rim. In *Crossroads of Continents: Cultures of Siberia and Alaska*, eds. W. Fitzhugh and A. Crowell. Washington, DC: Smithsonian Institution Press, pp. 294–312.

Fladmark, K. R. (1979). Routes: alternate migration corridors for early man in North America. *American Antiquity*, **44**, 55–69.

Flannery, T. (1995). *Mammals of New Guinea*. Sydney: Australian Museum/Reed Books.

Fleagle, J. G. (1998). *Primate Adaptation and Evolution*, 2nd edn. San Diego: Academic Press.

Fleisch, H., and Sanlaville, P. (1974). La plage de +52 m et son Acheuléen a Ras Beyrouth et a L'Oudai Aabet (Liban). *Paléorient*, **2**, 45–85.

Fogarty, M. E., and Smith, F. H. (1987). Late Pleistocene climatic reconstruction in North Africa and the emergence of modern Europeans. *Human Evolution*, **2**, 311–9.

Foley, R. A. (2002). Adaptive radiations and dispersals in hominin evolutionary ecology. *Evolutionary Anthropology*, Supplement **1**, 32–7.

Foley, R. A., and Lahr, M. M. (1997). Mode 3 technologies and the evolution of modern humans. *Cambridge Journal of Archaeology*, **7**, 3–36.

Fondon, J. W., III, Garner, H. R. (2004). Molecular origins of rapid and continuous morphological evolution. *Proceedings of the National Academy of Sciences*, **101**, 18058–63.

Forster, P. (2004). Ice ages and the mitochondrial DNA chronology of human dispersals: a review. *Philosophical Transactions of the Royal Society of London, Series B, Biological Sciences*, **359**, 255–64.

Forster, P., and Matsumura, S. (2005). Did early humans go north or south? *Science*, **308**, 965–6.

Fox, R. B. (1970). *The Tabon Caves*. Manila: National Museum.

Frayer, D., and Wolpoff, M. H. (1993). Response to Milo and Quiatt, glottogenesis and modern *Homo sapiens*. *Current Anthropology*, **34**, 582–4.

Friedman, J. (2008). *Anticipation Diseases*. Victoria: University of Victoria PhD thesis.

Gabunia, L., Antón, S. C., Lordkipanidze, D., Vekua, A., Justus, A., and Swisher, C. C. (2001). Dmanisi and dispersal. *Evolutionary Anthropology*, **10**, 158–70.

Gabunia, L., Vekua, A., and Lordkipanidze, D. (2000). The environmental contexts of early human occupation of Georgia (Transcaucasia). *Journal of Human Evolution*, **38**, 785–802.

Gabunia, L., Vekua, A., Lordkipanidze, D., Swisher, C. C., Ferring, R., Justus, A., Nioradze, M., Tvalchrelidze, M., Antón, S. C., Bosinski, G., Jöris, O., de

Lumley, M-A., Majsuradze, G., and Mouskhelishvili, A. (2000). Earliest Pleistocene hominid cranial remains from Dmanisi, Republic of Georgia: taxonomy, geological setting, and age. *Science*, **288**, 1019–25.

Galaburda, A. M., and Pandya, D. N. (1982). Role of architectonics and connections in the study of primate evolution. In *Primate Brain Evolution: Methods and Concepts*, eds. E. Armstrong and E. Falk. New York: Plenum, pp. 203–16.

Gamble, C. (2001). Modes, movement and moderns. *Quaternary International*, **75**, 5–10.

Gannon, P. J., Holloway, R. L., Broadfield, D. C., and Braun, A. R. (1998). Asymmetry of chimpanzee planum temporale: humanlike pattern of Wernicke's brain language area homolog. *Science*, **279**, 220.

Gao, X., and Norton, C. (2002). A critique of the Chinese 'Middle Palaeolithic.' *Antiquity*, **76**, 397–412.

Gathorne-Hardy, F. J., and Harcourt-Smith, W. E. H. (2003). The super-eruption of Toba, did it cause a human bottleneck? *Journal of Human Evolution*, **45**, 227–30.

Geoffroy de Saint-Hilaire, É. (1815–1822). *Philosophie anatomique*. Paris.

Geological Society of America. (2006). *New evidence of early horse domestication.* Press Release 06–49, 20 October, at http://www.geosociety.org/news/pr/06–49.htm.

Gibbons, A. (2001). Tools show humans reached Asia early. *Science*, **293**, 2368–9.

(2002a). New fossils raise molecular questions. *Science*, **295**, 1217.

(2002b). Humans' head start: new views of brain evolution. *Science*, **296**, 835–7.

Giuffra, E., Kijas, J. M. H., Amarger, V., Carlborg O., Jeon J.-T., Andersson, L. (2000). The origin of the domestic pig: independent domestication and subsequent introgression. *Genetics*, **154**, 1785–91.

Glantz, M. H. (1996). *Currents of Change: El Niño's impact on climate and society.* Cambridge: Cambridge University Press.

Goebel, T., Waters, M. R., and Dikova, M. (2003). The archaeology of Ushki Lake, Kamchatka, and the Pleistocene peopling of the Americas. *Science*, **301**, 501–5.

Goldschmidt, R. B. (1940). *The Material Basis of Evolution.* New Haven, CT: Yale University Press.

(1960). *In and Out of the Ivory Tower.* Seattle: University of Washington Press.

Golson, J. (1971). Australian Aboriginal food plants: some ecological and culture-historical implications. In *Aboriginal Man and Environment in Australia*, eds. D. J. Mulvaney and J. Golson. Canberra: Australian National University Press, pp. 196–238.

González, S., Huddart, D., Bennett, M. R., González-Huesca, A. (2006). Human footprints in Central Mexico older than 40,000 years. *Quaternary Science Reviews*, **25**, 201–22.

González-José, R., González-Martin, A., Hernández, M., Pucciarelli, H. M., Sardi, M., Rosales, A., and Van der Molen, S. (2003). Craniometric evidence for Palaeoamerican survival in Baja California. *Nature*, **425**, 62–5.

González-José, R., Hernández, M., Neves, W. A., Pucciarelli, H. M., and Correal, G. (2002). *Cráneos del Pleistoceno tardío-Holoceno tempramo de México en relación al patron morfológico paleoamericano.* Paper presented at the 7th Congress of the Latin American Association of Biological Anthropology, Mexico City.

González-Oliver, A., Ascunce, M. S., and Mulligan, C. J. (2004). Comparison of Y-chromosome and mitochondrial genetic diversity in Panamanian Amerinds. *American Journal of Physical Anthropology Suppl.*, **123**, 102 (Abstr.).

Goodwin, A. J. H. (1929). The Earlier Stone Age in South Africa IV: The Fauresmith industry. In *The Stone Age Cultures of South Africa: Annals of the South African Museum*, eds. A. J. H. Goodwin and C. Van Riet Lowe, **27**, 71–94.

Goodyear A. C. III (1999). The early Holocene occupation of the southeastern United States: a geoarcheological summary. In *Ice Age Peoples of North America*, eds.

R. Bonnichsen and K. L. Turnmire. Corvallis: Center for the Study of the First Americans, Oregon State University, pp. 432–81.

Goren-Inbar, B., and Speth, J. D., eds. (2004). *Human Paleoecology in the Levantine corridor*. Oxford: Oxbow Books.

Goren-Inbar, N., and Saragusti, I. (1996). An Acheulian biface assemblage from Gesher Benot Ya'aqov, Israel: indications of African affinities. *Journal of Field Archaeology*, **23**, 15–30.

Gould, S. J. (1977). *Ontogeny and Phylogeny*. Cambridge, MA: Belknap.

(1981). *The Mismeasure of Man*. New York: Norton.

Gowlett, J. A. J. (1984). Mental abilities of early man: a look at some hard evidence. In *Hominid Evolution and Community Ecology*, ed. R. A. Foley. London: Academic Press, pp.167–92.

Graham, R. (1998). *Mammal's eye view of environmental change in the United States at the end of the Pleistocene*. Paper presented at 63rd Annual Meeting of the Society for American Archaeology, Seattle.

Greenburg, J. H., Turner, C. G., II, and Zegura, S. L. (1986). The settlement of the Americas: a comparison of the linguistic, dental, and genetic evidence. *Current Anthropology*, **27**, 477–97.

Greenfield, P. M. (1991). Language, tools and brain: the ontogeny and phylogeny of hierarchically organized sequential behavior. *Behavioral and Brain Sciences*, **14**, 531–95.

Grichuk, V. P. (1997). Late Cenozoic changes of flora in extra-tropical Eurasia in the light of paleomagnetic stratigraphy. In *The Pleistocene Boundary and the Beginning of the Quaternary*, ed. J. A. Van Couvering. Cambridge: Cambridge University Press, pp. 104–13.

GRIP (Greenland Ice-core Project Members) (1993). Climate instability during the last interglacial period recorded in the GRIP ice core. *Nature*, **364**, 203–7.

Groube, L., Chappell, L., Muke, J., and Price, D. (1986). 40,000-year-old human occupation site at Huon Peninsula, Papua New Guinea. *Nature*, **324**, 453–5.

Grousset, F. E., Labaeyrie, L., Sinko, J. A., Cremer, M., Bond, G., Duprat, J., Cortigo, E., and Huon, S. (1993). Patterns of ice-rafted detritus in the glacial north Atlantic (40° – 55° N). *Paleoceanography*, **8**, 175–92.

Gruhn, R. (1988). Linguistic evidence in support of the coastal route of earliest entry into the New World. *Man (N.S.)*, **23**, 77–100.

(1994). The Pacific coast route of entry: an overview. In *Method and Theory for Investigating the Peopling of the Americas*, eds. R. Bonnichsen and D. G. Steele. Corvallis: Center for the Study of the First Americans, Oregon State University, pp. 249–56.

Grün, R., and Stringer, C. B. (2000). Tabun revisited: revised ESR chronology and new ESR and U-series analyses of dental material from Tabun C1. *Journal of Human Evolution*, **39**, 601–2.

Grün, R., and Thorne, A. (1997). Dating the Ngandong humans. *Science*, **276**, 1575–6.

Guidon, N. (1986). Las unidades culturales de São Raimundo Nonato – Dudeste del Estado de Piauí – Brasil. In *New Evidence for the Pleistocene Peopling of the Americas*, ed. A. L. Bryan. Orono, ME: Center for the Study of the First Americans, pp. 157–71.

Guidon, N., and Delibras, G. (1986). Carbon-14 dates point to man in the Americas 32,000 years ago. *Nature*, **321**, 769–71.

Gupta, A. K. (2004). Origin of agriculture and domestication of plants and animals linked to early Holocene climate amelioration. *Current Science*, **87**, 54–9.

Guthrie, R. D. (2001). Origin and causes of the mammoth steppe: a story of cloud cover, woolly mammoth tooth pits, buckles, and inside-out Beringia. *Quaternary Science Reviews*, **20**, 549–74.

(2003). Rapid body size decline in Alaskan Pleistocene horses before extinction. *Nature*, **426**, 169–71.

(2004). Radiocarbon evidence of mid-Holocene mammoths stranded on an Alaskan Bering Sea island. *Nature*, **429**, 746–9.

Haberle, S. G. (1998). Late Quaternary vegetation change in the Tari Basin, Papua New Guinea. *Palaeogeography, Palaeoclimatology, Palaeoecology*, **137**, 1–24.

Haberle, S. G., Hope, G. S., and DeFreytes, Y. (1991). Environmental change in the Baliem Valley Montane Irian Jaya Republic of Indonesia. *Journal of Biogeography*, **18**, 25–40.

Haile-Selassie, Y., Asfaw, B., and White, T. D. (2004). Hominid cranial remains from Upper Pleistocene deposits at Aduma, Middle Awash, Ethiopia. *America Journal of Physical Anthropology*, **123**, 1–10.

Hamilton, A. C. (1976). The significance of patterns of distribution shown by forest plants and animals in tropical Africa for the reconstruction of Upper Pleistocene palaeo-environments. A review. *Palaeoecology of Africa*, **9**, 63–97.

(1982). *Environmental History of East Africa: A Study of the Quaternary.* London: Academic Press.

Hammer, M. F., Karafet, T. M., Redd, A. J., Jarjanazi, H., Santachiara-Benerecetti, S., Soodyall, H., and Zegura, S. L. (2001). Hierarchical patterns of global human Y-chromosome diversity. *Molecular Biology and Evolution*, **18**, 1189–1203.

Hanihara, K. (1991). Dual structure model for the population history of the Japanese. *Japan Review*, **2**, 1–33.

Hansen, J. (2007). Huge Sea Level Rises Are Coming Unless We Act Now. *New Scientist*, **2614** (July 25, 2007), pp. 30–4. http://environment.newscientist.com/channel/earth/mg19526141.600-huge-sea-level-rises-are-coming–unless-we-act-now.html.

Harlan, J. A. (1967). A wild wheat harvest in Turkey. *Archaeology*, **19**, 197–201.

Harpending, H. C., Batzer, M. A., Gurven, M., Jorde, L. B., Rogers, A. R., and Sherry, S. T. (1998). Genetic traces of ancient demography. *Proceedings of the National Academy of Sciences of the United States of America*, **95**, 1961–7.

Harrison, T. (1958). The caves of Niah: a history of prehistory. *Sarawak Museum Journal*, **8** (n.s. 12), 549–95.

Haury, E. W., Sayles, E. B., and Wasley, W. W. (1959). The Lehner Mammoth site, southeastern Arizona. *American Antiquity*, **25**, 2–30.

Hawks, J. (2001). The Y chromosome and the replacement hypothesis. *Science*, **293**, 567a.

Hawks, J., and Wolpoff, M. H. (2001). The four faces of Eve: hypothesis compatibility and human origins. *Quaternary International*, **75**, 41–50.

Haynes, C. V. (1969). The earliest Americans. *Science*, **166**, 709–15.

(1992). Contributions of radiocarbon dating to the geochronology of the peopling of the New World. In *Radiocarbon after Four Decades*, eds. R. E. Taylor, A. Long, and R. S. Kra. New York: Springer-Verlag, pp. 355–74.

Heaney, L. R. (1991). A synopsis of climatic and vegetational change in Southeast Asia. *Climatic Change*, **19**, 53–61.

Heaton, T. H., Talbot, S. L., and Shields, G. F. (1996). An ice age refugium for large mammals in the Alexander Archipelago, southeastern Alaska. *Quaternary Research*, **46**, 186–92.

Hebda, R. J. (1997). Late Quaternary paleoecology of Brooks Peninsula. In *Brooks Peninsula: An Ice Age Refugium on Vancouver Island*, Occasional Paper No. 5, eds. R. J. Hebda and J. C. Haggarty. Victoria: Royal BC Museum, pp. 9.1–9.48.

Hellmann, I., Ebersberger, I., Ptak, S. E., Paabo, S., and Przeworski, M. (2003). A neutral explanation for the correlation of diversity with recombination rates in humans. *American Journal of Human Genetics*, **72**, 1527–35.

Hemmer, H. (2000). Out of Asia: a palaeoecological scenario of man and his carnivorous competitors in the European lower Pleistocene. In *Early Humans at the Gates of Europe: Études et Recherches Archeologiques de l'Université de Liege*, eds. D. Lordkipanidze, O. Bar-Yosef, and M. Otte. Liege: University of Liege, pp. 99–106.

Hemming, S. R. (2004). Heinrich events: massive late Pleistocene detritus layers of the North Atlantic and their global climate imprint. *Reviews of Geophysics*, **42**, RG1005, doi:10.1029/2003RG000128.

Henshilwood, C. S., and Marean, C. W. (2003). The origin of modern human behavior: critique of the models and their test implications. *Current Anthropology*, **44**, 627–51.

Hetherington, R., Barrie, J. V., MacLeod, R., and Wilson, M. (2004). Quest for the lost land. *Geotimes*, **49**, 20–3.

Hetherington, R., Barrie, J. V., Reid, R. G. B., MacLeod, R., and Smith, D. J. (2004). Paleogeography, glacially induced crustal displacement, and Late Quaternary coastlines on the continental shelf of British Columbia, Canada. *Quaternary Science Reviews*, **23**, 295–318.

Hetherington, R., Barrie, J. V., Reid, R. G. B., MacLeod, R., Smith, D. J., James, T. S., and Kung, R. (2003). Late Pleistocene coastal paleogeography of the Queen Charlotte Islands, British Columbia, Canada, and its implications for terrestrial biogeography and early postglacial human occupation. *Canadian Journal of Earth Sciences*, **40**, 1755–66.

Hetherington, R., and Reid, R. G. B. (2003). Malacological insights into the marine ecology and changing climate of the late Pleistocene-early Holocene Queen Charlotte Islands archipelago, western Canada, and implications for early peoples. *Canadian Journal of Zoology*, **81**, 626–61.

Hetherington, R., Weaver, A., and Montenegro, A. (2007). *Climate and the migration of early peoples into the Americas*, Special Paper **426**. Boulder, CO: Geological Association of America, pp. 113–32.

Hetherington, R., Wiebe, E., Weaver, A. J., Carto, S., Eby, M., and MacLeod, R. (2008). Climate, African and Beringian subaerial continental shelves, and migration of early peoples. *Quaternary International*, **183**, 83–101.

Heusser, C. J. (1960). *Late Pleistocene environments of Pacific North America*, Special Publication **35**. New York: American Geological Society.

Hey, J. (2005). On the number of New World founders: a population genetic portrait of the peopling of the Americas. *PLoS Biology*, **3**(6), e193, doi: 10.1371/journal.pbio.0030193.

Heyes, C. (2000). Evolutionary psychology in the round. In *The Evolution of Cognition*, eds. C. Heyes and L. Huber. London: MIT Press, pp. 3–22.

Higham, C. (1989). *The Archaeology of Mainland Southeast Asia*. Cambridge: Cambridge University Press.
  (1995). The transition to rice cultivation in Southeast Asia. In *Last Hunters, First Farmers: New Perspectives in the Prehistoric Transition to Agriculture*, eds. T. D. Price and A. B. Gebauer. Sante Fe, NM: School of American Research Press, pp. 127–56.

Hoáng, X. C. (1991). Faunal and cultural changes from Pleistocene to Holocene in Vietnam. *Indo-Pacific Prehistoric Association Bulletin*, **10**, 74–8.

Hodell, D. A., Curtis, J. H., Brenner, M. (1995). Possible role of climate in the collapse of classic Mayan civilisation. *Nature*, **375**, 341–7.

Holden, C. (1999). Archaeology: Were Spaniards among the first Americans? *Science*, **286**, 1467–8.

Hole, F. (1994). *Environmental Instabilities and Urban Origins: Chiefdoms and Early States in the Near East. The Organizational Dynamics of Complexity*, Monographs in World Archaeology 18. Madison WI: Prehistory Press.

Holliday, T. W. (1997a). Body proportions in Late Pleistocene Europe and modern human origins. *Journal of Human Evolution*, **32**, 423–47.

(1997b). Postcranial evidence of cold adaptation in European Neandertals. *America Journal of Physical Anthropology*, **104**, 245–58.

Holloway, R. L. (1985). The poor brain of *Homo sapiens neanderthalensis*: see what you please. In *Ancestors: The Hard Evidence*, ed. E. Delson. New York: Alan R. Liss, pp. 319–24.

Homer-Dixon, T., and Garrison, N., eds. (2009). *Carbon Shift: How the Twin Crises of Oil Depletion and Climate Change Will Define the Future*. Toronto: Random House.

Hooper, L. V., Wong, W. H., Thelin, A., Hansson, L., Falk, P. G., and Gordon, J. I. (2001). Molecular analysis of commensal host-microbial relationships in the intestine. *Science*, **291**, 881–4.

Hope, G. S., and Golson, J. (1995). Late Quaternary change in the mountains of New Guinea. In *Transitions*, eds. F. J. Allen and J. F. O'Connell. Antiquity, **69** (special no. 265), 818–30.

Hopkin, M. (2004). Early man steered clear of Neanderthal romance. *Nature Science Update*, March 15 (available at http://www.amren.com/news/news04/03/19/neanderthal.html).

(2005). Early African migrants made eastward exit. *Nature*, **12**, May 2005, doi: 10.1038/news050509–10.

Horai, S., Murayama, K., Hayasaka, K., Matsubayashi, S., Hattori, Y., Fucharoen, G., Harihara, S., Park, K.-S., Omoto, K., and Pan, I.-H. (1996). mtDNA polymorphism in East Asian populations, with special reference to the peopling of Japan. *American Journal of Human Genetics*, **59**, 579–90.

Höss, M. (2000). Neanderthal population genetics. *Nature*, **404**, 453–4.

Hou, Y., Potts, R., Baoyin, Y., Zhengtang, G., Deino, A., Wei, W., Clark, J., Guangmao, X., and Weiwen, H. (2000). Mid-Pleistocene Acheulean-like stone technology of the Bose basin, South China. *Science*, **287**, 1622–6.

Huang, W. P., Ciochon, R., Gu, Y. M., Fang, Q. R., Schwartz, H., Yonge, C., Devos, J., and Rink, W. (1995). Early *Homo* and associated artifacts from Asia. *Nature*, **378**, 275–8.

Huang, W. P., Si, X., Hou, Y., Miller-Antonio, S., and Schepartz, L. A. (1995). Excavations at Panxian Dadong, Guizhou Province, southern China. *Current Anthropology*, **36**, 844–6.

Huang, W. P., and Wang, D. (1995). La recherche récente sur le Paléolithique ancient en Chine. *L'Anthropologie*, **99**, 637–51.

Hublin, J-J., Barroso Ruiz, C., Medina Lara, P., Fontugne, M, and Reyss, J.-L. (1995). The Mousterian site of Zafarraya (Andalucia, Spain): dating and implications on the palaeolithic peopling processes of Western Europe. *Comptes Rendues de l'Académie de Sciences de Paris*, **321**, 931–7.

Hublin, J-J., Spoor, F., Braun, M., Zonneveld, F., and Condemi, S. (1996). A late Neanderthal associated with Upper Palaeolithic artefacts. *Nature*, **381**, 224–6.

Huffman, O. F., and Zaim, Y. (2003). Mojokerto Delta, East Jawa: paleoenvironment of *Homo modjokertensis* – first results. *Journal of Mineral Technology*, **10**, 1–9.

Hughes, D. J. (2009). The energy issue: a more urgent problem than climate change? In *Carbon Shift: How the Twin Crises of Oil Depletion and Climate Change Will Define the Future*, eds. T. Homer-Dixon, and N. Garrison. Toronto: Random House.

Huoponen, K., Torroni, A., Wickman, P. R., Sellitto, D., Gurley, D. S., Scozzari, R., and Wallace, D. C. (1997). Mitochondrial and Y chromosome-specific polymorphisms in the Seminole tribe of Florida. *Eurasian Journal of Human Genetics*, **5**, 25–34.

Hurcombe, L., and Dennell, R. W. (1992). A Pre-Acheulean in the Pabbi Hills, northern Pakistan? In *South Asian Archaeology 1989: Proceedings of the International Conference of South Asian Archaeologists in Western Europe, Paris, July 1989*, ed. C. Jarrige. Madison: Prehistory Press, pp. 133–6.

Hurtado de Mendoza, D., and Braginski, R. (1999). Y Chromosomes point to Native American Adam. *Science*, **283**, 1439–40.

Huxley, J. (1942). *Evolution: The Modern Synthesis*. 3rd edn., London: Allen & Unwin.

Huxley, T. H. (1863). *Evidence as to Man's Place in Nature*. London: Williams and Norgate.

Imamura, K. (1996). *Prehistoric Japan: New Perspectives on Insular East Asia*. Honolulu: University of Hawai'i Press.

Imbrie, J., Hays, J. D., Martinson, D. G., McIntyre, A., Mix, A. C., Morley, J. J., Pisias, N. G., Prell, W. L., and Shackleton, N. J. (1984). The orbital theory of Pleistocene climate: support from a revised chronology of the marine $\delta^{18}O$ record. In *Milankovitch and Climate, Part 1*, eds. A. L. Berger, J. Hays, G. Kukla, and B. Salzman. Norwell, MA: D. Reidel, pp. 269–305.

Ingman, M., Kaessmann, H., Pääbo, S., and Gyllensten, U. (2000). Mitochondrial genome variation and the origin of modern humans. *Nature*, **408**, 708–13.

Intergovernmental Panel on Climate Change (IPCC) (2001). *Climate Change 2001. Impacts, Adaptation and Vulnerability. Contribution of Working Group II to the Third Assessment Report of the Intergovernmental Panel on Climate Change*, eds. J. J. McCarthy, O. F. Canziani, N. A. Leary, D. J. Dokken, and K. S. White. Cambridge: Cambridge University Press, available online at www.grida.no/climate/ipcc_tar/wg2/180.htm.

    (2007a). *Climate Change 2007: Impacts, Adaptation and Vulnerability. Contribution of Working Group II to the Fourth Assessment Report of the Intergovernmental Panel on Climate Change*, eds. M. L. Parry, O. F. Canziani, J. P. Palutikof, P. J. van der Linden, and C. E. Hanson. Cambridge: Cambridge University Press, available online at www.gtp89.dial.pipex.com/chpt.htm.

    (2007b). *Climate Change 2007: The Physical Science Basis. Contribution of Working Group I to the Fourth Assessment Report of the Intergovernmental Panel on Climate Change*, eds. S. Solomon, D. Qin, M. Manning, Z. Chen, M. Marquis, K. B. Averyt, M. Tignor, and H. L. Miller. Cambridge: Cambridge University Press, available online at ipcc-wg1.ucar.edu/wg1/wg1-report.html.

Irwin, G. (1992). *The Prehistoric Exploration and Colonization of the Pacific*. Cambridge: Cambridge University Press.

Jablonka, E., and Lamb, M. J. (1995). *Epigenetic Inheritance and Evolution. The Lamarckian Dimension*. Oxford: Oxford University Press.

Jablonski, N. G. (1997). The relevance of environmental change and primate life histories to the problem of hominid evolution in East Asia. In *The Changing Face of East Asia during the Tertiary and Quaternary*, ed. N. Jablonski. Hong Kong: Centre of Asian Studies, University of Hong Kong, pp. 462–75.

Jackson, L. E., Jr., and Duk-Rodkin, A. (1996). Quaternary geology of the ice-free corridor: glacial controls on the peopling of the New World. In *Prehistoric Mongoloid Dispersals*, eds. T. Akazawa and E. J. E. Szathmary. New York: Oxford University Press, pp. 214–27.

Jackson, L. E., Jr., Phillips, F. M., Shimamura, K., and Little, E. C. (1997). Cosmogenic ^{36}Cl dating of the Foothills erratics train, Alberta, Canada. *Geology*, **25**, 195–8.

Jahns, S., Huls, M., and Sarnthein, N. (1998). Vegetation and climate history of west equatorial Africa based on a marine pollen record off Liberia (site GIK 16776) covering the last 400 000 years. *Review of Palaeobotany and Palynology*, **102**, 277–88.

James, H. V. A., and Petraglia, M. D. (2005). Modern human origins and the evolution of behavior in the later Pleistocene record of South Asia. *Current Anthropology*, **46**, S3–S27.

Jantz, R. L., and Owsley, D. W. (1997). Pathology, taphonomy, and cranial morphometrics of the Spirit Cave Mummy. *Nevada Historical Society Quarterly*, **40**, 57–61.

Jerison, H. J. (1973). *Evolution of the Brain and Human Intelligence*. New York: Academic Press.

Jia, L. P. (1989). Early Paleolithic industries in China. In *Paleoanthropology in China*, eds. R. K. Wu, X. Z. Wu, and S. S. Zhang. Beijing: Science Press, pp. 81–96.

Jia, L. P., and Wei, Q. (1987). Artefacts lithiques provenant du site Pleistocene ancien de Donggutuo près de Nihewan (Nihowan), province d'Hebei, Chine. *L'Anthropologie*, **91**, 727–32, cited in Dennell, R. (2004).

Jin, L., and Su, B. (2001). Response: the Y chromosome and the replacement hypothesis. *Science*, **293**, 567a.

Johannessen, C. (1998). Maize diffused to India before Columbus came to America. In *Across before Columbus? Evidence for Transoceanic Contact with the Americas Prior to 1492*, eds. D. Y. Gilmore and L. S. McElroy. Edgecomb, ME: New England Antiquities Research Association Publications, pp. 111–25.

Jones, R. (1973). Emerging picture of Pleistocene Australians. *Nature*, **246**, 278–81.

Jones, S., Martin, R., and Pilbeam, D., eds. (1992). *The Cambridge Encyclopaedia of Human Evolution*. Cambridge: Cambridge University Press.

Josenhans, H. W., Fedje, D. W., Conway, K. W., and Barrie, J. V. (1995). Post glacial sea levels on the western Canadian continental shelf: evidence for rapid change, extensive subaerial exposure, and early human habitation. *Marine Geology*, **125**, 73–94.

Jousse, H. (2006). What is the impact of Holocene climatic changes on human societies? Analysis of West African Neolithic populations dietary customs. *Quaternary International*, **151**, 63–73.

Justino, F. (2004). *The Influence of Boundary Conditions on the Last Glacial Maximum: Climate Dynamics*. Aachen: Shaker Verlag.

Kaifu, Y., Baba, H., Aziz, F., Indriati, E., Schrenk, F., and Jacob, T. (2005). Taxonomic affinities and evolutionary history of the Early Pleistocene hominids of Java: dentognathic evidence. *American Journal of Anthropology*, **128**, 709–26.

Kaplan, J. O. (2001). *Geophysical Applications of Vegetation Modeling*. PhD thesis, Lund University, Sweden.

Karafet, T. M., Zegura, S. L., Posukh, O., Osipova, L., Bergen, A., Long, J., and Goldman, D. (1999). Ancestral Asian source(s) of New World Y-chromosome founder haplotypes. *American Journal of Human Genetics*, **64**, 817–31.

Karafet, T. M., Zegura, S. L., Vuturo-Brady, J., Posukh, O., Osipova, L., Wiebe, V., and Romero, F. (1997). Y-chromosome markers and trans-Bering Strait dispersals. *American Journal of Physical Anthropology*, **102**, 301–14.

Kauffman, S. (1993). *The Origins of Order: Self-Organization and Selection in Evolution*. New York: Oxford University Press.

Kayser, M., Brauer, S., Weiss, G., Underhill, P. A., Roewer, L., Schiefenhövel, W., and Stoneking, M. (2000). Melanesian origin of Polynesian Y chromosomes. *Current Biology*, **10**, 1237–46.

Ke, Y., Su, B., Song, X., Lu, D., Chen, L., Li, H., Qi, C., Marzuki, S., Deka, R., Underhill, P., Xiao, C., Shriver, M., Lell, J., Wallace, D., Wells, R. S., Seielstad, M., Oefner, P., Zhu, D., Jin, J., Huang, W., Chakraborty, R., Chen, Z., and Jin, L. (2001). African origin of modern humans in East Asia: a tale of 12,000 Y chromosomes. *Science*, **292**, 1151–3.

Keates, S. G. (2004). Home range size in Middle Pleistocene China and human dispersal patterns in Eastern and Central Asia. *Asian Perspectives*, **43**, 227–47.

Keefer, D. K., deFrance, S. D., Moseley, M. E., Richardson, III J. B., Satterlee, D. R., and Day-Lewis, A. (1998). Early maritime economy and El Niño events at Quebrada Tacahuay, Peru. *Science*, **281**, 1833–5.

Keller, G., Adatte, T., Pardo Juez, A., Lopez-Oliva, L. (2009). New evidence concerning the age and biotic effects of the Chicxulub impact in Mexico. *Journal of the Geological Society*, **166**, 393–411.

Kemp, B. M., Malhi, R. S., McDonough, J., Bolnick, D. A., Eshleman, J. A., Rickards, O., Martinez-Labarga, C., Johnson, J. R., Lorenz, J. G., Dixon, E. J., Fifield, T. E., Heaton, T. H., Worl, R., and Smith, D. G. (2007). Genetic analysis of early Holocene skeletal remains from Alaska and its implications for the settlement of the Americas. *American Journal of Physical Anthropology*, **132**, 605–21.

Kempler, D. (1993). Disorders of language and tool use: neurological and cognitive links. In *Tools, Language and Cognition in Human Evolution*, eds. K. R. Gibson and T. Ingold. Cambridge: Cambridge University Press, pp. 193–215.

Kennedy, K. A. (2000). *Gods, Apes, and Fossil Men: Paleoanthropology in South Asia*. Ann Arbor: University of Michigan Press.

Kennedy, K. A., Sonakia, A., Chiment, J., and Verma, K. K. (1991). Is the Narmada hominid an Indian *Homo erectus*? *American Journal of Physical Anthropology*, **86**, 475–96.

Kennett, J. P., and Ingram, B. L. (1995). A 20,000 year record of ocean circulation and climate change from the Santa Barbara basin. *Nature*, **377**, 510–14.

Kimmins, H. (2008). Carbon and nutrient cycling: is the forest bioenergy sustainable? *Canadian Silviculture*, May, 12–16.

Kivisild, T., Kaldma, K., Metspalu, M., Parik, J., Papiha, S., and Villems, R. (1999). The place of the Indian mtDNA variants in the global network of maternal lineages and the peopling of the Old World. In *Genomic Diversity*, eds. R. Deka, and S. Papiha. New York: Plenum Publishers, pp. 135–52.

Klein, G. K. (1992). The archeology of modern human origins. *Evolutionary Anthropology*, **1**, 5–14.

Klein, R. G. (1999). *The Human Career*, 2nd edn. Chicago: Chicago University Press.
    (2000). Archeology and the evolution of human behaviour. *Evolutionary Anthropology*, **9**, 17–36.

Klein, R. G., Avery, G., Cruz-Aribe, K., Halkett, D., Hart, T., Milo, R. G., and Volman, T. P. (1999). Duinefontein 2: An Acheulean site in the Western Cape Province of South Africa. *Journal of Human Evolution*, **37**, 153–90.

Knight, A. (2003). The phylogenetic relationship of Neandertal and modern human mitochondrial DNAs based on informative nucleotide sites. *Journal of Human Evolution*, **44**, 627–32.

Koestler, A. (1971). *The Case of the Midwife Toad*. New York: Random House.

Kolman, C. J., and Bermingham, E. (1997). Mitochondrial and nuclear DNA diversity in the Choco and Chibcha Amerinds of Panama. *Genetics*, **147**, 1289–302.

Kolman, C. J., Sambuughin, N., and Bermingham, E. (1996). Mitochondrial DNA analysis of Mongolian populations and implications for the origin of New World founders. *Genetics*, **142**, 1321–34.

Korisettar, R. (2005). Comment to: James, H. V. A., Petraglia, M. D., 2005. Modern human origins and the evolution of behavior in the later Pleistocene record of South Asia. *Current Anthropology*, **46**, S19–S20.

Korisettar, R., and Rajaguru, S. N. (1998). Quaternary stratigraphy, palaeoclimate, and the Lower Palaeolithic of India. In *Early Human Behaviour in Global Context: The Rise and Diversity of the Lower Palaeolithic Record*, eds. M. D. Petraglia and R. Korisettar. London: Routledge, pp. 304–42.

Korisettar, R., and Ramesh, R. (2002). The Indian monsoon: roots, relations, and relevance. In *Indian Archaeology in Retrospect: Archaeology and Interactive Disciplines*, eds. S. Settar and R. Korisettar. New Delhi: Indian Council of Historical Research and Manohar Publishers, pp. 23–59.

Kotilainen, A. T., and Shackleton, N. J. (1995). Rapid climate variability in the North Pacific Ocean during the past 95,000 years. *Nature*, **377**, 323–6.

Kozlowski, J. K., ed. (1982). *Excavations in the Bacho Kiro Cave (Bulgaria): Final Report*. Warsaw: Państwowe Wydawnictwo Naukowe.

Kramer, A., Crummett, T. L., and Wolpoff, M. H. (2001). Out of Africa and into the Levant: replacement or admixture in Western Asia? *Quaternary International*, **75**, 51–63.

Kreiger, A. D. (1961). Review of 'Late Pleistocene environments of North Pacific North America,' by C. J. Heusser (1960). *American Antiquity*, **27**, 249–50.

Krings, M., Capelli, C., Tschentscher, F., Geisert, H., Meyer, S., von Haeseler, A., Grossschmidt, K., Possnert, G., Paunovic, M., and Pääbo, S. (2000). A view of Neandertal genetic diversity. *Nature Genetics*, **26**, 144–6.

Krings, M., Stone, A., Schmitz, R. W., Krainitzki, H., Stoneking, M., and Pääbo, S. (1997). Neandertal DNA sequences and the origin of modern humans. *Cell*, **90**, 19–30.

Ku, T.-L., Kimmel, M. A., Easton, W. H., and O'Neil, T. J. (1974). Eustatic sea level 120,000 years ago on Oahu, Hawaii. *Science*, **183**, 959–62.

Kukla, G., and Gavin, J. (2004). Milankovitch climate reinforcements. *Global and Planetary Change*, **40**, 27–48.

Kumar, N., Anderson, R. F., Mortlock, R. A., Froelich, P. N., Kubik, P., Dittrich-Hannen, B., and Suter, M. (1995). Increased biological productivity and export production in the glacial southern ocean. *Nature*, **378**, 675–80, doi: 10.1038/378675a0.

Kuper, R. (2004). La transition du desert libyque après 5000 BC. In *Proceeding of Climats, Cultures et Sociétiés*, Paris, 13–16 September 2004. Académie des Sciences, p.70.

Kurten, B. (1968). *Pleistocene Mammals of Europe*. Chicago: Aldine.

Kurz, W. A., Dymond, C. C., Stinson, G., Rampley, G. J., Neilson, E. T., Carroll, A. L., Ebata, T., and Safranyik, L. (2008). Mountain pine beetle and forest carbon feedback to climate change. *Nature*, **452**, 987–90, doi:10.1038/nature06777.

Lahr, M. M. (1995). Patterns of modern human diversification: implications for Amerindian origins. *Yearbook of Physical Anthropology*, **38**, 163–98.

Lahr, M. M., and Foley, R. (1994). Multiple dispersals and modern human origins. *Evolutionary Anthropology*, **3**, 48–60.

(1998). Towards a theory of modern human origins: geography, demography, and diversity in recent human evolution. *Yearbook of Physical Anthropology*, **41**, 137–76.

Lamarck, J-B. (1809). *Philosophie zoologique*. Paris.

(1815–1822). *L'Histoire naturelle des animaux sans vertebres*. Paris.

Lambeck, K., and Chappell, J. (2001). Sea level change through the last glacial cycle. *Science*, **292**, 679–86.

Lambeck, K. Esat, T. M., and Potter, E.-K. (2002). Links between climate and sea levels for the past three million years. *Nature*, **419**, 199–206.

Lang, C., Leuenberger, M., Schwander, J., and Johnsen, J. (1999). 16 °C rapid temperature variation in central Greenland 70,000 years ago. *Science*, **286**, 934–7.

Langbroek, M. (2003). *Out of Africa: A Study into the Earliest Occupation of the Old World*. Unpublished PhD dissertation, University of Leiden.

Langbroek, M., and Roebroeks, W. (2000). Extraterrestrial evidence on the age of the hominids from Java. *Journal of Human Evolution*, **38**, 595–600.

Larichev, V., Khol'ushkin, U., and Laricheva, I. (1987). Lower and Middle Paleolithic of Northern Asia: achievements, problems, and perspectives (translated from Russian by Inna Laricheva). *Journal of World Prehistory*, **1**, 415–64.

    (1992). The Upper Paleolithic of northern Asia: achievements, problems and perspectives 3. Northeastern Siberia and the Russian Far East. *Journal of World Prehistory*, **6**, 441–6.

Larick, R., Ciochon, R. L., Zaim, Y., Sudijono, Suminto, Rizal, Y., Aziz, F., Reagan, M., and Heizler, M. (2001). Early Pleistocene ^{40}Ar/^{39}Ar ages for Bapang Formation hominins, Central Jawa, Indonesia. *Proceedings of the National Academy of Sciences of the United States of America*, **98**, 4866–71.

Leakey, R. E., and Lewin, R. (1977). *Origins: What New Discoveries Reveal about the Emergence of Our Species and Its Possible Future*. Harmondsworth, UK: Penguin Books.

Lee, R. B. (1993). *The Dobe Ju/'hoansi*. Orlando, FL: Harcourt Brace College Publishers.

Lell, J. T., Brown, M. D., Schurr, T. G., Sukernik, R. I., Starikovskaya, Y. B., Torroni, A., Moore, L. G., Troup, G. M., and Wallace, D. C. (1997). Y chromosome polymorphisms in Native American and Siberian populations: identification of founding Native American Y chromosome haplotypes. *Human Genetics*, **100**, 536–43.

Lell, J. T., Sukernik, R. I., Starikovskaya, Y. B., Su, B., Jin, L., Schurr, T. G., Underhill, P. A., and Wallace, D. C. (2002). The dual origin and Siberian affinities of Native American Y chromosomes. *American Journal of Human Genetics*, **70**, 192–206.

Lemmen, D., Duk-Rodkin, A., and Bednarski, J. (1994). Late glacial drainage systems along the northwestern margin of the Laurentide ice sheet. *Quaternary Science Reviews*, **13**, 805–28.

Leng, J. (1992). *Early Paleolithic Technology in China and India*. Unpublished PhD dissertation, Washington University, St. Louis.

Lenton, T. M., Held, H., Kriegler, E., Hall, J. W., Lucht, W., Rahmstorf, S., and Joachim Schellnhuber, H. (2008). Tipping elements in the Earth's climate system. *Proceedings of the National Academy of Sciences of the United States of America*, **105**, 1786–93.

Lericolais, G., Popescu, I., Guichard, F., and Popescu, S. M. (2007). A Black Sea lowstand at 8500 yr BP indicated by a relict costal sand dune system at a depth of 90 m below sea level. *Geological Society of America* (Special Paper), **426**, 171–88.

Lévêque, F., and Vandermeersch, B. C. (1980). Paléontologie humaine: découverte de restes humains dans un niveau castelperronien. *Comptes Rendus Académie des Sciences de Paris*, **291**, 187–9.

Leverington, D. W., Mann, J. D., and Teller, J. T. (2000). *Modelling the bathymetry and dynamics of the early stages of glacial Lake Agassiz: GeoCanada 2000 joint conference (Abstract 108)*. Paper presented at the Geological Association of Canada/ Mineralogical Association of Canada Annual Conference, Calgary, Alberta.

Levine, D. (2001). *At the Dawn of Modernity: Biology, Culture, and Material Life in Europe after the Year 1000*. Berkeley: University of California Press.

Li, H. M., and Wang, J. D. (1982). Magnetostratigraphic study of several typical geologic sections in North China. In *Quaternary Geology and Environment of China*, ed. T. S. Lui. Beijing: China Ocean Press, pp. 33–8.

Lieberman, D. E., McBratney, B. M., and Krovitz, G. E. (2002). The evolution and development of cranial form in *Homo sapiens*. *Proceedings of the National Academy of Sciences of the United States of America*, **99**, 1134–9.

    (2003). Cranial base flexion and *H. Erectus* skulls. *Science*, **300**, 249.

Lieberman, P. (1989). The origins of some aspects of human language and cognition. In *The Human Revolution: Behavioural and Biological Perspectives on the Origins of*

Modern Humans, eds. P. Mellars and C. Stringer. Princeton, NJ: Princeton University Press, pp. 391–414.

Linsley, B. K. (1996). Oxygen-isotope record of sea level and climate variations in the Sulu Sea over the past 150,000 years. *Nature*, **380**, 234–7.

Lipinski, M. J., Lutz, F., Baysac, K. C., Billings, N. C., Leutenegger, C. L., Levy, A. M., Longeri, M., Nuni, T., Ozpinar, H., Slater, M. R., Pedersen, N. C., and Lyons, L. A. (2008). The ascent of cat breeds: genetic evaluations of breeds and worldwide random-bred populations. *Genomics*, **91**, 12–28.

Liu, C., Jin, Z. X., Zhu, R. X., and Yang, H. (1991). Chronological measurement of the earliest strata bearing *Homo* fossils in China: a magnetostratigrahic study on the lower Pleistocene in Wushan, (in Chinese with English abstract). Cited in Zhu, R., An, Z., Potts, R., and Hoffman, K. A. (2003). Magnetostratigraphic dating of early humans in China. *Earth-Science Reviews*, **61**, 341–59.

Liu, L. (1996). Settlement patterns of chiefdom variability, and the development of early states in North China. *Journal of Anthropological Archaeology*, **15**, 237–88.

Liu, T. S., and Ding, Z. L. (1999). Comparison of Plio-Pleistocene climatic changes in different monsoonal regions and implications for human evolution (in Chinese with English abstract). *Quaternary Science*, **4**, 289–98.

Lorenz, J. G., and Smith, D. G. (1996). Distribution of four founding mtDNA haplogroups among native North Americans. *American Journal of Physical Anthopology*, **101**, 307–23.

(1997). Distribution of sequence variations in the mtDNA control region of native North Americans. *Human Biology*, **69**, 749–76.

Løvtrup, S. (1974). *Epigenetics*. New York: Wiley.

Lowe, J. J. (2001). Climatic oscillations during the last glacial cycle – nature, causes and the case for synchronous effects. *Biology and Environment: Proceedings of the Royal Irish Academy*. **101B**, 19–33 (http://www.ria.ie/publications/journals/ProcBI/2001/PB101I1–2/PDF/101B1203.pdf).

Lowe, J. A., Gregory, J. M., Ridley, J., Huybrechts, P., Nicholls, R. J., and Collins, M. (2006). The role of sea-level rise and the Greenland ice sheet in dangerous climate change: implications for the stabilisation of climate. In *Avoiding Dangerous Climate Change*, eds. H. J. Schellnhuber, W. Cramer, N. Nakicenovic, T. Wigley, and G. Yohe. Cambridge: Cambridge University Press, pp. 29–36.

Lowell, T. V., Heusser, C. J., Andersen, B. G., Moreno, P. I., Hauser, A., Denton, G. H., Heusser, L. E., Schluchter, C., and Marchant, D. R. (1995). Interhemispheric symmetry of paleoclimatic events during the last glaciation. *Science*, **269**, 1541–9.

Lucotte, G. (1989). Evidence for the paternal ancestry of modern humans: evidence from a Y-chromosome specific sequence polymorphic DNA probe. In *The Human Revolution: Behavioural and Biological Perspectives on the Origins of Modern Humans*, eds. P. Mellars and C. Stringer. Princeton, NJ: Princeton University Press, pp. 39–46.

Luis, J. R., Rowold, D. J., Regueiro, M., Caeiro, B., Cinnioğlu, C., Roseman, C., Underhill, P. A., Cavalli-Sforza, L. L., and Herrera, R. J. (2004). The Levant versus the Horn of Africa: evidence for bidirectional corridors of human migrations. *American Journal of Human Genetics*, **74**, 532–44.

Lukacs, J. R. (2005). Comment to: James, H. V. A., and Petraglia, M. D., 2005. Modern human origins and the evolution of behavior in the later Pleistocene record of South Asia. *Current Anthropology*, **46**, S20–S21.

Lund, D. C., and Mix, A. C. (1998). Millennial-scale deep water oscillations: reflections of the North Atlantic in the deep Pacific from 10 to 60 ka. *Paleoceanography*, **13**, 10–19.

Lyell, C. (1830–33). *Principles of Geology.* London: John Murray.

Lynas, M. (2007). *Six Degrees: Our Future on a Hotter Planet.* London: Harper Collins.

Macaulay, V., Hill, C., Achilli, A., Rengo, C., Clarke, D., Meehan, W., Blackburn, J., Semino, O., Scozzari, R., Cruciani, F., Taha, A., Shaari, N. K., Raja, J. M., Ismail, P., Zainuddin, Z., Goodwin, W., Bulbeck, D., Bandelt, H-J., Oppenheimer, S., Torroni, A., and Richards, M. (2005). Single, rapid coastal settlement of Asia revealed by analysis of complete mitochondrial genomes. *Nature,* **308,** 1034–6.

Macgowan, K., and Hester, J. A.Jr., (1962). *Early Man in the New World.* New York: Doubleday.

MacPhee, R. D., and Marx, P. A. (1997). The 40 000-year plague: humans, hyperdisease, and first-contact extinctions. In *Natural Change and Human Impact in Madagascar,* eds. S. M. Goodman and B. D. Patterson. Washington, DC: Smithsonian Institution Press, pp. 169–217.

Maglio, V. C., and Cooke, H. B., eds. (1978). *Evolution of African Mammals.* Cambridge, MA: Harvard University Press.

Mahli, R. S., Mortensen, H. M., Eshleman, J. A., Kemp, B. M., Lorenz, J. G., Kaestle, F. A., Johnson, J. R., Gorodesky, C., and Smith, D. G. (2003). Native American mtDNA prehistory in the American west. *American Journal of Physical Anthropology,* **120,** 108–24.

Malhi, R. S., Schultz, B. A., and Smith, D. G. (2001). Distribution of mitochondrial lineages among Native American tribes of northeastern North America. *Human Biology,* **73,** 17–55.

Malthus, T. R. (1798). *An Essay on the Principle of Population as it Affects the Future Improvement of Society.* London: J. Johnson in St. Paul's Church-yard.

Manzi, G., Mellegni, F., and Ascenzi, A. (2001). A cranium for the earliest Europeans: phylogenetic position of the hominid from Ceprano, Italy. *Proceedings of the National Academy of Sciences of the United States of America,* **98,** 10011–6.

Marean, C. W., Bar-Matthews, M., Bernatchez, J., Fisher, E., Goldberg, P., Herries, A. I. R., Jacobs, Z., Jerardino, A., Karkanas, P., Minichillo, T., Nilssen, P. J., Thompson, E., Watts, I., and Williams, H. M. (2007). Early human use of marine resources and pigment in South Africa during the Middle Pleistocene. *Nature,* **449,** 905–9, doi:10.1038/nature06204.

Margulis L., and Sagan, D. (2002). *Acquiring Genomes. A Theory of the Origins of Species.* New York: Basic Books.

Marshall, F., and Hildebrand, E. (2002). Cattle before crops: the beginnings of food production in Africa. *Journal of World Prehistory,* **16,** 99–143.

Marshall, S. J., and Cuffey, K. M. (2000). Peregrinations of the Greenland ice sheet divide through the last glacial cycle and implications for disturbance of central Greenland ice cores. *Earth and Planetary Science Letters,* **179,** 73–90.

Marshall, S. J., and Koutnik, M. R. (2006). Ice sheet action vs. reaction: distinguishing between Heinrich events and Dansgaard-Oeschger cycles in the North Atlantic. *Paleoceanography,* **21,** PA2021, doi: 10.1029/2005PA001247.

Marshall, S. J., Tarasov, L., Clarke, G. K. C., and Peltier, W. R. (2000). Glaciological reconstruction of the Laurentide ice sheet: physical processes and modeling challenges. *Canadian Journal of Earth Sciences,* **37,** 769–93.

Martin, P. S. (1973). The discovery of America. *Science,* **179,** 969–74.

(1984). Prehistoric overkill: the global model. In *Quaternary Extinctions: A Prehistoric Revolution,* eds. P. S. Martin and R. G. Klein. Tucson: University of Arizona Press, pp. 354–403.

Martin, R. D. (1983). *Human Brain Evolution in an Ecological Context.* New York: American Museum of Natural History.

Martrat, B., Grimalt, J. O., Lopez-Martinez, C., Cacho, I., Sierro, F. J., Flores, J. A., Zahn, R., Canals, M., Curtis, J. H., and Hodell, D. A. (2004). Abrupt temperature changes in the Western Mediterranean over the past 250,000 yrs. *Science*, **306**, 1762–5.

Massone M. (1996). Hombre temprano y paleoambiente en la region de Magallanes: evaluacion critica y perspectiva. *Anales del Instituto de la Patagonia*, **24**, 81–98.

Maté, G. (2008). *In the Realm of Hungry Ghosts: Close Encounters with Addiction.* Toronto: Knopf Canada.

Mathewes, R. W. (1989). The Queen Charlotte Islands refugium: a paleoecological perspective. In *Quaternary Geology of Canada and Greenland.*, ed. R. J. Fulton. Ottawa: Geological Survey of Canada, pp. 486–91.

Matsuda, R. (1987). *Animal Evolution in Changing Environments, with Special Reference to Abnormal Metamorphosis.* New York: Wiley.

Matthews, H. D., Weaver, A. J., Eby, M., and Meissner, K. J. (2003). Radiative forcing of climate by historical land cover change. *Geophysical Research Letters*, **30** (2), 27:1–27:4, 1055.

Matthews, H. D., Weaver, A. J., Meissner, K. J., Gillett, N. P., and Eby, M. (2003). Natural and anthropogenic climate change: incorporating historical land cover change, vegetation dynamics and the global carbon cycle. *Climate Dynamics*, **22**, 461–79.

Matthews, J. V. (1982). East Beringia during late Wisconsin time: a review of biotic evidence. In *Paleoecology of Beringia*, eds. D. Hopkins, J. V. Matthews, C. E. Schweger, and S. B. Young. New York: Academic Press, pp. 127–50.

Matthews, R. (2003). *The Archaeology of Mesopotamia: Theories and Approaches.* London: Routledge.

Mayr, E. (1954). Change of genetic environment and evolution. In *Evolution as a Processs*, eds. J. S. Huxley, A. C. Hardy, and E. B. Ford. London: Allen and Unwin, pp. 157–80.

  (1963). *Animal Species, and Evolution.* Cambridge, MA: Belknap Press of Harvard University Press.

Mayr, E., and Provine, W., eds. (1980). *The Evolutionary Synthesis.* Cambridge: Harvard University Press.

McAvoy, J. M., Baker, J. C., Feathers, J. K., Hodges, R. L., McWeeney, L. J. *et al.*, (2000). Summary of research at the Cactus Hill Archeological Site 40SX02, Sussex County, Virgina. *Report to the National Geographic Society in Compliance with Stipulations of Grant #6345–98.*

McAvoy, J. M., and McAvoy, L. D., eds. (1997). *Archeological investigations of Site 44SX202, Cactus Hill, Sussex County, Virginia*, Research Report Series 8. Richmond, VA: Virginia Department of Historic Resources.

McBrearty, S. (1993). Reconstructing the environmental conditions surrounding the appearance of modern humans in East Africa. In *Culture and Environment: A Fragile Coexistence*, eds. R. Jamieson, S. Abonyi, and N. Mirau. Calgary: Chacmool Archaeological Association, pp. 145–54.

McBrearty, S., and Brooks, A. S. (2000). The revolution that wasn't: a new interpretation of the origin of modern human behaviour. *Journal of Human Evolution*, **39**, 453–563.

McCown, T. D., and Keith, A. (1939). The Stone Age of Mount Carmel. Vol. 2: *The fossil human remains from the Levalloiso-Mousterian.* Oxford: Clarendon Press.

McDougall, I., Brown, F. H., and Fleagle, J. G. (2005). Stratigraphic placement and age of modern humans from Kibish, Ethiopia. *Nature*, **433**, 733–6.

McIntosh, J. R. (2002). *A Peaceful Realm: The Rise and Fall of the Indus Civilization.* New York: Westview.

McManus, J. F., Bond, G. C., Broecker, W. J., Johnson, S., Labeyrie, L., and Higgins, S. (1994). High-resolution climate records from the North Atlantic during the last interglacial. *Nature*, **371**, 326–39.

McManus, J. F., Oppo, D. W., and Cullen, J. L. (1999). A 0.5 million-year record of millennial scale climate variability in the North Atlantic. *Science*, **283**, 971–5, doi: 10.1126/science.283.5404.971.

McMillan, H. J., and Wynne-Edwards, K. E. (1998). Evolutionary change in the endocrinology of behavioral receptivity: divergent roles for progesterone and prolactin within the genus *Phodopus*. *Biology of Reproduction*, **59**, 30–8.

(1999). Divergent reproductive endocrinology of the estrous cycle and pregnancy in dwarf hamsters (*Phodopus*). *Comparative Biochemistry and Physiology* A, **124**, 53–67.

Meggers, B. J. (1971). The transpacific origin of Mesoamerican civilization: a preliminary review of the evidence and its theoretical implications. *American Anthropology*, **77**, 1–27.

(1998a). Archaeological evidence for transpacific voyages from Asia since 6000 BP. *Estudios Atacaeños*, **15**, 107–24.

(1998b). Jomon-Valdivia similarities: convergence or contact? In *Across before Columbus? Evidence for Transoceanic Contact with the Americas Prior to 1492*, eds. D. Y. Gilmore and L. S. McElroy. Edgecomb, ME: NEARA Publications (New England Antiquities Research Association), pp. 16–17.

(1998c). Archaeological evidence for transpacific voyages from Asia since 6000 BP. *Estudios Atacameños*, **15**, pp. 107–20.

Meggers, B. J., Evans, C., and Estrada, E. (1965). *Early Formative Period of Coastal Ecuador: the Valdivida and Machalilla Phases*. Smithsonian Contributions to Anthropology, vol. 1. Washington, DC: Smithsonian Institution.

Mehlman, M. J. (1979). Mumba-Höhle revisited: the relevance of a forgotten excavation to some current issues in East African prehistory. *World Archaeology* **11**, 80–94.

(1987). Provenience, age and associations of archaic *Homo sapiens* crania from Lake Eyasi, Tanzania. *Journal of Archeological Science*, **14**, 133–62.

(1989). *Late Quaternary Archaeological Sequences in Northern Tanzania*. Unpublished PhD dissertation, University of Illinois, Urbana.

Meissner, K. J., Schmittner, A., Weaver, A. J., and Adkins, J. (2003). Ventilation of the North Atlantic Ocean during the last glacial maximum: a comparison between simulated and observed radiocarbon ages. *Paleoceanography*, **18**, 1023, doi: 10.1029/ 2002PA000762.

Meissner, K. J., Weaver, A. J., Matthews, H. D., and Cox, P. M. (2003). The role of land-surface dynamics in glacial inception: a study with the UVic Earth System Model. *Climate Dynamics*, **21**, 515–37, doi: 10.1007/s00382–003–0352–2.

Mellars, P. A. (1973). The character of the Middle-Upper Palaeolithic transition in southwest France. In *The Explanation of Culture Change*, ed. C. Rengrew. London: Duckworth, pp. 255–76.

(1992). Archaeology and the population-dispersal hypothesis of modern human origins in Europe. *Philosophical Transactions: Biological Sciences*, **337**, 225–34.

Mellars, P. (1996). The big transition. In *The Neanderthal Legacy: An Archaeological Perspective from Western Europe*. Princeton, NJ: Princeton University Press, pp. 392–419.

(2004a). Neanderthals and the modern human colonization of Europe. *Nature*, **43**, 461–5.

(2004b). Stage 3 climate and the Upper Palaeolithic revolution in Europe: evolutionary perspectives. In *Explaining Social Change: Studies in Honour of Colin Renfrew*, eds.

J. Cherry, C. Scarre, and S. Shennan. Cambridge: McDonald Institute for Archaeological Research.

(2005). The impossible coincidence: a single-species model for the origins of modern human behavior in Europe. *Evolutionary Anthropology*, **14**, 12–27.

Meltzer, D. J. (2002). What do you do when no one's been there before? Thoughts on the exploration and colonization of new lands. In *The First Americans: The Pleistocene Colonization of the New World*, ed. N. G. Jablonski. Berkeley: University of California Press, pp. 27–58.

Mercier, N., and Valladas, H. (2003). Reassessment of TL age estimates of burnt flints from the Paleolithic site of Tabun Cave, Israel. *Journal of Human Evolution*, **45**, 401–9.

Mercier, N., Valladas, H, Joron, J. L., Reyss, J. L., Lévêque, F., and Vandermeersch, B. (1991). Thermoluminescence dating of the late Neanderthal remains from Saint-Césaire. *Nature*, **351**, 737–9.

Midant-Reynes, G. (1992). *The Prehistory of Egypt*. Oxford: Blackwell.

Miklosi, A. (2008). *Dog Behaviour, Evolution and Cognition*. Oxford: Oxford University Press.

Milankovic, M. (1998). *Canon of Insolation and the Ice-Age Problem*. Royal Serbian Academy, special publications, v. **132**.

Miller, G. H., Magee, J. W., Johnson, B. J., Fogel, M. L., Spooner, N. A., McCulloch, M. T., and Ayliffe, L. K. (1999). Pleistocene extinction of *Genyornis newtoni*: human impact on Australian megafauna. *Science*, **283**, 205–8.

Miller, G. H., Magee, J. W., and Jull, A. J. (1997). Low-latitude glacial cooling in the southern hemisphere from amino-acid racemization in emu eggshells. *Nature*, **385**, 241–4.

Misra, V. N. (2005). Comment on James and Petraglia, modern human origins and the evolution of behaviour in the Later Pleistocene record of South Asia. *Current Anthropology*, **46**, S21.

Montenegro, A., Araujo, A., Eby, M., Ferreira, L. F., Hetherington, R., and Weaver, A. J. (2006). Parasites, paleoclimate, and the peopling of the Americas: using the hookworm to time the Clovis migration. *Current Anthropology*, **47**, 193–200.

Montenegro, A., Hetherington, R., Eby, M., and Weaver, A. J. (2006). Modelling pre-historic transoceanic crossings into the Americas. *Quaternary Science Reviews*, **25**, 1323–38.

Moratto, M. J. (1984). *California Archaeology*. New York: Academic Press.

Morton, O. (1997). Review of Steven Pinker's book *How the Mind Works*. *New Yorker*, 3 November, 102.

Morwood, M. J. (2001). Early hominid occupation of Flores, East Indonesia, and its wider significance. In *Faunal and Floral Migrations and Evolution in SE Asia-Australasia*, eds. I. Metcalfe, J. M. B. Smith, M. Morwood, and I. Davidson. Lisse, The Netherlands: Balkema, pp. 387–98.

Morwood, M. J., O'Sullivan, P., Aziz, G., and Raza, A. (1998). Fission-track ages of stone tools and fossils on the east Indonesian island of Flores. *Nature*, **392**, 173–6.

Morwood, M. J., Soejono, R. P., Roberts, R. G., Sutikna, T., Turney, C. S. M., Westaway, K. E., Rink, W. J., Zhao, J.-X, van den Bergh, G. D., Due, R. A., Hobbs, D. R., Moore, M. W., Bird, M. I., and Fifield, L. K. (2004). Archaeology and age of a new hominin from Flores in eastern Indonesia. *Nature*, **431**, 1087–91.

Mountain, J. L., Lin, A. A., Bowcock, A. M., and Cavalli-Sforza, L. L. (1993). Evolution of modern humans: evidence from nuclear DNA polymorphisms. In *The Origin of Modern Humans and the Impact of Chronometric Dating*, eds. M. J. Aitken, C. B. Stringer, and P. A. Mellars. Princeton, NJ: Princeton University Press, pp. 69–83.

386 References

Movius, H. L. (1944). Early man and Pleistocene stratigraphy in southern and eastern Asia. *Papers of the Peabody Museum, Harvard University*, **19**, 1–125.

Munford, D., Zanini, M. C., and Neves, W. A. (1995). Human cranial variation in South America: implications for the settlement of the New World. *Brazilian Journal of Genetics*, **18**, 673–88.

Nagorsen, D. W., and Keddie, G. (2002). Late Pleistocene mountain goats (*Oreamnos americanus*) from Vancouver Island: biogeographic implications. *Journal of Mammology*, **81**, 666–75.

Nakada, M., and Yokose, H. (1992). Ice age as a trigger of active Quaternary volcanism and tectonism. *Tectonophysics*, **212**, 321–9.

Neves, W. A., and Hubbe, M. (2005). Cranial morphology of early Americans from Lagoa Santa, Brazil: implications for the settlement of the New World. *Proceedings of the National Academy of Sciences of the United States of America*, **102**, 18309–14.

Neves, W. A., Hubbe, M., Mercedes, M., Okumura, M., González-José, R., Figuti, L., Eggers, S., and Dantas De Blasis, P. A. (2005). A new early Holocene human skeleton from Brazil: implications for the settlement of the New World. *Journal of Human Evolution*, **48**, 403–14.

Neves, W. A., Powell, J. F., and Ozolins, E. G. (1999a). Extra-continental morphological affinities of Lapa Vermelha IV, Hominid 1: A multivariate analysis with progressive numbers of variables. *Homo*, **50**, 263–82.

(1999b). Extra-continental morphological affinities of Palli Aike, southern Chile. *Interciência*, **24**, 258–63.

(1999c). Modern human origins as seen from the peripheries. *Journal of Human Evolution*, **34**, 96–105.

Neves, W. A., Prous, A., González-José, R., Kipnis, R., and Powell, J. (2003). Early Holocene human skeletal remains from Santana do Riacho, Brazil: implications for the settlement of the New World. *Journal of Human Evolution*, **45**, 19–42.

Neves, W. A., and Pucciarelli, H. M. (1989). Extra-continental biological relationships of early South American human remains: a multivariate analysis. *Ciência e Cultura*, **41**, 566–75.

Oakley, K. P. (1956). *Man the Tool-Maker*, 3rd edn. London: British Museum.

O'Brien, E. M. (1984). What was the Acheulean hand axe? *Natural History*, **93**, 20–3.

O'Brien, E. M., and Petters, C. R. (1999). Shifting paradigms: away from simplicity, towards dynamic complexity and testable hypotheses. In *The Environmental Background to Hominid Evolution in Africa*, INQUA XV International Congress, Book of Abstracts, eds. J. Lee-Thorp and H. Clift. Rondebosch. South Africa: University of Cape Town, p. 134.

O'Connell, J. F., and Allen, J. (2004). Dating the colonization of Sahul (Pleistocene Australia-New Guinea): a review of recent research. *Journal of Archaeological Science*, **31**, 835–53.

O'Connor, S., Aplin, K., Spriggs, M., Veth, P., and Ayliffe, L. (2001). From savannah to rainforest: changing environments and human occupation of Liang Lemdubu, Aru Islands, Maluku (Indonesia). In *Bridging Wallace's Line: Advances in GeoEcology*, vol. 34, eds. P. Kershaw, B. David, N. Tapper, D. Penny, and J. Brown. Reiskirchen: Catena Verlag, pp. 279–306.

O'Connor, S., and Veth, P. (2000). The world's first mariners: savannah dwellers in an island continent. In *East of Wallace's Line: Studies of Past and Present Maritime Cultures of the Indo-Pacific region*, eds. S. O'Connor and P. Veth. Rotterdam: Balkema, pp. 99–138.

Oeschger, H., Beer, J., Siegenthaler, U., Stauffer, B., Dansgaard, W., and Langway, C. C. (1984). Late glacial climate history from ice cores. In *Climate Processes and Climate*

Sensitivity, Geophysical Monograph 29, eds. J. E. Hansen and T. Takahashi. Washington, DC: American Geophysical Union, pp. 299–306.

Ohno, S. (1970). *Evolution by Gene Duplication*. Berlin: Springer.

Okladnikov, A. P., Muratov, V. M., Ovodov, N. D., and Fridenberg, E. O. (1973). *Peshchera Strashnaya-noviy pam'atnik paleolita Altaya* (Strashnaya cave: a new Paleolithic site in the Altai), Materiali po Arkheologii Sibiri i Dal'nego Vostoka. Novosibirsk: Nauka.

Olsen, J. W., and Ciochon, R. L. (1990). A review of evidence for postulated Middle Pleistocene occupations in Viet Nam. *Journal of Human Evolution*, **19**, 761–88.

Omoto, K., and Saitou, N. (1997). Genetic origins of the Japanese: a partial support for the dual structure hypothesis. *American Journal of Physical Anthropology*, **102**, 437–46.

Oms, O., Parés, J. M., Martinez-Navarro, B., Agusti, J., Toro, I., Martinez-Fernández, G., and Turq, A. (2000). Early human occupation of Western Europe: paleomagnetic dates for two Paleolithic sites in Spain. *Proceedings of the National Academy of Sciences of the United States of America*, **97**, 10666–70.

Opdyke, N. D. (1995). Mammalian migration and climate over the last seven million years. In *Paleoclimate and Evolution, with Emphasis on Human Origins*, eds. E. S. Vrba, G. H. Denton, T. C. Partridge, and L. H. Burckle. New Haven, CT: Yale University Press, pp. 109–14.

Oppenheimer, C. (2002). Limited global change due to the largest known Quaternary eruption, Toba ~74 kyr BP? *Quaternary Science Reviews*, **21**, 1593–1609.

Oppenheimer, S. (1998). *Eden in the East: The Drowned Continent of Southeast Asia*. London: Weidenfeld and Nicholson, 560 pp.

  (2003a). Journey of mankind interactive trail adapted from *Out of Eden/The Real Eve*, www.bradshawfoundation.com/journey/.

  (2003b). *Out of Eden: The Peopling of the World*. London: Constable.

Osborn, A. J. (1999). From global models to regional patterns: possible determinants of Folsom hunting weapon design diversity and complexity. In *Folsom Lithic Technology: Explorations in Structure and Variation*, ed. D. S. Amick. Ann Arbor: International Monographs in Prehistory, pp. 188–213.

Oschenius, C., and Gruhn, R., eds. (1979). *Taima-Taima: A Late Pleistocene Paleo-Indian Kill Site in Northernmost South America. Final Reports of the 1976 Excavations*. Germany: CIPICS/South American Quaternary Documentation Program.

O'Sullivan, P. B., Morwood, M., Hobbs, D., Aziz, F., Suminto, Situmorang, M., Raza, A., and Maas, R. (2001). Archaeological implications of the geology and chronology of the Soa basin, Flores, Indonesia. *Geology*, **29**, 607–10.

Oswalt, W. H. (1976). *An Anthropological Analysis of Food-getting Technology*. New York: Wiley.

Ovchinnikov, I. V., Götherström, A., Romanova, G. P., Kharitonov, V. M., Lidén, K., and Goodwin, W. (2000). Molecular analysis of Neanderthal DNA from the northern Caucasus. *Nature*, **404**, 490–3.

Overpeck, J., Anderson, D., Trumbore, S., and Prell, W. (1996). The southwest Indian monsoon over the last 18,000 years. *Climate Dynamics*, **12**, 213–25.

Oyama, S. (1985). *The Ontogeny of Information*. Durham, NC: Duke University Press.

Pääbo, S. (2003). The mosaic that is our genome. *Nature*, **421**, 409–12.

Paddayya, K. (1982). *The Acheulian Culture of the Hunsgi Valley, Peninsular India: A Settlement System Perspective*. Pune: Deccan College.

Parenti, F. (1993). Le Gisement Quaternaire de la Tocado Boqueirão da Pedra Furada (Piauí, Brésil) dans le Contexte de la Préhistoire Américaine. *Fouilles, Stratigraphique, Chronologie, Evolution Culturelle*. Thesis presented to l'Ecole de Hautes Etudes en Sciences Sociales, Paris.

(1996). Questions about the Upper Pleistocene prehistory in Northeastern Brazil: Pedra Furada Rockshelter in its regional context. In *Proceedings of the International Meeting on the Peopling of the Americas, São Raimundo Nonato, Piauí, Brasil*, vol. 1, ed. A-M. Pessis. São Raimundo Nonato, Brazil: Fundaçao Museu do Homen Americano, pp. 15–53.

(2001). *Le Gisement Quaternaire de Pedra Furada (Piauí, Brésil). Stratigraphie, Chronologie, Evolution Culturelle*. Paris: Editions Rècherche sur les Civilisations.

Parfitt, S. A., Barendregt, R. W., Breda, M., Candy, I., Collins, M. J., Coope, G. R., Durbidge, P., Field, M. H., Lee, J. R., Lister, A. M., Mutch, R., Penkman, K. E. H., Preece, R. C., Rose, J., Stringer, C. B., Symmons, R., Whittaker, J. E., Wymer, J. J., and Stuart, A. J. (2005). The earliest record of human activity in northern Europe. *Nature*, **438**, 1008–12.

Parry, M. L., Canziani, O. F., Palutikof, J. P., and co-authors (2007). Technical summary. In *Climate Change 2007: Impacts, Adaptation and Vulnerability. Contribution of Working Group II to the Fourth Assessment Report of the Intergovernmental Panel on Climate Change*, eds. M. L. Parry, O. F. Canziani, J. P. Palutikof, P. J. van der Linden, and C. E. Hanson. Cambridge: Cambridge University Press, pp. 23–78.

Parsons, T. J., Muniec, D. S., Sullivan, K., Woodyatt, N., Alliston-Greiner, R., Wilson, M. R., Berry, D. L., Holland, K. A., Weedn, V. W., Gill, P., and Holland, M. M. (1997). A high observed substitution rate in the human mitochondrial DNA control region. *Nature Genetics*, **15**, 363–8.

Pavlides, C., and Gosden, C. (1994). 35,000-year-old sites in the rainforests of West New Britain, Papua New Guinea. *Antiquity*, **68**, 604–10.

Pavlov, P., Svendsen, J. I., and Indrelid, S. (2001). Human presence in the European Arctic nearly 40,000 years ago. *Nature*, **413**, 64–7.

Pearson, O. M. (2001). Postcranial remains and the origin of modern humans. *Evolutionary Anthropology*, **9**, 229–47.

Pease, C. M., Lande, R., and Bull, J. J. (1989). A model of population growth, dispersal and evolution in a changing environment. *Ecology*, **70**, 1657–64.

Perreault, M. (2003). 85 milliards d'humans plus tard: évolution du nombre des humains depuis 65000 ans. *Populations et Sociétés*, **394**.

Pessis, A-M., ed. (1996). *Proceedings of the International Meeting on the Peopling of the Americas. São Raimundo Nonato, Piauí, Brasil*, vol. 1. São Raimundo Nonato, Brazil: Fundaçao Museu do Homen Americano.

Petit, J. R., Jouzel, J., Raynaud, D., Barkov, N. I., Barnola, J.-M., Basile, I., Bender, M., Chappellaz, J., Davis, M., Delayque, G., Delmotte, M., Kotlyakov, V. M., Legrand, M., Lipenkov, V. Y., Lorius, C., Pépin, L, Ritz, C., Saltzman, E., and Stievenard, M. (1999). Climate and atmospheric history of the past 420,000 years from the Vostok ice core, Antarctica. *Nature*, **399**, 429–36.

Petraglia, M. D. (1998). The Lower Palaeolithic of India and its bearing on the Asian record. In *Early Human Behaviour in Global Context: The Rise and Diversity of the Lower Palaeolithic Record*, eds. M. D. Petraglia and R. Korisettar. London: Routledge, pp. 343–90.

(2003). The Lower Paleolithic of the Arabian Peninsula: occupations, adaptations, and dispersals. *Journal of World Prehistory*, **17**, 141–79.

Petraglia, M. D., LaPorta, P., and Paddayya, K. (1999). The first Acheulian quarry in India: stone tool manufacture, biface morphology, and behaviors. *Journal of Anthropological Research*, **55**, 39–70.

Petraglia, M. D., Schuldenrein, D. J., and Korisettar, R. (2003). Landscapes, activity, and the Acheulean to Middle Paleolithic transition in the Kaladgi Basin, India. *Eurasian Prehistory*, **1**, 3–24.

Pierson, L. J. (1979). *New evidence of Asiatic shipwrecks off the California coast.* Unpublished manuscript. University of San Diego.

Pierson, L. J., and Moriarty, J. R. (1980). Stone anchors: Asiatic shipwrecks off the California coast. *Anthropological Journal of Canada*, **18**, 17–23.

Pinker, S. (1997). *How the Mind Works.* New York: Norton.

Pitulko, V. V., Nikolsky, P. A., Girya, E. Yu., Basilyan, A. E., Tumskoy, V. E., Koulakov, S. A., Astakhov, S. N., Pavlova, E. Yu., and Anisimov, M. A. (2004). The Yana RHS site: humans in the Arctic before the last glacial maximum. *Science*, **303**, 52–6.

Plunket, P., and Uruneula, G. (2006). Social and cultural consequences of a late Holocene eruption of Popocatepetl in central Mexico. *Quaternary International*, **151**, 19–28.

Ponce de León, M. S., and Zollikofer, C. P. E. (2001). Neanderthal cranial ontogeny and its implications for late hominid diversity. *Nature*, **412**, 534–8.

Pope, G. G. (1989). Bamboo and human evolution. *Natural History*, **98**, 49–57.
  (1995). The influence of climate and geography on the biocultural evolution of the Far Eastern hominids. In *Paleoclimate and Evolution with Emphasis on Human Origins*, eds. E. S. Vrba, G. H. Denton, T. C. Partridge, and L. H. Burckle. New Haven, CT: Yale University Press, pp. 493–506.

Pope, G. G., Barr, S., Macdonald, A., and Nakabanlang, S. (1986). Earliest radiometrically dated artifacts from Southeast Asia. *Current Anthropology*, **27**, 275–9.

Pope, G. G., and Keates, S. G. (1994). The evolution of human cognition and cultural capacity: a view from the Far East. In *Integrative Pathways to the Past*, eds. R. L. Ciochon and R. Corruccini. Englewood Cliffs, NJ: Prentice Hall, pp. 531–67.

Pope, K. O., Pohl, M. E. D., Jones, J. G., Lentz, D. L., von Nagy, C., Vega, F. J., and Quitmyer, I. R. (2001). Origin and environmental setting of ancient agriculture in the lowlands of Mesoamerica. *Science*, **292**, 1370–3.

Porch, N., and Allen, J. (1995). Tasmania: archaeological and paleoecological perspectives. *Antiquity*, **69**, 714–32.

Porter, S. C., and An, Z. S. (1995). Correlation between climate events in the North Atlantic and China during the last glaciation. *Nature*, **375**, 305–8.

Possehl, G. L. (1990). A short history of archaeological discovery at Harappa. In *Harappa Excavations 1986–1990, Monographs on World Archaeology*, ed. R. H. Meadow. Madison WI: Prehistory Press.
  (2002). *The Indus Civilization: A Contemporary Perspective.* Oxford: Oxford University Press.

Potts, R. (1996a). *Humanity's Descent: The Consequences of Ecological Instability.* New York: Morrow.
  (1996b). *Evolution and climate variability. Science*, **273**, 922–3.

Powell, J. F., and Neves, W. A. (1999). Craniofacial morphology of the first Americans: pattern and processes in the peopling of the New World. *Yearbook of Physical Anthropology*, **42**, 153–98.

Powell, J. F., Neves, W. A., Ozolins, E., and Pucciarelli, H. M. (1999). Afinidades biológicas extra-continentales de los dos esqueletos más antiguos de América: implicaciones para el poblamiento del Nuevo Mundo. *Atropoligia Fisica Latinoamericana*, **2**, 7–22.

Powell, J. F., and Rose, J. C. (1999). Report on the osteological assessment of the 'Kennewick Man' skeleton (CENWW.97.Kennewick). National Park Service, Archeology Program, http://www.cr.nps.gov/aad/kennewick/powell_rose.htm.

Pray, L. (2004). Epigenetics. Genome meet your environment. *The Scientist*, **18**, 14–20.

Price, G., and Sobbe, I. (2005). Pleistocene palaeoecology and environmental change on the Darling Downs, Southeastern Queensland, Australia. *Memoirs of the Queensland Museum*, **51**, 171–201.

Qiu, Z. X. (2000). Nihewan fauna and Q/N boundary in China (in Chinese with English abstract). *Quaternary Science*, **20**, 142–54.

Quade, J., Cerling, T. E., Bowman, J. R., and Jah, A. (1993). Paleoecologic reconstruction of floodplain environments using palaeosols from Upper Siwalik Group sediments, northern Pakistan. In *Himalaya to the Sea: Geology, Geomorphology and the Quaternary*, ed. J. F. Schroder. London: Routledge, pp. 213–26.

Quam, R. M., Arsuaga, J-L., de Castro, J-M. B., Díez, J. C., Lorenzo, C., Carretero, J. M., Garcia, N., and Ortega, A. I. (2001). Human remains from Veldegoba Cave (Huérmeces, Burgos, Spain). *Journal of Human Evolution*, **41**, 385–435.

Rampino, M. R., and Ambrose, S. H. (2000). Volcanic winter in the Garden of Eden: the Toba super-eruption and the Late Pleistocene human population crash. In *Volcanic Hazards and Disasters in Human Antiquity, Special Paper 345*, eds. F. W. McCoy and G. Heiken. Boulder, CO: Geological Society of America, pp. 71–82.

Rampino, M. R., and Self, S. (1992). Volcanic winter and accelerated glaciation following the Toba super-eruption. *Nature*, **359**, 50–2.

Rampino, M. R., Self, S., and Stothers, R. B. (1988). Volcanic winters. *Annual Review of Earth and Planetary Science*, **16**, 73–99.

Rampino, M. R., and Stothers, R. B. (1988). Flood basalt volcanism during the past 250 million years. *Science*, **241**, 663–8.

Rasmussen, T. L., and Thomsen, E. (2004). The role of the North Atlantic Drift in the millennial timescale glacial climate fluctuations. *Palaeogeography, Palaeoclimatology, Palaeoecology*, **210**, 101–16.

Ravelo, C., Andreasen, D., Lyle, M., Lyle, A. O., and Wara, M. W. (2004). Regional climate shifts caused by gradual global cooling in the Pliocene epoch. *Nature*, **429**, 263.

Reid, R. G. B. (1985). *Evolutionary Theory: The Unfinished Synthesis*. New York: Cornell University Press.

  (2007). *Biological Emergences: Evolution by Natural Experiment*, Vienna Series in Theoretical Biology. Cambridge, MA: MIT Press.

Reitz, E. J., and Sandweiss, D. A. (2001). Environmental change at Ostra base camp. *Journal of Archaeological Science*, **28**, 1085–100.

Relethford, J. H., and Harpending, H. C. (1994). Craniometric variation, genetic theory, and modern human origins. *American Journal of Physical Anthropology*, **95**, 249.

Rendell, H. M., Hailwood, E. A., and Dennell, R. W. (1987). Magnetic polarity stratigraphy of Upper Siwalik Sub-Group, Soan Valley, Pakistan: implications for early human occupance of Asia. *Earth and Planetary Science Letters*, **85**, 488–96.

Renne, P. R., Feinberg, J. M., Waters, M. R., Arroyo-Cabrales, J., Ochoa-Castillo, P., Perez-Campa, N., and Knight, K. B. (2005). Geochronology: age of Mexican ash with alleged footprints. *Nature*, **438**, E7–E8, doi: 10.1038/nature04425.

Reynolds, T. E. G. (1990). Problems in the stone age of Thailand. *Journal of the Siam Society*, **48**, 109–14.

Ricaut, F-X., Fedoseeva, A., Keyser-Tracqui, C., Crubézy, E., and Ludes, B. (2005). Ancient DNA analysis of human Neolithic remains found in northeastern Siberia. *American Journal of Physical Anthropology*, **26**, 458–62.

Richerson, P. J., and Boyd, R. (2000). Climate, culture, and the evolution of cognition. In *The Evolution of Cognition*, eds. C. Heyes and L. Huber. Cambridge, MA: Massachusetts Institute of Technology, pp. 329–46.

Rightmire, G. P. (1998). Evidence from facial morphology for similarity of Asian and African representatives of *Homo erectus. American Journal of Physical Anthropology*, **106**, 61–85.

(2001). Patterns of hominid evolution and dispersal in the Middle Pleistocene. *Quaternary International*, **75**, 77–84.

Riley, D. (1978). Developmental psychology, biology and Marxism. *Ideology and Consciousness*, **4**, 73–91.

Rink, W. J., Schepartz, L. A., Miller-Antonio, S., Huang, W., Hou, Y. Bakken, D., Richter, D., and Jones, H. L. (2003). Electron spin resonance (ESR) dating of mammalian tooth enamel at Panxian Dadong Cave, Guizhou, China. In *Current Research in Chinese Pleistocene Archaeology*, British Archaeology Reports, International Series, eds. S. Chen and S. G. Keates, **1179**, 111–8.

Rink, W. J., Schwarcz, H. P., Ronen, A., and Tsatskin, A. (2004). Confirmation of a near 400 ka age for the Yabrudian industry at Tabun Cave, Israel. *Journal of Archaeological Science*, **31**, 15–20.

Roberts, M., and Parfitt, S. (1999). *Boxgrove: A Middle Pleistocene Hominid Site at Eartham Quarry, Boxgrove, West Sussex*. London: English Heritage.

Roberts, R. G., Flannery, T. F., Ayliffe, L. K., Yoshida, H., Olley, J. M., Prideaux, G. J., Laslett, G. M., Baynes, A., Smith, M. A., Jones, R., and Smith, B. L. (2001). New ages for the last Australian megafauna: continent-wide extinction about 46,000 years ago. *Science*, **292**, 1888–92.

Roberts, R. G., Jones, R., and Smith, M. A. (1990). Thermoluminescence dating of a 50,000-year-old human occupation site in northern Australia. *Nature*, **345**, 153–6.

(1994). Beyond the radiocarbon barrier in Australia. *Antiquity*, **68**, 611–6.

Rodbell, D. T., Seltzer, G. O., Anderson, D. M., Abbott, M. B., Enfield, D. B., and Newman, J. H. (1999). An ~15,000-year record of El Niño-driven alluviation in southwestern Ecuador. *Science*, **283**, 516–20.

Rodríguez, J. (1997). *Paleoecología del Pleistoceno de Atapuerca*. PhD dissertation, Universidad Autónoma de Madrid.

Roebroeks, W. (2001). Hominid behaviour and the earliest occupation of Europe: an explanation. *Journal of Human Evolution*, **41**, 437–61.

Rogers, R. A. (1985a). Glacial geography and native North American languages. *Quaternary Research*, **23**, 130–7.

(1985b). Wisconsinan glaciation and the dispersal of native ethnic groups in North America. In *Woman, Poet, Scientist: Essays in New World Anthropology Honoring Dr. Emma Lou Davis*, ed. T. C. Blackburn. Los Altos, CA: Ballena Press, pp. 104–13.

Rogers, R. A., Martin, L. D., and Nicklas, T. D. (1990). Ice Age geography and the distribution of native North American languages. *Journal of Biogeography*, **17**, 131–43.

Rogers, S. L. (1963). *The Physical Characteristics of the Aboriginal La Jollan Population of Southern California*, San Diego Museum Papers. No. 4. San Diego: Museum of Man.

Rolland, N. (1998). The Lower Paleolithic settlement of Eurasia, with specific reference to Europe. In *Early Human Behaviour in Global Context: The Rise and Diversity of the Lower Palaeolithic Record*, eds. M. D. Petraglia and R. Korisettar. London: Routledge, pp. 187–220.

(2004). Was the emergence of home bases and domestic fire a punctuated event? A review of the Middle Pleistocene record in Eurasia. *Asian Perspectives*, **43**, 248–80.

Rolland, N., and Crockford, S. (2005). Late Pleistocene dwarf *Stegodon* from Flores, Indonesia. *Antiquity*, **79**, antiquity.ac.uk/projgall/rolland/index.html.

Rollo, D. (1994). *Phenotypes: Their Epigenetics, Ecology and Evolution*. London: Chapman and Hall.

Ron, H., and Levi, S. (2001). When did hominids first leave Africa? New high-resolution paleomagnetic evidence from the Erk-El-Ahmar formation, Israel. *Geology*, **29**, 887–90.

Roosevelt, A. C., da Costa, M. L., Machado, C. L., Michab, M., Mercier, N. *et al.* (1996). Paleoindian cave dwellers in the Amazon: the peopling of the Americas. *Science*, **272**, 373–84.

Roots, E. F. (1989a). Climate change: high-latitude regions. *Climate Change*, **15**, 223–53.
 (1989b). Environmental issues related to climate change in northern high latitudes. In *Arctic and Global Change: Proceedings of the Symposium on the Arctic and Global Change*, ed. J. A. W. McCulloch. Washington, DC: The Climate Institute, pp. 6–31.
 (1993). Population, 'carrying capacity', and environmental processes. In *Human Rights in the Twenty-first Century: A Global Challenge*, eds. K. E. Mahoney and P. Mahoney. Dordrecht: Kluwer Academic Publishers, pp. 529–49.

Rosas, A. (1995). Seventeen new mandibular specimens from the Atapuerca/Ibeas Middle Pleistocene Hominids sample (1985–1992). *Journal of Human Evolution*, **28**, 533–59.
 (1997). A gradient of size and shape for the Atapuerca sample and Middle Pleistocene hominid variability. *Journal of Human Evolution*, **33**, 319–31.

Rostek, F., Ruhland, G., Bassinot, F., Müller, P. J., Labeyrie, L., Lancelot, Y., and Bard, E. (1993). Reconstructing sea surface temperature and salinity using $\delta^{18}O$ and alkenone records. *Nature*, **364**, 319–21, doi: 10.1038/364319a0.

Rubicz, R., Schurr, T. G., Babb, P. L., and Crawford, M. H. (2003). Mitochondrial DNA diversity in modern Aleuts, and their genetic relationship with other circumarctic populations. *Human Biology*, **75**, 809–35.

Rudenko, S. I. (1960). Ust'-Kanskaya peshchernaya paleoliticheskaya stoyanka (The Ust'-Kanskaya Paleolithic cave-site). *Paleolit i neolit*, **4** (Materiali i issledovaniya po arkheologii SSSR, No. 79), 104–25.

Ruff, C. B. (1994). Morphological adaptation to climate in modern and fossil hominids. *Yearbook of Physical Anthropology*, **37**, 65–107.

Ryan, W., and Pitman, W. (1999). *Noah's Flood. The New Scientific Discoveries about the Event that Changed History*. New York: Simon and Schuster.

Sachidanandam, R., Weissman, D., Schmidt, S. C., Kakol, J. M., Stein, L. D., Marth, G., Sherry, S., Mullikin, J. C., Mortimore, B. J., Willey, D. L., Hunt, S. E., Cole, C. G., Coggill, P. C., Rice, C. M., Ning, Z., Rogers, J., Bentley, D. R., Kwok, P. Y., Mardis, E. R., Yeh, R. T., Schultz, B., Cook, L., Davenport, R., Dante, M., Fulton, L., Hillier, L., Waterston, R. H., McPherson, J. D., Gilman, B., Schaffner, S., Van Etten, W. J., Reich, D., Higgins, J., Daly, M. J., Blumenstiel, B., Baldwin, J., Stange-Thomann, N., Zody, M. C., Linton, L., Lander, E. S., Altshuler, D., and International SNP Map Working Group (2001). A map of human genome sequence variation containing 1.42 million single nucleotide polymorphisms. *Nature*, **409**, 928–33.

Sachs, I. (1976). Environment and styles of development. In *Outer Limits and Human Needs*, ed. W. Mathews. Uppsala, Sweden: Dag Hammarskjold Foundation.

Saillard, J., Forster, P., Lynnerup, N., Bandelt, H.-J., and Norby, S. (2000). mtDNA variation among Greenland Eskimos : the edge of the Beringian expansion. *American Journal of Human Genetics*, **67**, 718–26.

Sánchez Goñi, M. F., Vanhaeren, M., d'Errico, F., Valladas, H., Grousset, F., Cortijo, E., Turon, J.-L., Malaizé, B., Courty, M. A., and Tisnerat-laborde, N. (2001). *Proposition d'un nouveau cadre chronoclimatique pour les technocomplexes du Paléolithique supérieur d'après le message des glaces, des océans, des lacs et des sites archéologiques*. Presented at the 14th Congress of the International Union for Prehistoric and Protohistoric Sciences, Liège.

Sánchez-Marco, A. (1999). Implications of the avian fauna for paleoecology in the Early Pleistocene of the Iberian Peninsula. *Journal of Human Evolution*, **37**, 375–88.

Sandweiss, D. H., Maasch, K. A., and Anderson, D. G. (2001). Variations in Holocene El Niño frequencies: climate records and cultural consequences in Ancient Peru. *Geology*, **7**, 603–6.

Santos, F. R., Pandya, A., Tyler-Smith, C., Pena, S. D., Schanfield, M., Leonard, W. R., Osipova, L., Crawford, M. H., and Mitchell, R. J. (1999). The central Siberian origin for Native American Y chromosomes. *American Journal of Human Genetics*, **64**, 619–28.

Santos, M. R., Ward, R. H., and Barrantes, R. (1994). mtDNA variation in the Chibcha Amerindian Huetar from Costa Rica. *Human Biology*, **66**, 963–77.

Saragusti, I., and Goren-Inbar, N. (2001). The biface assemblage from Gesher Benot Ya'aqov, Israel: illuminating patterns in 'Out of Africa' dispersal. *Quaternary International*, **75**, 85–9.

Sarnthein, M. (1978). Sand deserts during glacial maximum and climatic optimum. *Nature*, **272**, 43–6.

Sarnthein, M., Janse, E., Weinelt, M., Arnold, M., Duplessy, J.-C., Erlenkeuser, H., Flatoy, A., Johannessen, G., Johanessen, T., Jung, S., Koç, N., Labeyrie, L., Masin, M., Pflaumann, U., and Schulz. H. (1995). Variations in Atlantic Ocean paleoceanography, 50°–85° N: a time-slice record of the last 30,000 years. *Paleocoeanography*, **10**, 1063–94.

Sarnthein, M., and Koopman, B. (1980). Late Quaternary deep-sea record on northwest African dust supply and wind circulation. *Palaeoecology of Africa*, **12**, 238–53.

Sawicki, O., and Smith, D. G. (1992). Glacial Lake Invermere, upper Columbia River Valley, British Columbia: a paleogeographic reconstruction. *Canadian Journal of Earth Sciences*, **29**, 687–92.

Schefuß, E., Schouten, S., Jansen, J. H. F., and Sinninghe Damsté, J. S. (2003). African vegetation controlled by tropical sea surface temperatures in the mid-Pleistocene period. *Nature*, **422**, 418–21.

Schepartz, L. A., Miller-Antonio, S., and Bakken, D. A. (2000). Upland resources and the early Palaeolithic occupation of Southern China, Vietnam, Laos, Thailand and Burma. *World Archaeology*, **32**, 1–13.

Schick, K. D., and Dong, Z. (1993). Early Paleolithic of China and Eastern Asia. *Evolutionary Anthropology*, **2**, 22–35.

Schmittner, A., Meissner, K. J., Eby, M., and Weaver, A. J. (2002). Forcing of the deep circulation in simulations of the last glacial maximum. *Paleooceanography*, **17**, 5:1–5:15, doi: 10.1029/2001/PA000633.

Schmittner, A., Saenko, O. A., and Weaver, A. J. (2003). Coupling of the hemispheres in observations and simulations of glacial climate change. *Quaternary Science Reviews*, **22**, 659–71.

Schmitz, R. W., Serre, D., Bonani, G., Feine, S., Hillgruber, F., Krainitzki, H., Pääbo, S., and Smith, F. H. (2002). The Neandertal type site revisited: interdisciplinary investigations of skeletal remains from the Neander Valley, Germany. *Proceedings of the National Academy of Sciences of the United States of America*, **99**, 13342–7.

Schneider, R. R., Müller, P. J., and Ruhland, G. (1995). Late Quaternary surface circulation in the eastern equatorial South Atlantic: evidence from alkenone sea surface temperature. *Paleoceanography*, **10**, 197–219.

Schneider, R. R., Müller, P. J., Ruhland, G., Meinecke, G., Schmidt., H., and Wefer, G. (1996). Late Quaternary surface temperatures and productivity in east-equatorial South Atlantic: response to changes in trade/monsoon wind forcing and surface water advection. In *The South Atlantic: Present and Past Circulation*, eds. G. Wefer, W. H. Berger, G. Siedler, and D. J. Webb. Berlin: Springer-Verlag, pp. 527–51.

Scholz, C. A., Johnson, T. C., Cohen, A. S., King, J. W., Peck, J. A., Overpeck, J. T., Talbot, M. R., Brown, E. T., Kalindekafe, L., Amoako, P. Y. O., Lyons, R. P., Shanahan, T. M., Castañeda. I. S., Heil, C. W., Forman, S. L., McHargue, L. R., Beuning, K. R., Gomez, J., and Pierson, J. (2007). East African megadroughts between 135 and 75 thousand years ago and bearing on early-modern human origins. *Proceedings of the National Academy of Sciences of the United States of America*, **104**, 16416–21.

Schulz, H., von Rad, U., and Erlenkeuser, H. (1998). Correlation between Arabian Sea and Greenland climate oscillations of the past 111,000 years. *Nature*, **393**, 54–7.

Schulz, M., Timmermann, A., and Paul, A. (2002). On the 1470-year pacing of Dansgaard-Oeschger warm events. *Paleoceanography*, **17**, doi: 10.1029/2000PA000571.

Schurr, T. G. (2003). Molecular genetic diversity of Siberian populations : implications for ancient DNA studies of archeological populations from the Cis-Baikal region. In *Prehistoric Foragers of the Cis-Baikal, Siberia; Proceedings of the First Conference of the Baikal Archaeology Project*, eds. A. Weber and H. McKenzie. Edmonton, AB: Canadian Circumpolar Institute, pp. 155–86.

(2004). The peopling of the New World: perspectives from molecular anthropology. *Annual Review of Anthropology*, **33**, 551–83.

Schurr, T. G., Ballinger, S. W., Gan, Y.-Y., Hodge, J. A., Merriwether, D. A., Lawrence, D. N., Knowler, W. C., Weiss, K. M., and Wallace. D. C. (1990). Amerindian mitochondrial DNAs have rare Asian mutations at high frequencies, suggesting they derived from four primary maternal lineages. *American Journal of Human Genetics*, **46**, 613–23.

Schurr, T. G., Starikovskaya, Y. B., Sukernik, R. I., Torroni, A., and Wallace, D. C. (2000). Mitochondrial DNA diversity in lower Amur River populations, and its implications for the genetic history of the North Pacific and the New World. *American Journal of Physical Anthropology Suppl.*, **30**, 274–5. (Abstr.)

Schurr, T. G., and Wallace, D. C. (2003). Genetic prehistory of Paleoasiatic-speaking peoples of northeastern Siberia and their links to Native American populations. In *Constructing Cultures Then and Now: Celebrating Franz Boas and the Jesup North Pacific Expedition*, eds. L. Kendall and I. Krupnik. Baltimore: Smithsonian Institution Press, pp. 239–58.

Schurr, T. G., Zhadanov, S. I., and Osipova, L. P. (2004). mtDNA variation in indigenous Altaians, and their genetic relationships with Siberian and Mongolian populations. *American Journal of Physical Anthropology Suppl.*, **123**, 176. (Abstr.).

Schwartz, J. H. (1999). *Sudden Origins: Fossils, Genes, and the Emergence of Species*. New York: Wiley.

Schwartz, J. H., and Tattersall, I. (1996). Whose teeth? *Nature*, **381**, 201.

Schwartz, M. K., Mills, L. S., McKelvey, K. S., Ruggiero, L. F., and Allendorf, F. W. (2002). DNA reveals high dispersal synchronizing the population dynamics of Canada lynx. *Nature*, **415**, 520–2.

Scozzari, R., Cruciani, F., Santolamazza, P., Sellitto, D., Cole, D. E., Rubin, L. A., Labuda, D., Marini, E., Succa, V., Vona, G., and Torroni, A. (1997). mtDNA and Y-chromosome-specific polymorphisms in modern Ojibwa : implications about the origin of their gene pool. *American Journal of Human Genetics*, **60**, 241–4.

Seager, R., Battisti, D.., Yin, J., Gordon, N., Naik, N., Clement, A. C., and Cane, M. A. (2002). Is the Gulf Stream responsible for Europe's mild winters? *Quarterly Journal of the Royal Meteorological Society*, **128**, 2563–86.

Sellards, E. H. (1952). *Early Man in America: A Study in Prehistory.* Austin: University of Texas Press.

Sémah, F., Saleki, H., Falguères, C., Féraud, G., and Djubiantono, T. (2000). Did early man reach Java during the Late Pliocene? *Journal of Archaeological Science*, **27**, 763–9.

Semino, O., Magri, C., Benuzzi, G., Lin, A. A., Al-Zahery, N., Battaglia, V., Maccioni, L., Triantaphyllidis, C., Shen, P., Oefner, P. J., Zhivotovsky, L. A., King, R., Torroni, A., Cavalli-Sforza, L. L., Underhill, P. A., and Santachiara-Benerecetti, A. S. (2004). Origin, diffusion, and differentiation of Y-chormosome haplogroups E and J: inferences on the Neolithization of Europe and later migratory events in the Mediterranean Area. *American Journal of Human Genetics*, **74**, 1023–34.

Semino, O., Passarino, G., Oefner, P. J., Lin, A. A., Arbuzova, S., Beckman, L. E., De Benedictis, G., Francalacci, P., Kouvatsi, A., Limborska, S., Marcikiæ, M., Mika, A., Mika, B., Primorac, D., Santachiara-Benerecetti, A. S., Cavalli-Sforza, L. L., and Underhill, P. A. (2000). The genetic legacy of paleolithic *Homo sapiens sapiens* in extant Europeans: a Y chromosome perspective. *Science*, **290**, 1155–9.

Serpell, J., ed. (1995). *The Domestic Dog: Its Evolution, Behaviour and Interactions with People.* Cambridge: Cambridge University Press.

Serre, D., Langaney, A., Chech, M., Teschler-Nicola, M., Paunovic, M., Mennecier, P., Hofreiter, M., Possnert, G., and Pääbo, S. (2004). No evidence of Neandertal mtDNA contribution to early modern humans. *PLoS Biology*, **2**, 0313–7.

Shackleton, N. J. (1987). Oxygen isotopes, ice volume and sca level. *Quaternary Science Reviews*, **6**, 183–90.

Shaw, J. (2005). Geomorphic evidence of postglacial terrestrial environments on Atlantic continental shelves. *Géographie physique et Quaternaire*, Victor Prest special issue, **59**, 141–54.

Shaw, J., Piper, D. J. W., Fader, G. B. J., King, E. L., Todd, B. J., Bell, T., Batterson, M. J., and Liverman, D. G. E. (2006). A conceptual model of the deglaciation of Atlantic Canada. *Quaternary Science Reviews*, **25**, 2059–81.

Shea, J., Fleagle, J. G., Brown, F., Assefa, Z., Feibel, C., McDougall, I., Bender, L., and Jagich, A. (2004). *Archaeology of the Kibish Formation, Lower Omo Valley, Ethiopia.* Paper presented at the Paleoanthropology Society Annual Meeting, Montreal, Quebec.

Shen, G. (1993). Uranium-series ages of speleothems from Guizhou paleolithic sites and their paleoclimatic implications. In *Evolving Landscapes and Evolving Biotas of East Asia since the Mid-Tertiary*, ed. N. Jablonski. Hong Kong: Centre of Asian Studies, University of Hong Kong, pp. 275–82.

Shen, G., Liu, J., and Lin, L. (1997). Preliminary results on U-series dating of Panxian Dadong in Guizhou Province, S-W China. *Acta Anthropologica Sinica*, **16**, 221–30.

Shennan, S. (2002). *Genes, Memes and Human History: Darwinian Archaeology and Cultural Evolution.* London: Thames and Hudson.

Sherry, S. T., Harpending, H. C., Batzer, M. A., and Stoneking, M. (1997). Alu evolution in human populations: using the coalescent to estimate effective population size. *Genetics*, **147**, 1977–82.

Sherry, S. T., Rogers, A. R., Harpending, H. C., Soodyall, H., Jenkins, T., and Stoneking, M. (1994). Mismatch distributions of mtDNA reveal recent human population expansions. *Human Biology*, **66**, 761–75.

Shettleworth, S. (2000). Modularity and the evolution of cognition. In *The Evolution of Cognition*, eds. C. Heyes and L. Huber. London: MIT Press, pp. 43–60.

Shields, G. F., Schmiechen, A. M., Frazier, B. L., Redd, A., Voevoda, M. I., Reed, J. K., and Ward, R. H. (1993). mtDNA sequences suggest a recent evolutionary divergence for Beringian and northern North American populations. *American Journal of Human Genetics*, **53**, 549–62.

Siegenthaler, U., Stocker, T. F., Monnin, E., Lüthi, D., Schwander, J., Stauffer, B., Raynaud, D., Barnola, J-M., Fischer, H., Masson-Delmotte, V., and Jouzel, J. (2005). Stable carbon cycle-climate relationship during the Late Pleistocene. *Science*, **310**, 1313–7.

Sinha, C. (1987). Adaptation and representation: the role of ontogenesis in human evolution and development. In *Behaviour as One of the Main Factors of Evolution*, eds. B. Leonoviká and V. Novak. Prague: Czechoslovak Academy of Sciences.

Skinner, A. R., Blackwell, B. A. B., Martin, S., Ortega, A., Blickstein, J. I. B., Golovanova, L. V., and Doronichev, V. B. (2005). ESR dating at Mezmaiskaya Cave, Russia. *Applied Radiation and Isotopes*, **62**, 219–24.

Slijper, E. J. (1942a). Biologic-anatomical investigations on the bipedal gait and upright posture in mammals with special reference to a little goat, born without forelegs. *I, Akademie Van Wetenschappen, Amsterdam, Afdeeling, Proceedings of the Section of Sciences*, **45**, 288–95.

  (1942b). Biologic-anatomical investigations on the bipedal gait and upright posture in mammals with special reference to a little goat, born without forelegs. *II. Akademie Van Wetenschappen, Amsterdam, Afdeeling, Proceedings of the Section of Sciences*, **45**, 407–15.

Smith, B. H. (1994). Patterns of dental development in *Homo, Australopithecus, Pan,* and *Gorilla. American Journal of Physical Anthropology*, **94**, 307–25.

Smith, D. G. (1995). Glacial Lake McConnell: paleogeography, age, duration, and associated river deltas, Mackenzie River basin, western Canada. *Quaternary Science Reviews*, **13**, 829–43.

Smith, D. G., Malhi, R. S., Eshleman, J. A., Lorenz, J. G., and Kaestle, F. A. (1999). Distribution of haplogroup X among native North Americans. *American Journal of Physical Anthropology*, **110**, 271–84.

Smith, D. G., Malhi, R. S., Eshleman, J. A., and Schultz, B. A. (2000). *A study of mtDNA of early Holocene North American skeletons*. Presented at the 99th Annual Meeting of the American Anthropological Association, San Francisco.

Smith, F. H. (2003). The fate of the Neandertals. *Scientific American*, **13**, 106–7.

Smith, F. H., Trinkaus, E., Pettitt, P. B., Karavanić, I., and Paunović, M. (1999). Direct radiocarbon dates for Vindija G₁ and Velicka Pećina Late Pleistocene hominid remains. *Proceedings of the National Academy of Sciences of the United States of America*, **22**, 12281–6.

Smith, P. E. L. (1986). *Palaeolithic Archaeology in Iran*. Philadelphia: University Museum.

Smith, S. L., and Harrold, F. B. (1997). A paradigm's worth of difference? Understanding the impasse over modern human origins. *Yearbook of Physical Anthropology*, **40**, 113–38.

Smocovitis, V. B. (1996). *Unifying Biology. The Evolutionary Synthesis and Evolutionary Biology*. Princeton: Princeton University Press.

Soffer, O. (1994). Ancestral lifeways in Eurasia: the Middle and Upper Palaeolithic records. In *Origins of Anatomically Modern Humans*, eds. M. H. Nitecki and D. V. Nitecki. New York: Plenum, pp. 101–19.

Soffer, O., Adovasio, J. M., Illingworth, J. S., Amirkhanov, H. A., Praslov, N. D., and Street, M. (2000). Palaeolithic perishables made permanent. *Antiquity*, **74**, 812–21.

Sonakia, A., and Kennedy, K. A. R. (1985). Skull cap of an early man from the Narmada Valley Alluvium (Pleistocene) of Central India. *American Anthropologist*, **87**, 612–6.

Soriano, S., Villa, P., and Wadley, L. (2007). Blade technology and tool forms in the Middle Stone Age of South Africa: the Howiesons Poort and post-Howiesons Poort at Rose Cottage Cave. *Journal of Archaeological Science*, **34**, 681–703.

Spahni, R., Chappellaz, J., Stocker, T. F., Loulergue, L., Hausammann, G., Kawamura, K., Flückiger, J., Schwander, J., Raynaud, D., Masson-Delmotte, V., and Jouzel, J. (2005). Atmospheric methane and nitrous oxide of the Late Pleistocene from Antarctic ice cores. *Science*, **310**, 1317–21.

Srivastava, P., Singh, I. B., Sharma, M., and Singhvi, A. K. (2003). Luminescence chronometry and late Quaternary geomorphic history of the Ganga Plain, India. *Paleogeography, Palaeoclimatology, Palaeoecology*, **197**, 15–41.

Stanford, D., and Bradley, B. (2000). The Solutrean solution – did some ancient Americans come from Europe? *Scientific American Discovering Archaeology*, **2**, 54–55.
  (2002). Ocean trails and prairie paths? In *The First Americans: The Pleistocene Colonization of the New World*, ed. N. G. Jablonski. Berkeley: University of California Press, pp. 255–71.

Starikovskaya, Y. B., Sukernik, R. I., Schurr, T. G., Kogelnik, A. M., and Wallace, D. C. (1998). Mitochondrial DNA diversity in Chukchi and Siberian Eskimos: implications for the genetic prehistory of ancient Beringia. *American Journal of Human Genetics*, **63**, 1473–91.

Stedman, H. H., Kozyak, B. W., Nelson, A., Thesier, D. M., Su, L. T., Low, D. W., Bridges, C. R., Shrager, J. B., Minugh-Purvis, N., and Mitchell, M. A. (2004). Myosin gene mutation correlates with anatomical changes in the human lineage. *Nature*, **428**, 415–8.

Steele, D. G., and Powell, J. F. (1992). The peopling of the Americas: the paleobiological evidence. *Human Biology*, **63**, 301–6.
  (1994). Paleobiological evidence of the peopling of the Americas: A morphometric view. In *Method and Theory for Investigating the Peopling of the Americas*, eds. R. Bonnichsen and D. G. Steele. Corvallis: Center for the Study of the First Americans, Oregon State University, pp. 141–63.
  (2002). Facing the past: a view of the North American human fossil record. In *The First Americans: The Pleistocene Colonization of the New World*, ed. N. G. Jablonski. Berkeley: University of California Press, pp. 93–122.

Steig, E. J. (1999). Mid-Holocene climate change. *Science*, **286**, 1485–7.
  (2006). Climate change: the south-north connection. *Nature* **444**, 152–3.

Stein, A. D., and Lumey, L. H. (2000). The relationship between maternal and offspring birth weights after maternal prenatal famine exposure: the Dutch famine birth cohort study. *Human Biology*, **72**, 641–54.

Stenger, A. (1991). Japanese-influenced ceramics in precontact Washington Sate: a view of the wares and their possible origin. In *The New World Figurine Project*, vol. 1, ed. T. Stocker. Provo: Research Press, pp. 111–22.

Stewart, I., and Cohen, J. (1997). *Figments of Reality: The Evolution of the Curious Mind*. Cambridge: Cambridge University Press.

Stocker, T. F. (1998). The seesaw effect. *Science*, **282**, 61–2.

Stokes, S., Thomas, D. S. G., and Washington, R. (1997). Multiple episodes of aridity in southern Africa since the last interglacial period. *Nature*, **388**, 154–8.

Stoneking, M., and Cann, R. L. (1989). African origin of human mitochondrial DNA. In *The Human Revolution: Behavioural and Biological Perspectives on the Origins of*

Modern Humans, eds. P. Mellars and C. Stringer. Princeton, NJ: Princeton University Press, pp. 17–30.

Straus, L. G. (1989). On early hominid use of fire. *Current Anthropology*, **30**, 488–91.

(1997). The Iberian situation between 40,000 and 30,000 BP in light of European models of migration and convergence. In *Conceptual Issues in Modern Human Origins Research*, eds. G. A. Clark and C. M. Willermet. New York: Aldine de Gruyter, pp. 235–52.

(2000a). Human adaptations to the reforestation of the south coast of the Bay of Biscay: 13–9 ka bp. *Préhistoire Européene*, **16–17**, 271–9.

(2000b). Solutrean settlement of North America? A review of reality. *American Antiquity*, **65**, 219–26.

(2001). Africa and Iberia in the Pleistocene. *Quaternary International*, **75**, 91–102.

Straus, L. G., Meltzer, D. J., and Goebel, T. (2005). Ice Age Atlantis? Exploring the Solutrean-Clovis 'connection'. *World Archaeology*, **37**, 507–32.

Strauss, E. (1999). Can mitochondrial clocks keep time? *Science*, **283**, 1435.

Street-Perrott, F. A., Marchand, D. S., Roberts, N., and Harrison, S. P. (1989). *Global Lake-Level Variations from 18,000 to 0 Years Ago: A Palaeoclimatic Analysis*. Technical Report 46. Washington, DC: US Dept of Energy.

Stringer, C. B. (1989). The origin of early modern humans: a comparison of the European and non-European evidence. In *The Human Revolution: Behavioural and Biological Perspectives on the Origins of Modern Humans*, eds. P. Mellars and C. Stringer. Princeton, NJ: Princeton University Press, pp. 232–44.

(1990). The emergence of modern humans. *Scientific American*, **263**, 98–104.

Stringer, C. B., and Andrews, P. (1988). Genetic and fossil evidence for the origin of modern humans. *Science*, **239**, 1263–8.

Stringer, C. B., and Davies, W. (2001). Those elusive Neanderthals. *Nature*, **413**, 791–2.

Su, B., Xiao, J., Underhill, P., Deka, R., Zhang, W., Akey, J., Huang, W., Shen, D., Lu, D., Luo, J., Chu, J., Tan, J., Shen, P., Davis, R., Cavalli-Sforrza, L., Chakraborty, R., Xiong M., Du, R., Oefner, P., Chen, Z., and Jin, L. (1999). Y-Chromosome evidence for a northward migration of modern humans into Eastern Asia during the last ice age. *American Journal of Human Genetics*, **65**, 1718–24.

Surovell, T. A. (2003). Simulating coastal migration in New World colonization. *Current Anthropology*, **44**, 580–91.

Swisher, C. C., Curtis, C., Jacob, T., Getty, A., Sceprijo, A., and Widiamostro (1994). Age of the earliest known hominids in Java, Indonesia. *Science*, **263**, 1118–21.

Swisher, C. C., Rink, W. J., Antón, S. C., Schwarcz, H. P., Curtis, G. H., Suprijo, A., and Widiasmoro (1996). Latest *Homo erectus* of Java: potential contemporaneity with *Homo sapiens* in Southeast Asia. *Science*, **274**, 1870–4.

Sykes, B. (2001). *The Seven Daughters of Eve: The Science that Reveals our Genetic Ancestry*. New York: W. W. Norton & Company, Inc.

Tanaka, M., Cabrera, V. M., Gonzalez, A. M., Larruga, J. M., Takeyasu, T., Fuku, N., Guo, L. J., Hirose, R., Fugita, Y., Kurata, M., Shinoda, K., Umetsu, K, Yamada, Y., Oshida, Y., Sato, Y., Hattori, N., Mizuno, Y., Arai, Y., Hirose, N., Ohta, S., Ogawa, O., Tanaka, Y., Kawamori, R., Shamoto-Nagai, M., Maruyama, W., Shimokata, H., Suzuki, R., and Shimodaira, H. (2004). Mitochondrial genome variation in Eastern Asia and the peopling of Japan. *Genome Research*, **14**, 1832–50.

Tang, Y. J., Li, Y., and Chen, W. Y. (1995). Mammalian fossil and the age of Xiaochangliang Paleolithic site of Yangyuan, Hebei (in Chinese with English abstract). *Vertebr. Pal-Asiatica*, **33**, 74–83.

Tardieu, C. (1998). Short adolescence in early hominids: infantile and adolescent growth of the human femur. *American Journal of Physical Anthropology*, **107**, 163–78.

Tattersall, I. (1998). *Becoming Human: Evolution and Human Uniqueness*. New York: Harcourt Brace.

(2000). Paleoanthropology: the last half-century. *Evolutionary Anthropology*, **9**, 2–16.

(2001). How we came to be human. *Scientific American*, **285**, 56–63.

(2003). Once we were not alone. *Scientific American*, **13**, 20–7.

Tattersall, I., and Schwartz, J. (1999). Hominds and hybrids: the place of Neanderthals in human evolution. *Proceedings of the National Academy of Sciences of the United States of America*, **96**, 7117.

Tchernov, E. (1987). The age of the 'Ubeidiya Formation', an early Pleistocene hominid site in the Jordan Valley, Israel. *Israel Journal of Earth Science*, **36**, 3–30.

(1988). Biochronology of the Middle Palaeolithic and dispersal events of hominids in the Levant. *L'Homme de Neandertal*, **2**, 153–68.

(1992a). The Afro-Arabian component in the Levantine mammalian fauna – a short biogeographic review. *Israel Journal of Zoology*, **38**, 155–92.

(1992b). Biochronology, paleoecology, and dispersal events of hominids in the southern Levant. In *The Evolution and Dispersal of Modern Humans in Asia*, eds. T. Akazawa, K. Aoki, and T. Kimura. Tokyo: Hokusan-Shen, pp. 149–88.

(1992c). Eurasian-African biotic exchanges through the Levantine corridor during the Neogene and Quaternary. *Courier Forschungs-Institut Senckenberg*, **153**, 103–23.

Teilhard de Chardin, P. (1941). *Early Man in China*. Peking: Institute of Geo-Biology.

Thangaraj, K., Chaubey, G., Kivisild, T., Reddy, A. G., Kumar Singh, V., Rasalkar, A., and Singh, L. (2005). Reconstructing the origin of Andaman Islanders. *Science*, **308**, 996.

Thiel, B. (1987). Early settlement of the Philippines, eastern Indonesia, and Australia-New Guinea: a new hypothesis. *Current Anthropology*, **28**, 236–41B.

Thieme, H. (1997). Lower Palaeolithic hunting spears from Germany. *Nature*, **385**, 807–10.

Thomas, C. D., Cameron, A., Green, R. E., Bakkenes, M., Beaumont, L. J., Collingham, Y. C., Erasmus, B. F. N., Ferreira de Siqueira, M., Grainger, A., Hannah, L., Hughes, L., Huntley, B., van Jaarsveld, A. S., Midgley, G. F., Miles, L., Ortega-Huerta, M. A., Peterson, A. T., Phillips, O. L., and Williams, S. E. (2004). Extinction risk from climate change. *Nature*, **427**, 145–8, doi: 10.1038/nature02121.

Thomas, M. F. (2000). Late Quaternary environmental changes and the alluvial record in humid tropical environments. *Quaternary International*, **72**, 23–36.

Thorne, A., Grün, R., Mortimer, G., Spooner, N. A., Simpson, J. J., McCulloch, M., Taylor, L., and Curnoe, D. (1999). Australia's oldest human remains: age of the Lake Mungo 3 skeleton. *Journal of Human Evolution*, **36**, 591–612.

Thorne, A. G., and Wolpoff, M. H. (1992). The multiregional evolution of humans. *Scientific American*, **266**, 28–33.

Timmermann, A., and An, S.-I. (2005). ENSO suppression due to weakening of the North Atlantic thermohaline circulation. *Journal of Climate*, **18**, 3122–39.

Tishkoff, S. A., and Kidd, K. K. (2004). Implications of biogeography of human populations for 'race' and medicine. *Nature Genetics Supplement*, **36**, S21–S27.

Tomasello, M. (1999). *The Cultural Origins of Human Cognition*. Cambridge, MA: Harvard University Press.

(2000). Two hypotheses about primate cognition. In *The Evolution of Cognition*, eds. C. Heyes and L. Huber. London: MIT Press, pp. 165–83.

Tomasello, M., and Call, J. (1997). *Primate Cognition*. New York: Oxford University Press.

Tong, G. B., and Shao, S. X. (1991). The evolution of Quatenary climate in China. In *The Quaternary of China*, eds. Z. H. Zhang and S. X. Shao. Beijing: China Ocean Press, pp. 42–76.

Torgerson, T., Luly, J., De Dekker, P., Jones, M. R., Seale, D. E., Chivas, A. R., and Ullman, W. J. (1988). Late Quaternary environments of the Carpentaria Basin, Australia. *Palaeogeography, Palaeoclimatology, Palaeoecology*, **67**, 245–61.

Torrence, R. (1989). Re-tooling: towards a behavioral theory of stone tools. In *Time, Energy and Stone Tools*, ed. R. Torrence. Cambridge: Cambridge University Press, pp. 57–66.

   (2000). Hunter-gatherer technology: macro- and microscale approaches. In *Hunter-gatherers: An Interdisciplinary Perspective*, eds. C. Panter-Brick, R. H. Layton, and P. Rowley-Conwy. Cambridge: Cambridge University Press, 99–143.

Torroni, A., Chen, Y-S., Semino, O., Santachiara-Beneceretti, A. S., Scott, C. R. et al. (1994a). MtDNA and Y-chromosome polymorphisms in four Native American populations from southern Mexico. *American Journal of Human Genetics*, **54**, 308–18.

Torroni, A., Neel, J. V., Barrantes, R., Schurr, T. G., and Wallace, D. C. (1994b). A mitochondrial DNA 'clock' for the Amerinds and its implications for timing their entry into North America. *Proceedings of the National Academy of Sciences of the United States of America*, **91**, 1158–62.

Torroni, A., Schurr, T. G., Cabell, M. F., Brown, M. D., Neel, J. V., Larsen, M., Smith, D. G., Vullo, C. M., and Wallace, D. C. (1993a). Asian affinities and the continental radiation of the four founding Native American mtDNAs. *American Journal of Human Genetics*, **53**, 563–90.

Torroni, A, Schurr, T. G., Yang, C.-C., Szathmary, E. J. E., Williams, R. C., Schanield, M. S., Troup, G. A., Knowlcr, W. C., Lawrcncc, D. N., Wciss, K. M., and Wallace, D. C. (1992). Native American mitochondrial DNA analysis indicates that the Amerind and the Na-Dene populations were founded by two independent migrations. *Genetics*, **130**, 153–62.

Torroni, A., Sukernik, R. I., Schurr, T. G., Starikovskaya, Y. B., Cabell, M. F., Crawford, M. H., Comuzzie, A. G., and Wallace, D. C. (1993b). mtDNA variation of aboriginal Siberians reveals distinct genetic affinities with Native Americans. *American Journal of Human Genetics*, **53**, 591–608.

Trauth, M. H., Maslin, M. A., Deino, A., and Strecker, M. R. (2005). Late Cenozoic moisture history of East Africa. *Science*, **309**, 2051–3.

Trinkaus, E. (1981). Neandertal limb proportions and cold adaptation. In *Aspects of Human Evolution*, ed. C. Stringer. London: Taylor and Francis, pp. 187–224.

Trinkaus, E., and Duarte C. (2003). The hybrid child from Portugal. *Scientific American*, **13**, 33.

Trinkaus, E., Moldovan, O., Milota, S., Bilgar, A., Sarcina, L., Athreya, S., Bailey, S. E., Rodgrigo, R., Mircea, G., Higham, T., Bronk Ramsey C., and van der Plicht, J. (2003). An early modern human from the Pestera cu Oase, Romania. *Proceedings of the National Academy of Sciences of the United States of America*, **100**, 11231–6.

Trinkaus, E., Ruff, C. B., and Churchill, S. E. (1998). Upper limb versus lower limb loading patterns among near eastern Middle Paleolithic Hominids. In *Neandertals and Modern Humans in Western Asia*, eds. T. Akazawa, K. Aoki, and O. Bar-Yosef. New York: Plenum Press, pp. 391–404.

Trueman, C. N. G., Field, J. H., Dortch, J., Charles, B., and Wroe, S. (2005). Prolonged coexistence of humans and megafauna in Pleistocene Australia. *Proceedings of the National Academy of Sciences of the United States of America*, **101**, 8381–5.

Tudhope, A. W., Chilcott, C. P., McCulloch, M. T., Cook, E. R., Chappell, J., Ellam, R. M., Lea, D. W., Lough, J. M., and Shimmield, G. B. (2001). Variability in the El Niño-Southern Oscillation through a glacial-interglacial cycle. *Science*, **291**, 1511–7.

Turner, A. (1992). Large carnivores and earliest European hominids: changing determinants of resource availability during the Lower and Middle Pleistocene. *Journal of Human Evolution*, **22**, 109–26.

(1999). Assessing earliest human settlement of Eurasia: Late Pliocene dispersions from Africa. *Antiquity*, **73**, 563–70.

Turner, C. G. (1983). Dental evidence for the peopling of the Americas. In *Early Man in the New World*, ed. R. Shutler Jr. Beverly Hills: Sage Publications, pp. 147–57.

(1990). Major features of sundadonty and sinodonty including suggestions about east Asian microevolution, population history, and late Pleistocene relationships with Australian aboriginals. *American Journal of Physical Anthropology*, **82**, 295–317.

(2002). Teeth, needles, dogs, and Siberia: bioarchaeological evidence for the colonization of the New World. In *The First Americans: The Pleistocene Colonization of the New World*, ed. N. G. Jablonski. Berkeley: University of California Press, pp. 123–58.

Turney, C. S. M., and Brown, H. (2007). Catastrophic early Holocene sea level rise, human migration and the Neolithic transition in Europe. *Quaternary Science Reviews*, **26**, 2036–41.

Turney, C. S. M., Kershaw, A. P., Moss, P., Bird, M. I., Fifield, L. K., Cresswell, R. G., Santos, G. M., di Tada, M. L., Hausladen, P. A., and Zhou, Y. (2001). Redating the onset of burning at Lynch's Crater (North Queensland): implications for human settlement in Australia. *Journal of Quaternary Science*, **16**, 767–71.

Tzedakis, P. C., Andrieu, V., de Beaulieu, J.-L., Crowhurst, S., Follieri, M., Hooghiemstra, H., Magri, D., Reille, M., Sadori, L., Shackleton, N. J., and Wijmstra, T. A. (1997). Comparison of terrestrial and marine records of changing climate of the last 500 000 years. *Earth and Planetary Science Letters*, **150**, 171–6.

Underhill, P. A., Jin, L., Lin, A. A., Mehdi, S. Q., Jenkins, T., Vollrath, D., Davis, R. W., Cavalli-Sforza, L. L., and Oefner, P. J. (1997). Detection of numerous Y chromosome biallelic polymorphisms by denaturing high performance liquid chromatography. *Genome Research*, **7**, 996–1005.

Underhill, P. A., Jin, L., Zemans, R., Oefner, P. J., Cavalli-Sforza, L. L. (1996). A pre-Columbian Y chromosome-specific transition and its implications for human evolutionary history. *Proceedings of the National Academy of Sciences of the United States of America*, **93**, 196–200.

Underhill, P. A., Passarino, G., Lin, A. A., Shen, P., Lahr, M. M., Foley, R. A., Oefner, P. J., and Cavalli-Sforza, L. L. (2001). The phylogeography of Y chromosome binary haplotypes and the origins of modern human populations. *Annals of Human Genetics*, **65**, 43–62.

Underhill, P. A., Shen, P., Lin, A. A., Jin, L., Passarino, G., Yang, W. H., Kauffman, E., Bonné-Tamir, B., Bertranpetit, J., Francalacci, P., Ibrahim, M., Jenkins, T., Kidd, J. R., Mehdi, S. Q., Seielstad, M. T., Wells, R. S., Piazza, A., Davis, R. W., Feldman, M. W., Cavalli-Sforza, L. L., and Oefner, P. J. (2000). Y chromosome sequence variation and the history of human populations. *Nature*, **26**, 358–61.

United Nations Environment Programme. (2007). *Global Environment Outlook 4: Environment for Development*. Valletta, Malta: Progress Press.

Ushijima, Y. (1954). The human skeletal remains from the Mitsu site, Saga Prefecture, a site associated with the 'Yayoishiki' period of prehistoric Japan. *Quarterly Journal of Anthropology*, **1**, 273–303.

Valladas, H., Mercier, N., Joron, J., and Reyss, J. (1998). GIF Laboratory dates for Middle Paleolithic Levant. In *Neandertals and Modern Humans in Western Asia*, eds. T. Akazawa, K. Aoki, and O. Bar-Yosef. New York: Plenum Press, pp. 69–75.

van den Bergh, G. D., de Vos, J., Aziz, F., and Morwood, M. J. (2001). Elephantoidea in the Indonesian region: new *Stegodon* findings from Flores. In *The World of Elephants: Proceedings of the First International Congress, Rome*, eds. G. Cavaretta, P. Gioia, M. Mussi, and M. R. Palombo. Rome: Consiglio Nazionale delle Ricerche, pp. 623–7.

van der Made, J. (2001). Les ongulés d'Atapuerca: stratigraphie et biogéographie. *L'Anthropologie*, **105**, 95–113.

van der Made, J., Aguirre, E., Bastir, M., Fernández Jalvo, Y., Huguet, R., Laplana, C., Márquez, B., Martínez, C., Martinón, M., Rosas, A., Rodríguez, J., Sánchez, A., (2003). El registro paleontológico y arqueológico de los yacimientos dc la Trinchera del Ferrocarril en la Sierra de Atapuvverca. *Coloquios de Paleontologia Volumen Extraordinario*, **1**, 345–72.

Van Neer, W. (2002). Food security in western and central Africa during the late Holocene: the role of domestic stock keeping, hunting and fishing. In *Droughts, Food and Culture: Ecological Change and Food Security in Africa's Later Prehistory*, ed. F. A. Hassan. Dordrecht: Kluwer Academic Press, pp. 251–74.

Van Vark, G. N., Kuizenga, D., and L'Engle Williams, F. (2003). Kennewick and Luzia: lessons from the European Upper Paleolithic. *American Journal of Physical Anthropology*, **121**, 181–4.

Vekua, A., Lordkipanidze, D., Rightmire, G. P., Agusti, J., Ferring, R., Maisuradze, G., Mouskhelishvili, A., Nioradze, M., Ponce de Leon, M., Tappen, M., Tvalchrelidze, M., and Zollikofer, C. (2002). A new skull of early *Homo* from Dmanisi, Georgia. *Science*, **297**, 85–9.

Verosub, K. K., Goren-Inbar, N., Feibel, C., and Saragusti, I. (1998). Location of the Matuyama/Brunhes boundary in the Gesher Benot Ya'aqov archaeological site, Israel. *Journal of Human Evolution*, **34**, A22.

Vettoretti, G., Peltier, W. R., and McFarlane, N. A. (2000). Global water balance and atmospheric water vapour transport at Last Glacial Maximum: climate simulations with the CCCma atmospheric general circulation model. *Canadian Journal of Earth Sciences*, **37**, 695–723.

Vignaud, P., Duringer, P., Mackaye, H. T., Likius, A., Blondel, C., Boisserie, J-R., de Bonis, L., Eisenmann, V., Etienne, M-E., Geraads, D., Guy, F., Lehmann, T., Lihoreau, F., Lopez-Martinez, N., Mourer-Chauviré, C., Otero, O., Rage, J-C., Schuster, M., Viriot, L., Zazzo, A., and Brunet, M. (2002). Geology and palaeontology of the Upper Miocene Toros-Menalla hominid locality, Chad. *Nature*, **418**, 152–5.

Villa, P. (2001). Early Italy and the colonization of Western Europe. *Quaternary International*, **75**, 113–30.

Visbeck, M. (2007). Oceanography: power of pull. *Nature*, **447**, 383.

Von Grafenstein, U., Erlenkeuser, H., Müller, J., Jouzel, J., and Johnsen, S. (1998). The short cold period 8,200 years ago documented in oxygen isotope records of precipitation in Europe and Greenland. *Climate Dynamics*, **14**, 73–81.

Vrba, E. S. (1995). The fossil record of African antelopes relative to human evolution. In *Paleoclimate and Evolution, with Emphasis on Human Origins*, eds. E. S. Vrba, G. H. Denton, T. C. Partridge, and L. H. Burckle. New Haven, CT: Yale University Press, pp. 385–424.

Wainscoat, J. S., Hill, A. V. S., Thein, S. L., Flint, J., Chapman, J. C., Weatherall, D. J., Clegg, J. B., and Higgs, D. R. (1989). Geographic distribution of alpha- and beta-globin gene cluster ploymorphisms. In *The Human Revolution: Behavioural and Biological Perspectives on the Origins of Modern Humans*, eds. P. Mellars and C. Stringer. Princeton, NJ: Princeton University Press, pp. 31–8.

Walker, A., and Shipman, P. (1996). *The Wisdom of the Bones: In Search of Human Origins.* New York: Knopf.

Wallace, A. R. (1890). *The Malay Archipelago: The Land of the Orang-utan, and the Bird of Paradise. A Narrative of Travel with Studies of Man and Nature.* London: Macmillan.

Walter, R. C., Buffler, R. T., Bruggemann, J. H., Guillaume, M. M. M., Berhe, S. M., Negassi, B., Libsekal, Y., Cheng, H., Edwards, R. L., Cosel, R., Néraudeau, D., and Gagnon, M. (2000). Early human occupation of the Red Sea coast of Eritrea during the last interglacial. *Nature*, **405**, 65–9.

Wang, H., Ambrose, S. H., Liu, J. C-L., and Follmer, L. R. (1997). Paleosol stable isotope evidence for early hominid occupation of East Asian temperate environments. *Quaternary Research*, **48**, 228–38.

Wang, P. (1999). Response of Western Pacific marginal seas to glacial cycles: paleoceanographic and sedimentological features. *Marine Geology*, **156**, 5–39.

Ward, B. C., Wilson, M. C., Nagorsen, D. W., Nelson, D. E., Driver J. C., and Wigen, R. J. (2003). Port Eliza Cave: North American west coast interstadial environment and implications for human migrations. *Quaternary Science Reviews*, **22**, 1383–8.

Warner, B. G., Mathewes, R. W., and Clague, J. J. (1982). Ice-free conditions of the Queen Charlotte Islands, British Columbia, at the height of Late Wisconsinan glaciations. *Science*, **218**, 675–7.

Warren, R. (2006). Impacts of global climate change at different annual mean global temperature increases. In *Avoiding Dangerous Climate Change*, eds. H. J. Schellnhuber, W. Cramer, N. Nakicenovic, T. Wigley, and G. Yohe. Cambridge: Cambridge University Press, pp. 93–131.

Watanabe, H. (1985). The chopper-chopping tool complex of eastern Asia: an ethnoarchaeological-ecological reexamination. *Journal of Anthropological Archaeology*, **4**, 1–18.

Waters, M. R., Forman, S. L., and Pierson, J. M. (1997). Diring Yuriakh: a Lower Paleolithic site in Central Siberia. *Science*, **275**, 1281–4.

Watson, A. J., Bakker, D. C. E., Ridgwell, A. J., Boyd, P. W., and Law, C. S. (2000). Effect of iron supply on southern ocean CO_2 uptake and implications for glacial atmospheric CO_2. *Nature*, **407**, 730–3, doi: 10.1038/35037561.

Weaver, A. J. (2004). The UVic Earth system climate model and the thermohaline circulation in past, present, and future climates. In *State of the Planet: Frontiers and Challenges in Geophysics, Geophysical Monograph 150*, eds. R. Stephen, J. Sparks, and J. C. Hawkesworth. Washington, DC: American Geophysical Union, pp. 279–96.

(2008). *Keeping Our Cool: Canada in a Warming World*. Toronto: Penguin Group, Canada.

Weaver, A. J., Eby, M., Fanning, A. F., and Wiebe, E. C. (1998). Simulated influence of carbon dioxide, orbital forcing and ice sheets on the climate of the last glacial maximum. *Nature*, **394**, 847–53, doi: 10.1038/29695.

Weaver, A. J., Eby, M., Kienast, M., and Saenko, O. A. (2007). Response of the Atlantic meriodional overturning circulation to increasing atmospheric CO_2: sensitivity to mean climate state. *Geophysical Research Letters*, **34**, L05078, doi: 10.1029/2006GL028756.

Weaver, A. J., Eby, M., Wiebe, E. C., Bitz, C. M., Duffy, P. B., Ewen, T. L., Fanning, A. F., Holland, M. M., MacFadyen, A., Matthews, H. D., Meissner, K. J., Saenko, O., Schmittner, A., Wang, H., and Yoshimori, M. (2001). The UVic Earth system climate model: model description, climatology and application to past, present and future climates. *Atmosphere-Ocean*, **39**, 361–428.

Weaver, A. J., Saenko, O. A., Clark, P. U., and Mitrovica, J. X. (2003). Meltwater pulse 1A
    from Antarctica as a trigger of the Bølling-Allerød warm interval. *Science*, **299**, 1709–13.
Wei, Q. (1997). The framework of archaeological geology of the Nihewan Basin (in Chinese
    with English abstract). In *Evidence for Evolution: Essays in Honor of Prof.
    Chungchien Yong on the Hundredth Anniversary of His Birth*, eds. Y. Tong *et al*.
    Beijing: Ocean Press, pp. 193–207.
  (1999). Paleolithic archaeological sites from the Lower Pleistocene in China. In *From
    Sozudai to Kamitakamori: World Views on the Early and Middle Palaeolithic in Japan*,
    ed. C. Serizawa. Tokyo: Tohoku Fukushi University, pp. 123–4.
Weidenreich, F. (1936a). The mandibles of *Sinanthropus pekinensis*: a comparative study.
    Palaeontologia Sinica Series D, **7**, 1–162.
  (1936b). Observations on the form and proportions of the endocranial casts of
    Sinanthropus pekinensis, other hominids, and the great apes: a comparative study of
    brain size. *Palaeontologia Sinica Series D*, **7**, 1–50, 1937.
  (1939). Six lectures on *Sinanthropus pekinensis* and related problems. *Bulletin of
    Geological Society of China*, **19**, 1–110.
  (1941). The extremity bones of *Sinanthropus pekinensis*. *Palaeontologia Sinica (New
    Series D)*, **5**, 1–150.
  (1943). The skull of *Sinanthropus pekinensis*: a comparative study of a primitive hominid
    skull. *Palaeontologia Sinica (New Series D)*, **10**, 1–298.
Wells, R. S., Yuldashev, N., Ruzibakiev, R., Underhill, P. A., Evseeva, I., Blue-Smith, J.,
    Jin, L., Su, B., Pitchappan, R., Shanmugalakshmi, S., Balakrishnan, K., Read, M.,
    Pearson, N. M., Zerjal, T., Webster, M. T., Zholoshvili, I., Jamarjashvili, E.,
    Gambarov, S., Nikbin, B., Dostiev, A., Aknazarov, O., Zalloua, P., Tsoy, I., Kitaev, M.,
    Mirrakhimov, M., Chariev, A., and Bodmera, W. F. (2001). The Eurasian heartland: a
    continental perspective on Y-chromosome diversity. *Proceedings of the National
    Academy of Sciences of the United States of America*, **98**, 10244–9.
West-Eberhard, M. J. (2003). *Developmental Plasticity and Evolution*. Oxford: Oxford
    University Press.
Whalen, N. M., Davis, W. P., and Pease, D. W. (1989). Early Pleistocene migrations into
    Saudi Arabia. *Atlal*, **12**, 59–75.
Whallon, R. (1989). Elements of cultural change in the Later Palaeolithic. In *The Human
    Revolution: Behavioural and Biological Perspectives on the Origins of Modern
    Humans*, eds. P. Mellars and C. Stringer. Princeton, NJ: Princeton University Press,
    pp. 433–54.
White, J. M., Mathewes, R. W., and Mathews, W. H. (1985). Late Pleistocene chronology
    and environment of the 'ice-free corridor' of northwestern Alberta. *Quaternary
    Research*, **24**, 173–86.
White, J. P., Crook, K. A. W., and Buxton, B. P. (1970). Kosipe: a late Pleistocene site in the
    Papuan highlands. *Proceedings of the Late Prehistory Society*, **36**, 152–70.
White, J. W. C. (1993). Don't touch that dial. *Nature*, **364**, 186, doi: 10.1038/364186a0.
White, T. D., Asfaw, B., Beyene, Y., Haile-Selassie, Y., Lovejoy, C. O., Suwa, G., and
    WoldeGabriel, G. (2009). *Ardipithecus ramidus* and the paleobiology of early
    hominids. *Science*, **326**, 64, 75–86.
White, T. D., Asfaw, B., DeGusta, D., Gilbert, H., Richards, G. D., Suwa, G., and
    Howell, F. C. (2003). Pleistocene *Homo sapiens* from Middle Awash, Ethiopia. *Nature*,
    423, 742–7.
Whiten, A., Goodall, J., McGrew, W. C., Nishida, T., Reynolds, V., Sugiyama, Y.,
    Tutin, C. E. G., Wrangham, R. W., and Boesch, C. (1999). Cultures in chimpanzees.
    Nature, **399**, 682–5.

Wilkinson, T. (1999). *Early Dynastic Egypt*. London: Routledge.

(2003). *Genesis of the Pharaohs*. London: Thames and Hudson.

Willoughby, P. R. (1993). Culture, environment and the emergence of *Homo sapiens* in East Africa. In *Culture and Environment: A Fragile Coexistence*, eds. R. Jamieson, S. Abonyi, and N. Mirau. Calgary: Chacmool Archaeological Association, pp. 135–43.

(2007). *The Evolution of Modern Humans in Africa: A Comprehensive Guide*. Lanham, MD: Altamira Press.

Wilson, A. C., and Cann, R. L. (1992). The recent African genesis of humans. *Scientific American*, **266**, 22–7.

Wintle, A. G. (1996). Archaeologically-relevant dating techniques for the next century: small, hot and identified by acronyms. *Journal of Archaeological Science*, **23**, 123–38.

Wolpoff, M. H. (1989). Multiregional evolution: the fossil alternative to Eden. In *The Human Revolution: Behavioural and Biological Perspectives on the Origins of Modern Humans*, eds. P. Mellars and C. Stringer. Princeton, NJ: Princeton University Press, pp. 62–108.

Wong, K. (2003). Who were the Neandertals? *Scientific American*, **13**, 28–37.

Woo, J. K. (1966). The skull of Lantian man. *Current Anthropology*, **7**, 83–6.

Wood, B., and Collard, M. (1999). The human genus. *Science*, **284**, 65–71.

Wood, B., and Turner, A. (1995). Out of Africa and into Asia. *Nature*, **378**, 239–40.

Wu, S. (2004). On the origin of modern humans in China. *Quaternary International*, **117**, 131–40.

Wu, W., and Liu, T. (2004). Possible role of the 'Holocene Event' on the collapse of Neolithic cultures around the Central Plain of China. *Quaternary International*, **117**, 153–66.

Wu, X. (2000). Longgupo mandible belongs to ape. *Acta Anthropológica Sinica*, **19**, 1–10.

Wu, X., and Poirier, F. E. (1995). *Human Evolution in China*. New York: Oxford University Press.

Wynn, J. G. (2004). Influence of Plio-Pleistocene aridification on human evolution: evidence from paleosols of the Turkana Basin, Kenya. *American Journal of Physical Anthropology*, **123**, 106–18.

Wynn, T. (1991). Tools, grammar and the archaeology of cognition. *Cambridge Archaeological Journal*, **1**, 191–206.

Wynn, T., and Tierson, F. (1990). Regional comparison of the shapes of later Acheulean handaxes. *American Anthropology*, **92**, 73–84.

Wynne-Edwards, K. E. (1998). Evolution of parental care in *Phodopus*: conflict between adaptations for survival and adaptations for rapid reproduction. *American Zoologist*, **38**, 238–50.

Wynne-Edwards, K. E.,, Surov, A. V., and Telitina, A. Y. (1999). Differences in endogenous activity within the genus *Phodopus*. *Journal of Mammalogy*, **80**, 855–65.

Yamei, H., Potts, R., Baoyin, Y., Zhengtang, G., Deino, A., Wei, W., Clark, J., Guangmao, X., and Weiwen, H. (2000). Mid-Pleistocene Acheulean-like stone technology of the Bose Basin, South China. *Science*, **287**, 1622–6.

Yeni-Komshian, G. H., and Benson, D. A. (1976). Anatomical study of cerebral asymmetry in humans, chimpanzees and rhesus monkeys. *Science*, **192**, 387–9.

Yoffee, N. (2005). *Myths of the Archaic State: Evolution of the Earliest Cities, States and Civilizations*. Cambridge: Cambridge University Press.

Yokoyama, Y., De Dekker, P., Lambeck, K., Johnston, P., and Fifield, L. K. (2001). Sea level at the Last Glacial Maximum: evidence from north-western Australia to constrain ice-volumes for oxygen-isotope stage 2. *Palaeogeography, Palaeoclimatology, Palaeoecology*, **165**, 281–97.

Yokoyama, Y., Purcell, A., Lambeck, K., and Johnston, P. (2001). Shoreline reconstruction around Australia during the Last Glacial Maximum and Late Glacial Stage. *Quaternary International*, **83–85**, 9–18.

Yoshimori, M., Reader, M. C., Weaver, A. J., and MacFarlane, N. A. (2002). On the causes of glacial inception at 116Ka BP. *Climate Dynamics*, **18**, 383–402.

You, Y. Z., Tang, Y. J., and Li, Y. (1980). Discovery of the Palaeoliths from the Nihewan Formation (in Chinese). *Quaternary Sinica*, **5**, 1–11.

Yu, G., Chen, X., Ni, J., Cheddadi, R., Guiot, J., Han, H., Harrison, S. P., Huang, C., Ke, M., Kong, Z., Li, S., Li, W., Liew, P., Liu, G., Liu, J., Liu, Q., Liu, K.-B., Prentice, I. C., Qui, W., Ren, G., Song, C., Sugita, S., Sun, X., Tang, L., Van Campo, E., Xia, Y., Xu, Q., Yan, S., Yang, X., Zhao, J., and Zheng, Z. (2000). Palaeovegetation of China: a pollen data-based synthesis for the mid-Holocene and last glacial maximum. *Journal of Biogeography*, **27**, 635–64.

Yuan, S., Chen, T., and Goa, S. (1986). Uranium series chronological sequence of some palaeolithic sites in South China (in Chinese with English abstract). *Acta Anthropológica Sinica*, **5**, 179–90.

Zagwijn, W. H. (1992). The beginning of the Ice Age in Europe and its major subdivisions. *Quaternary Science Reviews*, **11**, 583–91.

Zalasiewicz, J., Williams, M., Smith, A., Barry, T. L., Coe, A. L., Bown, P. R., Brenchley, P., Cantrill, D., Gale, A., Gibbard, P., Gregory, F. J., Hounslow, M. W., Kerr, A. C., Pearson, P., Knox, R., Powell, J., Waters, C., Marshall, J., Oates, M., Rawson, P., and Stone, P. (2008). Are we now living in the Anthropocene? *GSA Today*, **18**, 4–8.

Zazula, G. D., Froese, D. G., Schweger, C. E., Mathewes, R. W., Beaudoin, A. B., Telka, A. M., Harington, C. R., and Westgate, J. A. (2003). Ice-age steppe vegetation in east Beringia. *Nature*, **423**, 603.

Zeitlin, S. M. (1979). *Geologiya paleolita Severnoy Azii* (Geology of the Paleolithic of Northern Asia). Moscow: Nauka.

Zhang, H. C., Ma, Y. Z., Wünneman, B., and Pachur, H.-J. (2000). A Holocene climatic record from arid northwest China. *Palaeogeography, Palaeoclimatology, Palaeoecology*, **162**, 389–401.

Zhou, C., Wang, Y., Cheng, H., and Liu, Z. (1999). Discussion of Nanjing man's age (in Chinese with English abstract). *Acta Anthropológica Sinica*, **18**, 255–62.

Zhu, R. X., An, Z., Potts, R., and Hoffman, K. A. (2003). Magnetostratigraphic dating of early humans in China (in Chinese). *Earth-Science Reviews*, **61**, 341–59.

Zhu, R. X., Hoffman, K. A., Potts, R., Deng, C. L., Pan, Y. X., Guo, B., Shi, C. D., Guo, Z. T., Yuan, B. Y., Hou, Y. M., and Huang, W. W. (2001). Earliest presence of humans in northeast Asia. *Nature*, **413**, 413–7.

Zhu, R. X., Potts, R., Xie, F., Hoffman, K. A., Deng, C. L., Shi, C. D., Pan, Y. X., Wang, H. Q., Shi, R. P., Wang, Y. C., Shi, G. H., and Wu, N. Q. (2004). New evidence on the earliest human presence at high northern latitudes in northeast Asia. *Nature*, **431**, 559–62.

Zielinski, G. A. (2000). Use of paleo-records in determining variability within the volcanism-climate system. *Quaternary Science Reviews*, **19**, 417–38.

Zielinski, G. A., Mayewski, P. A., Meeker, L. D., Whitlow, S., and Twickler, M. S. (1996a). A 110,000-yr record of explosive volcanism from the GISP2 (Greenland) ice core. *Quaternary Research*, **45**, 109–18.

(1996b). Potential atmospheric impact of the Toba mega-eruption ~71,000 years ago. *Geophysical Research Letters*, **23**, 837–40.

Index

Printed in the United States
by Baker & Taylor Publisher Services